ELECTRONIC CIRCUIT ANALYSIS
basic principles

ELECTRONIC CIRCUIT ANALYSIS
basic principles

Roy A. Colclaser

Donald A. Neamen

Charles F. Hawkins

University of New Mexico

John Wiley & Sons

New York   Chichester   Brisbane   Toronto   Singapore

Cover photo by Geoffry Gove

Library of Congress Cataloging in Publication Data:

Colclaser, Roy A.
  Electronic circuit analysis.

  Includes index.
    1. Electronic circuits. 2. Electronic apparatus and
appliances. I. Neamen, Donald A. II. Hawkins, Charles F.
III. Title.
TK7867.C57   1984   621.3815'3   83-21743
ISBN 0-471-86626-1

Printed in the United States of America

10 9 8 7 6 5 4 3 2 1

to our families

preface _____

Electronics is one of the fastest growing and most rapidly changing disciplines of engineering. To keep pace with these developments, it is important to reassess on a regular basis the manner in which students are introduced to the subject. Historically, analog electronics dominated the applications of electronic devices. The introduction of the electronic computer in the 1940s was the forerunner of a major change in the emphasis of electronics. The widespread acceptance of the bipolar transistor in the 1950s and the integrated circuit in the 1960s also changed the electronics curriculum. The emergence of very large-scale integrated circuits in the 1970s and 1980s has been responsible for new pressures to change the approach for the teaching of electronics.

A typical electrical engineering curriculum includes two semesters or three quarters of electronic circuits, usually during the junior year. These courses are required of all students and serve as prerequisite material for advanced elective courses in the subject area. Laboratory courses are frequently used to supplement the circuits courses. In many instances, the material is introduced in an order that parallels the history of electronics—analog applications followed by digital applications. Recently, several textbooks have appeared that reverse the order of presentation. In addition to the pedagogical arguments for these changes, there is a new need which must be met at a number of universities. This need is the requirement for a single-semester course in electronics for majors in computer engineering. The interests of these students demand that a major part of the first course be devoted to digital circuits. However, it is necessary to provide a background in semiconductors, diodes, and bipolar and field effect transistors before digital circuits can be introduced. In this introductory material, both digital and analog concepts are presented.

This book was intended for a two-semester course sequence in which both computer and electrical engineering students would take the first course. Predominantly electrical engineering students would take the second course that is devoted to analog electronics. This approach allows more class time for the more difficult concepts of analog electronics such as feedback, frequency response, and differential amplifiers. Our approach has been to select a limited number of circuits and analyze them in detail, rather than trying to cover all possible circuit configurations. It is our opinion that advanced topics shouldbe in elective courses with specialized textbooks.

The book is divided into three parts: Introduction to Device Concepts, Digital Circuits, and Analog Circuits. It is intended that the first two parts be covered in the first semester, followed by the third part in the second semester. Because of the independence of the subject matter, it is possible to cover the material in a different order, presenting Parts I and III before Part II. In this case, it is not wise to try and finish Part III in the first course.

For a three-quarter sequence: Part I (without Chapter 5) plus Chapter 6 could be presented in the first quarter. Chapters 7, 5, and 8–11, would be a reasonable division for the second quarter. The remainder of the book would be left for the third quarter.

It is assumed that the students will have had an introduction to dc and ac steady-state circuit analysis and transient analysis through second-order systems. The mathematical background should include calculus and differential equations, while the physics background should include electrical theory and some elementary concepts of modern physics.

Electronics is a branch of engineering that has been expanding at an ever increasing rate. In the past, analog circuits dominated, and were constructed from discrete components. The terminal characteristics of the individual electronic devices were investigated, and models were developed to represent these devices. The models were used in conjunction with circuit theory to simulate the operation of the circuits. To analyze a complex electronic system like a television receiver, the usual practice was to break down the system into functional blocks that were more readily understood. When digital electronics became important, topics concerned with certain aspects of analog electronics gradually were replaced. The emergence of integrated circuits made it difficult to decide what topics should be covered in an introductory electronics text.

There are three tasks that may face electronics engineers in the 1980s. Some individuals will have the opportunity to design integrated circuits. A much larger number will be designing electronic systems in which integrated circuits are an important part. In some cases, engineers will be involved in the design of electronic circuits based on discrete components.

To prepare the reader for all three of these possibilities, it is necessary to investigate the operation of the individual devices, as well as that of standard integrated circuits. Part I introduces the reader to a variety of devices, junction diodes, bipolar junction transistors, junction field effect transistors, and metal–oxide–semiconductor field effect transistors. There is an emphasis on developing the appropriate models for each device for both digital and analog applications. In Part II, certain standard small-scale integrated logic circuits are examined. Each of these circuits represents a different electronic principle. The last chapter in this part provides some insight into specific circuits that are more complicated than those described in the preceding chapters. Part III is devoted to the principles of analog electronic circuits. Integrated circuit techniques are emphasized and compared with those used in discrete component circuitry.

We thank the many students who suffered through the early drafts of this text for their many constructive suggestions. We particularly acknowledge theassistance of Eliot Smyrl, Mark McLaughlin, Ross Dettmer, William Curtis, Mike Lang, and Adam Bryant for their help with the figures, and Hunter Snevily for his general assistance. We also thank Merrill Floyd for his guidance and patience during the evolution of this project.

Roy A. Colclaser
Donald A. Neamen
Charles F. Hawkins

contents _____

part I _____
INTRODUCTION TO DEVICE CONCEPTS   1

chapter 1 **INTRODUCTION**   3

1.1 A HISTORICAL PERSPECTIVE   **4**
1.2 ELECTRONICS IN THE 1980s   **8**
1.3 BASIC CONCEPTS   **9**
1.4 NOTATION   **12**

chapter 2 **SEMICONDUCTORS**   14

2.1 INTRINSIC SEMICONDUCTORS   **14**
2.2 EXTRINSIC SEMICONDUCTORS   **17**
2.3 DRIFT AND DIFFUSION   **20**
2.4 EXCESS CARRIER LIFETIME   **22**
2.5 DEVICE FABRICATION   **23**
    2.5.1 Crystal Growth   **23**
    2.5.2 Wafer Preparation   **24**
    2.5.3 Oxidation   **24**
    2.5.4 Photolithography   **25**
    2.5.5 Diffusion and Ion Implantation   **25**
    2.5.6 Chemical Vapor Deposition   **28**
    2.5.7 Metal Deposition   **28**
    2.5.8 Fabrication Summary   **28**
2.6 SUMMARY   **29**
    PROBLEMS   **29**

chapter 3 **SEMICONDUCTOR DIODES**   31

3.1 DIODE ELECTRICAL CHARACTERISTICS   **31**
3.2 JUNCTION THEORY   **32**

3.2.1 The Abrupt Junction in Equilibrium **33**
3.2.2 The Forward-Biased *pn* Junction **36**
3.2.3 Small-Signal Operation Under Forward Bias **39**
3.2.4 The Reverse-Biased *pn* Junction **41**
3.2.5 The Diode On–Off Switching Transient **43**
3.3 SCHOTTKY BARRIER DIODES **43**
3.4 DIODE CIRCUIT MODELS **45**
3.5 DIODE APPLICATIONS **49**
3.5.1 Rectifiers **49**
3.5.2 Detectors **51**
3.5.3 Voltage Regulators **53**
3.5.4 Clipping and Clamping **54**
3.5.5 Diode Logic Circuits **56**
3.6 SUMMARY **57**
PROBLEMS **58**

chapter 4 **THE BIPOLAR TRANSISTOR** **65**

4.1 BASIC TRANSISTOR AMPLIFYING MECHANISM **66**
4.2 TRANSISTOR CURRENTS AND OPERATING REGIONS **68**
4.3 TRANSISTOR CURRENT–VOLTAGE CHARACTERISTICS **74**
4.4 AN ELEMENTARY DEVICE MODEL FOR ACTIVE TRANSISTOR OPERATION **80**
4.5 SMALL–SIGNAL AMPLIFIER EXAMPLES **85**
4.5.1 Common-Emitter Amplifier **86**
4.5.2 Common-Emitter Amplifier with a Coupling Capacitor **89**
4.5.3 Common-Collector Circuit **91**
4.6 EXAMPLES OF AMPLIFIER dc CALCULATIONS **93**
4.6.1 Common-Emitter Bias Calculations **94**
4.6.2 Voltage Divider Bias for the Common-Emitter Amplifier **96**
4.6.3 Transistor Operating State **97**
4.7 INVERTER SWITCHING AND TIME DELAYS **100**
4.8 SUMMARY **106**
PROBLEMS **106**

chapter 5 **FIELD EFFECT TRANSISTORS** **114**

5.1 JFETS AND MESFETS **115**
5.1.1 Fabrication **116**
5.1.2 Current–Voltage Characteristics **117**

5.1.3 Transconductance   **120**
5.1.4 Switching and Frequency Characteristics   **121**
5.1.5 Applications   **122**
5.2 SURFACE FIELD EFFECT DEVICES   **122**
5.2.1 MOS Capacitor   **122**
5.2.2 MOS Field Effect Transistors   **124**
5.2.3 An MOS Inverter Amplifier   **129**
5.2.4 p-Channel MOSFETs   **131**
5.2.5 The Body Effect   **132**
5.2.6 Switching and Frequency Effects   **133**
5.3 SUMMARY   **136**
PROBLEMS   **137**

part II
DIGITAL ELECTRONICS   **143**

chapter 6 **TRANSISTOR–TRANSISTOR LOGIC CIRCUITS**   **145**

6.1 DIODE–TRANSISTOR LOGIC (DTL)   **145**
6.1.1 Basic DTL NAND Circuit   **146**
6.1.2 Current–Voltage Analysis   **146**
6.1.3 Minimum β **148**
6.1.4 The NAND Operation   **149**
6.1.5 Voltage Transfer Characteristic   **149**
6.1.6 Pull-Down Resistor   **150**
6.1.7 Fan-Out   **151**
6.1.8 Noise Margin   **155**
6.1.9 Propagation Delay Time   **156**
6.2 TRANSISTOR–TRANSISTOR LOGIC (TTL)   **157**
6.2.1 The Input Transistor   **158**
6.2.2 A Basic TTL NAND Circuit   **161**
6.2.3 TTL Output Stages   **170**
6.2.4 Tristate Output   **176**
6.3 SCHOTTKY TRANSISTOR–TRANSISTOR LOGIC (S-TTL)   **178**
6.3.1 The Schottky Clamped Transistor   **178**
6.3.2 A Schottky TTL NAND Circuit   **181**
6.3.3 Low-Power Schottky TTL (LS-TTL)   **182**
6.4 ADVANCED SCHOTTKY TTL LOGIC CIRCUITS   **185**
6.5 WIRED-AND LOGIC CONFIGURATIONS   **188**
6.6 SUMMARY   **189**
PROBLEMS   **192**

chapter 7 **BIPOLAR CURRENT MODE DIGITAL LOGIC CIRCUITS**   **202**

7.1 CONVENTIONAL INTEGRATED INJECTION LOGIC   **202**
    7.1.1 The Basic Inverter   **203**
    7.1.2 I^2L Logic Circuits   **210**
    7.1.3 Integrated Injection Logic Fan-out   **213**
7.2 SCHOTTKY INTEGRATED INJECTION LOGIC   **213**
    7.2.1 Schottky-Coupled Transistor Logic (SCTL)   **214**
    7.2.2 Schottky Transistor Logic (STL)   **216**
7.3 EMITTER COUPLED LOGIC (ECL)   **218**
    7.3.1 The Differential Amplifier   **218**
    7.3.2 The Basic ECL Gate   **220**
    7.3.3 Power Dissipation   **224**
    7.3.4 Transfer Characteristics   **224**
    7.3.5 Noise Margin   **226**
    7.3.6 Propagation Delay Time   **227**
    7.3.7 Fan-Out   **227**
    7.3.8 The Negative Supply Voltage   **229**
7.4 MODIFIED ECL CIRCUIT CONFIGURATIONS   **230**
7.5 SUMMARY   **234**
    PROBLEMS   **236**

chapter 8 **FET DIGITAL LOGIC CIRCUITS**   **242**

8.1 NMOS CIRCUITS   **243**
    8.1.1 NMOS Inverters   **243**
    8.1.2 NMOS Transmission Gate   **256**
    8.1.3 NMOS Digital Logic Circuits   **259**
8.2 CMOS CIRCUITS   **263**
    8.2.1 The CMOS Inverter   **263**
    8.2.2 The CMOS Transmission Gate   **268**
    8.2.3 CMOS Digital Logic Circuits   **269**
8.3 LIMITATIONS OF SPEED, POWER, AND SIZE IN MOSFETS   **270**
8.4 SUMMARY   **274**
    PROBLEMS   **274**

chapter 9 **MEMORY CIRCUITS**   **280**

9.1 ADDRESS DECODERS   **283**
9.2 CONTROL CIRCUITS   **284**
9.3 STATIC RANDOM ACCESS MEMORY CELLS   **287**
    9.3.1 TTL Static RAM Cell   **289**

9.3.2 NMOS Static RAM Cell   **291**

9.3.3 CMOS Static RAM Cell   **293**

9.4 DYNAMIC RANDOM ACCESS MEMORY CELLS   **293**

9.5 READ-ONLY MEMORY CELLS   **297**

9.5.1 Mask-Programmed ROM   **298**

9.5.2 User-Programmed ROM   **299**

9.5.3 Electrically Erasable PROM (EEPROM)   **300**

9.6 SUMMARY   **302**

part III _____

ANALOG ELECTRONICS   **305**

chapter 10 **TRANSISTOR BIASING PRINCIPLES**   **307**

10.1 BIPOLAR BASE CURRENT BIAS   **307**

10.2 BIPOLAR COMMON-EMITTER BIASING   **310**

10.3 MOSFET COMMON-SOURCE BIASING   **313**

10.4 STABILITY ANALYSIS   **315**

10.5 CURRENT SOURCES   **319**

10.5.1 A Simple Bipolar Current Source   **319**

10.5.2 A Two-Transistor Current Source   **320**

10.5.3 A Three-Transistor Current Source   **322**

10.5.4 Widlar Current Source   **323**

10.5.5 A Simple MOS Current Source   **328**

10.6 SUMMARY   **331**

PROBLEMS   **331**

chapter 11 **SMALL-SIGNAL TRANSISTOR MODELS**   **339**

11.1 TWO-PORT NETWORK MODELS   **340**

11.1.1 Hybrid Parameters for Bipolar Transistors   **343**

11.2 HIGH-FREQUENCY MODELS FOR BIPOLAR TRANSISTORS   **349**

11.2.1 The Hybrid-π Model   **349**

11.2.2 Frequency-Dependent Hybrid Parameter Model   **354**

11.2.3 Cutoff Frequency Calculations   **355**

11.2.4 Models for RF Transistors   **356**

11.3 FET MODELS   **357**

11.4 COMPUTER MODELS   **362**

11.5 SUMMARY   **362**

PROBLEMS   **363**

chapter 12 **FREQUENCY RESPONSE OF TRANSISTOR CIRCUITS** **367**

12.1 THE COMMON-EMITTER AMPLIFIER HIGH-FREQUENCY RESPONSE **371**
    12.1.1 Gain-Bandwidth Product **375**
12.2 THE COMMON-BASE AMPLIFIER **376**
12.3 MOSFET AMPLIFIER FREQUENCY RESPONSE **379**
12.4 THE LOW-FREQUENCY RESPONSE OF AN AMPLIFIER **381**
12.5 TUNED AMPLIFIERS **386**
12.6 SUMMARY **392**
    PROBLEMS **393**

chapter 13 **MULTIPLE-TRANSISTOR CIRCUITS** **398**

13.1 THE BASIC DIFFERENTIAL AMPLIFIER **398**
    13.1.1 Common-Mode and Differential-Mode Inputs **399**
    13.1.2 DC Analysis **399**
    13.1.3 Bipolar Differential Amplifier Small-Signal Analysis **405**
    13.1.4 MOSFET Differential Amplifier Small-Signal Analysis **408**
    13.1.5 Common-Mode Rejection Ratio **410**
    13.1.6 Input Resistance **411**
    13.1.7 The Differential Amplifier with Constant-Current Source **414**
13.2 MULTIPLE-STAGE TRANSISTOR CIRCUITS **417**
    13.2.1 Cascade Transistor Amplifiers **417**
    13.2.2 Darlington Pair **425**
    13.2.3 The Level-Shifting Amplifier **427**
13.3 A SIMPLIFIED OPERATIONAL AMPLIFIER **429**
    13.3.1 DC Analysis **430**
    13.3.2 Small-Signal Analysis **432**
    13.3.3 Bipolar Current Sources as Active Loads **434**
    13.3.4 MOS Current Sources as Active Loads **436**
13.4 SUMMARY **439**
    PROBLEMS **439**

chapter 14 **FEEDBACK** **446**

14.1 BASIC THEORY **447**
    14.1.1 The General Feedback Expressions **448**
    14.1.2 Bandwidth Modification **450**
    14.1.3 Gain Sensitivity **451**
    14.1.4 The Effect of Feedback on Distortion **452**
14.2 THE BASIC FEEDBACK CONNECTIONS **453**
    14.2.1 The Transfer-Impedance Amplifier **453**

14.2.2 The Transfer-Admittance Amplifier **460**

14.2.3 The Voltage Amplifier **464**

14.2.4 The Current Amplifier **467**

14.2.5 Summary of Single-Stage Negative Feedback Amplifier Configurations **469**

14.3 MULTISTAGE FEEDBACK AMPLIFIERS **470**

14.4 SUMMARY **475**

PROBLEMS **476**

chapter 15 **POWER AMPLIFIERS, TUNED AMPLIFIERS, AND OSCILLATORS** **482**

15.1 POWER AMPLIFIERS **483**

15.1.1 Amplifier Classifications **483**

15.1.2 Common-Collector Output Stage **485**

15.1.3 Class B Broadband Power Amplifiers **488**

15.2 TUNED AMPLIFIERS **492**

15.2.1 Class C Amplifiers **492**

15.2.2 Tuned Amplifier Characteristics **494**

15.2.3 Tuned Amplifier Stability **498**

15.2.4 Multistage Tuned Amplifiers **501**

15.2.5 Graphical Design of Tuned Amplifiers **504**

15.2.6 Stagger-Tuned Amplifiers **506**

15.3 OSCILLATORS **512**

15.3.1 The Phase Shift Oscillators **512**

15.3.2 The Colpitts Oscillator **514**

15.4 SUMMARY **519**

PROBLEMS **520**

chapter 16 **ANALOG BIPOLAR INTEGRATED CIRCUIT EXAMPLES** **523**

16.1 A BIPOLAR OPERATIONAL AMPLIFIER **523**

16.1.1 DC Analysis **527**

16.1.2 AC Analysis **530**

16.2 VOLTAGE REGULATORS **533**

16.3 SUMMARY **540**

chapter 17 **ANALOG MOSFET INTEGRATED CIRCUITS** **541**

17.1 ANALOG MOSFET DEVICE PROPERTIES **542**

17.2 INVERTER ANALYSIS **544**

17.3 VOLTAGE DIVIDERS **549**

17.4 CURRENT SOURCES **550**

17.5 DIFFERENTIAL AMPLIFIER **553**

17.6 AN NMOS OPERATIONAL AMPLIFIER **554**

    17.6.1 OP AMP Input Stage **555**

    17.6.2 OP AMP Voltage Level Shift Stage **557**

    17.6.3 OP AMP Voltage Gain Stage **560**

    17.6.4 OP AMP Output Stage **562**

17.7 SUMMARY **563**

    PROBLEMS **564**

ANSWERS TO SELECTED PROBLEMS **563**

INDEX **569**

part I

INTRODUCTION TO DEVICE CONCEPTS

chapter 1 _____
INTRODUCTION

Electronics is the field of science and engineering that deals with devices in which conduction is principally by electrons moving through a vacuum, gas, or semiconductor and with the utilization of these devices. It is an exciting part of electrical engineering because it permits us to construct and analyze circuitry containing components capable of performing marvelous operations on electrical signals. The most interesting of these operations is called amplification. In this process, the output of the electronic circuit is an enlarged version of the input. As far as the signal is concerned, the circuit appears to be active, in that the power delivered to the output terminals is greater than the input signal power to the circuit. It is important to recognize that electronic amplifiers do not violate that fundamental law of nature, "you cannot get something for nothing." An increase in the power carried by a signal is accompanied by an increase in the total power delivered to the circuit, usually from a dc power source. A familiar example of an electronic amplifier is the high fidelity music system found in many homes. In this system, the weak signal source may be the mechanical vibration of a stylus in contact with a record, a magnetic pattern on a plastic tape, or a radio wave arriving at the system through the air. In each case, the signal, if applied directly to a speaker, would not be able to provide enough power to reproduce the intended sound. An intermediate electronic system is placed between the weak input signal source and the speaker to convert the signal into an enhanced version containing the same information, but containing enough power to drive the speaker and provide enjoyment for the listener. The reader is also familiar with the controls on the front of the amplifier that enable the user to change the loudness and frequency characteristics of the system. The bass and treble controls allow the listener to alter the shape of the signal so that it sounds more pleasing to the ear. These controls are an example of selective amplification of different portions of the frequency spectrum.

The primary functions of electronic systems are associated with the transmission, processing, and manipulation of information. There are two forms of electronic signals capable of containing information. Continuous signals transmit information in the level or variation of the signal. Discrete signals transmit information in a pattern represented by the presence or absence of a voltage or current. The former type of signal is called analog, and the latter is called digital. They represent the two principal branches of electronics.

In the following sections, we look at electronics from several viewpoints. The first of these presents a brief history of the development of electronics from vacuum tubes to integrated circuits. The next sections discuss electronics in the 1980s and the basic concepts necessary for the applications and design of modern electronic circuits. Some of the terms in this chapter may be unfamiliar to the reader. No attempt has been made to describe them at this point, because they are treated in detail later in the book.

1.1 AN HISTORICAL PERSPECTIVE

Electronics began rather inauspiciously in 1883, when Thomas Edison discovered the vacuum diode as part of his research on materials for a practical electric light. This device consisted of an incandescent filament and a metal plate in an evacuated glass globe. When the plate was made positive with respect to the filament, current was observed in the external circuit. If the polarity was reversed, however, no current was observed in the external circuit. Edison did not apply for a patent on this device until the next year. An interesting aspect of the "Edison effect" is that the electron was not isolated as an individual particle until several years later.

This first electronic device exhibited a nonlinear, unilateral electrical characteristic but was not capable of producing amplification of a signal. This important step was taken independently in England by Fleming in 1905 and in the United States by DeForest in 1906. The placement of a third electrode, an open grid of fine wires between the plate and the filament, made it possible for a voltage between the grid and filament to control the current in the external plate–filament circuit. The voltage across an appropriately sized resistor in series with the plate could then be an enlarged replica of a signal voltage between the grid and the filament, resulting in a voltage amplification. We see here an important feature of all amplifying devices in the relative physical positions of the elements. The mechanical advantage of a lever is increased by placing the fulcrum closer to the load. Similarly, placing the grid closer to the filament increases the amplification factor of a vacuum triode. A geometric factor is important in the operation of every amplifying device.

The vacuum triode and its related devices, the tetrode and the pentode, provided the foundation for electronics for half a century. The most widespread applications of vacuum tubes during that time period were in the communications industry, first in radio and later in television. This period spanned two world wars, the development of radar and sonar, the golden age of radio, and the development of both black-and-white and color television.

Until nearly the end of the era of vacuum tube electronics, most of the applications involved an increase in the power contained in a signal with very little distortion in the waveform of the voltage or current. This type of amplification is called "analog" or "linear."

If the input signal to an amplifier is large enough, the output signal level reaches limits established by the power supply and is unable to duplicate the waveform of the input signal. Under these conditions, the device may be allowed to "switch" from one extreme to the other. If one extreme is labeled 1 and the other extreme is labeled 0, the device can be considered to be a switching element in a "digital" circuit.

The first large digital electronic system was a special-purpose vacuum tube circuit called the electronic numerical integrator and computer (ENIAC) built for the U.S. Army at the University of Pennsylvania in 1946. The ENIAC was the forerunner of the computer industry, but it also focused attention on some of the problems associated with vacuum tubes. For vacuum tubes to function, it is necessary for the filament or a cylinder surrounding the filament, called the cathode, to be heated to a temperature sufficient to boil large quantities of electrons from the material. The free electrons are necessary for the operation of the device, but the power and heat used to generate the electrons are unwanted. The reliability of vacuum tubes was poor. Large numbers of these devices in a system made it difficult to run even a short program without significant risk. It should also be recognized that even miniature vacuum tubes occupied a considerable volume when assembled into an electronic computer. These problems contributed to the rapid decline in the use of vacuum tubes when a semiconductor device was invented that could perform many of the functions previously associated exclusively with vacuum tubes.

The bipolar transistor was invented in 1948, but it was not the first application of semiconductors to electronics. The widespread acceptance of radio in the 1920s was due to the metal–semiconductor diode called the crystal detector. This device made a significant contribution to electronics, but it was not an amplifying device. The advantages of fabricating amplifying devices in semiconductor materials were apparent even as early as 1925. In that year, Lilienfeld proposed a solid-state device very similar to the surface field-effect transistors, which did not appear commercially until nearly 40 years later. The proposed solid-state amplifying device had no filament to consume unwanted power, since it had a sufficient supply of electrons at room temperature. Conduction was to take place within a single piece of material, making this device much more reliable than the multielement devices in evacuated glass globes of the vacuum tube era. The metallic alloys developed in the early vacuum tube days, with thermal expansion coefficients matching those of glass, are still used today to provide hermetic seals for many of the packages that contain semi-conductor devices.

The point contact or "cat's whisker" crystal detectors were the forerunners of the radar detectors of World War II. To improve the quality of these high-frequency rectifiers, research was initiated on two semiconductor materials, germanium and silicon. The results of this research set the stage for the dramatic events of 1947 and 1948.

A talented team of scientists from Bell Laboratories, led by Walter Brattain and John Bardeen, used purified germanium samples to investigate techniques for controlling the characteristics of semiconductor surfaces. Late in 1947. they discovered what was later to be named the "transistor" effect. This first solid-state amplifier consisted of two gold foil electrodes in pressure contact with a germanium base. The separation between the electrodes was less than 50 μm. This crude device was called the "point contact transistor." Its characteristics were not outstanding, but it amplified a sound signal and was used as the active element in an oscillator. This latter experiment demonstrated conclusively that the device had a power gain greater than 1. The point contact transistor inspired William Shockley, another Bell Laboratories scientist, to propose and supervise the fabrication of the bipolar transistor in the early months of 1948. The bipolar transistor exhibited superior electrical performance and reliability compared to the point-contact device, and quickly dominated commercial market.

The bipolar transistor uses an input current signal to control an output current signal, in contrast to the vacuum triode, which uses an input voltage signal to control an output current signal. An appropriate selection of input and output resistors can result in a transistor voltage amplifier with characteristics similar to those of vacuum tube amplifiers. Using circuit design techniques that differed considerably from those used in the past, transistor circuits soon replaced many of the commercial vacuum tube circuits. These circuits performed the electronic functions as well as their predecessors did, but they did not require filament power. They were also smaller and more reliable than their vacuum tube counterparts.

The early transistors were made from germanium. The most visible (and audible) application of these devices was in small, portable AM broadcast receivers. These transistor radios were very inexpensive and weighed a small fraction of their vacuum tube predecessors. For industrial and military applications, the required range of temperatures for the operation of equipment is larger than that for consumer applications. Germanium devices cease to function properly at package temperatures in the vicinity of 100°C. Silicon devices can operate successfully at higher package temperatures and exhibit better temperature stability at low temperatures. For these reasons, silicon transistors began to replace germanium transistors in the late 1950s.

The techniques developed for the fabrication of silicon bipolar transistors made possible the next revolutionary step in electronics. There was a great need in the space program, the military, and the fledgling computer industry for electronic circuits that were extremely small and ultrareliable. The response of the electronics industry to these needs was the introduction of bipolar integrated circuits.

The commercial success of the integrated circuit industry was based on standard products representing digital logic families. The early forms of integrated logic circuits were similar in circuit configuration to the discrete component versions which performed the same electronic functions. The next logic family to be developed included structures that were unique to integrated circuits. The transistor–transistor logic family, making use of multiple-emitter transistors, became the most successful bipolar logic technology in the history of integrated circuits.

Analog bipolar integrated circuits use another feature of integrated circuit fabrication technology. Devices located very close together on the same substrate frequently have very similar electrical characteristics, and matched electronic devices are useful in balanced circuits that make use of differential inputs. The most popular of these circuits is the operational amplifier. Integrated circuits allowed the widespread use of these quality circuits at very low cost.

The sophisticated processes developed for the fabrication of bipolar integrated circuits finally made it possible for Lilienfeld's idea of the 1920s, the surface field effect transistor, to become a reality. The surface of these devices is controlled by a voltage applied to a metal gate separated from the semiconductor substrate by a silicon dioxide insulating layer. These metal–oxide–semiconductor (MOS) integrated circuits rapidly established themselves as strong competitors with bipolar integrated circuits in the digital logic field. By 1970, the use of MOS random access memories had become widespread. The integrated circuit industry was moving from the era of small- and medium-scale circuits to large-scale integration (LSI).

The electronics industry was now equipped to make yet another giant step forward. The microprocessor was developed by the Intel Corporation in the early 1970s. The first microprocessor was a 4-bit device that evolved from specifications for a custom integrated circuit. Within a short time, 8-bit microprocessors appeared on the market from a number of manufacturers. Using a microprocessor as the central processing unit and a handful of support integrated circuits, the microcomputer became one of the most important electronic systems of the decade. The progress of semiconductor memories during the 1970s was phenomenal. Random access memories increased in capacity from 1024 bits to 65,536 bits with a significant decrease in the cost per bit. The combination of low-cost memory and the microprocessor made it possible for the widespread of computers in small businesses and homes.

As the decade of the 1970s came to a close, a new era in integrated circuits was beginning. This era is characterized by the inclusion of larger and larger numbers of components in a single circuit, and it is called very large-scale integration (VLSI). The first evidences of VLSI were new microprocessors that incorporated some of the functions previously requiring support circuits. These included clock generators, scratch-pad memories, read-only memories, and input/output ports. These circuits are microcomputers on a chip. Other examples of the early VLSI circuits are 16-bit microprocessors and 262,144-bit memories. These circuits are only the beginning. The technology has already been demonstrated to make devices small enough to include millions of them in a single integrated circuit.

1.2 ELECTRONICS IN THE 1980s

Electronics has become a pervasive influence in the civilized world. The communictions industry is heavily dependent on electronics. Even the printed word is usually typed, stored, and reproduced by an electronic computer system called a "word processor." The computer industry is entirely dependent upon electronics. Electronics has also invaded the automotive industry. Microcomputer-controlled ignition and fuel injection are electronic weapons in the war against air polution. In addition electronically controlled industrial robots are involved in the assembly of automobiles. Electronic instrumentation, supplemented with microcomputers, is used extensively in virtually every industry. The defense and aerospace industries also rely heavily on electronics. Consumer electronics is no longer limited to radio and television receivers. Inexpensive electronic calculators and timepieces are everywhere. Microwave ovens have changed the cooking habits of the nation. Perhaps the most sophisticated example of electronic equipment to enter the home is the personal computer. Competition and improved technology have resulted in an impressive decrease in the prices of integrated circuits, even in times of significant inflation. The personal computer of 1980 is comparable in performance to the main-frame business computer of 1960 with a price that is a factor of 100 lower than that of the earlier model.

The devices available for use in modern electronic systems represent a wide choice. Vacuum tubes have been displaced by the solid-state challenge in all but a few applications. The most obvious exception to this trend is the cathode ray tube. This device is popular as a display in television receivers computer teminals, and oscilloscopes. The other major application of vacuum tubes is in high-power radio transmitters. Discrete transistors and semiconductor diodes are available in a wide variety of power ratings. Discrete transistors include the bipolar transistor, the junction field effect transistor (JFET), the metal–oxide–semiconductor field effect transistor (MOSFET), and the metal–Schottky barrier—semiconductor field effect transistor (MESFET). Discrete diode types include the point contact, junction, Zener, avalanche, Schottky barrier, and tunnel. Most of these devices are made from silicon, but some are made from germanium or gallium arsenide when special characteristics are needed. This vast array of discrete devices is overshadowed by an even more impressive list of standard integrated circuits. These are available in both bipolar and MOS forms, for analog and digital circuitry. In some cases, the technologies are combined to provide improved performance, and, in the case of the analog microprocessor, the circuit functions are combined. The complexity of standard integrated circuits ranges from a 3-pin package containing a voltage regulator to a 64-pin package containing a microprocessor. The electronics engineer of the 1980s has access to a magnificent set of tools. It is a challenge to select the proper device or circuit to perform a particular electronic function.

In the era of integrated circuits, a designer must decide whether to build a new electronic system based on standard integrated circuits or to develop a custom integrated circuit. This is usually a decision based on economics. For an integrated circuit design to be profitable, there must be a need for 10,000 or more circuits, and six months or more may be required for the design. There are alternative approaches to this problem. One of these makes use of an array of devices on a wafer. This "breadboard" circuit is used to demonstrate the feasibility of connecting some of the devices together to perform the desired functions. If the designer is satisfied, an intraconnection mask set can be generated resulting in a custom integrated circuit at low cost in a very short time span. The introduction of the microprocessor provided another solution to the problem, at least for digital applications. In many cases, it is possible to trade hardware for software in logic operations, or to use a programmable logic array to replace custom integrated or standard integrated circuits. It must be emphasized again that the electronics engineers of the 1980s must be capable of choosing the appropriate devices or circuits to satisfy their functional requirements.

1.3 BASIC CONCEPTS

The study of electronic circuits is based on the same principles that are used in circuit theory. These include superposition and Kirchhoff's voltage and current laws. In most cases, it is possible to describe the operation of complicated electronic devices by models that consist of more familiar circuit components such as resistors, capacitors, inductors, independent sources, and dependent sources. It is important to recognize that these device models are dependent on a set of conditions called the *operating point* of the device. All amplifying devices require a dc energy source for activation. The dc energy state is required to establish the internal electric fields necessary for the operation of any active device. The particular dc levels in the device define the operating point.

In most electronic circuits, there are two sets of "input" terminals. One set makes use of a dc power supply to establish the operating point. The other set is used to introduce the signal into the circuit. This is shown in Figure 1.1. In this example, one terminal is used as a reference for the other terminals. It is called *ground* and is the zero of potential for the system. In an analog electronic circuit, the power supply imposes limits on the magnitude of the output signal. This indirectly places a limit on the magnitude of the input signal,

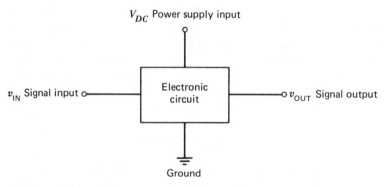

Figure 1.1. An electronic circuit that illustrates two sets of input terminals: the signal input v_{IN} and the power supply input V_{DC}.

since the output signal magnitude is expected to be an enhanced replica of the magnitude of the input signal. This is illustrated in Figure 1.2. As long as the input signal does not exceed the indirect limit, the power supply does not appear to have any effect on the output signal. However, if you remove the power supply, the output signal will probably disappear. For this particular example, with the power supply in place, the electronic circuit can be modeled as a voltage-controlled voltage source, as shown in Figure 1.3. Note that the power supply is not even shown in this figure; but, without the power supply to establish the operating point, the model would not be valid. The other function of the power supply is to act as a source for the apparent power gain associated

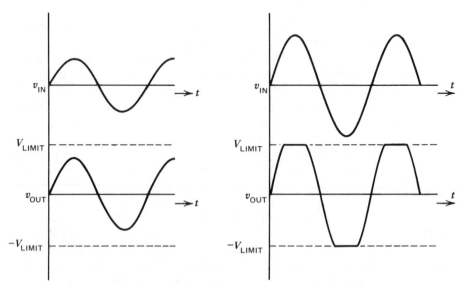

Figure 1.2. Limitations on analog signals imposed by the power supply.

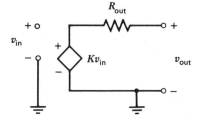

Figure 1.3. A model of the amplifier of Figure 1.1. using a voltage-controlled voltage source to represent the process of amplification for that particular circuit.

with the output signal as compared to the input signal. In most cases, the analysis of analog electronic circuits is divided into two parts, one dealing with the dc circuit to establish the operating point and the other concerning the signal path through the circuit.

In a digital electronic circuit, the operating point is usually established at one of the limits imposed by the power supply. It is also customary to connect the output of one circuit to an input of another circuit. To model a circuit that has two operating points and transitions between them, it may be necessary or useful to specify four different models, one for each region of operation. Using this technique, it is possible to change a complex circuit analysis problem into four relatively simple ones. A typical digital device modeled with electrical components is shown in Figure 1.4.

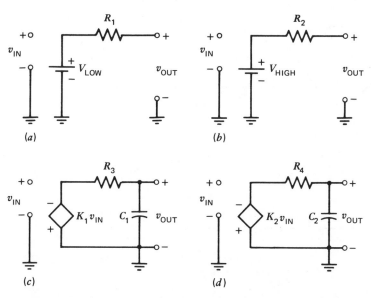

Figure 1.4. Models representing four electronic states of a digital logic circuit. (a) V_{OUT} = Low state. (b) V_{OUT} = High state. (c) Output voltage in transition from low to high. (d) Output voltage in transition from high to low.

Modeling is an important aspect of electronics. All models are based on approximations. The purpose of a model is to permit the mathematical prediction of performance that agrees with observation. The accuracy required of the model depends on the application. For example, a model for a transistor used in an audio-frequency amplifier does not have to be nearly as complicated as a model for the same transistor used in a radio-frequency amplifier. It should be obvious that the modeling of complicated circuits, either constructed from discrete components or in integrated form, requires the use of a computer. Computer software has been developed for the modeling of electronic circuits. Examples of these are the electronic circuit analysis program (ECAP) and the simulation program with integrated circuit emphasis (SPICE). The design of VLSI circuits relies heavily on circuit simulation programs.

1.4 NOTATION

The following notation is used throughout this text. Lower case letters like i and v are used to denote *instantaneous* values of time-varying functions. A lower case quantity with an upper case subscript indicates the total value, and

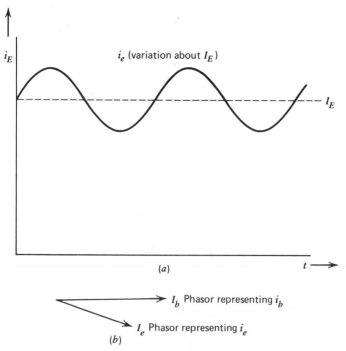

Figure 1.5. A summary of the notation. (*a*) Instantaneous. (*b*) Phasor.

a lower case quantity with a lower case subscript indicates the small-signal or ac value. An upper case letter with an upper case subscript represents the dc value of the quantity. An upper case letter with a lower case subscript represents the rms phasor value of the ac component of the quantity. For example, i_E denotes the total instantaneous emitter current, i_e is the instantaneous ac component of the emitter current, I_E represents the dc value of the emitter current, and I_e signifies the rms phasor value of the ac component of the emitter current. This notation is summarized in Figure 1.5.

chapter 2
SEMICONDUCTORS

Electronic amplifying devices can be made from several types of materials. Early devices were based on the transport of electrons through vacuum, usually in a fragile glass envelope. The physical structure of vacuum tubes is awkward, with metal electrodes separated at precise distances. When vibrations cause changes in the amplifying characteristics, the tubes are called "microphonic." Their operation is based on thermionic emission, which consists of supplying enough energy, in the form of heat, to boil electrons off the surface of one of the electrodes. Many of the problems associated with vacuum tubes vanished with the introduction of solid-state amplifying devices. These are fabricated within materials called *semiconductors*. In these materials, charge carriers are readily available, even at room temperature. Semiconductor devices are more rugged in construction and can withstand more severe environments. In this chapter, we investigate some of the properties of these unusual materials that make amplification possible.

2.1 INTRINSIC SEMICONDUCTORS

Semiconductors are the "in between" materials, bracketed by conductors and insulators. The best known conductors are metals such as copper, aluminum, and gold, having electrical resistivities on the order of 10^{-5} $\Omega \cdot$cm at room temperature. Insulators such as mica, diamond, quartz, and sapphire have electrical resistivities on the order of 10^{12} $\Omega \cdot$cm under the same conditions. Pure, single crystal semiconductors are similar to insulators, except that semiconductors have lower electrical resistivities. One characteristic of both insulators and semiconductors is that an increase in temperature results in an increase in electrical conductivity.

Pure semiconductors are called *intrinsic*, since their properties are controlled by the materials themselves. The room-temperature electrical resistivity of intrinsic silicon is 2.27×10^5 Ω·cm, that for intrinsic germanium is 45 Ω·cm, and that for intrinsic gallium arsenide is 7.94×10^7 Ω·cm. Silicon and germanium are elemental semiconductors. Compounds formed between elements from Groups III and V or Groups II and VI of the periodic table can also be semiconductors. Examples of compound semiconductors are gallium arsenide and cadmium sulfide. The compound semiconductors are quite useful for optical diodes and for certain very fast logic circuits, but the vast majority of semiconductor devices are made from silicon.

The electronic structures of insulators and intrinsic semiconductors are identical. They are both characterized by bands of allowed energy states separated by bands of forbidden energy levels. It should be recognized that a completely filled band or a completely empty band does not contribute to conduction. This is due to a lack of empty energy states into which the charge carriers can be accelerated by a moderate electric field. The valence electrons completely fill one of the allowed bands (called the *valence band*) at the absolute zero of temperature. The next higher band of allowed states is called the *conduction band*. The separation between the valence band and the conduction band is called the *energy gap* and is designated E_G, as shown in Figure 2.1. At temperatures above 0°K, thermal energy absorbed in the material can excite electrons from the valence band into the conduction band, leaving empty states behind in the valence band. The concentration of electrons that is excited into the conduction band at a particular temperature is dependent on the energy gap of the material. The relation for this concentration is

$$n_i = BT^{3/2} \exp\left[\frac{-E_G}{2kT}\right] \tag{2.1}$$

where n_i is the equilibrium intrinsic electron concentration (in electrons per cubic centimeter), B is a constant that depends on the material, T is the temperature in °K, and k is Boltzmann's constant (86×10^{-6} eV/°K). The energy gaps for several materials are listed in Table 2.1. At any specific temperature,

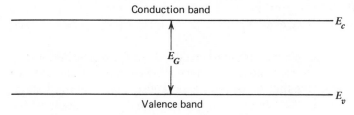

Figure 2.1. The energy band structure of a semiconductor.

Table 2.1. Energy Gaps for Semiconductors
and Insulators at Room Temperature

Material	E_G (eV)	B (cm^{-3}($^\circ$K)$^{-3/2}$)
Semiconductors		
Germanium	0.67	1.5×10^{15}
Silicon	1.1	3.3×10^{15}
Gallium arsenide	1.4	350×10^{12}
Gallium phosphide	2.2	1.3×10^{15}
Insulators		
Diamond	5.5	470×10^{12}
Silicon dioxide	8	—

insulators have much smaller intrinsic electron concentrations than semiconductors.

The values given in Table 2.1 are approximations that neglect the variation with temperature of B and E_G. If we calculate n_i for silicon at 300°K, we get

$$n_i = 3.3 \times 10^{15} \times (300)^{3/2} \exp\left[\frac{-1.1}{2 \times 86 \times 10^{-6} \times 300}\right]$$

$$= 9.5 \times 10^9 \text{ cm}^{-3}$$

This is slightly less than the accepted value of 1.5×10^{10} cm^{-3} that has been verified by a number of independent experimental measurements.

Conduction in an intrinsic semiconductor is due to the electrons in both the conduction band and the valence band. It was mentioned earlier that a full band does not contribute to conduction. The valence band is full of electrons at absolute zero, but there are no empty allowed energy states into which these electrons can be accelerated by an external electric field, except those in the conduction band. The excitation of an electron across the energy gap requires a much larger electric field than that usually applied to an electronic device. However, at temperatures above absolute zero, there is sufficient thermal energy to excite some electrons from the valence band to the conduction band. Under these circumstances, the valence band is not completely full and the conduction band is not completely empty. The electrons in the conduction band then have a multitude of empty allowed energy states, at almost the same energy level, into which they can be readily accelerated by an applied electric field.

The situation in the valence band is very different from that in the conduction band. The small concentration of empty allowed states in the valence band has energy levels near the top of the band. These empty allowed states do permit the electrons in the valence band to participate in conduction, but only to a limited extent. Since a full band does not contribute to conduction, it is reasonable to assume that an almost full band will not make a large contribution to conduction. The easiest way to account for the actions of all of the electrons remaining in the valence band is to assign a positive charge to the empty electron states and follow the motion of these empty states. The name assigned to these empty states is the *hole*. For each electron excited into the conduction band is an intrinsic semiconductor, a hole is left behind in the valence band. The process is called the *generation of an electron–hole pair*.

Conduction due to the drift of mobile charge carriers in an intrinsic semiconductor is ohmic in nature. This implies that electrons and holes respond, on the average, to an applied electric field E with an average drift velocity v_d that is proportional to the applied electric field. This is described by

$$v_d = \pm \mu E \qquad (2.2)$$

where μ is the mobility of the charge in a particular material. The sign in front of μ depends on the polarity of the charge carrier. Holes have a positive charge and drift in the same direction as the applied field. Electrons have a negative charge and drift in the same direction as the applied field. The current density in an intrinsic semiconductor due to drift J_{drift} is given by

$$J_{\text{drift}} = \sigma_i E \qquad (2.3)$$
$$= q(n_i \mu_n + p_i \mu_p)E$$

where σ_i is the intrinsic conductivity, q is the magnitude of the charge on an electron (1.6×10^{-19} C), $p_i = n_i$ is the intrinsic hole concentration, and μ_n and μ_p are the electron and hole mobilities, respectively. Values of these quantities for several semiconductors are given in Table 2.2. Note that S is the abbreviation for siemens, which is the unit for conductivity. $1 \text{ S} = (1 \ \Omega)^{-1}$.

2.2 EXTRINSIC SEMICONDUCTORS

Intrinsic semiconductors are interesting materials, but, from the viewpoint of electronics, the effects of incorporating controlled quantities of certain impurities in regions of single-crystal semiconductors are much more significant. The impurities of interest are those that enter the crystal lattice by substituting for the semiconductor atoms, even though they do not have the same valence electron structure as the host atoms. For the elemental semiconductors, which

Table 2.2. Some Important Data for Semiconductors at 300°K

Material	n_i (cm^{-3})	μ_n	μ_p	σ_i (S·cm^{-1})
		(cm²/V·s)		
Germanium	2.4×10^{13}	3900	1900	2.2×10^{-2}
Silicon	1.5×10^{10}	1350	480	4.4×10^{-6}
Gallium arsenide	9×10^6	8600	250	1.3×10^{-8}

come from the fourth group of the periodic table, the substitutional impurities are from Groups III and V. For the compound semiconductor made from one Group III element and one Group V element, substitutional impurities come from Groups II and VI.

Silicon is an elemental semiconductor that may use Group V impurities such as phosphorus and arsenic. When a phosphorus atom substitutes for a silicon atom in a single crystal of silicon, four of its valence electrons are involved in satisfying the bonding structure of the crystal. The fifth electron is loosely tied to the vicinity of the phosphorus atom when the temperature is low (below 100°K). If the temperature is raised toward room temperature, the thermal energy is sufficient to ionize the phosphorus atom, freeing the fifth valence electron to move throughout the crystal in the conduction band. Phosphorus is called a *donor* impurity because it readily "donates" electrons to the conduction band. The phosphorus ion is immobile and becomes positively charged. The electron that entered the conduction band came from the phosphorus atom, not from the valence band. This electron was generated without generating a hole. An important relationship for mobile charge carriers in semiconductors is given by

$$p_0 n_0 = n_i^2 \qquad (2.4)$$

where n_0 and p_0 are the equilibrium electron and hole concentrations, respectively. At room temperature, each donor atom supplies an electron to the conduction band and, if the donor concentration is N_d, then

$$n_0 \simeq N_d \qquad (2.5)$$

and

$$p_0 \simeq \frac{n_i^2}{N_d} \qquad (2.6)$$

For silicon, at room temperature, $n_i^2 = 2.25 \times 10^{20}$ cm^{-6}; and, if $N_d = 10^{18}$ cm^{-3}, $p_0 = 2.25 \times 10^2$ cm^{-3}. Clearly, electrons are the *majority* carriers and holes are the *minority* carriers in this case, making this material *n*-type. Phosphorus doping levels up to 10^{21} cm^{-3} are possible in silicon without destroying the silicon crystal structure. By controlling the doping level, it is possible to routinely vary the resistivity at room temperature for different samples of *n*-type silicon from 100 Ω·cm to 0.001, that is, a range of *five* orders of magnitude. Let us now look at the doping properties that occur in silicon from the Group III elements.

Boron is by far the most popular Group III element used for the doping of silicon. When a boron atom is substituted for a silicon atom in a silicon single crystal, its three valence electrons are involved in bonds with nearest-neighbor silicon atoms. Since there are only three valence electrons in the boron atom, one electron is lacking to satisfy the required bonding structure. At room temperature, this electron is supplied from the valence band, leaving a hole behind. The bound extra electron in the vicinity of the boron atom creates an immobile negative ion. The boron atoms are called *acceptors* because they "accept" electrons from the valence band. The equilibrium hole density is approximately equal to the acceptor density N_a, giving an equilibrium electron density of n_i^2/N_a. Silicon doped with Group III impurities is called p-type. The conductivity range for p-type silicon is slightly less than that for n-type silicon, because the maximum concentration of boron atoms that can be substituted for silicon atoms is not quite as large as that of phosphorus atoms. In addition, the mobility of holes is lower than that of electrons.

If an homogeneously doped sample contains both donors and acceptors, it is called a *compensated semiconductor*. If the two doping concentrations are equal, the material is totally compensated and acts in most ways like an intrinsic semiconductor. If the donor concentration exceeds the acceptor concentration, the material will be n-type. If the acceptor concentration is higher, the material will be p-type. In an uncompensated n-type semiconductor at room temperature, essentially all of the donor atoms are ionized, and the electrons that are removed from these atoms are elevated to the conduction band. In a compensated n-type semiconductor at room temperature, essentially all of the donor atoms are ionized, but some of the electrons from these atoms are involved in the ionization of the acceptor atoms, with the remainder going to the conduction band. The addition of acceptor atoms results in an effective reduction in the concentration of donor atoms to a new value given by $N_d - N_a$. A similar situation arises in a compensated p-type semiconductor.

In general, a charge balance equation can be written to describe a semiconductor sample in equilibrium. We can write a charge balance equation for the equilibrium concentrations of holes and electrons and for donor and acceptor dopant concentrations as

$$n_0 + N_a = p_0 + N_d \qquad (2.7)$$

that implies

$$n_0 \simeq N_d - N_a \quad \text{and} \quad p_0 \simeq \frac{n_i^2}{N_d - N_a}$$

for n-type material, and

$$p_0 \simeq N_a - N_d \quad \text{and} \quad n_0 \simeq \frac{n_i^2}{N_a - N_d}$$

for p-type material. For most cases, a compensated semiconductor can be treated as if it were uncompensated with a net doping density. The main effect of the compensation is a degradation in lifetime and mobility for the carriers. Every ionized impurity acts as a scattering center and, thus, reduces the mean free path between collisions for the carriers. This reduces the mobility. Charge centers also promote the recombination of excess carriers, as discussed below. The processes used for the fabrication of semiconductors are usually based on compensation techniques.

2.3 DRIFT AND DIFFUSION

Current in a semiconductor can be due to the drift of both electrons and holes, in an electric field; but diffusion is an additional mechanism for current that can be important in the operation of semiconductor devices. Diffusion is the process of particle flow from a region of high concentration to a region of lower concentration. This is a statistical phenomenon related to kinetic theory. The electrons and holes in a semiconductor are in continuous motion, randomized by interactions with the lattice atoms, with an average speed determined by the temperature. The direction of motion is random; and, therefore, in a region of high concentration, approximately half of the particles are directed away from the region at any particular instant. In regions of lesser concentration, there are smaller numbers of particles moving toward the high-concentration region. This results in a net flow of particles from the region of higher concentration to those regions of lower concentration. The concentration gradient can arise from a nonhomogeneous doping distribution or from the injection of a quantity of electrons or holes into a region.

The driving force for the diffusion of charge carriers is a concentration gradient. Since, by definition, the gradient is positive for an increasing gradient, the particle current for diffusion is proportional to the negative of the concentration gradient. Figure 2.2a illustrates this situation for a concentration gradient of holes. The gradient has a positive value for increasing distance x that causes a current in the opposite direction. Electrons respond in a similar manner to a positive concentration gradient; but, as shown in Figure 2.2b, the convention for positive current is shown now as the same direction as the gradient.

The electric current density for diffusion is the product of the carrier charge and the particle current density. For holes, the diffusion current density J_{diff} is given by

$$J_{\text{diff}} = -qD_p\frac{dp}{dx} \tag{2.8}$$

and, for electrons,

$$J_{\text{diff}} = +qD_n\frac{dn}{dx} \tag{2.9}$$

where D_p and D_n are the diffusion coefficients and p and n are the concentrations for holes and electrons, respectively.

The current in the semiconductor due to an electric field is called the drift current. An electric field can exist in a semiconductor either due to an applied voltage or because of an internal distribution of charge. Ionized impurities represent immobile charge, and electrons and holes represent mobile charge. If there is a region in the semiconductor where the mobile charge does not neutralize the immobile charge, a field will exist. If free, mobile carriers exist in this region, then a drift current density exists, given by Equation 2.3.

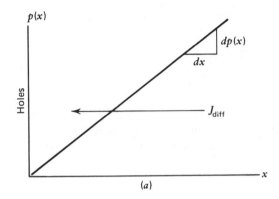

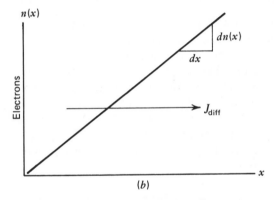

Figure 2.2. Concentration gradient diagrams for (a) holes and (b) electrons.

The total current density for a semiconductor is the sum of the individual hole and electron contributions due to drift and diffusion and is given by

$$ J = q(n\mu_n + p\mu_p)E + q\left(D_n\frac{dn}{dx} - D_p\frac{dp}{dx} \right) \tag{2.10} $$

In an extrinsic semiconductor, the terms in Equation 2.10 may have vastly different magnitudes. The majority carrier drift term is usually much larger than its minority carrier counterpart. The diffusion component for minority carriers can be significant since it depends on the gradient, not on the size, of the minority carrier concentration.

The physical mechanisms of drift and diffusion appear to be quite different, but they both result in the net flow of electric charge. For this reason, it is not surprising that the diffusion coefficient and the mobility for a carrier in a specific semiconductor are related. The ratio of these quantities is

$$\frac{D_n}{\mu_n} = \frac{D_p}{\mu_p} = \frac{kT}{q} \tag{2.11}$$

which is called Einstein's relation.

The carrier mobility is a function of the doping level. In silicon, the electron mobility varies from 1350 cm²/V·s for intrinsic material to 100 cm²/V·s for a doping level of 10^{19} cm^{-3} or more. Hole mobility varies from 480 to 50 cm²/V·s over the same doping range.

2.4 EXCESS CARRIER LIFETIME

Nonequilibrium mobile charge conditions may occur by carrier injection from a contact or electron–hole pair generation due to illumination. This creates a concentration of electrons and holes in the semiconductor that exceeds their equilibrium values. When the source of excess carriers is removed, the carrier densities "relax" to their equilibrium values through the process of recombination. The average time required for a recombination event to take place is called the *excess carrier lifetime* τ. The subscript for the lifetime is associated with the minority carrier, even though the recombination event eliminates both an electron and a hole. For p-type material, the excess electron concentration n', as a function of time t, is given by

$$n'(t) = n'(0) \exp\left[\frac{-t}{\tau_n}\right] \tag{2.12}$$

Lifetime in silicon is strongly influenced by impurity density. The impurity atoms participate in the recombination process, so that an increase in impurity density results in a decrease in lifetime. Excess carrier lifetime in relatively pure silicon can be several seconds, while in heavily doped silicon the lifetime is on the order of a microsecond.

A characteristic length associated with the diffusion of minority carriers in a semiconductor is called the *diffusion length L*. This is the average distance that a minority carrier travels by diffusion before recombination takes place. For electrons in p-type material, the diffusion length is

$$L_n = (D_n\tau_n)^{1/2} \tag{2.13}$$

Similarly, for holes in n-type material,

$$L_p = (D_p\tau_p)^{1/2} \tag{2.14}$$

Diffusion lengths in silicon range from 1500 μm to less than 1 μm as the doping density ranges from 10^{14} to 10^{19} cm^{-3}.

2.5 **DEVICE FABRICATION**

Large, nearly perfect single crystals are needed to make electronic devices out of very highly purified semiconductor materials. The important processes include the formation of junctions between regions with opposite impurity types, the deposition of conducting layers, and the photographic etching of patterns in the various layers. The processes for the fabrication of semiconductor devices and integrated circuits are discussed at length in the text *Microelectronics: Processing and Device Design* (R. A. Colclaser, Wiley, New York, 1980) and other similar treatments. In the next few paragraphs, we provide an introduction to these processes so that the reader can understand the basic device structures introduced in the ensuing chapters.

2.5.1 **Crystal Growth**

Silicon compounds are abundant in the earth's crust. Unfortunately, these materials must undergo extensive purification processes before they can be used to make electronic devices. The resulting high-purity chemicals are usually in the form of silicon tetrachloride ($SiCl_4$), silane (SiH_4), or one of the chlorosilanes such as $SiCl_2H_2$. These chemicals can be decomposed at high temperatures to grow polycrystalline ingots of high-purity silicon with resistivities greater than 100 $\Omega \cdot cm$. The polycrystalline silicon can then be melted (1404°C) and mixed with carefully controlled amounts of impurity to serve as a source for the growth of single-crystal silicon with a specified impurity concentration and type. The doping concentration for integrated circuits is often around 10^{15} boron atoms/cm^3 (p-type) or a similar concentration of phosphorus atoms/cm^3 for n-type. This provides crystals with resistivities between 10 and 20 $\Omega \cdot cm$. In some cases, this phosphorus concentration is much higher, typically 10^{20} atoms/cm^3, because the semiconductor crystal is to be used as a good conductor for specific applications. The resistivity is below 0.001 $\Omega \cdot cm$ for this material, and it is referred to as n^+-doped silicon.

The single crystal is grown by inserting a small single crystal called a seed crystal into the melt and slowly withdrawing it at a controlled rate. There is usually rotational motion of the seed relative to the melt, and the circular geometry of the crucible containing the melt is responsible for the cylindrical shape of the resulting crystals. This process was originally developed by Czochralski for other materials and adapted for silicon by Teal and Little. Many refinements have been made in the Czochralski process over a period of 30 years, and it is now possible to routinely grow essentially perfect silicon single crystals with diameters in excess of 125 mm. One characteristic of Czochralski-grown silicon is that it usually contains a high concentration of oxygen (10^{18} atoms/cm^3) due to the use of a silicon dioxide crucible liner. Molten silicon is so corrosive that it is difficult to find a container for it. If the oxygen problem cannot be overcome by other methods, a different technique, called float zone, is used for crystal growth, whereby the molten silicon does not contact the container. In most cases, the effects of the oxygen can be neutralized, and Czochralski material is used.

2.5.2 Wafer Preparation

After crystal growth, the boules of silicon are processed on a centerless grinder until the desired diameter is attained. At this point, the crystal is oriented by X-ray diffraction techniques and an orientation flat is ground on the crystal. This flat allows the manufacturer to make use of natural cleavage planes in a subsequent processing step when the individual chips on the wafer are to be separated. The orientation flat is the larger flat on the crystal. Another flat may be added to help in identifying the impurity type and crystal orientation.

The crystal is then sliced into wafers using a diamond-impregnated saw. The thickness of the wafers depends on the diameter, since the weight of the wafer can cause distortion and crystal deformation during subsequent high-temperature processing. A typical 100-mm wafer is 625 μm thick, after lapping and polishing to remove the saw damage.

The final steps in wafer preparation include a chemical-mechanical polish that leaves a microscopically smooth mirror finish on one side of the wafer.

2.5.3 Oxidation

One of the properties of silicon that has made it *the* outstanding device material is that it forms a stable oxide that can be used to protect its surface. Germanium and gallium arsenide do not have this property. Exposing silicon to an oxidizing atmosphere, even at room temperatures (or below), results in the immediate formation of an oxide layer. The SiO_2 chemical reaction takes place at the interface between the silicon and the oxide layer, requiring the oxidizing species to penetrate the existing layer for further growth. At room temperature, the "native" oxide stops growing after it has reached about 50 angstroms in thickness. Placing the wafers in oxygen or steam at temperatures in the 900–1200°C range results in repeatable oxide growths from 500 to 10,000 angstroms (1 μm) for typical device processes.

There are several uses for clean oxide layers in device fabrication. One feature of the oxide is that it retards the penetration of certain impurities. As we will see below, an important way of introducing impurities into single-crystal silicon is to surround the wafers at high temperature with a gas containing a high concentration of the impurity. The impurity atoms migrate into the semi-conductor by diffusion. If we have an oxide layer over part of the wafer and none (or a much thinner one) over the rest of the wafer, the impurity will be introduced much more readily into the uncovered part. The oxide is said to serve as a mask for diffusion. Silicon dioxide is a mask for boron, phosphorus, arsenic, and antimony.

Another important aspect of the oxide is that it can stabilize the surface of the silicon. If the oxide is grown using extreme care and cleanliness (often in the presence of HCl gas), it is possible to grow an oxide with very low concentrations of particular impurities, the most undesirable of which is sodium. If the oxide is clean, the resulting surface characteristics can be reproduced on a production basis. This is important for all types of devices, but it is essential for the operation of surface field effect transistors (see Chapter 5). The oxide also serves as an excellent insulating material, making it possible to run conducting paths between various parts of integrated circuits with very little electrical interactions. One of the best features of silicon dioxide is that it can be readily etched by hydrofluoric acid, a chemical that does not etch silicon.

2.5.4 **Photolithography**

Photographic techniques are essential in the mass production of semiconductor devices. The basic process, called photolithography, consists of coating the substrate (usually oxidized silicon) with a photosensitive organic polymer, exposing the polymer to ultraviolet radiation through a glass mask containing a pattern, rinsing off the unexposed polymer, harden the exposed polymer, and etching. The polymers used in this process are called photoresists because they are photosensitive and resist chemical attack. The process outlined above uses "negative" photoresist. It is also possible to use "positive" photoresist, in which the exposed polymer, rather than the unexposed, rinses off. This process has been refined to the point that patterns including linewidths approaching 1.0 μm can be reproduced in production. Unfortunately, we are reaching the limits of this simple process, and each small advance is gained by a great effort and expense. New techniques such as computer-controlled electron-beam lithography and X-ray lithography are needed for major advances in this area.

As mentioned above, silicon dioxide is readily patterned and etched using hydrofluoric acid. Other materials, including aluminum, silicon nitride, and polycrystalline silicon, can be patterned and etched in a similar manner with different chemicals. Advanced processes using plasmas for dry etching have been developed for fine line pattern definition. The final step in photolithography is the removal of the photoresist. Obviously, the better photoresists are more difficult to remove. This can be done either by oxidizing chemicals or by an oxygen plasma.

2.5.5 **Diffusion and Ion Implantation**

Semiconductor devices rely on the formation of pn junctions within single crystals of semiconductors. One technique starts with a uniformly doped n-type wafer that has been oxidized and patterned by photolithography. Next, a p-type impurity is introduced into the exposed regions. This is indicated in Figures 2.3a and 2.3b, where phosphorus and boron represent the n- and p-

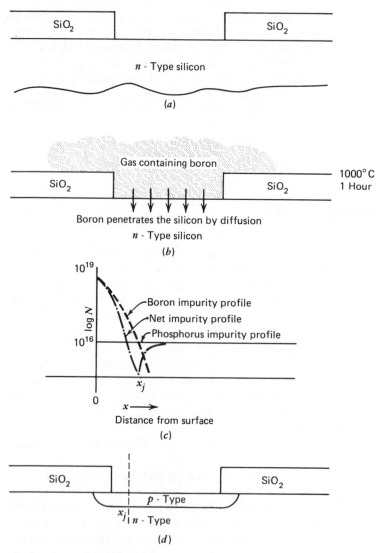

Figure 2.3. The formation of a *pn* junction. (*a*) The SiO_2 must be in place over an *n*-type silicon material. (*b*) Boron gas diffused into the silicon at high temperature. (*c*) Diffusion profiles. (*d*) Resultant *pn* junction formed in silicon material.

dopants, respectively. Boron enters the silicon by diffusion because the concentration of boron outside the silicon is greater than the concentration of boron inside the crystal. In a typical example, assume that the initial concentration of phosphorus inside the silicon is 10^{16} atoms/cm^3 and the concentration of the boron is maintained at 10^{19} atoms/cm^3 at the interface between the silicon and the gas containing the boron. The process takes place at 1000°C for 1 hour. The mathematics of diffusion indicates that the boron profile inside the semiconductor will be a complementary error function and that, when the wafer has cooled to room temperature, there will be a distance x_j (see Figures 2.3c and 2.3d) at which the concentration of the two impurities are equal. From our previous discussion of compensation, the dominant species from $x = 0$ to $x = x_j$ is boron (p-type), and the dominant species at distances greater than x_j is phosphorus. We have formed a pn junction, located at x_j, by diffusion and compensation. This is the basic configuration of the junction diode discussed in Chapter 3.

A repeat of the process, using a gas containing phosphorus and a new window in the oxide, results in the bipolar transistor structure (see Chapter 4) shown in Figure 2.4a. Using two junctions side by side with metal over the oxide between the two junction areas results in the MOS transistor structure (see Chapter 5) shown in Figure 2.4b.

Another approach to impurity insertion into silicon bombards the surface with high-energy ions of the desired impurity. The oxide layer can act as a mask for this process if it has the proper thickness to stop most of the ions before they penetrate to the silicon. In general, unless the energy of the ions is very high, implanted ions only penetrate a few tenths of a micrometer into the silicon. It is important to note that the ions cause physical damage when they enter the silicon and this damage must be annealed. One process sequence

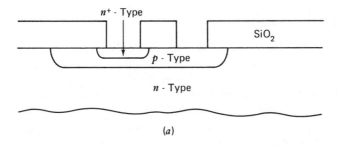

(a)

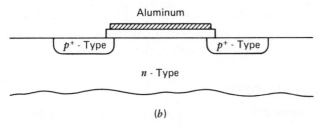

(b)

Figure 2.4. Transistor configurations. (a) Bipolar transistor structure. (b) MOS transistor structure.

that is used frequently inserts the impurity atoms by implantation and redistributes them by a subsequent high-temperature cycle during which the junction moves from close to the surface to the desired location by diffusion.

2.5.6 Chemical Vapor Deposition

Sometimes it is necessary to deposit a layer of silicon, silicon dioxide, or silicon nitride on a substrate. This is done by a process called chemical vapor deposition. The process is based on the thermal decomposition of compounds or temperature-dependent chemical reactions.

Most bipolar integrated circuits are fabricated in a thin, single crystal layer near the surface. This layer is grown as an extension of the substrate crystal lattice by a process called epitaxy. It is thus possible to grow abrupt junctions without compensation.

Polycrystalline silicon is used extensively in MOS integrated circuits as an electrode material and as a conductor. It is deposited readily on silicon dioxide by the same process as epitaxy. Since the oxide layer is not single crystal in structure, the deposited silicon atoms find a variety of nucleating sites, and the resulting layer is a fine-grain structure.

Silicon dioxide and silicon nitride are excellent insulator materials that can be deposited by chemical vapor deposition. These materials can be deposited on a variety of materials without difficulty. These dielectrics are used for crossover insulators in multilevel conductor systems.

2.5.7 Metal Deposition

Conductor materials are deposited on semiconductor devices for contacts and intraconnections. The depositions can be performed by vacuum evaporation or sputtering. The most common metal used for these purposes is aluminum, which is patterned using photolithography. Aluminum presents two problems related to the reliability of the devices. One is that it is relatively easily attacked by corrosion, and the second is its susceptibility to a failure mechanism called "electromigration." This is a process by which aluminum atoms are transported with the current under highly stressed conditions. The eventual creation of voids in the metal may be a true failure mechanism for aluminum-metallized integrated circuits. Other metal combinations have been demonstrated to have less susceptibility to electromigration, but the fabrication difficulties of trilayered metal systems such as TiPtAu or TiWAu have put more emphasis on alloying copper with silicon to reduce the problem.

2.5.8 Fabrication Summary

Silicon devices are fabricated by forming *pn* junctions in single-crystal wafers by a combination of oxidation, photolithography, diffusion, and/or ion implantation and the deposition of insulator and conductor layers. A wide variety of devices and circuits, from discrete diodes and transistors to 32-bit microprocessors and 256k RAMs, can be fabricated with the same basic processes.

2.6 SUMMARY

Most modern electronic devices are made from semiconductors. In their pure form, these materials have electrical properties somewhere between those of insulators and conductors but, in most cases, closer to those of insulators. Silicon is the most popular semiconductor. When a Group III impurity such as boron is incorporated in single-crystal silicon, the electrical properties of the material are dependent on the density of the impurity. The material is *p*-type with holes as the majority carriers, and the conductivity can be controlled over a wide range. If a Group V impurity such as phosphorus is used in place of the boron, the material becomes *n*-type, with electrons as the majority carriers. Conduction in semiconductors is by both drift and diffusion. Excess carriers have a doping density-dependent lifetime, mobility, diffusion coefficient, and diffusion length. The basic fabrication processes for the fabrication of semiconductor devices and integrated circuits, crystal growth, wafer preparation, oxidation, diffusion and ion implantation, chemical vapor deposition, and metal deposition have been described.

REFERENCES ───────────────────────────────

1. R. B. Adler, A. C. Smith, and R. L. Longini, *Introduction to Semiconductor Physics* (New York: Wiley, 1964).

2. R. A. Colclaser, *Microelectronics: Processing and Device Design* (New York: Wiley, 1980).

PROBLEMS ───────────────────────────────

2.1. (*a*) Calculate the density of free electrons and holes in a sample of silicon (Si) at 300°K that has a concentration of donor atoms equal to 2×10^{15} atoms/cm³. Is this *p*-type or *n*-type silicon? (*b*) Repeat part (*a*) for germanium (Ge).

2.2. (*a*) Determine the concentration of free electrons and holes in a sample of silicon at 300°K that has a concentration of donor atoms equal to 2×10^{15} atoms/cm³ and a concentration of acceptor atoms equal to 3×10^{15} atoms/cm³. Is this *p*-type or *n*-type silicon? (*b*) Repeat part (*a*) for germanium.

2.3. (*a*) Calculate the density of holes and electrons in *n*-type gallium arsenide at 300°K if the resistivity is 0.01 Ω·cm. (*b*) Repeat part (*a*) for *p*-type silicon if the resistivity is 10 Ω·cm.

2.4. Calculate the resistance of a bar of intrinsic silicon at 300°K which has a length of 2 cm and a cross section of 1 mm by 3 mm.

2.5. A sample of silicon is doped with 50×10^{15} donor atoms/cm^3 and 5×10^{15} acceptor atoms/cm^3. If the drift current density is 50 A/cm^2 at 300°K, find the electric field.

2.6. The conductivity of a silicon sample is 5 S/cm at 300°K. Calculate the electron and hole concentrations if the silicon is (*a*) *n*-type, (*b*) *p*-type.

2.7. Calculate the resistance of a bar of silicon at 300°K that has a length of 0.5 cm and a cross-sectional area of 10^{-4} cm^2 if the silicon has (*a*) a donor concentration of 1.0×10^{16} atoms/cm^3, (*b*) a donor concentration of 1.0×10^{15} atoms/cm^3 and an acceptor concentration of 1.0×10^{16} atoms/cm^3.

2.8. If a 6.0-V battery is connected across each of the samples in Problem 2.7, (*a*) what is the current through each sample, and (*b*) what is the average drift velocity of the majority carriers in each sample?

2.9. In an *n*-type sample of GaAs at 300°K, the electron concentration varies from 1.0×10^{18} to 7.0×10^{17} electrons/cm^3 over a distance of 0.10 cm. Calculate the diffusion current density.

2.10. The electrons in a *p*-type silicon semiconductor have an excess carrier lifetime of $\tau = 1.0$ μs. What is the diffusion length of the minority carriers? If the lifetime is 0.1 μs, what is the diffusion length? Repeat the problem for the case of minority carrier holes in an *n*-type silicon semiconductor.

chapter 3
SEMICONDUCTOR DIODES

The fundamental device of semiconductor electronics is the *pn* junction diode. A single-junction diode not only serves as a rectifier but is the basic building block in essentially all the semiconductor devices used today. These include the bipolar transistor, the junction field effect transistor (JFET), the metal-oxide–semiconductor field effect transistor (MOSFET), and the thyristor. In this book, you will be introduced to the operation and applications of all these devices except the thyristor. Thyristors deal with the control of power and are beyond the scope of this text.

In this chapter, we briefly consider the physical mechanisms responsible for the nonlinear, unilateral nature of the junction diode. A similar device, the Schottky barrier rectifier, will also be considered. The important models used to analyze diodes in a circuit are examined. We conclude the chapter with applications of these devices as rectifiers and as components in logic gates.

3.1 DIODE ELECTRICAL CHARACTERISTICS

The junction diode is a two-terminal device that can pass a large current with one voltage polarity and virtually no current in the reverse voltage polarity. Functionally, an ideal diode would be a short circuit with one polarity and an open circuit with the opposite polarity, as shown in Figure 3.1*a*. In practice, there is a nonlinear resistance in the forward voltage direction and a small current that is essentially independent of voltage in the reverse voltage direc-

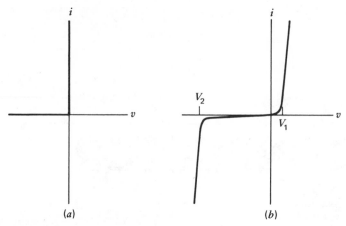

(a)   *(b)*

Figure 3.1. Current–voltage characteristics for diodes. (*a*)
Ideal. (*b*) Real.

tion. As the reverse voltage value is increased, a point is reached at V_2 where
the current increases rapidly with almost no increase in voltage, as shown in
Figure 3.1 *b*. V_1 represents the voltage at which the forward current first becomes significant. This is a relative concept, since "significant" depends on
the other components in the circuit. For a silicon diode, V_1 is between 0.5 and
0.7 V.

The word "bias" is used to denote a dc voltage across or a dc current
through a diode. The expressions "reverse bias voltage" and "forward bias
voltage" are usually shortened to just "reverse bias" and "forward bias."
These terms are frequently used to describe the states of diodes and transistors.

When a diode is reverse biased to a magnitude betweeen zero and V_2,
the diode behaves as a small capacitor. The value of the capacitance depends
on the bias voltage.

There is also a charge storage effect associated with a forward-biased
junction. This effect is most obvious when an attempt is made to rapidly switch
the voltage on a diode from forward to reverse bias. The current can change
direction instantaneously, but the voltage cannot switch until a certain amount
of charge has been removed from the device. This is a distinct capacitive effect
associated with a forward-biased *pn* junction and directly affects the switching
time of the junction. It is called the *diffusion capacitance* and is discussed
later. This effect is not observed in metal-semiconductor diodes called Schottky
barrier devices. Schottky diodes are used instead of junction diodes when fast
switching is necessary.

3.2 JUNCTION THEORY

In this section, a physical model of the *pn* junction is developed that predicts
the electrical characteristics described in the preceding section. The develop-

ment is largely qualitative, with more detailed models left for texts on device physics.

3.2.1 **The Abrupt Junction in Equilibrium**

A *pn* junction is formed within a single crystal of semiconductor material at the plane where the dominant impurity type changes from acceptors to donors. For simplicity, we shall use a device with a constant doping density on each side of the junction. This ideal device has N_a acceptor atoms/cm^3 for $x < 0$ and abruptly changes to N_d donor atoms/cm^3 for $x > 0$, as shown in Figure 3.2. In the figure, $x = 0$ represents the metallurgical junction, where the impurity concentration changes from *n*-type to *p*-type and $-x_1$ and x_2 represent the ends of the device on the *p*-side and *n*-side, respectively. Assume that metallic contacts are made to the semiconductor at $-x_1$ and x_2 in such a manner that no barrier to current exists from the metal to the semiconductor or from the semiconductor to the metal. This type of ideal metal-to-semiconductor contact is called *ohmic* and is difficult to obtain in practice.

If equilibrium conditions exist with an open circuit between the wires connected at $-x_1$ and x_2, no current exists through the device and no voltage can be measured between the wires. This is probably what you would have expected since the diode is a passive device, not a battery. A close analysis of Chapter 2 and, in particular, Equation 2.10 (repeated here for convenience),

$$J = q(n\mu_n + p\mu_p)E + q\left(D_n\frac{dn}{dx} - D_p\frac{dp}{dx}\right) \qquad (2.10)$$

may lead us to a different conclusion.

We begin this analysis by assuming (which we later show to be incorrect) that the values of the carrier concentrations are those listed in Table 3.1. From these values, it is clear that a very large increase in the electron density exists

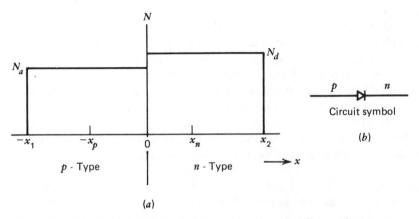

(a)

Figure 3.2. (*a*) The geometry and doping concentration for an abrupt *pn* junction and (*b*) Diode circuit symbol.

Table 3.1. Assumed Carrier Concentrations for an Abrupt Diode with $N_d = 10^{17}$ and $N_a = 10^{16}/cm^3$

Species	Region	
	$-x_1$ to 0	0 to x_2
n_0 (cm^{-3})	2.25×10^4	10^{17}
p_0 (cm^{-3})	10^{16}	2.25×10^3

in a vanishingly small region in the vicinity of $x = 0$. Similarly, there is a very large decrease in the hole density over the same very small region. The diffusion terms in Equation 2.10 indicate, however, that gradients in opposite directions for electrons and holes result in currents in the same direction. Thus, our assumption leads to the conclusion that a large current exists in the vicinity of $x = 0$. This is an unsatisfactory conclusion, since the device is effectively a node; and to satisfy Kirchhoff's current law, the same current must flow in the external leads. The external leads, however, are open circuited, invalidating our assumption concerning the carrier concentrations.

How is this enormous diffusion current stopped? The answer lies in the drift terms of Equation 2.10. If an electric field can be created that is just large enough and has the proper polarity to produce a *drift* current equal in magnitude but opposite in direction to the *diffusion* current, the total current will be zero. Unfortunately, the existence of an electric field implies that a voltage should appear between the external leads. This is not the case. There is a difference between the chemical energies in the two sides of the junction due to the presence of the impurities. This energy difference is exactly the right amount to counterbalance the voltage due to the built-in electric field.

We now have a way to satisfy the equilibrium conditions if a mechanism for creating an electric field within the device can be found. This is accomplished by the interaction of the impurity ions fixed in place within the crystal lattice, and the electrons and holes that are free to move about the lattice. The free electrons on the *n*-side may diffuse into the *p*-side and recombine with holes. The free holes on the *p*-side may diffuse into the *n*-side and recombine with electrons. Each electron or hole crossing the junction leaves behind a charge imbalance. The charge imbalance is seen as the fixed donor or acceptor atoms from which the free charges originate. Thus, powerful diffusion forces drive the free carriers across the junction where they recombine. The fixed charges left on both sides of the junction then create a large electric field across the *pn* junction between $-x_p$ and x_n, pointing from the *n*-side to the *p*-side. This region is called the *space charge, depletion,* or *transition region*.

The bound charge distribution may now be represented by Figure 3.3. In the regions from $-x_1$ to $-x_p$ and from x_n to x_2, the semiconductor material is charge neutral, and the assumptions from Table 3.1 apply. No field exists in these regions. The region from $-x_p$ to x_n is the proper width to produce the

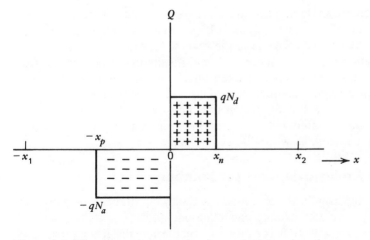

Figure 3.3. The space charge region in an abrupt *pn* junction.

voltage necessary to exactly balance the diffusion currents within the device. The voltage is called the *built-in voltage* across the junction, V_0. It is a function of the impurity densities on both sides of the junction, and it is given by

$$V_0 = \frac{kT}{q} \ln \left(\frac{N_d N_a}{n_i^2} \right) \tag{3.1}$$

At room temperature, the range of V_0 is 0.6 to 1.2 V in silicon. Remember that V_0 is the internal voltage necessary to balance the diffusion currents and that it cannot be measured at the external terminals. Note that the polarity of V_0 is positive from the *n*-side with respect to the *p*-side.

Example. If $N_d = 10^{19}/\text{cm}^3$ and $N_a = 10^{18}/\text{cm}^3$, find V_0 at room temperature in silicon.

Solution. The quantity $kT/q = 0.026$ V at room temperature, and $n_i^2 = 2.3 \times 10^{20}$ cm^{-6} for silicon. Then,

$$V_0 = 0.026 \ln \left(\frac{10^{19} \times 10^{18}}{2.3 \times 10^{20}} \right)$$

$$= 0.99 \text{ V}$$

There is another feature of the space-charge region that has an important effect on the electrical characteristics of the junction diode. The space-charge region contains two regions of fixed charge of opposite sign, like the dielectric that separates the plates of a capacitor. There is a capacitance per unit area, c_0, associated with the *pn* junction under equilibrium conditions. The capacitance value is influenced by the impurity densities on both sides of the junction.

In general, an increase in doping density increases the junction capacitance.

The magnitude of the reverse bias voltage affects the physical width of the space charge region. An increase in the reverse bias increases the electric field in the junction region. This increased field results in the removal of free carriers from an increased number of fixed donor and acceptor charges. The space charge region is measureably altered in size by the magnitude of the reverse bias voltage. This size alteration not only affects the capacitance of the diode but also has an affect on the amplifying properties of bipolar junction and field effect transistors, as we shall see in Chapters 4 and 5.

3.2.2 The Forward-Biased *pn* Junction

The equilibrium condition for a junction diode represents a delicate balance between large currents exactly canceling one another. This balance can be upset by applying a voltage across the external terminals. If the polarity of the applied voltage is such that the *p*-region is positive with respect to the *n*-region, the diode is said to be *forward* biased. If the *n*-side is biased positive with respect to the *p*-side, the diode is *reverse* biased.

Under forward bias conditions, the applied voltage v is opposite in polarity to the built-in voltage V_0. The effect is a reduction in the drift currents that balanced the diffusion currents under zero bias conditions. The resulting diffusion currents of holes into the *n*-side and electrons into the *p*-side can be significant, even for small values of v.

The minority carrier concentrations at the edges of the space charge region are controlled by the applied voltage v. The relationships are

$$n(-x_p) = n_0(-x_p) \exp\left(\frac{qv}{kT}\right) \tag{3.2}$$

and

$$p(x_n) = p_0(x_n) \exp\left(\frac{qv}{kT}\right) \tag{3.3}$$

where the subscript 0 signifies the equilibrium value. These relationships are often called "the law of the junction." The excess carrier concentrations at the planes, $n'(-x_p)$ and $p'(x_n)$ (see Figure 3.4), are given by

$$n'(-x_p) = n(-x_p) - n_0(-x_p)$$
$$= n_0(-x_p)\left[\exp\left(\frac{qv}{kT}\right) - 1\right] \tag{3.4}$$

and

$$p'(x_n) = p(x_n) - p_0(x_n)$$
$$= p_0(x_n)\left[\exp\left(\frac{qv}{kT}\right) - 1\right] \tag{3.5}$$

At $-x_1$ and x_2, the excess carrier concentrations are zero because ohmic contacts are located at these planes. If the lengths $x_2 - x_n = W_2$ and $(-x_p) - (-x_1) = W_1$ are small compared to the diffusion lengths for minority carriers in these regions, the distribution of excess minority carriers is approximately linear, as shown in Figure 3.4. If the conditions on W_1 and W_2 are not met, the distribution of excess carriers is more complicated, but the resulting current expression is similar to the one for this case.

The current in a junction diode is given by the diode equation,

$$i = I_s \left[\exp \left(\frac{qv}{kT} \right) - 1 \right] \tag{3.6}$$

where I_s is called the *reverse saturation current*. This terminology arises when v is large and negative, resulting in i $\simeq -I_s$. This constant depends on a number of factors, including the material from which the device is made, the doping densitites, the diffusion coefficients, and the dimensions of the device. In general, there are two terms in I_s, one for the diffusion current due to holes in the n-side and the other for the diffusion current due to electrons in the p-side. These terms are much larger than drift terms in most diodes. The diffusion terms are proportional to the excess minority carrier densities at the edges of the space charge region, given by Equations 3.4 and 3.5. The hole diffusion current on the n-side is proportional to $p_0(x_n) = n_i^2/N_d$, and the electron diffusion current on the p-side is proportional to $n_0(-x_p) = n_i^2/N_a$. If $N_d >> N_a$, $n_0(-x_p)$ will be very much greater than $p_0(x_n)$. This implies that the current in this nonsymmetrical doped diode will be dominated by electron diffusion from the heavily doped n-side into the lightly doped p-side, with a very small hole flow in the other direction. This is referred to as a one-sided diode, and it is frequently used in the construction of bipolar junction transistors.

The diode equation is plotted on several different scales in Figures 3.5*a–d*, where the diode current is plotted as multiples of I_s. The physical factors determining I_s have a strong effect on the diode current–voltage char-

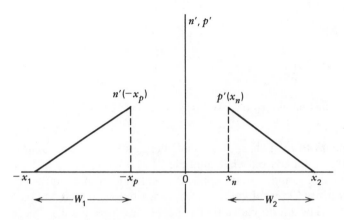

Figure 3.4. The excess minority carrier distributions on both sides of an abrupt *pn* junction with $W_2 << L_p$ and $W_1 << L_n$.

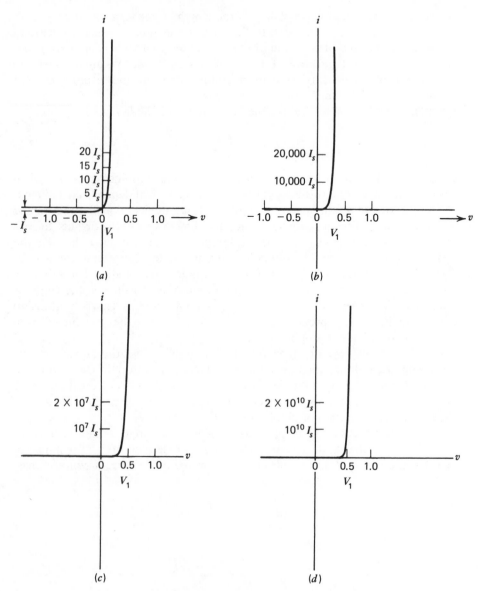

Figure 3.5. The diode equation plotted on four different current scales.

acteristic. What appears to be a sharp "turn-on" voltage V_1 in Figure 3.5d is a relative concept that depends on what multiple of I_s is considered to be a significant current level. In many circuit applications, the idea of a "turn-on" voltage is a useful concept. Typical values for V_1 range between 0.60 and 0.75 V for a silicon diode. The diode capacitive concepts are examined in more detail in Chapter 11.

The transition region described in the preceding section is reduced in width under forward bias conditions, resulting in a slight increase in capacitance with increasing forward bias. However, another capacitance effect, due to the storage of minority carriers in the device, usually dominates over the space-charge region capacitance in the forward-biased diode.

3.2.3 Small-Signal Operation Under Forward Bias

A typical operating condition for a junction diode occurs when a dc forward bias voltage and a small-signal variation are applied simultaneously to the device. This can be represented by

$$
\begin{aligned}
i &= I_s \left[\exp\left(\frac{qv}{kT}\right) - 1 \right] \\
&= I_s \left\{ \exp\left[\frac{q(V_{DC} + \Delta v)}{kT}\right] - 1 \right\} \\
&= I_s \left[\exp\left(\frac{qV_{DC}}{kT}\right) \exp\left(\frac{q\Delta v}{kT}\right) - 1 \right]
\end{aligned}
\tag{3.7}
$$

where i and v are the total instantaneous current and voltage, respectively; V_{DC} is the dc bias voltage; and Δv is the variation signal around the dc operating point. The ac or small-signal resistance of the diode may be derived in the following manner. If $\Delta v \ll kT/q$, Equation 3.7 can be expanded in a series and approximated by

$$
\begin{aligned}
i &\simeq I_s \left[\exp\left(\frac{qV_{DC}}{kT}\right)\left(1 + \frac{q\Delta v}{kT}\right) - 1 \right] \\
&\simeq I_s \exp\left(\frac{qV_{DC}}{kT}\right) + I_s\left(\frac{q\Delta v}{kT}\right)\exp\left(\frac{qV_{DC}}{kT}\right) - I_s
\end{aligned}
\tag{3.8}
$$

The current at the operating point I_{DC} is given by

$$
I_{DC} = I_s \left[\exp\left(\frac{qV_{DC}}{kT}\right) - 1 \right]
$$

Then,

$$
i = I_{DC} + \Delta i
$$

and

$$
\Delta i \simeq (I_{DC} + I_s)\frac{q\Delta v}{kT}
$$

resulting in

$$
r \simeq \frac{\Delta v}{\Delta i}
$$

$$r \simeq \frac{kT}{q(I_{DC} + I_s)} \tag{3.9}$$

where r is the small signal or ac resistance of the diode. At room temperature, $kT/q = 26$ mV. For a typical diode, $I_s << I_{DC}$ at bias voltages above 0.60 V, so that

$$r \simeq \frac{26}{I_{DC}}$$

with I_{DC} in milliamperes. The value of r can also be determined from the slope of the characteristic curve.

Example. If the dc bias current is 13 mA, find r. Repeat for $I_{DC} = 75$ μA.

Solution. $r = 26/I_{DC}$. In the first case, $r = 2$ Ω. In the second case, $r = 347$ Ω.

There is a small-signal capacitive effect in junction diodes that is important in the study of high-frequency limitations of the diode. This is due to the nature of the current in these devices, where the primary current mechanism is the diffusion of minority carriers. For simplicity, we consider a "one-sided" diode with the p-side very heavily doped with respect to the n-side. This results in an excess hole distribution in the n-side as shown in Figure 3.6a, if $x_2 - x_n = W$ is much less than the diffusion length for holes in this region. The diffusion current I_{diff} for this distribution is given by

$$
\begin{aligned}
I_{\text{diff}} &= -qAD_p \frac{dp'}{dx} \\
&= qAD_p \frac{p'(x_n)}{W}
\end{aligned}
\tag{3.10}
$$

where A is the cross-sectional area of the device. The total charge stored

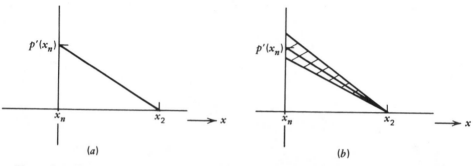

Figure 3.6. Excess hole distribution in the n-side of a junction diode. (*a*) dc condition. (*b*) dc condition with a small variation.

in this region due to holes Q can be found by integrating from x_n to x_2. The result is

$$Q = qp'(x_n)\,\frac{WA}{2}$$

Substituting this result in Equation 3.10 yields

$$I_{\text{diff}} = Q\,\frac{2D_p}{W^2}$$

The quantity $W^2/2D_p$ has the dimensions of time and represents the time required for a hole to traverse the region between x_n and x_2. It is designated τ_{tr}. Then,

$$I_{\text{diff}} = \frac{Q}{\tau_{tr}} \tag{3.11}$$

When the diode is forward biased to a dc operating point designated by I_{DC} and V_{DC} and a small signal voltage variation is applied, the charge due to excess minority carriers in the n-region varies along with the signal. A typical variation of charge is indicated in Figure 3.6b. This change in stored charge in response to a change in voltage represents a capacitance called the *diffusion capacitance* c_d. A detailed analysis indicates that the value of the diffusion capacitance is directly proportional to the dc bias current. The small-signal effects on a forward-biased pn junction can be modeled as a resistor in parallel with two capacitors, as shown in Figure 3.7. The value for the resistor, r, is $kT/[q(I_{DC} + I_s)]$, as given in Equation 3.9. The capacitors c_d and c_{sc}, represent the diffusion and space-charge region capacitors, respectively.

3.2.4 The Reverse-Biased *pn* Junction

The reverse bias condition for a junction diode results in an almost negligible current for voltages up to the breakdown voltage. Earlier, the balance

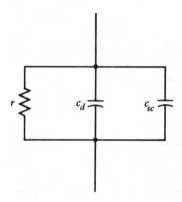

Figure 3.7. A small-signal model for a forward-biased *pn* junction diode.

of drift and diffusion currents in a diode under equilibrium conditions was described. A reverse bias voltage adds to the built-in voltage. This implies that reverse currents are due to drift rather than diffusion. The polarity of the voltage is such that electrons on the p-side and holes on the n-side cross the space-charge region to the opposite sides. The minority carriers swept out of the n- and p-regions are continually supplied by the thermal generation of electron–hole pairs at the edge of the space-charge region and by diffusion from farther away from the junction. The result is a reverse current that is very small and relatively independent of bias voltage. In a real diode, there is an observable increase in reverse current with an increase in reverse voltage. This is due to the thermal generation of carriers within the space charge region and the variation of the width of that region with increasing reverse bias. For most circuit applications, the diode equation, as stated in Equation 3.6, may be used for both forward and reverse bias situations.

An interesting feature of a reverse-biased diode is the variation of the small-signal space charge region capacitance as a function of bias voltage. If the zero bias capacitance per unit area is c_0, the junction capacitance is given by

$$c_{sc} = c_0 A \left(1 - \frac{V_{DC}}{V_0}\right)^{-1/n} \tag{3.12}$$

where V_0 is the built-in voltage, V_{DC} is the dc bias voltage (which is a *negative* quantity for reverse bias), and n is between 2 and 3. A typical value for c_0 is between 200 and 1000 pF/mm². The capacitance–voltage characteristic of a reverse-biased diode is used in electrically tunable resonant circuits. Diodes made specifically for this purpose are called varactors.

The diode equation does not apply to reverse bias conditions when breakdown occurs. Breakdown is characterized by a significant increase in current with very little change in voltage. There are two physical mechanisms that can participate in breakdown. The first of these is the *Zener* effect, which is based on a principle in quantum mechanics called *tunneling*. This can only happen in junctions with heavy doping concentrations on both sides, resulting in narrow space charge regions. Diodes exhibiting only the Zener effect exhibit breakdown at voltages less than 4 V. The second basic mechanism for breakdown is the *avalanche* effect. The electrons and holes are accelerated through the space charge region of a reverse-biased pn junction by the electric field of the built-in voltage and the applied voltage. If these carriers reach a high enough energy, they produce electron–hole pairs when they interact with the crystal lattice atoms in the space charge region. Since these new carriers are generated in a high-field region, they are also accelerated to high energy. If the newly produced carriers interact with lattice atoms before they leave the space charge region, even more carriers are generated. Once the voltage necessary to produce avalanche has been reached, the current is limited by external components in the circuit. For diodes that breakdown at voltages higher than 7 V, avalanche is the dominant mechanism. Those diodes with breakdown voltages between 4 and 7 V have both avalanche and Zener mechanisms contributing to the breakdown.

Breakdown can be either destructive or useful, depending on the circuit configuration. Power dissipation is the determining factor. Since breakdown can result in a combination of high voltage and relatively high current, the power dissipation can be sufficient to melt portions of the device. If the resistance in series with the diode limits the current and power dissipation to a value within the capabilities of the device, the breakdown mode can be used as a constant voltage in circuit applications. Of particular interest in this regard is the 5.6-V reference diode. The mixture of avalanche and Zener mechanisms causing this particular breakdown phenomenon produces a breakdown voltage that is almost independent of temperature.

The voltage across a diode under breakdown conditions does increase slightly with current. This results in a dynamic resistance r_d, which is a function of the operating point. This resistance can be determined by measuring the slope of the current–voltage characteristic.

3.2.5 **The Diode On–Off Switching Transient**

A forward-biased diode in series with a resistor will establish a steady-state charge distribution of minority carriers near the pn junction, as shown in Figure 3.6. If the voltage polarity is switched at t = 0, as shown in Figure 3.8a, the current direction changes at once, but the voltage across the device does not change immediately. This is shown in Figure 3.8b. The reverse current for the time interval between 0 and t_s is only limited by the resistance in the circuit. If the reverse voltage is increased, or the resistance is decreased, the reverse current during the initial transient is increased. The "storage" time t_s is the time necessary to reduce the excess carrier densities at the edges of the space charge region to zero. This time interval is a function of the magnitude of the reverse current, with a larger reverse current resulting in a shorter storage time. The remaining part of the transient, for $t > t_s$, is associated with the recombination of excess carriers within the diode. For a one-sided junction, this part of the transient is an exponential with a time constant equal to the excess carrier lifetime in the lightly doped side. The reduction of the storage time t_s is a major concern in both switching diodes and bipolar junction transistors used in digital switching applications.

3.3 **SCHOTTKY BARRIER DIODES**

Metal–semiconductor diodes have current–voltage characteristics that are very similar to pn junction diodes. The diode equation can describe the characteristics of both types of devices, but the physical mechanisms for operation of the two devices are not the same. When a metal and a semiconductor are brought together, there is an energy barrier to the transfer of electrons across

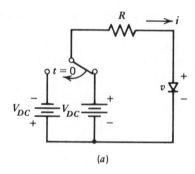

(a)

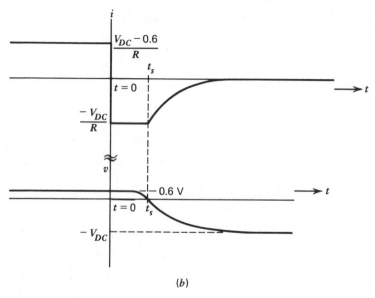

(b)

Figure 3.8. The diode turn-off transient. (*a*) Circuit. (*b*) Waveforms.

the junction unless great care is exercised to eliminate the barrier. This is called a *Schottky barrier*. Contacts with no barrier are called ohmic. In general, the barrier height depends on the type of metal and the doping concentration and type in the semiconductor. For example, aluminum forms an ohmic contact to heavily doped *n*-type and *p*-type silicon but forms a Schottky barrier to moderately doped *n*-type silicon.

The primary difference between metal–semiconductor and *pn* junction diodes is concerned with charge storage. Current in a junction diode is based on the diffusion of minority carriers. In a Schottky barrier diode, current is due to the drift of majority carriers. There is essentially no minority carrier

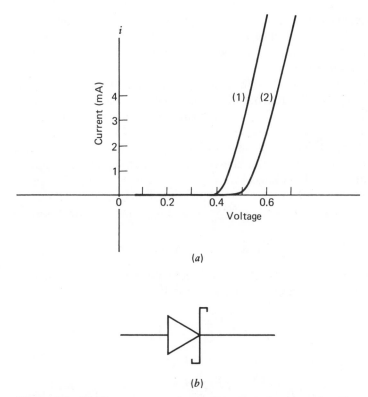

Figure 3.9. (*a*) The current–voltage characteristic curves for (1) a Schottky barrier diode and (2) a *pn* junction diode under forward bias. (*b*) The electronic symbol for a Schottky barrier diode.

storage in a Schottky barrier diode, so that the switching time from forward bias to reverse bias is very short compared to that of a junction diode.

The reverse saturation current I_s in a Schottky barrier diode is larger than that of a junction diode made from the same semiconductor. The diode equation (Equation 3.6) indicates that a larger I_s results in a larger forward current for the same forward bias voltage. Figure 3.9*a* shows a plot of the forward characteristics of a typical Schottky barrier diode and a junction diode on the same scale. Figure 3.9*b* shows the symbol for this diode.

3.4 DIODE CIRCUIT MODELS

To aid in the analysis of a circuit containing electronic devices, it is important to select an appropriate device model. The best model to select is the least complicated one that yields results consistent with measurements. For example, it makes no sense to include capacitors in a model for dc operation, but to ignore them in switching transients leads to erroneous results. It is important to recognize that the same device can be modeled in several different ways, depending on the application.

Three different modeling techniques will be used for the diode circuit of Figure 3.10. Each approach presents different insights, accuracies, and ease of usage. The three analytical techniques are (a) iterative solutions to a nonlinear Kirchhoff law equation, (b) simultaneous graphical solution of I_D-versus-V_D curves for the diode and the circuit, and (c) linear approximation of the diode for large currents. The object of the analysis is to determine the dc value of the current I_D and the diode voltage V_D.

The first approach uses the theoretical junction diode equation (Equation 3.6) in a Kirchhoff's law statement. Kirchhoff's voltage law for the circuit of Figure 3.10 yields

$$5.0 = 1.0 \times 10^3 I_D + V_D \tag{3.13}$$

This circuit equation has two unknowns. A direct solution can be obtained by substituting the diode relation of Equation 3.6 for V_D:

$$I_D = I_s \left[\exp \left(\frac{qV_D}{kT} \right) - 1 \right]$$

or

$$V_D = \frac{kT}{q} \ln \left(\frac{I_D + I_s}{I_s} \right) \tag{3.14}$$

Then,

$$5.0 = 1.0 \times 10^3 I_D + \frac{kT}{q} \ln \left(\frac{I_D + I_s}{I_s} \right) \tag{3.15}$$

Equation 3.15 is a transcendental equation without an obvious solution for I_D. However, the solution may be approximated rather quickly using an iterative method and a calculator. The approach is to substitute various values of I_D into the right-hand side of Equation 3.15 until a value of I_D is found that closely balances the equation. A reasonable initial trial value for I_D is the maximum possible value that it could have if there were no voltage drop across the diode. If V_D is zero, $I_D(\text{max}) = 5.0/1.0 \times 10^3 = 5.0$ mA. Substituting I_D = 5.0 mA and $I_s = 5.0 \times 10^{-12}$ A into Equation 3.15 results in an imbalance

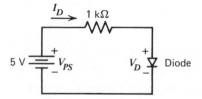

Figure 3.10. A steady-state diode circuit.

of the equation of 5.0 V to 5.33 V. By substituting smaller values for I_D in this trial and error solution, we find that I_D = 4.46 mA results in an imbalance of 0.004 V and is an acceptable answer. The voltage across the diode may then be calculated directly from Equation 3.14:

$$V_D = \frac{kT}{q} \ln \left(\frac{I_D + I_s}{I_s} \right)$$

$$= 0.536 \text{ V}$$

The second method of analyzing the circuit of Figure 3.10 uses a graphical approach to match the diode device and circuit properties. A diode character-istic curve can be readily obtained from a diode curve tracer instrument and could be plotted as shown in Figure 3.11. This curve represents the true mea-sured I_D-versus-V_D properties of a particular device. The Kirchhoff's law state-ment (Equation 3.13) for the circuit also relates I_D and V_D:

$$5.0 = 1.0 \times 10^3 I_D + V_D \tag{3.13}$$

This equation is a straight-line function referred to as the circuit "load line." The load line is also plotted in Figure 3.11, and the intersection of the two curves is the circuit solution. The current $I_D = I_{DQ}$ = 4.45 mA and the diode voltage $V_D = V_{DQ}$ = 0.534 V are the values at the dc operating point. The subscript Q is frequently used to designate the operating point and is the ab-breviation for "quiescent."

The load line is a frequently used concept in electronic circuits. It is a

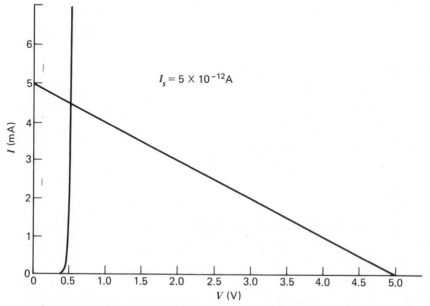

Figure 3.11. A load line solution for the diode current.

Kirchhoff's law statement of the circuit and is usually a straight-line function relating the current and voltage of a device. The load line (Equation 3.13) can be drawn by first locating the intercepts of the abscissa and ordinate. The abscissa intercept for Equation 3.13 is $V_D = 5.0$ V, and the ordinate intercept is 5.0 mA. The graphical method of solution is accurate but somewhat cumbersome for practical use. The concept associated with the graphical solution, however, is frequently used when discussing electronic circuit operating points and ac variations about the operating point. The concept is stressed frequently in the discussions of transistor circuits in later chapters.

The third approach is the easiest but least accurate. The diode equation is an excellent model of the diode but can lead to complex solutions, as indicated above. Therefore, it is desirable to develop other models that yield similar results. A simple but useful model for the forward-biased junction diode is a battery with a value between 0.50 and 0.75 V, depending on the current level. Referring to the preceding example, replacing the diode with a 0.60-V battery leads to a current of 4.4 mA. A similar approach is called a "piecewise" linear model. This method simulates or models the diode characteristic with straight-line segments, as shown in Figure 3.12. For example, in Figure 3.11, r is approximately 5 Ω and V_1 is 0.55 V. Since the circuit resistance is very large compared to r, this model yields very similar results to the previous model. The effect of r becomes more significant if the circuit resistance is reduced. The major source of error in the battery model is the choice of voltage for the forward-biased diode.

The reverse-biased diode can be modeled in several ways. In low-frequency applications, it can be represented as an open circuit. An alternate

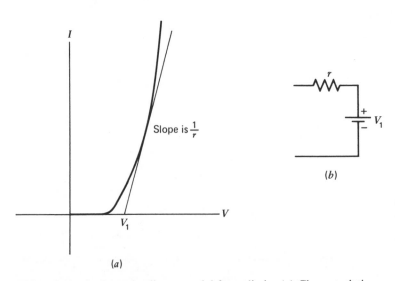

(a)

(b)

Figure 3.12. A piecewise linear model for a diode. (*a*) Characteristic. (*b*) Model.

model uses the space charge region capacitance discussed previously. The reverse saturation current can be represented as a constant current source in parallel with a large-valued resistor. A battery or a battery in series with a small resistor is an appropriate model for a diode in breakdown.

3.5 DIODE APPLICATIONS

Discrete diodes are commercially available in a wide variety of current and breakdown voltage ratings. Power diodes capable of forward currents up to 1000 A and reverse voltages up to 1000 V are readily available. These contrast with signal diodes having current ratings in milliamperes and breakdown voltages less than 10 V. In this section, we consider several typical applications of diodes. These are not the only applications for diodes but provide an introduction to the use of these versatile devices. Diodes that are built into the working structure of bipolar and junction field effect transistors are no longer called discrete diodes. This subject is developed in Chapters 4 and 5.

3.5.1 Rectifiers

A significant use of power diodes is the conversion of power from ac to dc for use as power supplies for electronic equipment. The electronic power supply is a sophisticated electronic system, but at this point we are not prepared to discuss all of the components of one of these complex systems. Instead, we concentrate on the "rectification" process.

A single diode is placed in series with a resistor and an ac power source in Figure 3.13a, producing the current waveform shown in Figure 3.13b. The shape of the current i is also the shape of the voltage across the resistor and is almost one half-cycle of a sine wave. During the other half-cycle, the current is the reverse bias saturation current and is essentially zero. The departure from a sinusoid during the first half-cycle is due to the nonlinear current–voltage characteristic of the forward-biased diode. If the peak voltage of the power source, V_{PS}, is very large compared to a forward-biased diode voltage drop, the current appears to be very nearly sinusoidal during the positive half-cycles.

The voltage across the resistor, v_R, represents the "output" of this circuit. Since our goal is to convert the sinusoidal ac input to a smooth dc output, it is clear that an efficient filter is required to remove most of the variation in the output. If a large-valued capacitor is connected in parallel with the resistor such that the RC time constant is large compared to the period of the sine wave, the output voltage v_{RC} becomes that shown in Figure 3.14.

A more efficient way to reduce the ripple from a rectifier circuit is to use a full-wave rectifier arrangement. Two popular full-wave rectifiers are shown in Figure 3.15. These circuits are used extensively in power supplies. In a power supply circuit, a final smoothing occurs on the waveform of Figure 3.15, using a voltage regulator circuit. The modern voltage regulator is an inexpensive

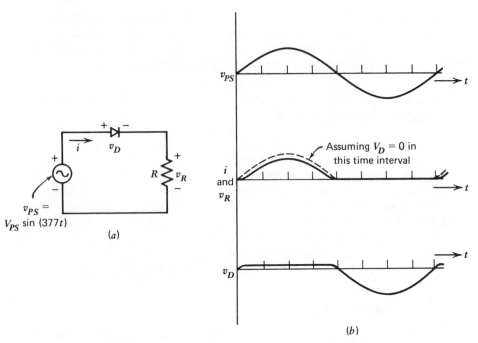

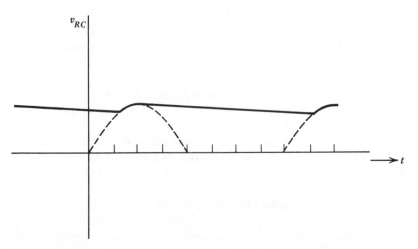

Figure 3.13. A half-wave rectifier circuit (*a*) and the resulting waveforms (*b*).

Figure 3.14. The steady-state output voltage of an RC filter excited by a half-wave rectifier.

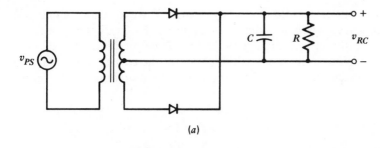

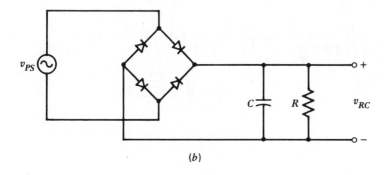

(a)

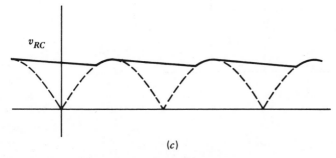

(b)

(c)

Figure 3.15. Full-wave rectifier circuits. (*a*) A center-tapped transformer full-wave rectifier circuit. (*b*) A bridge full-wave rectifier circuit. (*c*) A typical output from either circuit.

three-terminal integrated circuit available for a large range of input and output voltages.

3.5.2 **Detectors**

One of the first applications of semiconductor diodes was a "detector" in receivers for amplitude-modulated (AM) radio waves. The amplitude-modulated signal consists of a radio-frequency "carrier" wave whose amplitude varies at an audible frequency, as indicated in Figure 3.16a. The detector circuit, shown in Figure 3.16b, is very similar to the half-wave rectifier. The RC time constant is approximately the same as the period of the carrier in this case so that the output voltage can follow the variation in the amplitude of the input.

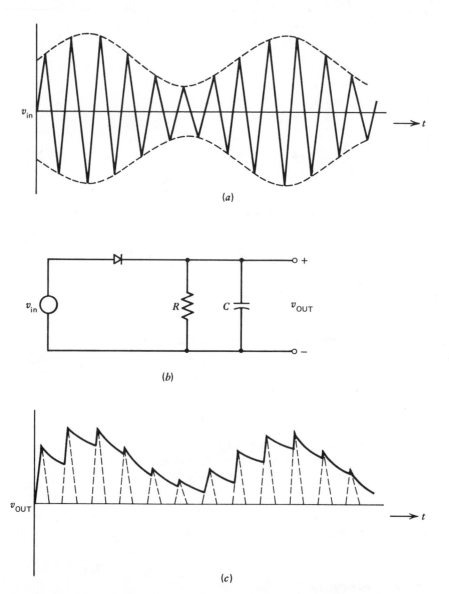

Figure 3.16. The signals and circuit for demodulation of an amplitude-modulated signal. (a) The amplitude-modulated input signal. (b) The detector circuit. (c) The demodulated output voltage.

The output voltage of the circuit is coupled to an amplifier through a capacitor to remove the dc level from the signal. The output of the amplifier is then coupled to a speaker to provide the acoustic vibration that matches the audio-frequency electrical signal from the AM radio wave.

3.5.3 **Voltage Regulators**

An interesting use of the breakdown characteristic of a diode is in the simple voltage regulator. The object of this circuit is to maintain a constant output voltage over a range of load resistance. The basic circuit is shown in Figure 3.17. Proper circuit operation requires that the diode breakdown voltage V_D be less than V_{PS}. The resistor R is selected to provide the proper voltage drop when the load resistance is at its minimum value. The diode must be capable of dissipating enough power when the load resistance is at its maximum value. It must be recognized that the power dissipated in an avalanche diode under breakdown conditions is the product of a large voltage and a relatively large current.

Example. If $V_{PS} = 20$ V and $V_D = 10$ V in Figure 3.17, find R and the maximum power dissipated in the diode when the load resistance varies from 100 to 500 Ω.

Solution. If $R_L = 100\ \Omega$ and $R = 100\ \Omega$, the voltage drop across R_L would be 10 V, even if the diode were not present. With the diode in place, under these circumstance, $I_D = 0$. In practice, it is desirable to have at least a small current in the diode when the load resistance is at its minimum value. Therefore, we select $R = 90\ \Omega$, slightly smaller than the maximum value that it could have and still satisfy the circuit criteria. To provide a 10-V drop across R, I must be 0.11 A. $I_L = 10/100 = 0.10$ A. Then $I_D = 0.01$ A under these conditions. If the load resistance is increased to 500 Ω, I_L decreases to $10/500 = 0.02$ A but the voltage drop across R must stay the same. Therefore, I stays at 0.11 A and I_D increases to $0.11 - 0.02 = 0.09$ A. The power dissipated in the diode under these circum-

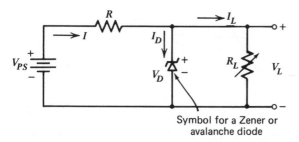

Figure 3.17. A basic voltage regulator circuit.

stances is $P_D = V_D I_D = 10 \times 0.09 = 0.90$ W. This is the maximum power dissipated in the diode when the load resistance stays within the specified range. The same circuit configuration can be used for a fixed load and a variable supply voltage, or for both load and supply variations.

3.5.4 Clipping and Clamping

Diodes are frequently used in wave-shaping circuits, since diodes and batteries (or avalanche diodes) can be used to limit voltage excursions. These circuits are called clipping circuits and an example is shown in Figure 3.18. The diode in the circuit is reverse biased when v_{in} is less than the battery voltage V. Effectively, no current exists in the circuit when the diode is reverse biased. The zero-current condition forces v_{OUT} to be equal to v_{in}, since the voltage drop across the resistor is zero. When v_{in} becomes greater than V, the diode becomes forward biased and conduction occurs. The output voltage is then $v_{OUT} = V_D + V$. This voltage condition holds when the diode is forward biased and is independent of the magnitude of v_{in} as long as $v_{in} > V$. As v_{in} becomes less than V, the diode becomes reverse biased again, and $v_{OUT} = v_{in}$. Other clipping variations of the circuit in Figure 3.18 can be obtained by reversing the polarity of the diode, the battery, or both.

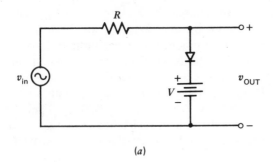

(a)

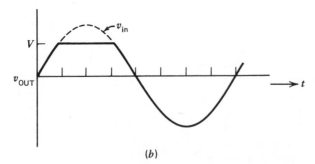

(b)

Figure 3.18. A diode clipping circuit. (*a*) Circuit. (*b*) Waveform.

Figure 3.19*a* illustrates a two-level clipping circuit whose response to a sinusoidal input is shown in Figure 3.19*b*. For $v_{in} < V$, the diode in series with V_1 is forward biased, and the output voltage is $v_{OUT} = V_1 - V_D \approx V_1$. For $V_1 < v_{in} < V_2$, both diodes are reverse biased and $v_{OUT} = v_{in}$. As v_{in} becomes greater than V_2, the diode in series with V_2 conducts, and $v_{OUT} = V_D + V_2 \approx V_2$. This is just one example of the versatility in the various combinations of diodes, batteries, and resistors.

A similar application for diodes is used to clamp a voltage so that it never goes positive (or negative). Figure 3.20 shows a negative clamping circuit that provides a charging path for a capacitor but uses a diode to prevent the discharge. This has the effect of providing a dc voltage shift of the waveform. The operation of this circuit is more subtle than that of the clipping circuit. During the first 90° of the input waveform, the ideal diode is forward biased, and the capacitor charges to a voltage equal to v_{in}. At $v_{in} = V_p$, the capacitor acquires a voltage $v_C = V_P$. As v_{in} reduces its amplitude from V_P, the capacitor holds its voltage V_P, and the diode acquires a reverse bias, shutting it off. No current exists in the circuit, and

$$v_{OUT} = -v_C + v_{in}$$

The waveforms for v_C and v_{in} are shown in Figure 3.20. The output voltage v_{OUT} is the algebraic sum of v_C and v_{in} and is shown at the bottom of the figure.

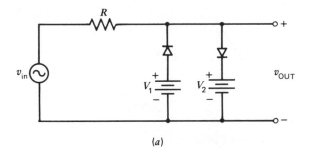

(*a*)

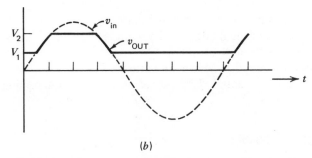

(*b*)

Figure 3.19. A two-level clipping circuit. (*a*) Circuit. (*b*) Waveform.

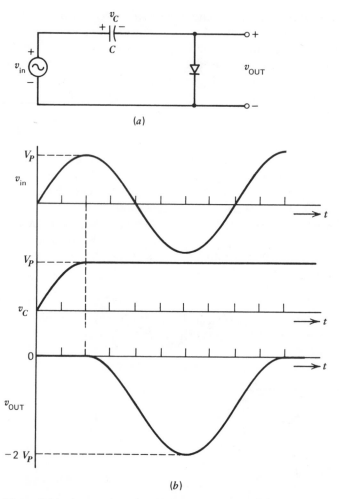

(a)

(b)

Figure 3.20. A negative clamping circuit. (a) Circuit. (b) Waveform.

The result is a sinusoid of the same amplitude and frequency of v_{in} but with a dc level shift of $-V_P$. A reversal of the diode polarity would produce a sinusoid v_{in} also of zero volts but with a dc shift of $+V_P$.

This section is a brief introduction to a wide variety of circuits for wave shaping. Space does not permit a detailed investigation of these interesting circuits.

3.5.5 Diode Logic Circuits

Diodes can be used to perform certain logic functions. In Figure 3.21, a diode AND gate and its truth table are shown. When either input terminal is connected to ground, the diode in series with that input is forward biased, and

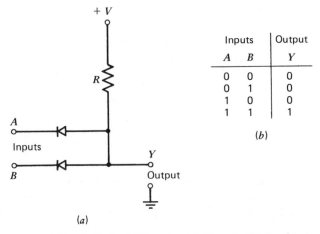

Inputs		Output
A	B	Y
0	0	0
0	1	0
1	0	0
1	1	1

(b)

(a)

Figure 3.21. A diode AND gate. (a) Circuit. (b) Truth table.

the output voltage is one forward-biased diode drop above ground. This is interpreted as a logical "0". It is only when both inputs are connected to $+V$ that both diodes are zero biased, leaving the output voltage at $+V$. This is interpreted as a logical "1." A similar analysis can be applied to the diode OR gate shown in Figure 3.22.

3.6 SUMMARY

The junction diode is an interesting and useful device. Under forward bias conditions, the diode has a current that increases exponentially with increasing voltage. The reverse bias condition produces a very small current that is essentially independent of bias voltage as long as that voltage is less than the breakdown voltage. Under breakdown conditions, the current increases drastically with almost no increase in voltage. The junction diode has capacitance

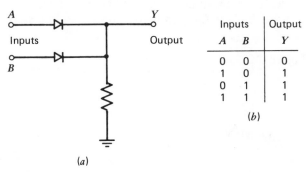

Inputs		Output
A	B	Y
0	0	0
1	0	1
0	1	1
1	1	1

(b)

(a)

Figure 3.22. A diode OR gate. (a) Circuit. (b) Truth table.

characteristics under both forward and reverse bias. Charge storage under forward bias makes it difficult to turn off a junction diode in a very short time. Schottky barrier diodes do not store charge when forward biased and can be switched off much more rapidly than junction diodes. The diode can be represented by a number of different circuit models, depending on the circuit application.

PROBLEMS

3.1. (a) Determine the applied voltage to achieve a forward current of 0.45 μA in a *pn* junction silicon diode at $T = 300°$K if the reverse saturation current is 1.0 nA (1.0×10^{-9}). (b) If the reverse saturation current in a germanium diode is 10 μA, what current would result if the voltage in part (a) were applied in the forward direction?

3.2. (a) A silicon *pn* junction diode at $T = 300°$K conducts 4.0 mA at 0.65 V in the forward direction. Calculate the reverse saturation current. (b) If the forward voltage is increased to 0.75 V, calculate the diode current.

3.3. If the reverse saturation current in a *pn* junction diode at 300°K is 10 nA, calculate the forward currents corresponding to voltages of 0.5, 0.6, and 0.7 V.

3.4. (a) Calculate the voltage at which the reverse current in a *pn* junction diode at $T = 300°$K reaches 90% of its reverse saturation current. (b) If the reverse saturation current in a germanium diode at $T = 300°$K is 5.0 μA, determine the current for a reverse voltage of 50 mV. Also, determine the current for a forward voltage of 50 mV.

3.5. (a) Consider a silicon *pn* junction diode at 300°K operating in the forward bias region. Calculate the change in the voltage that will cause a factor of 10 increase in the current. (b) Repeat part (a) for a factor of 100 increase in current.

3.6. (a) A silicon diode is in series with a 10 MΩ resistor and a 1.5-V power supply. Assuming that the reverse saturation current is 35 nA, determine the voltage across the diode if the diode is forward biased. (b) Repeat part (a) if the diode is reverse biased.

3.7. In the circuit of Figure 3.23, find V_{OUT}.

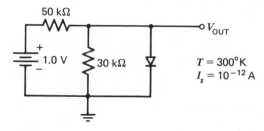

Figure 3.23. Circuit for Problem 3.7.

3.8. In the circuit of Figure 3.24, find I.

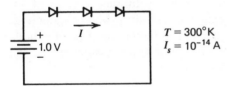

$T = 300°\text{K}$
$I_s = 10^{-14}\,\text{A}$

Figure 3.24. Circuit for Problem 3.8.

3.9. In the circuit of Figure 3.25, (a) at what input voltage will V_{OUT} be 600 mV? (b) What are the dynamic resistances of D_1, D_2, and D_3?

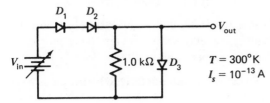

$T = 300°\text{K}$
$I_s = 10^{-13}\,\text{A}$

Figure 3.25. Circuit for Problem 3.9.

3.10. In the circuit of Figure 3.26, (a) find V so that the current I is 500 μA, and (b) what is the diode voltage?

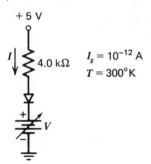

$I_s = 10^{-12}\,\text{A}$
$T = 300°\text{K}$

Figure 3.26. Circuit for Problem 3.10.

3.11. In the circuit of Figure 3.27, what is V_{DC} if the ac output voltage $v_{OUT} = v_1/2$ and $I_s = 1.0 \times 10^{-14}\,\text{A}$?

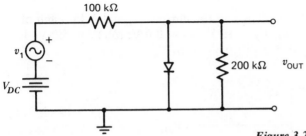

Figure 3.27. Circuit for Problem 3.11.

3.12. Consider a silicon diode that is in series with a 4.7-kΩ resistor and a 10-V power supply. Calculate the current in the circuit if a piecewise linear diode model is assumed and the diode is forward biased. Assume that the voltage difference across the diode is 0.60 V when conducting.

3.13. A *pn* junction diode at $T = 300°K$ is in series with a 1.2-kΩ resistor and a 30-V power supply. The diode has a reverse saturation current of 5.0 μA, a breakdown voltage of 85 V, and negligible resistance when conducting. Using the diode equation, determine the current in the circuit if (*a*) the diode is forward biased and (*b*) the diode is reverse biased. (*c*) Assuming that the breakdown voltage is 15 V, repeat parts (*a*) and (*b*).

3.14. A particular device whose *I*-versus-*V* characteristics are given by

$$I = (V^2 - 5.0)/220 \qquad \text{for } V \geq (5.0)^{1/2}$$
$$I = 0 \qquad \text{for } V < (5.0)^{1/2}$$

is connected in series with a 10 Ω resistor and a 25-V power supply. Find the current *I* in the device and the voltage *V* across the device using graphical techniques. Assume that *I* and *V* are in amperes and volts.

3.15. The diodes in the following circuit are characterized by the piecewise linear parameters:

$$D_1: V_{D1} = 0.3 \text{ V}, \qquad r_1 = 40 \text{ Ω}$$
$$D_2: V_{D2} = 0.6 \text{ V}, \qquad r_2 = 20 \text{ Ω}$$

Calculate the current in the circuit in Figure 3.28 and the voltage across each diode for (*a*) $R = 100 \text{ Ω}$ and (*b*) $R = 1.0 \text{ kΩ}$.

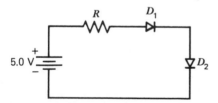

Figure 3.28. Circuit for Problem 3.15.

3.16. In the circuit of Figure 3.29, the diodes are characterized by the piecewise linear parameters:

$$V_D = 0.6 \text{ V}, \qquad r = 0.0 \text{ Ω}$$

Determine the output voltage V_{OUT} and the currents I_1 and I_2 for each of the following input conditions. (*a*) $V_1 = 10 \text{ V}$, $V_2 = 0.0 \text{ V}$; (*b*) $V_1 = 5.0 \text{ V}$, $V_2 = 0.0 \text{ V}$; (*c*) $V_1 = 10 \text{ V}$, $V_2 = 5.0 \text{ V}$; (*d*) $V_1 = 10 \text{ V}$, $V_2 = 10 \text{ V}$.

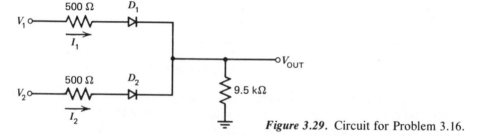

Figure 3.29. Circuit for Problem 3.16.

3.17. For the circuit of Figure 3.30, assuming the same piecewise linear parameters for the diodes as those in Problem 3.16, calculate the output voltage V_{OUT} and the currents I_1, I_2, and I_3 for each of the following input conditions: (a) $V_1 = V_2 = 5.0$ V; (b) $V_1 = 5.0$ V, $V_2 = 0.0$ V; (c) $V_1 = 5.0$ V, $V_2 = 2.0$ V; and (d) $V_1 = V_2 = 0.0$ V.

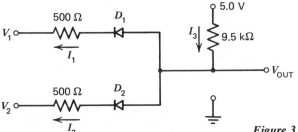

Figure 3.30. Circuit for Problem 3.17.

3.18. For the circuit of Figure 3.31, assuming the same piecewise linear parameters for the diodes as those in Problem 3.16, calculate the output voltage V_{OUT} and the currents I_1, I_2, I_3, and I_4 for each of the following input conditions: (a) $V_1 = V_2 = 5.0$ V; (b) $V_1 = 5.0$ V, $V_2 = 0.0$ V; (c) $V_1 = 5.0$ V, $V_2 = 2.0$ V; (d) $V_1 = V_2 = 0.0$ V.

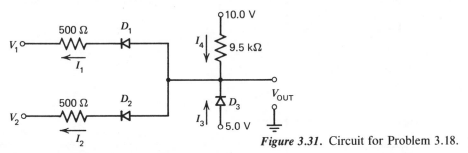

Figure 3.31. Circuit for Problem 3.18.

3.19. For the circuit of Figure 3.32, find V_1. Assume that $V_D = 0.6$ V when sufficiently forward biased.

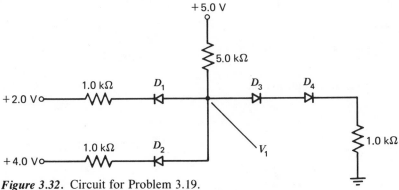

Figure 3.32. Circuit for Problem 3.19.

3.20. For the circuits in Figure 3.33, sketch the waveform of v_{OUT}.

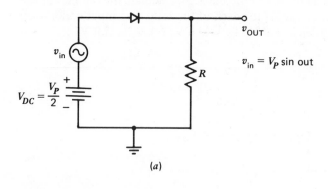

$v_{in} = V_P \sin$ out

(a)

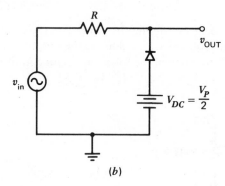

(b)

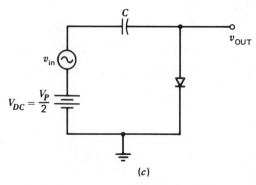

(c)

Figure 3.33. Circuits for Problem 3.20.

3.21. For the circuits in Figure 3.34, sketch the waveform of v_{OUT} for two cycles of the input voltage. The input voltage is a square wave with a peak amplitude of V_m and has a frequency of 250 Hz. Assume that the diode is ideal.

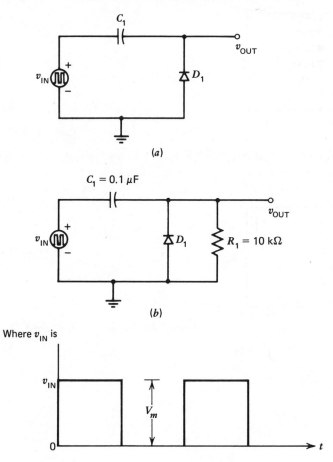

Figure 3.34. Circuits for Problem 3.21.

3.22. If the maximum allowable power dissipation in the 5.3-V Zener diode in Figure 3.35 is 800 mW, find the minimum value of R_L.

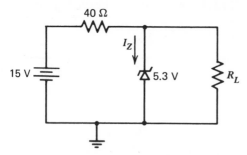

Figure 3.35. Circuit for Problem 3.22.

3.23. The load current in the circuit of Figure 3.36 varies from 35 to 50 mA. Find R such that $I_Z \leq 35$ mA.

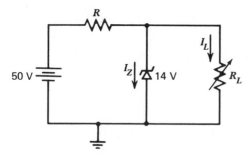

Figure 3.36. Circuit for Problem 3.23.

chapter 4
THE BIPOLAR TRANSISTOR

Historically, the bipolar transistor has been the dominant force in the semi-conductor era of electronics. This transistor contains two *pn* junctions placed in physical proximity. It is called bipolar because of the participation of both hole and electron carriers in the operation of the device. The bipolar transistor is distinct from the other major transistor known as the field effect or unipolar transistor, which is discussed in Chapter 5. There is intense industrial competition between advocates of bipolar and field effect transistors over which device can most economically perform the varieties of electronic functions required in the present age of electronics. In this chapter, the bipolar transistor is frequently referred to as, simply, the transistor.

The bipolar transistor finds extensive applications in small-signal linear amplification, digital logic circuitry, and large power amplification. The transistor is introduced as it applies to the small-signal amplification process, stressing the qualitative physical concepts and related current–voltage properties. The concept of bias and operating states is presented. The states of the transistor used in digital applications, called *cutoff* and *saturation*, are discussed in an analysis of the inverter circuit.

The role of circuit models for the transistor is stressed with two important examples. The small-signal hybrid parameter model is presented for simple analog circuitry, and a large-signal model is introduced to emphasize the speed limitations of a transistor during digital switching operation.

4.1 BASIC TRANSISTOR AMPLIFYING MECHANISM

The bipolar transistor operation is described using the concepts of the *pn* junction discussed in Chapter 3. Two *pn* junctions lying in close physical proximity within a single crystal are required for successful transistor operation. A highly idealized transistor structure, shown in Figure 4.1*a*, has two *pn* junctions and three separately doped regions. The device in Figure 4.1*a* is called an *npn*

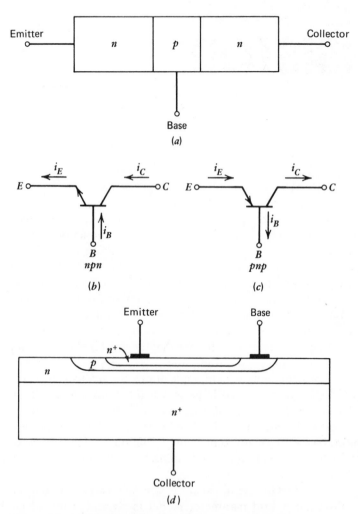

Figure 4.1. (*a*) Idealized *npn* bipolar junction transistor. (*b*) Circuit symbol for an *npn* transistor. (*c*) Circuit symbol for a *pnp* transistor. (*d*) Cross section illustrating modern transistor construction.

transistor. A *pnp* transistor is constructed in a similar way. The symbols and dc current conventions for the *npn* and *pnp* transistors are shown in Figures 4.1*b* and 4.1*c*, respectively. Most of the *npn* transistors produced after 1960 have a cross-sectional structure similar to that in Figure 4.1*d*. The heavily doped n^+-region between the collector contact and the more lightly doped *n*-type region provides physical strength without adding significant resistance to the device.

The *n*-doped regions of Figure 4.1*a* are labeled the emitter and the collector, and the *p*-doped region between the emitter and collector is called the base. The emitter and collector regions differ in geometry and doping concentration. The emitter is more heavily doped and possesses a much smaller surface area than the collector. The base region has a very narrow dimension between the emitter and collector. All of these conditions are critical for good transistor operation.

When the transistor is to be used as an analog amplifying device, the base–emitter junction is forward biased and the base–collector junction is reverse biased. These conditions establish a distribution of electrons (minority carriers) in the *p*-type base that is essential to the amplification process. The forward-biased base–emitter junction supplies a concentration of electrons at the emitter edge of the base that is exponentially related to the bias voltage. The nonsymmetrical doping of this junction, with the emitter doped much more heavily than the base, assures that the flow of electrons from the emitter to the base represents a much larger fraction of the emitter current than the flow of holes from the base to the emitter. The reverse bias of the base–collector junction provides a large electric field in a direction to sweep electrons that reach the collector edge of the base region across the base–collector junction into the collector. This maintains the electron concentration at nearly zero along the collector edge of the base. The combination of the two boundary conditions and a narrow base width [compared to a diffusion length $(D_n \tau_n)^{1/2}$] determines that the flow of minority carriers (electrons) across the base is primarily due to diffusion.

In a high-quality transistor, over 99% of the electrons injected across the base–emitter junction make their way to the base–collector junction. Some of the electrons undergo recombination with holes in the base before they complete this transit and do not contribute to the amplification process of the transistor. The base region is designed to be quite narrow compared to a diffusion length, reducing the probability for recombination.

An important role for transistors is the analog amplification of weak input signals, but the transistor action described thus far does not explain how this is accomplished. We find that analog amplification cannot be accomplished by the transistor alone. As with all amplifying devices in the transistor or vacuum tube family, external components are required. An amplifier consists of a properly biased transistor and a set of external components with the values selected in the correct ratios. Insight into the transistor contribution to the circuit amplification can be obtained by noting in Figure 4.1*b* that an input current $-i_E$

enters the transistor through the low impedance of the forward-biased base–emitter junction, but the output current i_C leaves the transistor from the very high impedance of the reverse-biased base–collector junction. Note that our convention for current directions makes it more convenient to discuss circuit properties but introduces minus signs into our physical discussion because of the importance of electron flow in the operation of the *npn* transistor structure. For a *pnp* transistor, the emitter and collector currents are positive in this kind of discussion. Since the input and output currents are approximately equal, a power gain for the signal occurs within the device.

There are fewer *pnp* transistors in electronic circuits than *npn* transistors, due primarily to the higher mobility of electrons compared to that of holes. This results in higher operating frequencies for *npn* transistors. There is, however, an important role for *pnp* transistors in circuitry where its characteristics combine with those of *npn* transistors to accomplish a particular electronic function more easily than could be done by either type alone. These applications are referred to as uses of complementary pairs of transistors. The operation of *pnp* transistors is essentially the same as that of *npn* transistors, except for a reversal in the roles of the electrons and holes.

4.2 TRANSISTOR CURRENTS AND OPERATING REGIONS

In this section, a more detailed examination is made of the charge transport processes that occur in transistors. The distribution of the minority carriers in the base region is the critical factor in determining how well a transistor will perform a specific function. This distribution is discussed for the active or linear mode of operation of the transistor as well as those modes associated with digital functions. We then interpret certain terminal properties of the transistor in view of the minority charge distribution.

The important current components of an *npn* transistor biased in its linear mode of operation are indicated pictorially in Figure 4.2. Electrons are injected into the base (1) across the forward-biased base–emitter junction. As these electrons diffuse across the base, some (4) recombine with the hole majority carriers. However, the largest fraction of electrons arrive at the base–collector junction and are swept into the collector by the electric field in the space charge region (3). There is a hole current injected into the emitter (2) from the forward-biased base–emitter junction. This hole flow is opposite in direction to the electron flow (1) and much smaller in magnitude. The asymmetric flow of electrons and holes across the base–emitter junction arises from the heavier doping of the emitter compared to the base. The hole current from the base to the emitter reduces the emitter efficiency of the transistor, and device designers attempt to make it as small as possible. Another source of current exists at the reverse-biased base–collector junction. A reverse saturation current arises due to thermally generated hole–electron pairs swept out of the junction by the electric field of the space-charge region. For a typical low-power transistor,

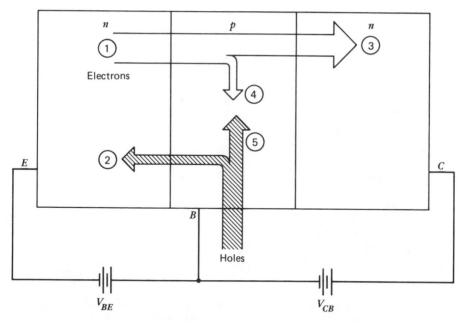

Figure 4.2. Electron and hole currents in an *npn* transistor biased in the forward active state. (1) Electrons injected from the emitter into the base. (2) Holes injected from the base into the emitter. (3) Electrons from the emitter that reach the collector. (4) Electrons from the emitter that recombine in the base. (5) Holes from the base contact that recombine in the base.

the saturation current of the base–collector junction is on the order of nanoamperes.

The components shown in Figure 4.2 that contribute to the base terminal current are loss factors for the transistor. The current at the base terminal consists of the holes injected into the emitter from the base (2) and the recombination loss of holes that must be replaced in the base from the power supply (5). There is also a reverse bias saturation current associated with the base–collector junction. The relative values of these components depend on the geometry and doping concentrations used in the fabrication of the transistor.

The base current is small compared to the emitter and collector currents, but it can be treated as if it controls these larger currents. If the emitter of an *npn* transistor is grounded, as indicated in Figure 4.3, the base current i_B *effectively determines the value for the collector current* i_C. This can be explained as follows. Assume that the dominant component of the base current is due to recombination, so that we may ignore the base–collector saturation current and the hole current from the base into the emitter. A minority carrier (electron) crosses the base from the emitter to the collector in an average transit time τ_{tr}. The width of the base, W_b, is constructed much smaller than the average

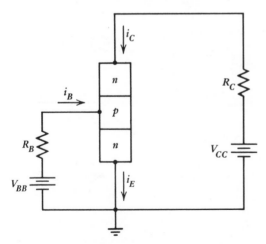

Figure 4.3. An *npn* transistor circuit in the grounded or common-emitter configuration.

distance L_n that an electron travels before recombination occurs. This construction guarantees that most of the electrons arrive at the base–collector junction, avoiding recombination. The collector current may then be related to the excess electron charge in the base, q_{Bn} and the base transit time by

$$i_C = \frac{q_{Bn}}{\tau_{tr}} \qquad (4.1)$$

Even though the effect is small, there is a finite amount of recombination in the base. The average lifetime of an excess hole or electron in the base before recombination occurs is τ_p or τ_n. Since holes and electrons recombine with each other, $\tau_p = \tau_n$. As indicated above, $\tau_{tr} \ll \tau_p$. This implies that many electrons cross the base region during the lifetime of a hole in the base. Assuming again that the major current supplied by the base terminal is used to supply holes lost by recombination, we may write

$$i_B = \frac{q_{Bp}}{\tau_p} \qquad (4.2)$$

where q_{Bp} is the charge due to excess holes in the base. The excess hole concentration in the base disappears by recombination at a rate of q_{Bp}/τ_p, while the base terminal current is the source for replenishing these holes. Charge neutrality must be maintained in the base region, since there are mobile charges in the region and an imbalance would cause large electric fields, resulting in a rapid redistribution of the charges until the field essentially disappears. This means that the excess hole concentration in the base is equal to the excess electron concentration in the base, or

$$q_{Bn} = q_{Bp} \qquad (4.3)$$

Combining these equations results in

$$\frac{i_C}{i_B} = \frac{q_{Bn}/\tau_{tr}}{q_{Bp}/\tau_p} = \frac{\tau_p}{\tau_{tr}} = \beta \qquad (4.4)$$

where β is the common-emitter forward current transfer ratio. We see here the close relationships between the collector and base currents and the recombination and transit times of the transistor. The recombination and transit times are properties of the doping concentration and geometry of the transistor. The tradeoffs in transistor design generally limit the β to values less than 600, with typical values between 50 and 150 in integrated circuits.

The β factor is useful in relating the three terminal currents in a transistor biased in the active region of operation. Neglecting saturation currents,

$$i_C = \beta i_B \qquad (4.5)$$

$$i_C + i_B = i_E \qquad (4.6)$$

and, by substitution,

$$i_E = (1 + \beta)i_B \qquad (4.7)$$

$$i_C = \frac{\beta}{1 + \beta}i_E \qquad (4.8)$$

These four equations are used so frequently for transistors biased in the active mode that they should be memorized. Equation 4.6 is applicable to the transistor under any condition, since it is a statement of Kirchhoff's law of charge conservation.

The discussion thus far has described current mechanisms in the *npn* transistor biased in the *forward active* mode. Under these circumstances, the emitter–base junction is forward biased while the base–collector junction is reverse biased. Three other bias possibilities exist with the two *pn* junctions:

1. Both junctions are reverse biased.
2. Both junctions are forward biased.
3. The emitter–base junction is reverse biased while the base–collector junction is forward biased.

These other bias conditions are not only possibilities but are found extensively in digital applications.

A transistor with both junctions reverse biased will inject no minority carriers into the base. The collector current is essentially zero, and the transistor is said to be in the *cutoff* mode. A transistor with both junctions forward biased will inject large quantities of minority carriers into the base from both the emitter and collector. The collector current is large, but finite, and limited by the circuit configuration. This situation, with both *pn* junctions forward biased, is called the *saturation* mode. Saturation usually occurs by increasing the base current when the device is in the forward active mode. The collector current tries to increase in proportion to the base current. As the collector current increases, the voltage drop across the resistor in series with the collector also increases, reducing the voltage on the collector until the collector is at a lower potential than the base. The base–collector junction is now forward biased; and, beyond this point, the collector current cannot continue to increase in proportion to the base current. Instead, the collector current remains almost constant (increasing slightly) when the base current increases. The collector–emitter voltage approaches a constant value, often as low as 0.1 V (depending on the fabrication of the device). Note that the base–emitter voltage is always larger than the base–collector voltage when saturation is reached from the forward active bias configuration.

The remaining transistor state to be considered has a forward-biased base–collector junction and a reverse-biased emitter–base junction and will inject minority carriers into the base from the collector. The injected electrons diffuse across the base and are swept into the emitter by the field in the space-charge region of the reverse-biased emitter–base junction. This biasing arrangement puts the transistor in the *inverse active* mode. The current amplification of a transistor in the inverse active mode is much smaller than that of the same transistor in the forward active mode due to asymmetries in the geometric layout and doping of the device. If the device goes into saturation from the inverse active configuration, the base–collector bias voltage is larger than the base–emitter bias voltage.

Another way to visualize the four possible bias modes of a transistor is to examine the minority carrier charge distributions in the base. In a transistor biased in the forward active mode, the minority carrier charge density in the base is a maximum at the emitter–base junction and essentially zero at the space charge region of the base–collector junction. The precise charge distribution or profile across the base can be obtained from a detailed mathematical analysis, but this is beyond the scope of the text. For base widths that are small compared to a diffusion length, the result is closely approximated by a straight line, as shown in Figure 4.4a. The minority carrier profile for the saturation mode in Figure 4.4b can be redrawn as the sum of the injected minority carriers of each junction, as shown in Figure 4.5. The minority carrier distribution in the cutoff mode in Figure 4.4c shows a small amount of thermally generated carriers in the base region. These minority carriers are swept out of the base by one of the reverse-biased *pn* junctions. The saturation and cutoff modes of the transistor are operational states in many types of digital circuits. The two states represent extremes of transistor operation and are convenient representations of the binary 0 and 1 in logic circuits. The inverse mode minority carrier profile in Figure 4.4d reflects the lower efficiency of injected minority carriers from the collector. One reason for this is the relatively large emitting area of the base–collector junction compared to the collecting area of the emitter–base junction, as shown in Figure 4.1d. Another reason is that the doping ratio of the collector to base is smaller than that of the emitter to base,

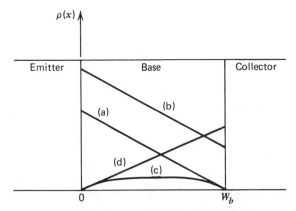

Figure 4.4. Minority carrier charge distribution
$\rho(x)$ as a function of one-dimensional distance
across the base region. (*a*) Normal mode. (*b*) Saturation mode. (*c*) Cutoff mode. (*d*) Inverse mode.

resulting in a lower minority carrier emitting efficiency from the collector. The
inverse active mode occurs in the operation of certain types of digital circuits.

The minority carrier charge density profiles of Figure 4.4 are useful not
only for visualizing transistor operation but can also be used to estimate transistor currents. If the drift term is negligible, the current density due to minority
carriers in the base is given by

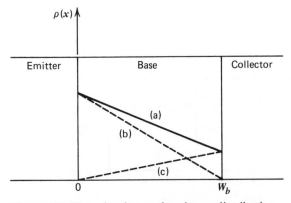

Figure 4.5. The minority carrier charge distribution
of a saturated transistor (*a*) is the sum of the emitter-
base injected carriers (*b*) and the collector-base injected carriers (*c*).

$$J_n(x) = +qD_n\frac{dn(x)}{dx} \qquad (4.9)$$

where these quantities have the same meaning as in Chapter 2. It is important to note that the current density is directly proportional to the *slope* of the charge distribution. Current at any particular location in the base may be obtained by multiplying the current density by the cross-sectional area of the active base, which is approximately equal to the area of the emitter for forward active bias conditions.

4.3 TRANSISTOR CURRENT–VOLTAGE CHARACTERISTICS

The physical description of the previous sections is now used to describe some of the measureable current–voltage characteristics of the transistor. Figure 4.6 illustrates an *npn* transistor, indicating the notation and current convention used in this section. The terminal currents are shown as well as the fraction of collected minority carriers αi_E and the base–collector reverse bias saturation current I_s. These current components can be expressed as

$$i_E = i_B + i_C \qquad (4.6)$$

$$i_C = \alpha i_E + I_s \qquad (4.10)$$

The grounded- or common-base transistor amplification factor is defined by

$$\alpha = \left.\frac{\partial i_C}{\partial i_E}\right|_{v_{CB}=\text{constant}} \qquad (4.11)$$

The value of α is slightly less than unity for a quality transistor. Substitution of Equation 4.6 into Equation 4.10 yields

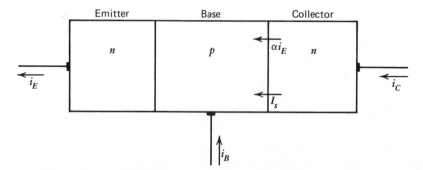

Figure 4.6. An *npn* transistor with the fraction of collected minority carriers αi_E and saturation current I_s. Note the postive convention of current flow in the *npn* transistor.

$$i_C = \frac{\alpha}{1 - \alpha} i_B + \frac{I_s}{1 - \alpha} \tag{4.12}$$

The change in collector current for a change in base current provides the current amplification factor β of the common emitter circuit. This quantity can be related to α by

$$\beta = \frac{\partial i_C}{\partial i_B}\bigg|_{v_{CE} = \text{constant}} = \frac{\alpha}{1 - \alpha} \tag{4.13}$$

Example. If $\alpha = 0.995$, $I_s = 1$ nA, and $i_B = 10$ μA, calculate i_C and β.

Solution. From Equation 4.12,

$$i_C = \frac{0.995}{(1 - 0.995)} (1 \times 10^{-5}) + \frac{1 \times 10^{-9}}{1 - 0.995}$$

$$= 1.99 \text{ mA}$$

$$\beta = \frac{\alpha}{1 - \alpha} = \frac{0.995}{1 - 0.995} = 199$$

Note that the reverse bias saturation current makes a negligible contribution to the collector current in this example. This is generally true, and Equation 4.12 may be written

$$i_C \simeq \frac{\alpha}{1 - \alpha} i_B \tag{4.14}$$

$$i_C \simeq \beta i_B$$

The transistor is often treated from the viewpoint of its terminal voltage and current characteristics. Figure 4.7a shows a circuit with a grounded- or common-base connection having variable voltage sources V_{EE} and V_{CC}. The V_{EE} source provides forward bias voltage to the base–emitter junction and permits control of the dc emitter current I_E. Assume that I_E is set to a value on the characteristic curves of Figure 4.7b. Then, observe the value of I_C as V_{CB} is varied via V_{CC} from a small negative to a positive value. V_{CB} is obtained from

$$V_{CB} = V_{CC} - I_C R_C$$

For $V_{CB} < 0$, the collector base junction is forward biased. In this region of operation, the transistor has both junctions forward biased and is in saturation. As V_{CB} changes polarity, the base–collector junction becomes reverse biased and the transistor enters the forward active mode of operation. In this region, the emitter and collector currents are essentially equal. The vast majority of electrons injected from the emitter into the base are "collected" at the base–collector junction. The bottom trace of Figure 4.7b is the response of the transistor with the emitter terminal open. For $I_E = 0$, but with $V_{CB} > 0$, the reverse bias saturation current I_{CBo} is seen.

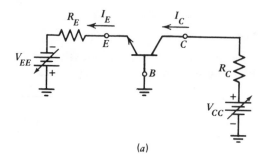

(a)

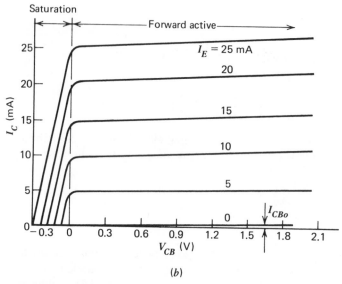

(b)

Figure 4.7. (a) Common-base circuit with variable power supplies V_{EE} and V_{CC}. (b) Common-base collector characteristics for *npn* transistors.

The low-frequency α of the transistor can be estimated from Figure 4.7b; but because it is very close to unity, it is difficult to accurately determine. The α is measured by taking

$$\alpha \simeq \frac{\Delta I_C}{\Delta I_E}\bigg|_{v_{CB}=\text{constant}} \tag{4.15}$$

It should be emphasized that the general characteristic curves of Figure 4.7b represent static or dc measurements, depicting I_C as a function of V_{CB} with I_E as a parameter.

The common-emitter circuit provides a different set of characteristic curves. Figure 4.8a illustrates a common-emitter circuit, and Figure 4.8b shows the characteristic curves for the transistor in that circuit. The value of I_C is given as a function of V_{CE} with I_B as a parameter. Select a fixed value of I_B in the circuit of Figure 4.8a and observe I_C as V_{CE} is altered by varying V_{CC}. For the base terminal open, the base–collector junction is reverse biased and the base–emitter junction is forward biased, resulting in a leakage current I_{CEo} that is approximately β times the leakage current in the common-base case. For I_B

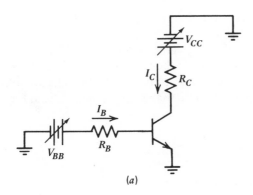

(a)

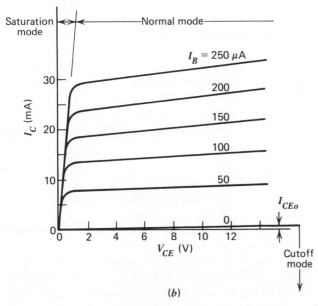

(b)

Figure 4.8. (a) Common-emitter circuit with variable base and collector power supplies. (b) Transistor general characteristic curves for common-emitter configuration.

> 0, the collector current curves rise steeply together before flattening out at the separate, nearly constant values of I_C. At low values of V_{CE} (< 0.4 V), the base–collector junction is forward biased since the emitter-base forward bias voltage is typically 0.7 V. With both junctions forward biased, the transistor is in saturation. Applying Kirchhoff's voltage law around the transistor terminals, we obtain

$$V_{BC} = V_{BE} - V_{CE} \tag{4.16}$$

A slight increase in V_{CE} reverses the polarity of V_{BC} and the device enters the forward active region of operation.

The collector current is relatively independent of V_{CE} for values greater than or equal to 0.7 V. The transistor is then in the forward active mode and acts essentially as a constant-current source at the collector. The low-frequency common-emitter current amplification factor can be estimated from Figure 4.8*b* using the approximation

$$\beta \simeq \left. \frac{\Delta I_C}{\Delta I_B} \right|_{v_{CE} = \text{constant}} \tag{4.17}$$

From the curves of Figure 4.8*b*, for $V_{CE} = 5$ V,

$$\beta \simeq \frac{20 - 14}{0.015 - 0.010} = 120$$

Example. Given the circuit and transistor characteristic curves of Figure 4.9, estimate I_B, I_C, and V_{CE}.

Solution. We use an approach that seeks a graphical simultaneous solution using Figure 4.9*b*. This figure is a plot of I_C versus V_{CE} with I_B as a parameter. Kirchhoff's voltage law is used to write an analytical expression for $I_C = f(V_{CE})$. For the loop containing V_{CC} and the collector–emitter terminals,

$$V_{CC} = I_C R_C + V_{CE} \tag{4.18}$$

or

$$I_C = \frac{V_{CC}}{R_C} - \frac{V_{CE}}{R_C} \tag{4.19}$$

Equation 4.19 can be plotted as a straight-line function with a slope of $-1/R_C$ and with I_C and V_{CE} intercepts of V_{CC}/R_C and V_{CC}. This equation is plotted on the same axes as the characteristic curves in Figure 4.9*b*. This straight-line statement of Kirchhoff's voltage law is known as a dc load line. It relates I_C, V_{CE}, and the circuit resistor R_C. The transistor is also precisely defined by its characteristic curves showing $I_C = f(V_{CE}, I_B)$, where f designates a function. Thus, a solution exists at some point

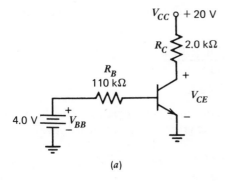

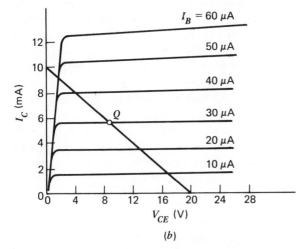

Figure 4.9. (*a*) Common-emitter circuit. (*b*) Transistor characteristic curves with circuit load line
superimposed.

on the load line. When that point is found, then all three circuit parameters
can be read from the graph of Figure 4.9*b*.

Kirchhoff's voltage law for the loop containing V_{BB} and the
base–emitter terminals of the transistor results in

$$V_{BB} = I_B R_B + V_{BE} \tag{4.20}$$

or

$$I_B = \frac{V_{BB} - V_{BE}}{R_B} \tag{4.21}$$

The values for V_{BB} and R_B are given. The value for V_{BE} may be estimated from the voltage of a forward-biased silicon *pn* junction that is approximately 0.70 V. Then,

$$I_B = \frac{4.0 - 0.70}{110 \times 10^3}$$

$$= 30 \ \mu A$$

This value is marked on the load line as Q for quiescent point. The values of I_C and V_{CE} are estimated from the graph as 6 mA and 8 V, respectively.

The concepts and insights gained by examination of the characteristic curves, load line, and circuit of Figure 4.9 are usually of more value than the graphical solution technique. If the base current is increased by increasing V_{BB}, the Q-point will rise along the load line toward the saturation region. Conversely, a drop in the base current will decrease the Q-point along the load line toward cutoff. The circuit designer has the option to locate the quiescent point of the circuit to achieve a specific function. For example, the circuit Q-point may be located in the forward active region to achieve analog amplification; or, for digital circuit applications, the circuit may be biased in either the cutoff or saturation mode. Note from Figure 4.9b that V_{CE} will have a value of 20 V in cutoff and a value less than 0.4 V in saturation. This large difference in V_{CE} between cutoff and saturation states is used as the electrical equivalent to the binary 1 and 0. Lastly, it can be observed in Figure 4.9b that, if the circuit were suddenly switched from the cutoff to the saturation state, the transistor would follow the load line through the forward active region.

4.4 AN ELEMENTARY DEVICE MODEL FOR ACTIVE TRANSISTOR OPERATION

The transistor discussions thus far have led from physical concepts to the terminal voltage–current properties. In this section, a circuit model for the transistor in the forward active mode of operation is developed and applied to a small-signal ac amplifier. The purpose of a small-signal amplifier is to amplify weak input signals or to provide an impedance match between other circuits. Figure 4.10 shows a common-emitter amplifier, transistor characteristic curves, and the superimposed circuit load line. The circuit differs slightly from previous configurations since, in addition to biasing or dc voltage sources, a small-signal voltage source v_{in} is present.

Two processes may be extracted from the circuit of Figure 4.10a. First, the dc or quiescent currents and voltages of the transistor are established by V_{CC}, V_{BB}, R_C, and R_B with $v_{in} = 0$. For this example, $I_{CQ} = 20$ mA, $I_{BQ} = 150 \ \mu A$, and $V_{CEQ} = 13.4$ V. Secondly, the ac signal is superimposed on these dc values. As i_B increases or decreases in value, the transistor operating point moves up or down along the load line from the Q-point, as shown in Figure

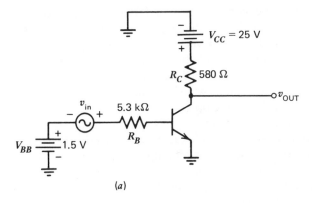

(a)

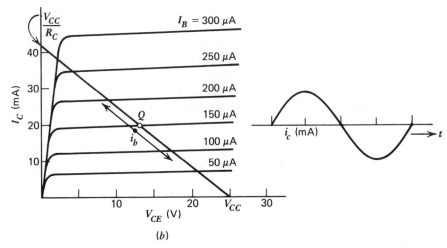

(b)

Figure 4.10. (*a*) Common-emitter circuit with ac signal sources v_i. (*b*) Transistor characteristic curves, circuit load line, and graphical transformation of signal currents, i_b, i_c.

4.10*b*. The corresponding larger value of i_C at any base current can be read from the vertical scale. In addition, the output voltage variation can be obtained from the horizontal scale. This variation represents the output signal voltage v_{OUT}. This graphical technique provides some visualization of the interaction of the ac and dc components of the currents and voltages in the circuit. However, the graphical technique is clumsy when used to analyze the circuit; and, for this reason, we seek more satisfactory analytical methods of analyzing transistor circuits.

Electronic circuits can be more easily analyzed by manual or computer analysis when a model representation is used for the transistor. The model may take the form of analytical equations relating transistor currents and voltages, or the model may be combinations of familiar circuit components such as resistors, capacitors, and controlled signal sources representing the transistor. In this section, we develop a very popular low-frequency model called the hybrid or *h*-parameter model of the transistor. The name arises because the units associated with the different parameters are not the same. The model applies to that frequency range where the internal capacitances of the transistor have a negligible effect on the operation of the circuit. The *h*-parameter model of the transistor is a combination of resistors and dependent signal sources that allow linear circuit analysis techniques to be used. This model permits approximate solutions to the important amplifier parameters such as voltage and current gain and input and output impedance. The mathematical form of the circuit ac equations provides important design guides for the engineer to achieve specific goals for the amplifier.

The *h*-parameter model is one of several circuit models that can be derived from two-port network theory. Figure 4.11 shows the standard configuration for a two-port network. This representation applies only to linear circuits, implying that the model will apply to a transistor operating in the forward active mode. Network theory states that a linear two-port network can be completely described in terms of its terminal voltages, terminal currents, and a set of constants. There are six possible sets of constants that might describe a particular two-port network. The hybrid parameters are one of these sets and are defined in phasor notation by

$$V_1 = h_{11}I_1 + h_{12}V_2 \tag{4.22}$$

$$I_2 = h_{21}I_1 + h_{22}V_2 \tag{4.23}$$

These two equations can be interpreted as statements of Kirchhoff's voltage and current laws whose electrical circuit representations are shown in Figure 4.12*a* and 4.12*b*. The two circuits can be combined as a single circuit, as indicated in Figure 4.12*c*.

The circuit representation of Figure 4.12*c* has four terminals. However, the bottom two terminals are directly connected, making an equivalent three terminals. The *h*-parameter model may then be applied to any linear three-terminal device such as the transistor when it is biased in the forward active mode. Small-signal models are treated in more detail in Chapter 11. Other types of models have been developed that include all of the operating modes of the transistor, and these models are called large-signal or nonlinear models.

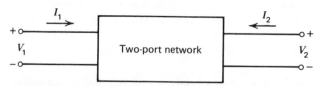

Figure 4.11. The two-port network representation showing the terminal current and voltage conventions using phasor notation.

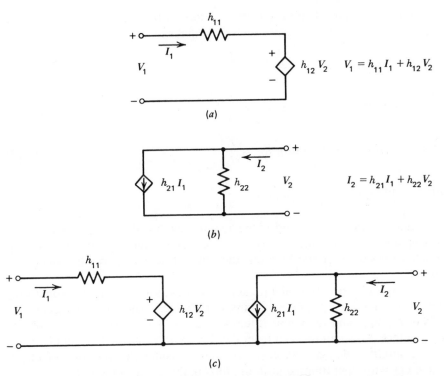

$$V_1 = h_{11}I_1 + h_{12}V_2$$

$$I_2 = h_{21}I_1 + h_{22}V_2$$

Figure 4.12. (*a*) Kirchhoff voltage law and equivalent circuit for the *h*-parameter Equation 4.22. (*b*) Kirchhoff current law and equivalent circuit for the *h*-parameter Equation 4.23. (*c*) Combined equivalent circuits representing electrical equivalent circuit to the two-port network *h*-parameters.

The application of the *h*-parameter small-signal model to the transistor is accomplished by substituting the equivalent circuit of Figure 4.12*c* for a given transistor in a circuit. The model is appropriate for both *npn* and *pnp* transistors. The values of the individual *h*-parameters for a particular transistor must be known to determine specific numerical results. The *h*-parameters can be measured with a transistor curve tracer or approximated from typical values given by the manufacturer on the transistor data sheet.

The measurement of transistor *h*-parameters follows precise definitions. For a transistor in the common-emitter configuration, the set of *h*-parameter equations can be written as

$$V_{be} = h_{ie}I_b + h_{re}V_{ce} \tag{4.24}$$

$$I_c = h_{fe}I_b + h_{oe}V_{ce} \tag{4.25}$$

The subscript terminology shown here reflects industrial usage. The lower case *e* refers to a transistor in the common-emitter connection. The *i*, *r*, *f*, and *o* are symbols for input, reverse, forward, and output, respectively. Appropriate partial differentiation of these equations leads to the following definitions of the h-parameters in the common-emitter configuration:

Input impedance: $\qquad h_{ie} = \dfrac{V_{be}}{I_b}\bigg|_{V_{ce}=0}$ (4.26)

Reverse voltage transfer ratio: $h_{re} = \dfrac{V_{be}}{V_{ce}}\bigg|_{I_b=0}$ (4.27)

Forward current transfer ratio: $h_{fe} = \dfrac{I_c}{I_b}\bigg|_{V_{ce}=0}$ (4.28)

Output admittance: $\qquad h_{oe} = \dfrac{I_c}{V_{ce}}\bigg|_{I_b=0}$ (4.29)

The variables v_{BE}, v_{CE}, i_B, and i_C are readily displayed on a transistor curve tracer in any combination of two or three variables. The h-parameters are then estimated as slopes of the lines on the curve tracer at particular operating points.

Figure 4.13 shows typical h-parameter values as a function of the operating point for the medium-power npn transistor 2N2222. The particular operating levels of I_C, V_{CE}, and the ambient temperature T_A have a strong effect on the h-parameters. This can be seen from the three curves at the bottom of Figure 4.13. In amplifier design, the designer must consider the variation of the h-parameters with operating point in optimizing the circuit function.

While the h-parameter model was derived from network theory, a physical mechanism can be associated with each of the individual parameters. The h_{ie} can be visualized as the approximate small-signal resistance of the forward-biased base–emitter diode. This diode resistance is a variable depending on the

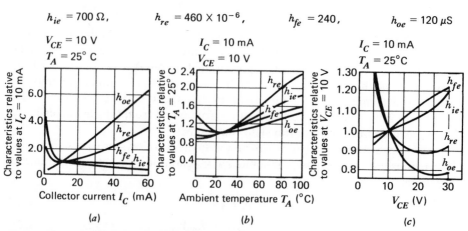

Figure 4.13. The variation of hybrid parameters with operating point for a typical 2N2222 bipolar transistor. The data in the curves is referenced to the h-parameter values given in the legend.

operating point of the base–emitter diode. The diode equation and its resulting current–voltage plot show that the diode forward-biased small-signal resistance is reduced as the current is increased. This is an important point in the design of transistor circuits in integrated circuits. A common method to achieve high-input resistance at the base of a transistor is to lower the emitter bias current to the microampere range. The h_{fe} parameter is approximately equal to the quantity β that we previously defined. This parameter describes the internal amplifying ability of the transistor and accounts for the multiplication of current from base to collector. The mechanisms for current amplification were discussed earlier in this chapter. The emitter efficiency and base width influence the value of h_{fe} for a particular device.

The h_{re} is a measure of the effectiveness of the transistor in isolating signal influences at the collector from the base terminal. It is related to the change in the effective width of the base when the collector–emitter voltage changes. This is called the *Early effect*. The value of h_{re} is normally very small for a silicon transistor. In subsequent circuit examples, h_{re} is assigned a value of zero.

The h_{oe} parameter is the output conductance and represents several physical processes associated with the conductance path between the collector and emitter. This parameter is often modeled as

$$h_{oe} = \frac{I_C}{V_A}$$

where V_A is called the *Early voltage*. It can be determined by locating the common intercept on the negative V_{CE} axis of the extrapolations of the collector current curves. This voltage has typical values between 50 and 200 V. An ideal transistor has a zero conductance at the collector–emitter terminals. Any finite value of h_{oe} degrades the ideality of the constant-current source property of the transistor. Typical values of h_{oe} lie in the tens of microsiemens, giving typical resistance values ($1/h_{oe}$) of 15 to 100 kΩ.

4.5 SMALL-SIGNAL AMPLIFIER EXAMPLES

We make use of the h-parameter model in a variety of examples in this section to analyze the small-signal or ac properties of transistor circuits. It should be emphasized that this is only one of several methods used to analyze linear or analog amplifier circuits.

In the examples presented below, we show how the ac or small-signal h-parameter model is used to calculate voltage and current gain and terminal impedances. The expressions for these circuit parameters provide the guidelines for satisfying circuit specifications. The technique for writing generalized circuit parameter functions or calculating numerical answers consists of:

1. Redrawing the circuit to include only the ac equivalent components.
2. Replacing the transistor by the h-parameter model.

3. Solving for the required parameter using the basic tools of linear circuit analysis.

The analysis of single-stage circuits using manual techniques is useful in learning the basic circuit laws involving transistors. Even with the availability of detailed computer solutions, the engineer must retain sufficient insights to estimate the accuracy of a computer solution. These insights come from a knowledge of the models for transistors and a knowledge of when assumptions are appropriate.

4.5.1 Common-Emitter Amplifier

Example. The circuit shown in Figure 4.14a is a common-emitter amplifier. Find the ac signal voltage gain, $A_v = V_{out}/V_{in}$. The quiescent currents and voltages are approximately

$$I_{BQ} = 20 \ \mu A, \qquad I_{CQ} = 1 \ mA, \qquad V_{CEQ} = 15 \ V$$

The h-parameter constants were measured on a transistor curve tracer at the operating point described above and were found to be

$$h_{ie} = 1.3 \ k\Omega, \qquad h_{re} = 0, \qquad h_{fe} = 50, \qquad h_{oe} = 25 \ \mu S$$

Solution. Just as the bias analysis of the transistor circuit in Figure 4.9 concerned itself only with the dc values of the circuit, the ac analysis concerns itself only with the ac equivalent circuit. The circuit of Figure 4.14 a is redrawn in Figure 4.14b showing only the ac values. Note that the dc power supplies (V_{CC}, V_{EE}) have a zero impedance and a zero value of ac voltage. Consequently, the dc power supplies are treated as ac grounds and eliminated in the ac circuit representation. As a final step, the transistor symbol in Figure 4.14b is replaced by the common-emitter h-parameter model, resulting in the ac equivalent circuit shown in Figure 4.14c.

In the following equations, a symbol $\|$ is used to indicate the parallel combination of impedances. For example,

$$Z_a\|Z_b\|Z_c = Z_{eq} = \left(\frac{1}{Z_a} + \frac{1}{Z_b} + \frac{1}{Z_c}\right)^{-1}$$

The ac voltage gain of the circuit is found from Figure 4.14c by the application of linear circuit analysis theory. Using rms phasor notation,

$$V_{out} = -(h_{fe}I_b)[(1/h_{oe})\|R_C] \qquad (4.30)$$

$$I_b = \frac{V_{in}}{R_B + h_{ie}} \qquad (4.31)$$

By substitution,

$$V_{\text{out}} = -\frac{h_{fe}V_{\text{in}}}{R_B + h_{ie}}[(1/h_{oe})\|R_C] \tag{4.32}$$

and

$$A_v = \frac{V_{\text{out}}}{V_{\text{in}}} \tag{4.33}$$

$$= -\frac{h_{fe}}{R_B + h_{ie}}[(1/h_{oe})\|R_C]$$

The numerical result is

$$A_v = -\frac{50}{220 \times 10^3 + 1.3 \times 10^3}[(1/25 \times 10^{-6})\|20 \times 10^3] \tag{4.34}$$

$$= -3.0$$

The negative sign in the gain expression (Equation 4.33) indicates that a 180° phase reversal occurs between the base and collector voltages in a common-emitter circuit. The small-signal equivalent circuit of Figure 4.14c shows this phase reversal property of the common-emitter circuit configuration. The phase reversal phenomenon can also be visualized in Figure 4.14a. If there were no base or collector current in that circuit, V_C would be equal to V_{CC} (+30V). As I_B increases, I_C also increases, and its direction is from the power supply to the transistor terminal. V_C is now less than V_{CC} by the amount I_CR_C. In other words, the collector voltage decreases as the base current increases. The base current increase is due to an increase in the input voltage.

The approach in analyzing ac voltage gain in this example was direct. The ac equivalent circuit was drawn (Figure 4.14c), and the ratio of V_{out} to V_{in} was obtained by substitution. This numerical calculation is useful, but the voltage gain expression (Equation 4.33) is probably more valuable to the design engineer than the numerical result (Equation 4.34). A designer must know what controls the circuit parameters. For example, the voltage gain of -3.0 for the single stage circuit of Figure 4.14 is quite low for an analog amplifier. Voltage gains of -50 to -1000 can easily be obtained with different resistor configurations. In this circuit, Equation 4.33 explains why the voltage gain is low. The hybrid parameters are essentially fixed by the selection of the operating point. The resistors, R_B and R_C, are somewhat under the control of the designer. Increasing R_C will increase the voltage gain. However, R_C is in parallel with $1/h_{oe}$, and the increase in gain is asymptotic. Resistor R_B has a very large value and lies in the denominator of Equation 4.33. Decreasing this resistor can have a significant effect on the gain. However, this resistor is involved in the biasing circuit; and to decrease it, an alternate biasing scheme must be devised. In a subsequent example, the base bias circuit is rearranged in a manner that essentially eliminates R_B from the voltage gain expression.

In addition to gain parameters, it is often of interest to examine impedances in various sections of the circuit by looking into a pair of terminals. These

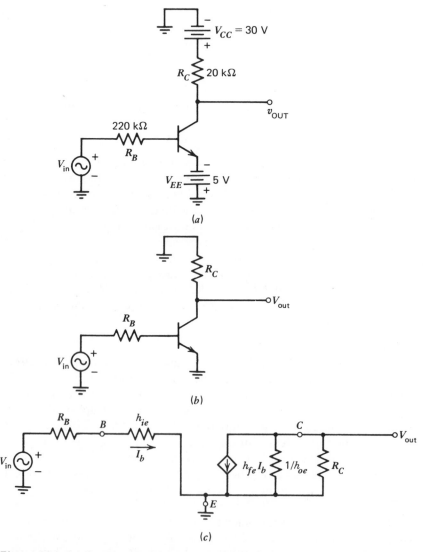

Figure 4.14. (*a*) Common-emitter circuit with the transistor in normal mode. (*b*) Circuit redrawn with power supplies V_{CC} and V_{EE} removed illustrating their zero contribution to the ac circuit. (*c*) Complete ac circuit representation using common-emitter h-parameter model for transistor.

are the Thevenin equivalent impedances that are determined by opening the circuit at the pair of terminals and finding the impedance with all *independent* sources disabled.

The relevant impedances of the circuit can be determined from Figure 4.14c. The input impedance looking into the transistor from the base terminal is

$$Z_B = h_{ie} = 1.3 \text{ k}\Omega \tag{4.35}$$

The impedance seen by the input signal source is

$$Z_{\text{in}} = R_B + h_{ie} = 221.3 \text{ k}\Omega \tag{4.36}$$

and the output impedance at the collector terminal is

$$Z_{\text{out}} = R_C \| (1/h_{oe}) = 13.3 \text{ k}\Omega \tag{4.37}$$

The analysis of this example by hand was relatively easy. The problem could have become much more tedious by the simple addition of a resistor in series with the emitter terminal. The computer simulation of electronic circuits by such programs as SPICE has taken the tedium from circuit calculations while providing more accurate answers. However, the circuit designers who use such programs must possess insight into the transistor and circuit properties in order to judge if the computer solutions are reasonable.

4.5.2 Common-Emitter Amplifier with a Coupling Capacitor

Example. Find the voltage gain for the circuit of Figure 4.15a. The bias arrangement has been altered to provide a nearly identical circuit, but with a greatly enhanced voltage gain. The circuit in Figure 4.15a contains a large *coupling capacitor* to isolate the ac and dc portions of the input signal and bias currents. A large capacitor is desired so that its capacitive reactance is negligible compared to the other impedances in the input circuit over the frequency range associated with the input signal.

The circuit quiescent and h-parameter values are the same as in the previous example:

$$I_{BQ} = 20 \ \mu\text{A}, \qquad I_{CQ} = 1.0 \text{ mA}, \qquad V_{\text{CEQ}} = 15 \text{ V}$$

$$h_{ie} = 1.3 \text{ k}\Omega, \qquad h_{re} = 0, \qquad h_{fe} = 50, \qquad h_{oe} = 25 \ \mu\text{S}$$

Solution. Figure 4.15b shows the ac equivalent circuit used for the voltage gain calculation. The capacitor C_C is eliminated since its capacitive reactance is small compared to circuit resistor values. The voltage gain is determined in the following manner:

$$V_{out} = -h_{fe}I_b[(1/h_{oe})\|R_C] \tag{4.38}$$

$$I_b = \frac{V_{in}}{h_{ie}} \tag{4.39}$$

By substitution,

$$V_{out} = -\frac{h_{fe}V_{in}}{h_{ie}}[(1/h_{oe})\|R_C] \tag{4.40}$$

and

$$A_v = \frac{V_{out}}{V_{in}} = -\frac{h_{fe}}{h_{ie}}[(1/h_{oe})\|R_C]$$

$$= -\frac{50}{1.3 \times 10^3}[(1/25 \times 10^{-6})\|20 \times 10^3] \tag{4.41}$$

$$= -513$$

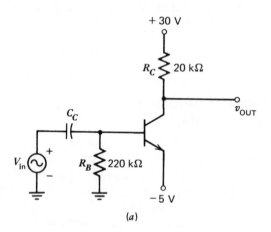

(a)

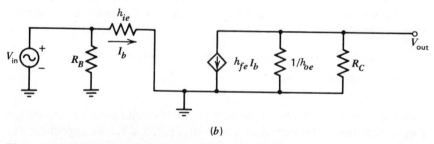

(b)

Figure 4.15. (*a*) Common-emitter amplifier with coupling capacitor C_c. (*b*) The small signal equivalent circuit using the *h*-parameter model for the transistor.

It is observed that the voltage gain in this example has increased from -3.0 to -513, even though component values and transistor parameters have not changed. The important point is that the gain expression (Equation 4.33) provided us with the specific direction to proceed for increasing A_v. The calculated numbers are important, but the effect of the circuit parameters on the gains and impedances tells the circuit designer where to look for optimizing performance.

4.5.3 Common-Collector Circuit

Example. Calculate the signal input and output impedance, the current gain, and the voltage gain for the common-collector circuit of Figure 4.16a. The circuit constants and quiescent values are

$$h_{ie} = 600\ \Omega, \qquad h_{fe} = 100, \qquad h_{re} = 0, \qquad h_{oe} = 0$$
$$I_{CQ} = 4.5\ \text{mA}, \qquad V_{CEQ} = 5.5\ \text{V}$$

The ac equivalent circuit is drawn in Figure 4.16b.

Solution. The signal input impedance seen from the voltage sources V_{in} is composed of the bias resistors in parallel with the base impedance of the transistor. The base terminal impedance is found from $V_b/I_b = Z_b$. The ratio of V_b to I_b is found from a voltage loop equation and by noting that the sum of I_b and $h_{fe}I_b$ comprises the load current. Then,

$$V_b = I_b h_{ie} + (1 + h_{fe})I_b R_E = [h_{ie} + (1 + h_{fe})R_E]I_b \qquad (4.42)$$

The impedance looking in at the base terminal is then

$$Z_b = \frac{V_b}{I_b} = h_{ie} + (1 + h_{fe})R_E$$
$$= 101.6\ \text{k}\Omega \qquad (4.43)$$

and the signal input impedance is

$$Z_{in} = R_1 \| R_2 \| Z_b = 18 \times 10^3 \| 24 \times 10^3 \| 101.6 \times 10^3 = 9.3\ \text{k}\Omega \qquad (4.44)$$

The signal output impedance is obtained by analyzing Figure 4.16b. To determine this impedance, let V_{in} go to zero and imagine a voltage generator V_{out} driving the circuit at the emitter or output terminal. The output impedance, $Z_{out} = V_{out}/I_x$, is obtained from a Kirchhoff current law statement,

$$I_x = \frac{V_{out}}{R_E} - I_b - h_{fe}I_b \qquad (4.45)$$

$$I_b = -\frac{V_{out}}{h_{ie}} \qquad (4.46)$$

By substitution,

$$I_x = \frac{V_{\text{out}}}{R_E} + \frac{V_{\text{out}}}{h_{ie}} + \frac{h_{fe}V_{\text{out}}}{h_{ie}} \tag{4.47}$$

and

$$
\begin{aligned}
Z_{\text{out}} &= \frac{V_{\text{out}}}{I_o} = \left[\frac{1}{R_E} + (1 + h_{fe})/h_{ie} \right]^{-1} \\
&= R_E \| [h_{ie}/(1 + h_{fe})] \\
&= (1.0 \times 10^3) \| (600/101) \\
&= 5.9 \ \Omega
\end{aligned}
\tag{4.48}
$$

The current gain, $A_i = I_e/I_{\text{in}}$, is also obtained from Figure 4.16b. The signal current I_{in} enters a current divider between the bias resistors R_1 and R_2 and the impedance at the base terminal. Therefore,

$$I_b = \{(R_1 \| R_2)/[(R_1 \| R_2) + Z_b]\} I_{\text{in}} \tag{4.49}$$

$$I_e = (1 + h_{fe}) I_b \tag{4.50}$$

By substitution,

$$
\begin{aligned}
A_i &= \frac{I_e}{I_{\text{in}}} \\
&= (1 + h_{fe})\{(R_1 \| R_2)/[(R_1 \| R_2) + Z_b]\} \\
&= 101 \ \{(18 \times 10^3 \| 24 \times 10^3)/[(18 \times 10^3 \| 24 \times 10^3) \\
&\quad + 101.6 \times 10^3]\} \\
&= 9.3
\end{aligned}
\tag{4.51}
$$

The voltage gain, $A_v = V_{\text{out}}/V_{\text{in}}$, is found by substitution:

$$V_{\text{out}} = (1 + h_{fe}) I_b R_E \tag{4.52}$$

$$I_b = \frac{V_b}{Z_b} = \frac{V_{\text{in}}}{Z_b} \tag{4.53}$$

Then,

$$
\begin{aligned}
A_v &= \frac{V_{\text{out}}}{V_{\text{in}}} = \frac{R_E(1 + h_{fe})}{Z_b} \\
&= \frac{(1.0 \times 10^3)(101)}{101.6 \times 10^3} \\
&= 0.99
\end{aligned}
\tag{4.54}
$$

In this example, the transistor collector, even though connected to the positive terminal of the power supply, is at ground potential in the ac circuit. This common-collector circuit is distinguished by properties shown in the cal-

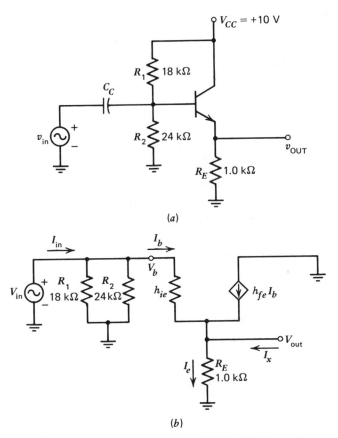

Figure 4.16. (a) Common-collector circuit. (b) Ac equivalent circuit of common-collector circuit in (a).

culations. The voltage gain is positive and less than but very close to unity.
The current gain is much higher than unity. The input impedance is high, and
the output impedance is very low. This type of circuit is primarily used to
provide a low source impedance for driving subsequent loads or for maximum
power transfer applications.

The methodology in using the small-signal equivalent circuit is straightforward. The ac circuit is drawn for the amplifier with the small-signal transistor
equivalent circuit inserted for the transistor symbol. Conventional linear circuit
analysis techniques are then used to calculate the amplifier parameters.

4.6 EXAMPLES OF AMPLIFIER dc CALCULATIONS

The dc calculations associated with electronic circuits have been introduced
in previous discussions. The method employs dc circuit analysis with assumptions made for any diode voltage drops encountered in a pathway. The critical

assumption in a transistor dc analysis is the voltage drop attributed to the forward-biased base–emitter junction. A typical silicon transistor drawing a few milliamperes of emitter current in the forward active mode will have a $V_{BE} \simeq 0.70$ V. The base–emitter junction of a transistor will "cut in" and start drawing a few microamperes of current when V_{BE}(cut-in) $\simeq 0.60$ V. The assignments for various conditions of the transistor states are approximate and will vary among transistor types. More exact models of diodes are used in computer simulation circuits but are too complicated to use here.

The estimation of dc values for a circuit is illustrated below. The four major steps employed in the analysis of most dc transistor circuits are:

1. A voltage loop equation encompassing the base–emitter terminals.
2. A voltage loop equation encompassing the collector–emitter terminals.
3. Assignment of a value for V_{BE} in the various transistor states as described above.
4. The use of the active mode relationships between h_{FE}, I_B, I_C, and I_E.

4.6.1 Common-Emitter Bias Calculations

Example. Estimate I_{EQ}, I_{CQ}, and V_{CEQ} for the circuit of Figure 4.17. The common-emitter forward current ratio is $\beta = 100$.

Solution. First decide whether the transistor is in the active or saturated state. V_{BE} is certainly positive in Figure 4.17. To determine the polarity of V_{BC}, assume that V_{BC} is negative and the transistor is then in the active mode. If the transistor is not in the active mode, the calculated active mode currents will produce a voltage drop across R_C and R_E that will make V_{CE} a few tenths of a volt, or even negative, and the problem must be recalculated assuming that the transistor is in the saturated mode.

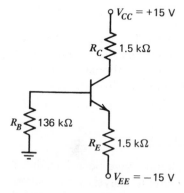

Figure 4.17. Transistor circuit used as example for dc analysis.

The collector and emitter currents are controlled by the base current so that the next step is to write a voltage loop equation including the base-emitter junction:

$$I_{BQ}R_B + V_{BE} + I_{EQ}R_E = V_{EE} \tag{4.55}$$

Substitute

$$I_{BQ} = \frac{I_{EQ}}{1 + h_{FE}} \tag{4.56}$$

and solve for I_{EQ}

$$I_{EQ} = \frac{V_{EE} - V_{BE}}{[R_B/(1 + h_{FE})] + R_E} \tag{4.57}$$

Using the active mode voltage approximation, $V_{BE} \simeq 0.70$ V, and substituting the given values,

$$
\begin{aligned}
I_{EQ} &= \frac{15 - 0.70}{(136 \times 10^3/101) + 1.5 \times 10^3} \\
&= 5.0 \text{ mA}
\end{aligned}
\tag{4.58}
$$

and

$$
\begin{aligned}
I_{CQ} &= \frac{I_{EQ}h_{FE}}{1 + h_{FE}} \\
&= \frac{5 \times 100}{101} \\
&\simeq 5.0 \text{ mA}
\end{aligned}
\tag{4.59}
$$

The V_{CEQ} value is obtained from the voltage loop equation that includes the collector–emitter terminals and the current pathways of I_C and I_E:

$$V_{CC} = I_{CQ}R_C + V_{CEQ} + I_{EQ}R_E - V_{EE} \tag{4.60}$$

Solving for V_{CEQ} yields

$$
\begin{aligned}
V_{CEQ} &= V_{CC} + V_{EE} - I_{CQ}R_C - I_{EQ}R_E \\
&= 15 + 15 - 5.0 \times 10^{-3} \times 1.5 \times 10^3 \\
&\quad - 5.0 \times 10^{-3} \times 1.5 \times 10^3 \\
&= 15 \text{ V}
\end{aligned}
\tag{4.61}
$$

Note the use of the base–emitter loop equation to calculate emitter current. The loop equation through the collector–emitter terminals and the I_C and I_E pathways determines V_{CE}. The value calculated for V_{CEQ}, 15 V, is well above the saturated value of $V_{CE}(\text{sat})$, which is typically 0.10 to 0.20 V. The assumption that the transistor was in the active mode was true, and the calculated active mode currents and voltages are correct.

4.6.2 **Voltage Divider Bias for the Common-Emitter Amplifier**

Example. Figure 4.18*a* shows a popular discrete circuit design of a common-emitter amplifier. In this example, we analyze the values of I_{EQ}, I_{CQ}, and V_{CEQ}. The amplifier is designed for an active mode transistor and $V_{BE}(\text{act}) \simeq 0.70$ V. The β is 200.

Solution. The loop equation through the base–emitter junction appears slightly complex in this example. However, the method in solving for the emitter current is to reduce the circuit portion to the left of the base terminal to a dc Thevenin circuit. Figure 4.18*b* shows the redrawn circuit. The dc Thevenin voltage to the left of the base terminal is

$$
\begin{aligned}
V_{BB} &= \frac{V_{CC}R_2}{R_1 + R_2} \\
&= 10 \times \frac{4.6 \times 10^3}{(32 \times 10^3) + (4.6 \times 10^3)} \\
&= 1.25 \text{ V}
\end{aligned} \tag{4.62}
$$

The dc Thevenin resistance to the left of the base terminal is

$$
\begin{aligned}
R_{BB} &= R_1 \| R_2 \\
&= 4.0 \text{ k}\Omega
\end{aligned} \tag{4.63}
$$

Now the base–emitter loop equation may be written from the circuit in Figure 4.18*b* as

$$
V_{BB} = I_{BQ}R_{BB} + V_{BE} + I_{EQ}R_E \tag{4.64}
$$

and, solving for I_{EQ},

$$
\begin{aligned}
I_{EQ} &= \frac{V_{BB} - V_{BE}}{[R_{BB}/(1 + \beta)] + R_E} \\
&= \frac{1.25 - 0.70}{(4 \times 10^3/201) + 200} \\
&= 2.5 \text{ mA}
\end{aligned} \tag{4.65}
$$

The collector–emitter loop equation yields

$$
\begin{aligned}
V_{CEQ} &= V_{CC} - I_{CQ}R_C - I_{EQ}R_E \\
&= 10 - (2.5 \times 10^{-3} \times 1.6 \times 10^3) - (2.5 \times 10^{-3} \times 200) \\
&= 5.5 \text{ V}
\end{aligned} \tag{4.66}
$$

This transistor is clearly in the active mode with $V_{CEQ} \gg 0.20$ V. This example illustrates the use of the dc Thevenin equivalent circuit in reducing an apparently complex base biasing network to a simple circuit.

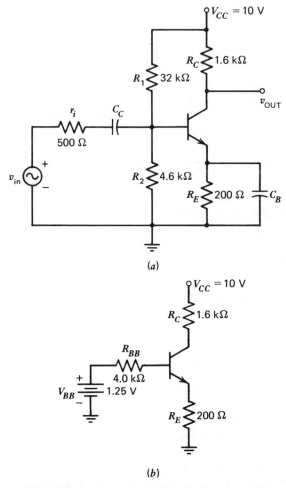

(a)

(b)

Figure 4.18. (a) Discrete design of common-emitter amplifier. (b) Circuit of part (a) with a dc Thevenin equivalent circuit to the left of the base terminal.

4.6.3 **Transistor Operating State**

Example. The common-emitter circuit of Figure 4.19 is used to illustrate the influence of the load resistor on the transistor operating state. The dc forward current gain is $\beta = 100$. Fill in Table 4.1 and draw conclusions concerning the transistor junction biasing and operational state.

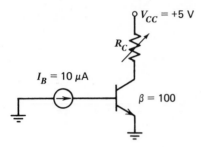

Figure 4.19. Common-emitter amplifier with a variable R_C that leads the transistor into the saturated state.

Solution. For the first row, with $R_C = 1.0 \text{ k}\Omega$, the collector current is

$$I_C = \beta I_B$$
$$= 100 \times 10 \times 10^{-6}$$
$$= 1.0 \text{ mA}$$

The $V_{BE}(\text{act}) \simeq 0.70$ V, and

$$V_{CE} = V_{CC} - I_C R_C$$
$$= 5 - (1.0 \times 10^{-3})(1.0 \times 10^{3})$$
$$= 4.0 \text{ V}$$

The V_{BC} is

$$V_{BC} = V_{BE} - V_{CE}$$
$$= 0.70 - 4.0$$
$$= -3.3 \text{ V}$$

As the collector load resistor R_C is increased to 4.0 kΩ, the bias voltages still hold the transistor in the active state. As R_C increases to 4.3 kΩ, V_{BC} goes to zero volts. Ideally, this defines the onset of saturation, with the base–collector junction becoming forward biased and injecting minority carriers into the base instead of collecting them. In practice, the transistor will not show the pronounced effects of saturation until V_{BC} exceeds 0.40 V. This is because significant minority carriers are not injected across the base–collector diode until the forward bias increases to this point.

Table 4.1. Circuit Values

R_C (kΩ)	I_C (mA)	V_{BE} (V)	V_{CE} (V)	V_{BC} (V)
1.0				
2.0				
4.0				
4.3				
4.6				
6.0				

As R_C increases to 6.0 kΩ, the transistor goes into the saturation mode. The collector–emitter voltage drops to a low value of $V_{CE}(\text{sat}) \simeq$ 0.10, V_{BE} decreases slightly and the collector current becomes

$$I_C = \frac{V_{CC} - V_{CE}(\text{sat})}{R_C}$$

$$= \frac{5.0 - 0.10}{6 \times 10^3}$$

$$= 0.82 \text{ mA}$$

and

$$V_{BC} = V_{BE} - V_{CE}$$

$$= 0.69 - 0.10$$

$$= +0.59 \text{ V}$$

The completed table of values is given in Table 4.2. Important observations should be noted. In the active mode, a change in R_C does not significantly affect I_C, since the transistor acts as a constant current source in the active mode. However, as R_C increases, a corresponding decrease in voltage occurs across the collector–emitter terminals. When the transistor enters saturation, the constant current nature of the transistor is lost with $V_{CE}(\text{sat})$ retaining a low value in the range of 0.10 to 0.20 V. The current through the base–emitter junction decreases as R_C gets larger as indicated by comparing I_C in Table 4.2 for $R_C = 4.6$ kΩ and $R_C = 6.0$ kΩ. The voltage across the base–emitter junction then drops slightly so that for this circuit, $V_{BE}(\text{sat}) < V_{BE}(\text{act})$.

In the example of Figure 4.19, the transistor was brought into saturation by increasing the load resistance for a fixed base current. In some digital circuits, an output transistor is brought into saturation by increasing the base current, using a fixed value of load resistance. The principle of transistor saturation is the same. The increase in base current causes an increasingly large collector current through the relation $I_C = \beta I_B$. When the voltage drop across

Table 4.2. Completed Values

R_C (kΩ)	I_C (mA)	V_{BE} (V)	V_{CE} (V)	V_{BC} (V)
1.0	1.0	0.70	+4.0	−3.30
2.0	1.0	0.70	+3.0	−2.30
4.0	1.0	0.70	+1.0	−0.30
4.3	1.0	0.70	+0.70	0.00
4.6	1.0	0.70	+0.10	+0.30
6.0	0.82	0.69	+0.10	+0.59

the load resistance causes the collector voltage to drop below that of the base, the transistor enters the saturation region. In this case, the current through the base–emitter junction increases such that, in this mode of saturation, the base–emitter voltage usually increases to $V_{BE}(\text{sat}) \simeq 0.8$ V. This mode of saturating a transistor is seen in the next section where we examine the switching time of an inverter circuit.

4.7 INVERTER SWITCHING AND TIME DELAYS

In Chapters 6 and 7, we analyze several different bipolar digital circuits. The purpose of these circuits is to propagate information by switching between two definite discrete voltage levels when directed to do so by an input signal. This input signal often has the same voltage swing as the output signal. Like most events occurring in our physical world, this switching cannot be accomplished in zero time. In this section, we examine the switching of a transistor inverting amplifier and the time delays associated with the transitions. We use this to develop a large-signal model of the bipolar transistor. This model is less accurate than the traditional models due to Ebers and Moll, Beufoy and Sparkes, Linvill, and Gummel and Poon that are used in computer simulations such as SPICE, but the model gives us qualitative insights into the operation of the bipolar transistor as a switch.

The circuit to be analyzed is shown in Figure 4.20. The drive signal to the base is chosen to be a *slow* current ramp starting at zero and increasing to a value sufficient to cause the transistor to go well into saturation. For $t < t_1$, the base–emitter junction draws no current and is in the high-resistance or reverse-bias condition. As $i_B(t)$ increases in the time range from t_1 to t_2, the transistor exists in the active region. As $i_B(t)$ increases beyond its value at t_2, the base–collector junction becomes forward biased and the transistor enters the saturation region.

These events can be followed in the circuit load line and transistor characteristic curve plots of Figure 4.20c. The transistor is cut off for $t < t_1$, and $v_{CE} = V_{CC} = 5$ V. As $i_B(t)$ increases from zero, the state of the transistor can be tracked on the load line. The transistor is in the active region for $t_1 < t < t_2$, and $i_C(t) = h_{FE}i_B(t)$. As both $i_B(t)$ and $i_C(t)$ increase in magnitude, the voltage drop across R_C also increases. The v_{CE} value drops, since $v_{CE} = V_{CC} - i_C R_C$. The base voltage remains essentially constant at $v_B = V_{BE}(\text{act}) \simeq 0.7$ V. When $v_{CE} = v_{BE}$, the v_{BC} must become zero. Further decreases in v_{CE} then cause significant forward bias of the base–collector junction, and the transistor enters saturation at $t = t_2$. A further increase in $i_B(t)$ to the point $t = t_3$ drives the transistor deeper into saturation. In deep saturation, v_{CE} remains constant at $V_{CE}(\text{sat}) \simeq 0.1$ V and v_{BE} increases to $V_{BE}(\text{sat}) \simeq 0.8$ V. In saturation, it is

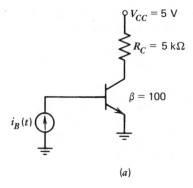

(a)

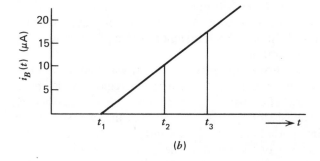

(b)

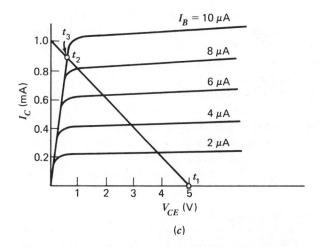

(c)

Figure 4.20. (*a*) Inverter switch. (*b*) Base current waveform. (*c*) Transistor characteristic curves with circuit
load line.

important to note that $I_C(\text{sat})$ remains constant, even though $i_B(t)$ continues to increase. The collector current under these conditions is

$$I_C(\text{sat}) \simeq \frac{V_{CC} - V_{CE}(\text{sat})}{R_C} \tag{4.67}$$

$$\simeq \frac{V_{CC}}{R_C} \tag{4.68}$$

For the circuit of Figure 4.20*a*, this becomes

$$I_C(\text{sat}) \simeq \frac{5.0 - 0.1}{5 \times 10^3}$$

$$= 0.98 \text{ mA}$$

Observe in Figure 4.20*c* that the two points on the load line, t_2 and t_3, are quite close together in spite of a large difference in the base currents associated with these times. It should be remembered that the family of i_B curves is very dense in this saturation region.

In this example, the transistor of the inverter circuit was taken from the cutoff region through the active region to the saturation region as the base current was *slowly* increased. The switching process we have described is sometimes referred to as "quasi-static." This implies that the changes in the bias conditions occur in time periods that are long compared to time constants associated with the components of the circuit. These time constants include the transit time across the base and the excess carrier lifetime in the base of the transistor and a number of RC time constants associated with parasitic capacitances in the transistor. In the discussion below, we see the effects of these time constants on the operation of the circuit.

If we change the wave shape of $i_B(t)$ so that it is a rectangular pulse with a magnitude of 15 μA and a duration of 10 μs, as shown in Figure 4.21*a*, the collector current cannot follow the base current. We represent the base–emitter junction as a series combination of a battery, an "ideal" diode, and a resistor with a capacitor in parallel with the series components, as shown in Figure 4.21*b*. The "ideal" diode refers to one that acts as an open circuit when reverse biased and a short circuit when forward biased. When the step of current is applied at t_1, the voltage from B to E cannot change instantaneously. This voltage is given by

$$v_{BE}(t) = \frac{1}{C} \int_{t_1}^{t} i_1(t') \, dt' + v_{BE}(t_1) \tag{4.69}$$

For this situation, $v_{BE}(t_1) = 0$, and $i_1(t) = i_B(t) = 15 \ \mu$A and $i_2(t) = 0$ until $v_{BE} = V_D$. If we let $V_D = 0.60$ V and $r = 22.2$ kΩ, the value of v_{BE} is 0.80 V when i_2 reaches 9.0 μA, which agrees with our previous value for the saturation voltage. We can now identify a time t_2, when $v_{BE} = V_D = 0.60$ V:

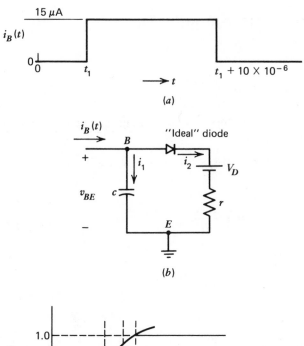

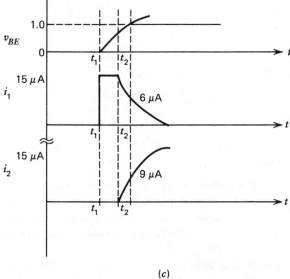

Figure 4.21. Model for the turn-on transient. (*a*) The base current pulse. (*b*) The model for the base-emitter junction. (*c*) Timing diagrams for the turn-on transient.

$$0.60 = \frac{1}{c}[15 \times 10^{-6}(t_2 - t_1)]$$

or

$$t_2 - t_1 = \frac{0.60c}{15 \times 10^{-6}}$$

For $c = 2.0$ pF, $t_2 - t_1 = 80$ ns. After this time, the voltage can be represented by the following relations:

$$i_1 + i_2 = 15 \times 10^{-6}$$

but

$$i_1 = c\frac{dv_{BE}}{dt}$$

and

$$i_2 = \frac{v_{BE} - 0.60}{r}$$

Then,

$$c\frac{dv_{BE}}{dt} + \frac{v_{BE} - 0.60}{r} = 15 \times 10^{-6} \tag{4.70}$$

The voltage tries to rise from 0.60 V to a final value of

$$0.60 + I_B r = 0.60 + (15 \times 10^{-6})(22.2 \times 10^3) = 0.93 \text{ V}$$

with a time constant given by

$$rc = (22.2 \times 10^3)(2.0 \times 10^{-12})$$
$$= 44.2 \text{ ns}$$

This is indicated in Figure 4.21c. The collector current is proportional to i_2 and does not begin until t_2. When v_{BE} reaches 0.80 V, the device goes into saturation, and the model is no longer valid. The collector current reaches 0.98 mA and remains constant while the base current continues to increase. The turn-on transient concludes when $i_2 = 15$ μA and $i_1 = 0$.

The base current in excess of that required to drive the transistor to the edge of saturation is responsible for a buildup of charge in the base, as indicated in Figure 4.22. To turn off the device, it is necessary to remove the excess charge from the device before the collector current begins to decrease. In our example, the base current is abruptly terminated 10 μs after it is initiated. More rapid switching is obtained if the base current is reversed for a short period of time. The time required to reduce the excess charge in the base to the point where the transistor reenters the active region is called the storage time t_s. In general, t_s is proportional to the excess carrier lifetime. In some cases, excess

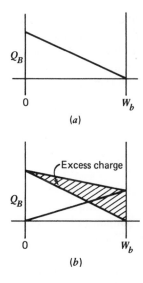

(a)

(b)

Figure 4.22. (a) Charge in base for forward active bias. (b) Charge in base for saturation.

carrier lifetime "killers" such as gold are used to reduce the storage time. This usually is much longer than the remainder of the turn-off transient.

The total transient is shown in Figure 4.23. The important concepts to be gained from this example are that there are several time delays associated with the switching of a bipolar transistor. These include a brief delay before the turn-on transient of the collector current can begin. This is followed by a one-minus exponential transition to the saturation point. After this transition, there is a delay associated with the storage of excess charge in the base region. When

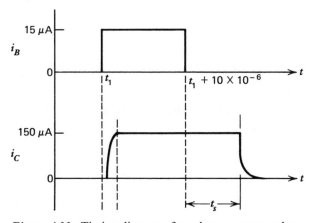

Figure 4.23. Timing diagrams for a base current pulse.

the turn-off transient is initiated, there is a time period, called the storage time, during which the excess charge beyond that needed to drive the transistor to the edge of saturation is removed from the base region. Following the storage interval, the device passes through the active region to cutoff. These delays are important in the digital circuits discussed in Chapters 6 and 7.

4.8 SUMMARY

The bipolar transistor is one of the two important "active" components used in solid state electronics. In Chapter 5, we introduce the other important contributor, the field-effect transistor. The current gain mechanism in the bipolar transistor is due to the motion of both electrons and holes. The most important contributions are related to the distribution of minority carriers in the base region of the device. Since the transitor has three terminals, the emitter, the base, and the collector, it is possible to connect it in circuits with the common terminal between the input signal and the output signal as any one of the terminals. The most popular configuration is the common-emitter circuit, in which both the current and voltage gains are greater than one. But the common-base connection, with a current gain less than one and a voltage gain greater than one, and the common-collector connection, with the current gain greater than one and the voltage gain less than one, have applications in electronic circuits. The bipolar transistor can be used as a small-signal amplifier, or as a large-signal switch. In Chapters 6 and 7, we discuss the switching circuit applications of the bipolar transistor and, in Part III, analog applications are presented.

PROBLEMS

Note: Note: Problems 4.1 to 4.5 are intended to review the analysis of dependent source networks.

4.1. Calculate the Thevenin equivalent circuit at the terminals of the circuit in Figure 4.24.

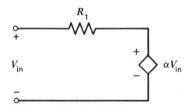

Figure 4.24. Circuit for Problem 4.1.

4.2. Calculate the Thevenin equivalent circuit at the terminals of the circuit in Figure 4.25.

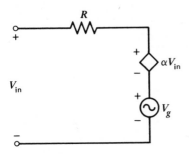

Figure 4.25. Circuit for Problem 4.2.

4.3. (*a*) Calculate the terminal impedance for the circuit in Figure 4.26. (*b*) Determine the ratio V_3/V_1 in the same circuit.

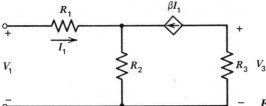

Figure 4.26. Circuit for Problem 4.3.

4.4. (*a*) Calculate the terminal impedance for the circuit in Figure 4.27. (*b*) Determine the ratio I_3/I_{in} for the same circuit.

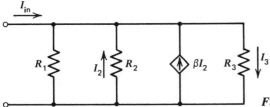

Figure 4.27. Circuit for Problem 4.4.

4.5. (*a*) Calculate the terminal impedance for the circuit in Figure 4.28. (*b*) Determine the ratio V_3/V_1 for the same circuit.

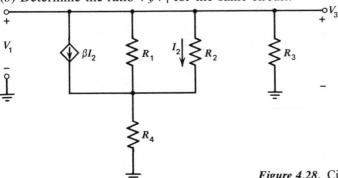

Figure 4.28. Circuit for Problem 4.5.

4.6. For the circuit of Figure 4.29, with $\beta = 100$, find I_{CQ} and V_{CEQ}.

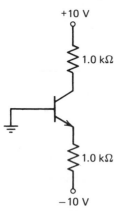

+10 V

1.0 kΩ

1.0 kΩ

−10 V

Figure 4.29. Circuit for Problem 4.6.

4.7. For the circuit of Figure 4.30, with $\beta = 100$, find I_{CQ}, V_{CEQ}, V_E, and V_{BC}.

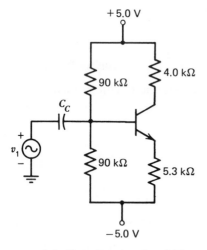

+5.0 V

90 kΩ

4.0 kΩ

C_C

v_1

90 kΩ

5.3 kΩ

−5.0 V

Figure 4.30. Circuit for Problem 4.7.

4.8. For the circuit of Figure 4.31, with $h_{fe} = 50$, $h_{oe} = 50\ \mu S$, and $h_{ie} = 1.8\ k\Omega$, calculate A_v, Z_{in}, Z_{out}, and $A_i = I_c/I_{\text{in}}$.

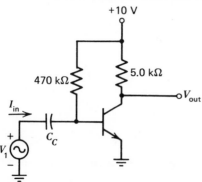

+10 V

470 kΩ

5.0 kΩ

V_{out}

I_{in}

C_C

V_1

Figure 4.31. Circuit for Problem 4.8.

4.9. Find the Thevenin equivalent circuit looking to the right of *ab* in Figure 4.32.

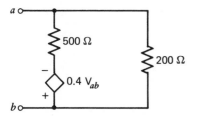

Figure 4.32. Circuit for Problem 4.9.

4.10. For the circuit of Figure 4.33, with $h_{fe} = 100$, $h_{oe} = 25$ μS, and h_{ie} 10 kΩ, find Z_C.

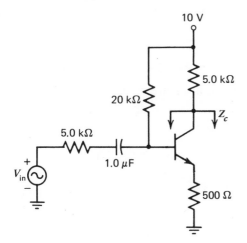

Figure 4.33. Circuit for Problem 4.10.

4.11. For the common-base circuit of Figure 4.34, with $\beta = 50$, $h_{oe} = 25$ μS, $h_{ie} = 2.0$ kΩ, and $R_C = 4.0$ kΩ, find $A_v = V_{\text{out}} / V_{\text{in}}$.

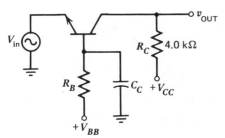

Figure 4.34. Circuit for Problem 4.11.

4.12. For the circuit of Figure 4.35, with $h_{fe} = \beta = 100$, $h_{re} = h_{oe} = 0$, and $h_{ie} = \beta \times 26 \times 10^{-3}/I_{CQ}$, find $A_v = V_{out} / V_{in}$.

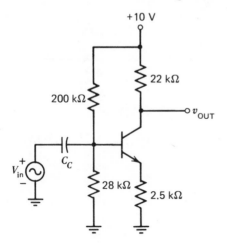

Figure 4.35. Circuit for Problem 4.12.

4.13. Find I_{CQ} and V_{CEQ} for the circuit in Figure 4.36.

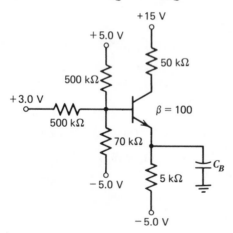

Figure 4.36. Circuit for Problem 4.13.

4.14. For the circuit and characteristics of Figure 4.37, with $\beta = 100$, (a) find I_{CQ} and V_{CEQ}, and (b) draw the load line and mark the operating point.

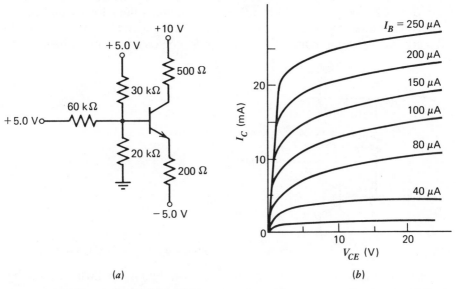

(a)   (b)

Figure 4.37. Circuit for Problem 4.14.

4.15. The common-emitter circuit in Figure 4.38 is to be biased at a quiescent point such that $V_{CEQ} = 5.0$ V. (a) Find a value of R_B to accomplish this if $h_{fe} = \beta = 200$. (b) If $h_{oe} = 10^{-5}$ S, find A_v. (c) Find the voltage gain and new value of R_B if the bias current I_{CQ} is reduced to half of the value in part (a). (d) How did the input resistance change when I_{CQ} was reduced by a factor of 2? To determine h_{ie} see comment in problem 4.12.

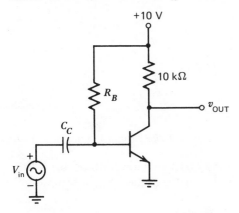

Figure 4.38. Circuit for Problem 4.15.

4.16. The common-emitter circuit in Figure 4.39 is to be designed for maximum voltage gain under conditions that the output voltage may vary by ± 1.0 V from the quiescent point and that R_C is at least 1.0 kΩ. Assume that $h_{fe} = \beta = 200$ and $h_{oe} = 20$ μS. (a) Select optimal values for R_B and R_C to achieve maximum voltage gain. (b) Draw a load line on an I_C-versus-V_{CE} graph and mark the operating point. To determine h_{ie} see comment in problem 4.12.

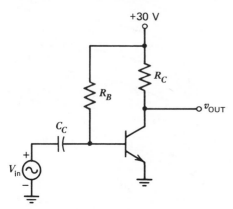

Figure 4.39. Circuit for Problem 4.16.

4.17. In Figure 4.40, $h_{fe} = 100$, $h_{oe} = 25$ μS, and $h_{ie} = 1500$ Ω. Find the resistance seen looking into the collector.

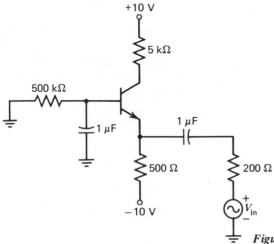

Figure 4.40. Circuit for Problem 4.17.

4.18. In Figure 4.41, $h_{fe} = 100$ and the onset of saturation occurs when $V_{CE} \le 0.4$ V. What value of I_B will put the transistor at the edge of saturation?

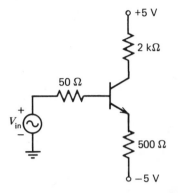

Figure 4.41. Circuit for Problem 4.18.

4.19. In Figure 4.42, what value of R_C will cause the transistor to enter the saturation state?

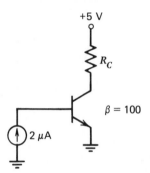

Figure 4.42. Circuit for Problem 4.19.

4.20. The inverter circuit shown in Figure 4.43 is driven at the base by a step current waveform of 10 μA. Using the model of Figure 4.21b and given that $c = 1.0$ pF and $r = 20$ kΩ, find the delay time from the start of the waveform until $V_{BE} = 0.8$ V.

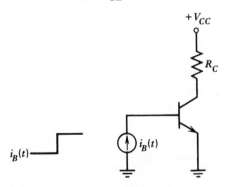

Figure 4.43. Circuit for Problem 4.20.

4.21. A 20-μA current step is applied to the inverter circuit of Figure 4.43. The delay time for V_{BE} to reach 0.6 V from the start of the current step is 60 ns, and the total delay time for V_{BE} to reach 0.8 V is 100 ns. Find the values for c and r in the model of Figure 4.21b.

chapter 5
FIELD EFFECT TRANSISTORS

The field effect transistor (FET) operates in a much different manner from the bipolar junction transistor. As discussed in Chapter 4, bipolar transistor operation depends on the transport of minority carriers across a narrow region called the base. It is often represented as a current-controlled dependent current source. One significant feature of the bipolar junction transistor is a low input impedance when the device is connected in a common-emitter or common-base configuration. In contrast, the operation of the FET is based on the control of majority carriers in a channel by an applied voltage. In some cases, the geometry of the channel is altered by the applied voltage, while in others, the number of charge carriers depends on the voltage. The FET is best represented as a voltage-controlled dependent current source. It has a very large input impedance compared to that of a bipolar junction transistor.

There are three basic FET physical configurations, as shown in Figure 5.1. In each configuration, there are three external terminals called the *source,* the *drain,* and the *gate.* In general, a voltage between the gate and the source controls the drain current. In the first structure shown, the gate is formed by creating a junction between the gate contact and the channel that connects the source and the drain. This device is called the junction field effect transistor (JFET). The second FET structure in the figure is similar to the first, except that the gate is formed by a rectifying metal–semiconductor contact. This Schottky barrier diode device is called a metal–Schottky barrier FET, or MESFET. The third device is the metal–oxide–semiconductor FET, or MOSFET. It is distinguishable from the other FET structures by a layer of silicon dioxide that

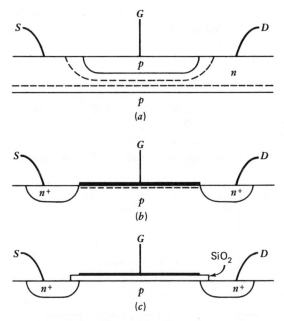

Figure 5.1. The three basic types of field effect transistors. (*a*) JFET. (*b*) MESFET. (*c*) MOSFET.

separates the gate from the channel. This insulator provides the MOSFET with the highest input resistance of the FET structures. Unfortunately, the gate oxide in a MOSFET is such a good insulator that it is susceptible to destruction by the buildup of static charge on the gate, and it is common practice to use gate protection diodes to provide a path for discharging the gate-to-substrate capacitor if the voltage between the gate and substrate becomes excessive. Precautions are necessary in the handling of MOSFETs to prevent destruction by static charge.

In this chapter, we examine the physical mechanisms responsible for the electrical characteristics of the three types of FETs. We also discuss the operation of the MOS capacitor, an essential part of the MOSFET. We conclude the chapter with an analysis of the FET inverter biasing properties and propagation delay.

5.1 JFETS AND MESFETS

There are many similarities in the basic theory of operation of JFETs and MESFETs. Both devices rely on the variation of the width of the space-charge region of a reverse-biased diode to control the area available for conduction between two ohmic contacts to a bar of semiconductor. As the names imply, the JFET has a *pn* junction gate, and the MESFET has a metal–semiconductor (Schottky barrier) gate region.

5.1.1 **Fabrication**

The *n*-channel JFET is usually fabricated in lightly doped *n*-type silicon by forming a heavily doped *p*-type region at the surface, as shown in Figure 5.2*a*. It is possible to fabricate this device with symmetrical gates on the top and bottom as indicated in Figure 5.2*b;* but for ease of manufacturing (partic-

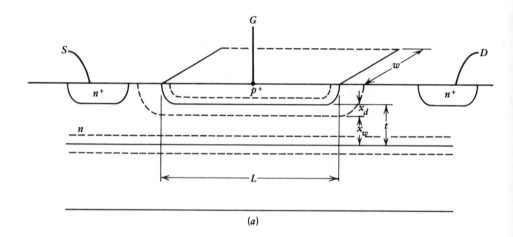

(*a*)

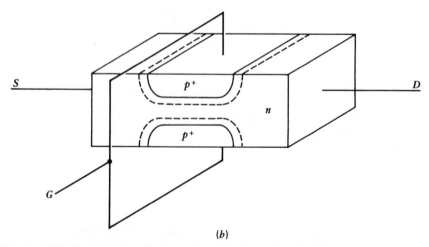

(*b*)

Figure 5.2. The dimensions of an *n*-channel junction FET. (*a*) JFET with a symmetric gate structure. (*b*) JFET with symmetric gate structure.

ularly in integrated circuits), it is more common to make an unsymmetrical structure using a lightly doped p-type substrate, an n-type epitaxial layer, and a diffused p^+-gate region on top. If the substrate and source are grounded, the substrate–channel junction does not have as much effect as the gate–channel junction on the operation of the device.

The MESFET can be made either in gallium arsenide or silicon, but the majority of these devices are gallium arsenide. They are used extensively in the frequency range above 1 gigahertz. A typical MESFET is formed by implanting selenium into semiinsulating (chromium-doped) gallium arsenide to form a lightly doped n-type channel. Aluminum is used for the Schottky barrier and gold–germanium is used for the ohmic contacts. In some cases, epitaxial layers of n-type material on semiinsulating substrates are used for the channels. Devices are isolated in epitaxial structures by etching away the material between devices to form mesas. As in the case of JFETs, the channel-to-substrate junction is less important than the gate diode in controlling the operation of the device. The electron mobility in gallium arsenide is five times that of silicon, but fabrication difficulties have prevented the development of GaAs-insulated gate structures.

5.1.2 Current–Voltage Characteristics

In a JFET or MESFET, a reverse bias voltage from gate to channel can be used to control the drain current. The effect of this voltage is to change the width of the space charge region of the diode, as discussed in Chapter 3. The width of the space charge region, x_d, is given by

$$x_d = x_{d0}\left(1 - \frac{V_{GC}}{V_0}\right)^{1/2} \tag{5.1}$$

where x_{d0} is the width with $V_{GC} = 0$, V_0 is the built-in voltage associated with the diode, and V_{GC} is the gate-to-channel voltage. V_{GC} is positive for forward bias and negative for reverse bias. In normal operation, the gate-to-channel voltage is zero at the source end of the channel and negative at the other points in the channel as we progress from the source to the drain.

For low values of drain-to-source voltage V_{DS}, the FET acts like a resistor whose cross-sectional area perpendicular to the carrier flow is controlled by the gate-to-source voltage. In this region of operation with a constant gate-to-source voltage, the drain current is directly proportional to the drain voltage, as expected for a passive resistor. In Figure 5.2a, we make the assumption that the width of the channel, x_w, is given by $t - x_d$ and the resistance of the channel is given by

$$R = \frac{\rho L}{x_w W} \tag{5.2}$$

where L is the channel length and W is the channel width. The resistivity of the channel, ρ, is given by

$$\rho = \frac{1}{q\mu_n N_d} \tag{5.3}$$

where q is the magnitude of the charge on the electron, μ_n is the electron mobility, and N_d is the donor density. The drain current is then

$$I_D = \frac{V_{DS}}{R} = \frac{q\mu_n N_d x_w V_{DS} W}{L} \tag{5.4}$$

Even though this equation is applicable over a limited range of V_{DS}, it indicates the important quantities that influence the magnitude of the drain current. The mobility μ_n is a materials property but also depends on the channel doping concentration N_d. The mobility decreases by a factor less than 20 when the doping concentration increases by a factor of 10^7, so it is less important than N_d in its influence on the drain current. The width-to-length ratio of the channel, W/L, is important and requires serpentine geometries for power devices. Although not indicated by this equation, the channel length L has a significant effect on the frequency response, since the time required for a carrier to traverse the length of the channel is a fundamental time limitation on the device. The factor $x_w = t - x_d$ contains not only the geometrical quantities but, from Equation 5.1, also includes the effect of the gate-to-channel voltage V_{GC} on the drain current. For this case, with V_{DS} low, $V_{GC} \simeq V_{GS}$. Since there is an increase in x_d for an increase in the magnitude of the reverse bias voltage between gate and channel, there is a voltage V_{GS}(off) for which $x_w = 0$, and the drain current is reduced to a negligible value. As discussed below, the voltage necessary to pinch off the channel is important in understanding the nonlinear characteristics of the JFET and MESFET.

Equation 5.4 was developed on the basis that V_{DS} was very small, so there was essentially no variation in V_{GC} along the channel. An increase in the magnitude of V_{DS} results in a voltage drop along the channel. For an n-channel device, V_{DS} is positive and V_{GS} is negative and less (in magnitude) than V_{GS}(off) for "forward active" operation. As V_{DS} increases, the gate-to-channel voltage near the drain represents a larger reverse bias than that near the grounded source. The area available for conduction perpendicular to carrier flow near the drain is then smaller than the similar area near the source, as indicated in Figure 5.3. If we consider the case of the gate connected to the source ($V_{GS} = 0$), we observe that I_D no longer responds in a linear manner as V_{DS} is increased. In fact, when $V_{DS} = -V_{GS}$(off), the channel area has been "pinched off" at the drain end of the channel. The effect on the drain current is to produce a saturation characteristic as shown in Figure 5.4. The electrons traverse the high field region of the pinched-off area in much the same manner as the electrons traverse the space charge region of the base–collector junction in a forward active-biased bipolar junction transistor. The voltage V_{GS}(off) is usually called the "pinch off" voltage and designated V_p. Further increases in V_{DS} beyond $|V_p|$ do not result in a significant increase in I_D, and this property defines the saturation region of operation for the JFET or MESFET. The small

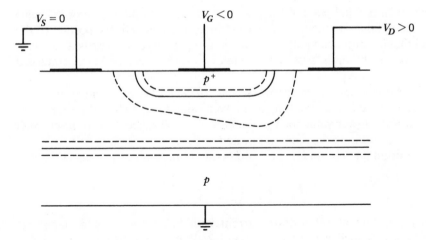

Figure 5.3. The JFET approaching pinch-off.

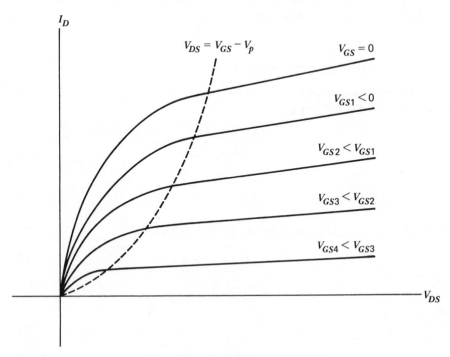

Figure 5.4. The drain characteristics of a JFET or MESFET.

increase in I_D beyond pinch-off is due to an increase in the length of that part of the channel that is pinched off. This reduces the length of the portion of the channel that is not pinched off, resulting in a slight decrease in resistance. This can be modeled by drain current by a factor $1 + \lambda V_{DS}$, where λ is an empirical parameter that is typically 1.0×10^{-4} V^{-1}.

If we now apply a small reverse bias between the gate and source, the channel area is reduced, resulting in a lower value for I_D. The saturation point now occurs at a lower value for V_{DS} than for $V_{GS} = 0$, since the gate-to-channel voltage at the drain is the sum of V_{GS} and $-V_{DS}$.

The drain current under saturation conditions can be approximated by

$$I_D \simeq I_{DSS} \left(1 - \frac{V_{GS}}{V_p} \right)^n \tag{5.5}$$

where I_{DSS} is the saturated drain current with $V_{GS} = 0$, V_p is the pinch-off voltage V_{GS}(off), and n is between $\frac{3}{2}$ and 2. We assume that $n = 2$ for our examples since there is good experimental agreement for this value of n.

Example. For $I_{DSS} = 2.0$ mA and $V_p = -2.0$ V, find and plot the drain characteristics for $V_{GS} = 0, -0.4, -0.8, -1.2, -1.6$, and -2.0 V. Also find the locus of saturation points ($V_{DS} = V_{GS} - V_p$).

Solution. Substitute the quantities in Equation 5.5, place the results in Table 5.1, and plot the results in Figure 5.5.

5.1.3 Transconductance

The gain parameter for the JFET or MESFET is called the transconductance g_m and is defined by

$$\begin{aligned} g_m &= \left. \frac{\partial I_D}{\partial V_{GS}} \right|_{V_{DS} = \text{constant}} \\ &= \frac{-2I_{DSS}}{V_p} \left(1 - \frac{V_{GS}}{V_p} \right) \end{aligned} \tag{5.6}$$

Table 5.1. Solution to Example

V_{GS} (V)	$V_{DS} = V_{GS} - V_p$ (V)	I_D (mA)
0	2.0	2.0
−0.4	1.6	1.28
−0.8	1.2	0.72
−1.2	0.8	0.32
−1.6	0.4	0.08
−2.0	0	0

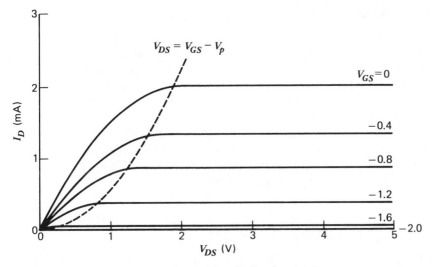

Figure 5.5. Solution to Example problem in Section 5.1.2.

For our example,

$$g_m = \frac{-2 \times 2.0 \times 10^{-3}}{-2.0}\left(1 - \frac{V_{GS}}{-2.0}\right)$$

$$= 2.0 \times 10^{-3}\left(1 + \frac{V_{GS}}{2}\right) \, S$$

Note that the transconductance is a function of V_{GS}. This is different from the current gain of the bipolar junction transistor, where the current gain is essentially independent of the base current over a wide range.

5.1.4 Switching and Frequency Characteristics

As in the case of the bipolar transistor, there are several time constants that determine the transient characteristics of JFETs and MESFETs. The fundamental limitation is associated with the transit time for majority carriers in the channel under the gate. For most devices, this transit time is so short that it can be neglected compared to the resistance-capacitance time constants in the device or circuit configuration. Since the gate is a reverse-biased diode, it has capacitances from gate to source and from gate to drain that are only a few picofarads. For low-power devices having high-valued resistors in the circuit, time constants are in the range from 1 to 10 ns. The gallium arsenide MESFET is one of the highest-frequency solid-state devices known, while JFETs are often used in radio-frequency amplifiers in the 100-MHz range.

5.1.5 **Applications**

JFETs and MESFETs are most often used in analog circuits, while MOS-FETs are predominantly used in digital circuits. Some logic circuits have been fabricated using gallium arsenide MESFETs, but the opposite polarities of the gate and drain voltages make it cumbersome to develop a system for interconnecting devices. It is possible to fabricate enhancement mode MESFETs by precisely implanting a channel that is the same thickness as the space charge region associated with the Schottky barrier diode. Applying a forward bias less than that required for significant forward current results in a reduction of the space charge region width and thus creates a channel. This type of circuit requires power supply voltages of less than a volt.

5.2 **SURFACE FIELD EFFECT DEVICES**

The MOS class of field effect devices has characteristics that are similar to the JFET but based on a different principle. In these devices, the semiconductor surface is controlled by a voltage between the semiconductor and the gate electrode separated from the semiconductor by an insulating layer. The gate can be either a metal or a polycrystalline semiconductor doped to act like a metal. The insulator is usually silicon dioxide, grown by the thermal oxidation process, but several other materials such as silicon nitride and titanium oxide have been used successfully. The most important metal–oxide–semiconductor (MOS) devices are the FETs, but we first examine the MOS capacitor, since it is a simpler device and is useful in understanding the more complicated structures.

5.2.1 **MOS Capacitor**

The basic structure of the MOS capacitor is shown in Figure 5.6. This particular device uses p-type silicon, but we should remember that there are complements of all of these devices with opposite impurity type and polarity of voltages. We also assume that there is no charge trapped in the oxide or at the interface between the silicon and the oxide. The presence of charge in either or both of these places results in a shift of the curves as if there were a voltage on the gate when the applied voltage is zero.

When a negative voltage is applied between the gate and the substrate, holes are attracted to the interface between the oxide and semiconductor where they are blocked by the oxide. Since the material is a p-type semiconductor, the effect of the negative gate voltage is an accumulation of holes at the interface. The small-signal capacitance measured between the gate and substrate is essentially the same as obtained at zero bias. The MOS capacitor, under zero or negative bias, can be treated as if it were a parallel-plate capacitor with two metal plates separated by the oxide serving as the dielectric. It is customary

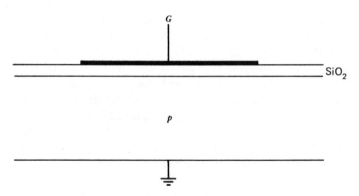

Figure 5.6. The construction of an MOS capacitor.

to designate this capacitance on a per-unit-area basis as C_{ox}. The capacitance is equal to WLC_{ox}, where W is the width and L is the length of the gate.

If, however, a small positive voltage is applied from gate to substrate, then a different effect is observed. The positive voltage repels holes from the interface, leaving a space charge of immobile, negatively charged acceptor ions. This region is very similar to the space-charge region associated with a reverse-biased junction diode. The capacitance per unit area of this device under the condition of small positive bias is the series combination of the oxide capacitance per unit area and the voltage-dependent space charge capacitance per unit area, $C_{SC}(V_G)$. This is given by

$$C = \frac{1}{1/C_{ox} + 1/C_{SC}(V_G)} \tag{5.7}$$

An increase in V_G results in an increase in the width of the space charge region and thus a reduction in the total capacitance.

A large positive gate potential will not only repel holes but attract free electrons that were created by thermal electron–hole pair generation. When the positive potential is large enough to convert the material from p-type to n-type at the insulator–semiconductor interface, the gate voltage loses its ability to increase the width of the space charge region. The interface is said to be "inverted" under these conditions. The voltage at which the interface concentration of electrons equals the hole concentration in the bulk is called the threshold voltage V_T. When V_G is increased beyond this value, the space charge region width stays the same and the electron concentration at the interface increases in response to the voltage. The capacitance per unit area then assumes a constant value given by

$$C_{min} = \frac{1}{1/C_{ox} + 1/C_{SC}(V_T)} \tag{5.8}$$

It has been assumed in this discussion that the gate voltage variations were made in a quasi-static manner. This means that the device was allowed to attain a stable steady-state situation before the measurements were made at any particular gate bias voltage. The frequency used for the measurement of small-signal capacitance of the MOS capacitor also has an influence on the results. At very low frequencies, there is a redistribution of charge between the edge of the space charge region and the inversion layer. Figure 5.7 shows a plot of small-signal capacitance-versus-gate voltage at high frequency (greater than 1.0 kHz) for a typical MOS capacitor made from p-type silicon. These interesting characteristics are primarily used to evaluate materials and device structures but have very little application in electronic circuits.

5.2.2 **MOS Field Effect Transistors**

The metal–oxide–semiconductor field effect transistor (MOSFET) is a versatile semiconductor device. It can be made with an n-channel or a p-channel, and either with or without a channel when the gate to source voltage is zero. The array of possible MOSFETs is indicated in Figure 5.8, with the arrow indicating the direction from p-type to n-type. The line connecting the source and drain is continuous for depletion mode devices (channel exists for zero gate-to-source bias) and discontinuous for enhancement mode devices (channel does not exist for zero gate-to-source bias). When MOSFETs are used as discrete devices, the substrate contact (labeled B for "body" in the figure) is usually connected to the source. In integrated circuits that use only one channel type, all of the transistors have a common substrate. Under these

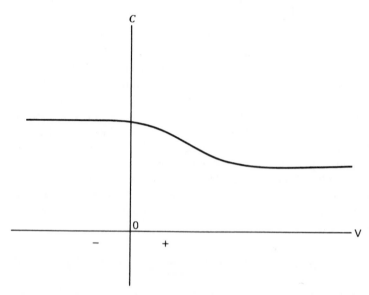

Figure 5.7. The current–voltage curve for an MOS capacitor.

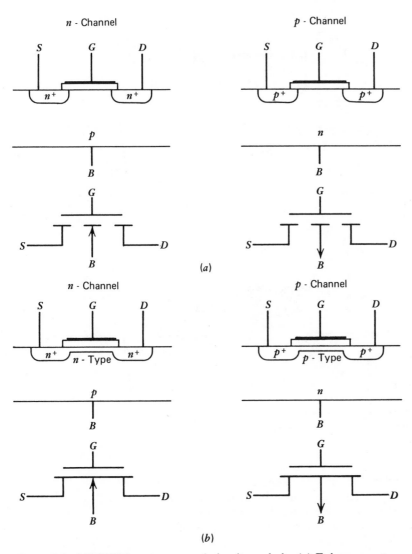

Figure 5.8. MOSFET structures and circuit symbols. (*a*) Enhancement mode. (*b*) Depletion mode.

circumstances, it is common practice for a significant number of these devices to have their source terminals connected to other circuit components, not to the substrate. This connection results in complicated relationships between the gate-to-source voltage and drain current. For most of the examples in this book, we assume that the source and substrate are connected together.

Consider the enhancement mode *n*-channel MOSFET shown in Figure 5.9. A positive voltage between the gate and source (connected to the substrate)

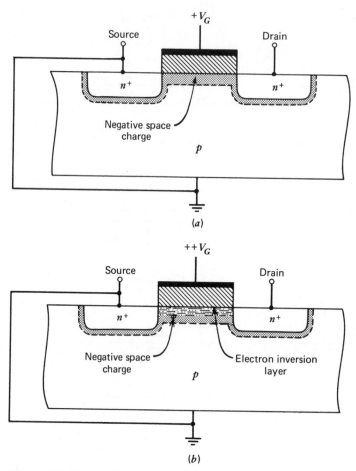

Figure 5.9. Enhancement mode *n*-channel MOSFET. (*a*) For V_G slightly less than V_T, a negative space charge region appears. (*b*) For V_G much larger than V_T, both an electron inversion and negative space charge regions are present.

that is greater than the threshold voltage inverts the interface in the same manner as in the MOS capacitor described above. In this device, however, the geometry is such that the inverted layer provides a conducting path between the two n^+-regions called the drain and source. There are also large quantities of mobile electrons in the source and drain that participate in conduction in the drain–source circuit.

The source and drain form *pn* junctions with the substrate. The space charge regions associated with these junctions are in equilibrium; and, when the source is tied to the substrate, the space charge region of the source–substrate junction remains unchanged. The gate-to-substrate voltage, when positive, produces a space charge region under the gate area. When $V_G > V_T$, both a channel and a space charge region appear under the gate, as shown in Figure 5.9b.

The space charge region due to the drain–substrate junction has a significant effect on the device operation. Since the source and substrate are connected, the combination of the gate-to-substrate voltage and the drain-to-source voltage controls the thickness of the channel in the vicinity of the drain. Increasing the drain-to-substrate voltage from a low value while maintaining a constant gate-to-substrate voltage decreases the gate-to-drain voltage and reduces the channel thickness. This has essentially the same result as the pinch-off in the JFET described above.

When $V_{DS} > V_{GS} - V_T$, the interface potential is not large enough to produce a channel at the drain end of the gate structure. Under these circumstances, the drain current remains almost constant. The current is said to "saturate" at this value. Note that "saturation" in a field effect transistor is current saturation, while "saturation" in a bipolar junction transistor is voltage saturation. The locus of points dividing the saturation region from the nonsaturation region is given by

$$V_{DS}(\text{sat}) = V_{GS} - V_T \qquad (5.9)$$

For digital applications, the devices operate on both sides of the curve specified by this equation. For analog applications, FETs are usually operated in saturation.

The channel in an enhancement mode MOSFET extends only 30 to 300 angstroms into the silicon from the oxide–silicon interface. Conduction in the channel occurs primarily by drift in the electric field caused by the drain-to-source voltage. It is characterized by a mobility that is considerably less than the bulk mobility described in Chapter 2. The reduced mobility is from surface scattering and the presence of charge trapped at the interface. The number of carriers in the channel is related to the electric field due to the gate voltage. The geometry of the device is also important in determining the drain current relationship.

The drain current versus drain-to-source voltage (with gate-to-source voltage as a parameter) characteristics can be divided into two sections, saturation and nonsaturation, as indicated in Figure 5.10a. In the saturation region, drain current is given by

$$I_D(\text{sat}) = k(V_{GS} - V_T)^2 \qquad (5.10)$$

where V_{GS} is the gate-to-source voltage and V_T is the threshold voltage. The conduction constant k is given by

$$k = \frac{W \mu C_{ox}}{2L} \qquad (5.11)$$

where W and L are the width and length of the channel, respectively; C_{ox} is the capacitance per unit area due to the gate oxide; and μ is the mobility of electrons in the channel. In the nonsaturation region,

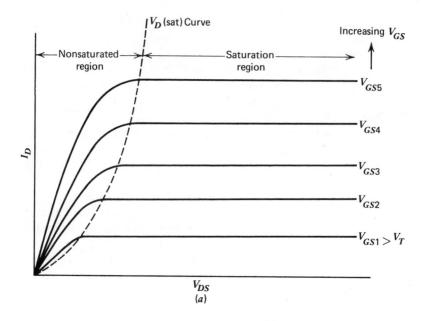

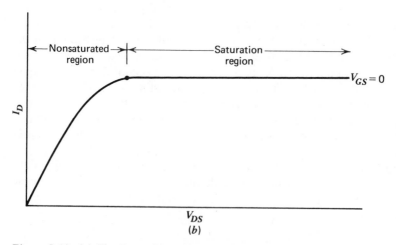

Figure 5.10. (a) The I_D-vs.-V_{DS} characteristics for an *n*-channel enhancement mode transistor. (b) Characteristic curves for I_D vs. V_{DS} for
an *n*-channel depletion mode transistor with $V_{GS} = 0$.

$$I_D(\text{nonsat}) = k[2(V_{GS} - V_T)V_{DS} - V_{DS}{}^2] \tag{5.12}$$

The curve that separates the two regions is given by Equation 5.9. As in the JFET, there is a small dependence of I_D on V_{DS} in the saturation due to a reduction in the channel length with increasing V_{DS}. This effect can be approximated by

$$I_D(\text{sat}) = k(V_{GS} - V_T)^2(1 + \lambda V_{DS}) \tag{5.13a}$$

where λ is an empirical parameter that has a typical value of $20 \times 10^{-3} \text{ V}^{-1}$. In some models, Equation 5.13a is written in the form

$$I_D(\text{sat}) = k(V_{GS} - V_T)^2\left[1 + \left(\frac{\lambda'}{L}\right)V_{DS}\right] \tag{5.13b}$$

where λ' is of the order of 10^{-7} m/V. Equations 5.10 and 5.12 also apply to n-channel depletion mode MOSFETs whose threshold voltages are negative. The drain characteristics for a depletion mode MOSFET are shown in Figure 5.10b.

The gain parameter for the MOSFET is the transconductance g_m, which may be derived for the saturation case using Equation 5.10,

$$\begin{aligned}g_m(\text{sat}) &= \left.\frac{\partial I_D(\text{sat})}{\partial V_{GS}}\right|_{V_{DS}=\text{constant}} \\ &= 2k(V_{GS} - V_T) \tag{5.14} \\ &= 2kV_{DS}(\text{sat}) = \frac{2I_D(\text{sat})}{V_{GS} - V_T}\end{aligned}$$

and for the nonsaturation case using Equation 5.12,

$$g_m(\text{nonsat}) = 2kV_{DS} \tag{5.15}$$

Note that $g_m(\text{sat})$ depends linearly on $V_{GS} - V_T$, which represents the bias voltage. Also, $g_m(\text{nonsat})$ is proportional to V_{DS}. Since k contains the ratio W/L, the gain of the MOSFET, once the fabrication process has been established, depends primarily on the surface geometry.

5.2.3 An MOS Inverter Amplifier

We now consider the MOSFET in an inverter circuit using a typical value for k as 50×10^{-6} A/V^2. The threshold voltage for an enhancement mode MOSFET is usually between 1.0 and 2.0 V, and we assume that $V_T = 1.0$ V for the circuit shown in Figure 5.11a. Kirchhoff's voltage law around the loop containing the drain and source yields

$$V_{DD} = i_D R_D + v_{DS} \tag{5.16}$$

In the gate circuit,

$$v_{GS} = V_{GG} + v_{\text{in}} \tag{5.17}$$

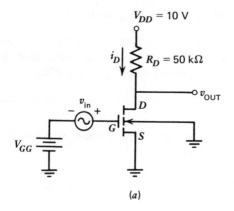

(a)

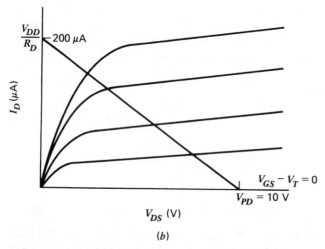

(b)

Figure 5.11. (a) An n-MOSFET inverting amplifier. (b) The load line superimposed on a family of curves for I_D vs. V_{DS}.

For $v_{in} = 0$, the dc relationships are

$$V_{DD} = I_D R_D + V_{DS}$$

and

$$V_{GS} = V_{GG}$$

If V_{GS} is greater than V_T, the drain current is given by either Equation 5.10 or 5.12, depending on the operating mode of the transistor.

Examination of Figure 5.11b indicates that a small voltage increment for V_{GS} greater than V_T places the transistor in the saturation region of operation. Then, I_D is given by

$$I_D = k(V_{GS} - V_T)^2$$

For $V_{GS} = 1.5$ V,

$$I_D = (50 \times 10^{-6})(1.5 - 1.0)^2$$
$$= 12.5 \ \mu A$$

Then,

$$V_{DS} = V_{DD} - I_D R_D$$
$$= 10 - (12.5 \times 10^{-6})(50 \times 10^3)$$
$$= 9.4 \ V$$

This verifies that the MOSFET is in saturation, since V_{DS} is greater than $V_{GS} - V_T$.

When V_{GS} increases to 2.0 V, $I_D = 50 \ \mu A$ and $V_{DS} = 7.5$ V. The operating point proceeds along the load line until $V_{DS} = V_{GS} - V_T$, which occurs for $V_{GS} = 2.81$ V and $V_{DS} = 1.81$ V.

When V_{GS} increases beyond 2.81 V, the device goes into the nonsaturation region of operation. For $V_{GS} = 3.0$ V,

$$I_D = k[2(V_{GS} - V_T)V_{DS} - V_{DS}^2]$$
$$= (50 \times 10^{-6})[2(3.0 - 1.0)V_{DS} - V_{DS}^2]$$

Substituting this into Equation 5.16 results in

$$V_{DD} = (50 \times 10^{-6})(4.0V_{DS} - V_{DS}^2)(50 \times 10^3) + V_{DS}$$
$$10 = 10V_{DS} - 2.5V_{DS}^2 + V_{DS}$$

The meaningful solution for the quadratic equation is $V_{DS} = 1.28$ V. Note that increasing V_{GS} to 4.0 V only changes V_{DS} to 0.70 V.

One interesting feature of the MOSFET inverter is the size of the load resistor. It is typically more that a factor of 10 greater than the load resistor for a bipolar inverting amplifier operating over the same voltage range. It is difficult to fabricate large-valued resistors in integrated circuits. To overcome this problem, MOSFETs connected as passive elements are used for load devices. These circuit configurations are discussed in Chapter 8 for digital applications and in Part III for analog applications.

5.2.4 p-Channel MOSFETs

MOSFETs can also be made with p-channels. For an enhancement mode device, the threshold voltage and the drain-to-source voltage are negative. The major disadvantage of the p-channel MOSFET is that the current carriers are holes, whose mobility is less than that of electrons by a factor between 2 and 3. The drain current under saturation conditions for a p-channel device is written

$$I_D(\text{sat}) = k(V_{SG} + V_T)^2 \tag{5.18}$$

and, for nonsaturation conditions,

$$I_D(\text{nonsat}) = k[2(V_{SG} + V_T)V_{SD} - V_{SD}^2] \tag{5.19}$$

In these equations, the drain current is positive when it is *out* of the drain.

The major application of p-channel MOSFETs is in conjunction with n-channel MOSFETs in the complementary MOS or CMOS configuration. In general, p-channel and n-channel devices are used in equal number in CMOS circuits. The basic CMOS inverter is shown in Figure 5.12. The magnitudes of the threshold voltages for the NMOS and PMOS transistors are approximately equal. When the input voltage is high (approximately V_{DD}), the NMOS device is on and the PMOS device is off. When the input is at ground potential, the situation is reversed. The outstanding feature of this configuration is that there is virtually no current in either steady-state condition. The only power dissipation in these circuits other than the small amount due to leakage currents is during the switching transient. CMOS digital circuits are discussed in Chapter 8.

5.2.5 **The Body Effect**

When the source and substrate are not connected, it is the gate-to-substrate voltage that establishes the channel rather than the gate-to-source voltage. The usual way of treating this effect is to continue to use the gate-to-source voltage in the relationships for I_D (Equations 5.10 and 5.12) but alter the value of the threshold voltage. The correction to the threshold voltage is proportional to the square root of the voltage between substrate and source. In most cases, this effect is difficult to predict because the substrate-to-source voltage varies with the operating point of the device.

A typical circuit illustrating this effect is shown in Figure 5.13. In this circuit, transistor M_1 is always in the saturation region of operation, since its gate and drain are tied together making $V_{GS1} = V_{DS1}$ and $V_{DS1} > V_{GS1} - V_{T1}$. The source voltage of M_1 varies relative to the substrate, depending on the

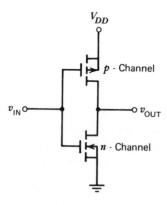

Figure 5.12. A CMOS inverter.

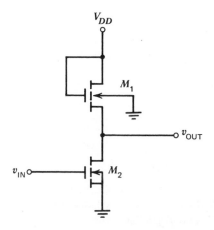

Figure 5.13. An inverter with an *n*-MOSFET load transistor illustrating differences in V_{GS} and V_{GB} for M_1.

input voltage v_{IN}. When v_{IN} is less than V_{T2}, M_2 is off, and $v_{OUT} = V_{S1} = V_{DD} - V_{T1}$. The voltage across M_2 is not zero because the device is in saturation and Equation 5.10 applies. This equation implies that $V_{GS1} = V_{DS1} = V_{T1}$ for $I_D = 0$. Under these circumstances, the voltage from gate to substrate for M_1 is V_{DD} and $V_{GS1} = V_{T1}$. When the input voltage v_{IN} is sufficient to drive M_2 into nonsaturation so that $V_{D2} \simeq 0$, the voltage from gate to substrate and the voltage from gate to source are equal. If we are to assume that the gate-to-source voltage, rather than the gate-to-substrate voltage, is the control voltage, the threshold voltage must be adjusted to take the substrate bias into account. Unfortunately, this bias changes with operating point. This can be approximated by

$$\Delta V_T \simeq K(|V_{BS}|)^{1/2} \tag{5.20}$$

where K is an empirical constant. In this equation, the substrate-to-source voltage V_{BS} varies over a wide range. A further approximation to simplify this expression is to represent V_{BS} by its average value.

5.2.6 **Switching and Frequency Effects**

The dominant frequency limitations for MOS devices are determined by the effective channel resistance and the capacitances in the circuit. In most cases, the transit time for carriers under the gate is so short that it can be ignored. This transit time T can be estimated by dividing the length of the channel by the velocity of the carriers in the channel. A simplified relation, which is strictly correct only for very low drain-to-source voltages, is

$$T = \frac{L}{\mu E} \simeq \frac{L^2}{\mu V_{DS}} \tag{5.21}$$

The capacitances that must be considered for the time constants include the gate capacitance of the following stage and stray capacitance associated

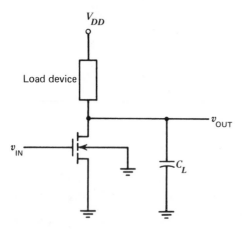

V_{DD}

Load device

v_{OUT}

v_{IN}

C_L

Figure 5.14. An *n*-MOSFET inverter circuit illustrating the load capacitance that must be charged or discharged during switching of voltage levels.

with the interconnection layout. In some large-scale integrated circuits, the interconnection capacitance is the dominant capacitance.

We can lump all of the load and stray capacitance into a single capacitor, C_L, from the output terminal to ground, as shown in Figure 5.14. It is the charging and discharging of this capacitor that primarily determines the propagation delay associated with the transfer of information through an MOS inverting amplifier or logic circuit.

THE DISCHARGE TRANSIENT. Consider the situation where v_{IN} is less than V_{T1}, so that M_1 is cut off. At this time, the voltage across C_L is the maximum allowed by the load. For most types of loads, this voltage is V_{DD}; but, as we saw above, if the load is a MOSFET in saturation, it is less than V_{DD} by the threshold voltage of the transistor. To include the general case, we designate the initial voltage across C_L as $V_{\text{out}H}$.

We now make an assumption that greatly simplifies the analysis by treating the current through the load i_L as negligible compared to the capacitor current $-i_C$ for the discharge transient. This implies that $i_{D1} = -i_C$. This assumption, while losing some accuracy, allows a solution that illustrates the sequence of events occurring in the transistor during the discharge transient. In the next section, we make the assumption that i_D is negligible when treating the charging transient.

The discharge transient is analyzed as a capacitor discharged by a MOS-FET, ignoring the current contributions from the load. If the input voltage is a step function of magnitude $V_{\text{out}H}$ ($>V_{T1}$) at $t = 0$, the output voltage, which is also v_{DS1}, starts at $V_{\text{out}H}$ and decreases as C_L discharges. Initially, M_1 is in saturation, since $v_{DS1} = V_{\text{out}H}$ is greater than $v_{GS1} - V_{T1} = V_{\text{out}H} - V_{T1}$. The current is a constant during this time period and is given by

$$i_{D1} = -i_C = k_1(V_{\text{out}H} - V_{T1})^2 = -C_L \frac{dv_{\text{OUT}}}{dt} \qquad (5.22)$$

for the time interval from zero to t_1. At t_1,

$$v_{DS1} = v_{OUT} = V_{outH} - V_{T1}$$

and the solution to Equation 5.22 with this boundary condition is

$$t_1 = \frac{C_L}{k_1} \frac{V_{T1}}{V_{outH} - V_{T1}^2} \tag{5.23}$$

At $t = t_1$, M_1 goes into nonsaturation. From t_1 until the end of the transient, the drain current is (from Equation 5.12) given by

$$i_D = k_1[2(V_{outH} - V_{T1})v_{out} - v_{OUT}^2]$$

But

$$i_D = -i_C = -C_L \frac{dv_{OUT}}{dt}$$

Then,

$$-\frac{k_1}{C_L} dt = \frac{dv_{OUT}}{2(V_{outH} - V_{T1})v_{OUT} - v_{OUT}^2}$$

The result of this, applying the t_1 boundary condition, is

$$t_2 - t_1 = \frac{C_L}{2k_1(V_{outH} - V_{T1})} \ln \frac{2(V_{outH} - V_{T1}) - V_{outL}}{V_{outL}} \tag{5.24}$$

where V_{outL} is the lowest level for the output voltage and is typically 0.10 V. Combining Equations 5.23 and 5.24 gives the total discharge time.

Example. For $k_1 = 3.3$ $\mu A/V^2$, $V_{outH} = 5.0$ V, $V_{outL} = 0.10$ V, and $V_{T1} = 1.0$ V, find the time that it takes to discharge a load capacitance of 2.0 pF.

Solution. t_1 can be calculated from Equation 5.20:

$$t_1 = \frac{C_L}{k_1} \frac{V_{T1}}{(V_{outH} - V_{T1})^2} = \frac{2.0 \times 10^{-12}}{3.3 \times 10^{-6}} \frac{1}{(5.0 - 1.0)^2}$$

$$= 38 \text{ ns}$$

Using Equation 5.21,

$$t_2 - t_1 = \frac{C_L}{2k_1(V_{outH} - V_{T1})} \ln \frac{2(V_{outH} - V_{T1}) - V_{outL}}{V_{outL}}$$

$$= \frac{2.0 \times 10^{-12}}{2(3.3 \times 10^{-6})(5.0 - 1.0)} \ln \frac{2(5.0 - 1.0) - 0.10}{0.10}$$

$$= 331 \text{ ns}$$

Thus, the total discharge time is 369 ns.

THE CHARGING TRANSIENT. The longest time constant associated with the MOS inverter is that for the charging of the load capacitance. We ignore the current in M_1 for this analysis giving $i_L = i_C = C_L dv_{OUT}/dt$. We assume an initial voltage V_{outL} and a final value of V_{outH}.

The detailed transient analysis depends on the nature of the load device. For a resistive load, the solution is the familiar one-minus exponential curve with a time constant of RC_L, approaching V_{DD} for a final value. For a saturated transistor load, the current is given by

$$i_D = k^2(v_{GS2} - V_{T2})^2 = k^2(V_{DD} - v_{OUT} - V_{T2})^2$$

and

$$i_C = C_L \frac{dv_{OUT}}{dt}$$

that leads to

$$t_3 = \frac{C_L}{k_2}\left[\frac{1}{V_{DD} - V_{T2} - V_{outH}} - \frac{1}{V_{DD} - V_{T2} - V_{outL}}\right] \tag{5.25}$$

Unfortunately, the first term in brackets is infinite, since $V_{outH} = V_{DD} - V_{T2}$ for the saturated load. This is not unusual for transient solutions, since all exponential solutions take an infinite time. Instead, we find the time for the solution to go to 90% of the final value. Using the numbers from the preceding example, with $k_2 = k_1/10$, to give a reasonable ratio for a digital circuit (see Chapter 8), we find that t_3' is 13.6 μs. If we use a 500-kΩ resistor, the charging time is reduced to 1.0 μs. Note that both of these times are very large compared to the discharge time for our example. It is interesting to compare these times to the transit time for the device under consideration. For a channel length of 7.0 μm and a mobility of 500 cm^2/V·s, Equation 5.21 gives

$$
\begin{aligned}
T &= \frac{L^2}{\mu V_{DS}} \\
&= \frac{(7.0 \times 10^{-4})^2}{500 \times 5.0} \\
&= 0.2 \text{ ns}
\end{aligned}
$$

This is at least a factor of 1000 smaller than the other time constants in the circuit.

5.3 SUMMARY

Field effect transistors come in three basic forms: JFETs, MESFETs, and MOSFETs. All of these devices share the common properties of high input impedance and majority carrier operation. JFETs and MESFETs have diode gate structures, and control of the channel dimensions and resistance is accomplished by changing the voltage between the gate and source. In contrast, MOSFETs have gates that are insulated from the substrate. In enhancement mode devices, a voltage from gate to source produces a channel for carriers between the source and the drain. In depletion mode devices, a channel already exists, and the voltage controls the conductivity of the channel.

MOS capacitors have interesting characteristics that depend on the polarity and magnitude of the bias voltage. The time required for achieving steady state is sufficient that temporary storage of information can be performed if reading and restoring of information is performed every two milliseconds.

Like bipolar transistors, FET devices can be used both for analog and digital applications. The gain characteristics for FETs are less favorable than those for bipolar devices, making digital circuits more important than analog circuits for FETs.

PROBLEMS

5.1. Consider the n-channel JFET circuit shown in Figure 5.15. Calculate the quiescent values I_{DQ} and V_{DSQ} and plot the quiescent point on the load line. Is the transistor operating in the saturation or nonsaturation region? Assume that $I_{DSS} = 0.50 \, \text{mA}$, $V_p = -1.0 \, \text{V}$, and $v_{IN} = 0.0 \, \text{V}$. Repeat the problem for $v_{IN} = -0.50 \, \text{V}$.

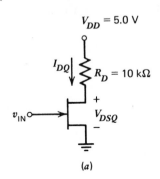

(a)

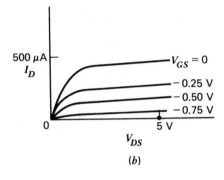

(b)

Figure 5.15. Circuit and I_D-vs.-V_{DS} characteristic curves for Problem 5.1.

5.2. In the circuit of Figure 5.15, let $I_{DSS} = 0.40$ mA and $V_p = -1.5$ V. The quiescent value of V_{DSQ} is to be 2.5 V. Calculate I_{DQ} and v_{IN} to achieve this result.

5.3. Calculate the transconductance at the quiescent point for Problem 5.2.

5.4. For the circuit of Figure 5.16, let $V_T = 1.5$ V and $k = 10$ μA/V^2. (*a*) Plot the voltage transfer characteristics (v_{OUT} versus v_{IN}) for $0 \le v_{IN} \le 5.0$ V and for $R_D = 20$ kΩ. Mark the values of v_{IN} and v_{OUT} at the transition point on the curve. (*b*) Repeat the problem for $R_D = 200$ kΩ.

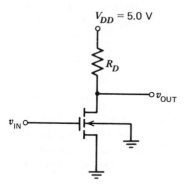

Figure 5.16. Circuit for Problem 5.4.

5.5. For the circuit of Figure 5.17, with $k = 1.0$ mA/V^2 and $V_T = 2.0$ V, find V_{OUT} and I_S.

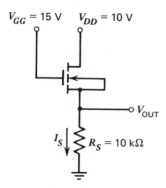

Figure 5.17. Circuit for Problem 5.5.

5.6. For the circuit of Figure 5.17, let $k = 500$ μA/V^2 and $V_T = 1.5$ V. Find the value of R_S that will result in $V_{OUT} = 4.5$ V.

5.7. For the circuit of Figure 5.18, let $V_{TL} = 1.5$ V, $k_L = 1.0$ μA/V², $V_{TD} = 1.5$ V, and $k_D = 20$ μA/V². Calculate (a) the values of V_{IN} and V_{OUT} when the driver is at its transition point, (b) V_{OUT} at $V_{IN} = 0$, (c) V_{OUT} at $V_{IN} = 3.5$ V, (d) V_{OUT} at $V_{IN} = 5.0$ V, and (e) I_D at $V_{IN} = 3.5$ V.

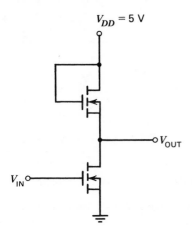

$V_{DD} = 5$ V

V_{OUT}

V_{IN}

Figure 5.18. Circuit for Problem 5.7.

5.8. For the circuit of Figure 5.19, let $k = 1.0$ mA/V² and $V_T = 2.0$ V. (a) Find V_{OUT} and V_{DS} if $R_S = 10$ kΩ. (b) Determine the value of R_S that will result in $V_{OUT} = 2.75$ V.

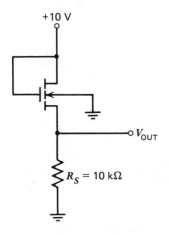

+10 V

V_{OUT}

$R_S = 10$ kΩ

Figure 5.19. Circuit for Problem 5.8.

5.9. For the depletion mode MOSFET shown in Figure 5.20, $k = 1.0$ mA/V^2 and $V_T = -3.0$ V. Find V_{OUT} and V_{DS}.

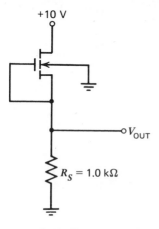

Figure 5.20. Circuit for Problem 5.9.

5.10. Repeat Problem 5.9 if the gate of the transistor is connected to (a) the 10-V power supply and (b) ground.

5.11. For the circuit of Figure 5.21, let $V_T = 2.0$ V and $k = 20$ μA/V^2 for all three transistors. Find V_{OUT} and I_D for the following conditions: (a) $V_x = V_y = V_z = 0.0$ V; (b) $V_x = 3.0$ V, $V_y = V_z = 0.0$ V; (c) $V_y = 4.0$ V, $V_x = V_z = 0.0$ V; (d) $V_z = 5.0$ V, $V_x = V_y = 0.0$ V; (e) $V_x = 3.0$ V, $V_y = 4.0$ V, and $V_z = 5.0$ V; and (f) $V_x = V_y = V_z = 8.0$ V.

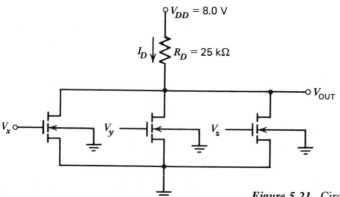

Figure 5.21. Circuit for Problem 5.11.

5.12. For the MOSFET circuit in Figure 5.11, let $k = 0.15$ mA/V^2, $V_T = 1.2$ V, $R_D = 24$ kΩ, and $V_{DD} = 5.0$ V. Calculate the quiescent values I_{DQ} and V_{DSQ} if $V_{GG} = 2.0$ V. Is the transistor biased in the saturation or nonsaturation region?

5.13. For the circuit of Figure 5.22, let $V_{TL} = -3.0$ V, $k_L = 10$ μA/V^2, $V_{TD} = 1.0$ V, and $k_D = 40$ μA/V^2. Investigate the body effect by determining the voltage transfer characteristics, including the voltage between the source of the load device and ground in your calculation.

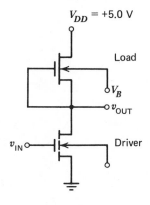

$V_{DD} = +5.0$ V

Load

V_B

v_{OUT}

v_{IN}

Driver

Figure 5.22. Circuit for Problem 5.13.

5.14. For the p-channel MOSFET circuit of Figure 5.23, let $V_{TL} = -1.5$ V, $k_L = 5.0$ μA/V^2, $V_{TD} = -1.5$ V, and $k_D = 125$ μA/V^2. (*a*) Determine the discharge time for the capacitor if $C_L = 5.0$ pF, and (*b*) determine the time it takes to charge the same capacitor to -4.5 V through the load transistor.

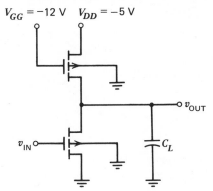

$V_{GG} = -12$ V $V_{DD} = -5$ V

v_{OUT}

v_{IN}

C_L

Figure 5.23. Circuit for Problem 5.14.

part II

DIGITAL CIRCUITS

chapter 6
TRANSISTOR–
TRANSISTOR LOGIC
CIRCUITS

Digital systems incorporate electronic logic circuits to perform the various required logic functions. Five basic logic circuits or logic gates are used extensively; they include the inverter (NOT), AND, NAND, OR, and NOR. The input and output levels of these logic gates are one of two binary states designated by the logic 1 or the logic 0. These binary states are represented in electronic circuits by either voltages or currents. In a positive logic system based on voltage, the more positive voltage of the two states represents a logic 1, and the more negative voltage represents a logic 0. Several types of circuits are available to implement the electronic logic gates. A combination of the diode AND circuit considered in Chapter 3 and the bipolar inverter circuit considered in Chapter 4 provides the basis for diode–transistor logic (DTL) that is considered first in this chapter. Other types of bipolar logic circuits have evolved, including transistor–transistor logic (TTL), emitter-coupled logic (ECL), and integrated-injection logic (I^2L). A detailed circuit analysis of these bipolar logic gates is given in this chapter and in Chapter 7.

6.1 DIODE–TRANSISTOR LOGIC (DTL)

The diode–transistor logic circuit provides an introduction to the study of the transistor–transistor logic circuit. Historically, TTL evolved from the DTL circuit. However, the basic DTL circuit has reappeared in the more advanced Schottky TTL circuits that are considered in the last sections of this chapter.

6.1.1 Basic DTL NAND Circuit

A basic DTL circuit that performs the NAND logic operation is shown in Figure 6.1. The input circuitry of DTL is a diode logic gate that feeds an output transistor. The circuit is designed so that the output transistor operates between cutoff and saturation. This provides the maximum output voltage swing, thus minimizing loading effects and improving the ability of the circuit to reject electronic noise.

A piecewise linear model for the diodes and transistor is assumed in the initial analysis of this DTL circuit. The piecewise linear parameters given in Table 6.1 are used, and the circuit is assumed to be in a static or quasi-static condition that ignores charge storage and transient effects. Upper case letters describe dc voltages and currents in the circuits, emphasizing the quasi-static nature of the analysis. The magnitude of the values for V_{BE} in Table 6.1 reflect the amount of emitter current normally found in the transistors of a DTL or TTL circuit. The V_{BE}(act) refers to the base–emitter voltage of a transistor in the forward active region, while V_{BE}(sat) and V_{CE}(sat) refer to the terminal voltages of the transistor in saturation.

6.1.2 Current–Voltage Analysis

If both input signals V_x and V_y are in a low- or zero-voltage state, then input diodes D_x and D_y are forward biased through resistor R_1 and voltage source V_{CC}. The input diodes conduct, and the voltage V_1 is clamped to a value

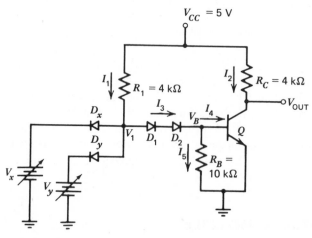

Figure 6.1. A basic DTL NAND logic circuit.

Table 6.1. Piecewise Linear Parameters for the Circuit in Figure 6.1

$$V_D(\text{on}) = 0.70 \text{ V}$$
$$V_D(\text{cut-in}) = 0.65 \text{ V}$$
$$V_{BE}(\text{cut-in}) = 0.65 \text{ V}$$
$$V_{BE}(\text{act}) = 0.70 \text{ V}$$
$$V_{BE}(\text{sat}) = 0.80 \text{ V}$$
$$V_{CE}(\text{sat}) = 0.10 \text{ V}$$

Note: The values given in Table 6.1 may differ for actual transistors in a given logic circuit but we will use these approximate values in our analyses in this chapter.

of one diode drop above the input voltage so that $V_1 = 0.70$ V. The current I_1 can be calculated since the voltage drop across R_1 is known. Hence,

$$I_1 = \frac{V_{CC} - V_1}{R_1}$$

$$= \frac{5.0 - 0.70}{4.0 \times 10^3} \tag{6.1}$$

$$= 1.08 \text{ mA}$$

The diodes D_1 and D_2 and the output transistor Q are nonconducting and in their off-state. If D_1 and D_2 were conducting, a 0.70 V drop would exist across each diode so that

$$V_B = V_1 - 2V_D$$

$$= 0.70 - 1.40 \tag{6.2}$$

$$= -0.70 \text{ V}$$

However, there is no mechanism for V_B to become negative and still have a forward-biased diode current. Thus, D_1 and D_2 are off, V_B is zero, and the output transistor Q is off. Under these circumstances, the collector current I_2 is zero and the output voltage V_{OUT} rises to 5.0 V.

Since diodes D_1 and D_2 are off, the current I_1 divides through the input diodes to the voltage sources V_x and V_y. If the input diodes have identical characteristics and the voltage sources V_x and V_y are identical, the current I_1 can be assumed to divide equally between diode D_x and D_y.

Now, consider the case when both input voltages are in a high state, where $V_x = V_y = 5.0$ V. Here, both input diodes are off, as there is no mechanism for V_1 to exceed 5.0 V. The diodes D_1 and D_2 and the base–emitter junction of Q become forward biased through R_1 and V_{CC}. Since the circuit is designed to drive the output transistor into saturation, the base voltage is $V_B = V_{BE}(\text{sat}) = 0.80$ V, forcing V_1 to be

$$V_1 = V_{BE}(\text{sat}) + 2V_D$$

$$= 0.80 + 2 \times 0.70 \tag{6.3}$$

$$= 2.2 \text{ V}$$

The voltage V_1 cannot increase further. Note that $V_x = V_y = 5.0$ V and $V_1 = 2.2$ V, so that indeed both input diodes D_x and D_y are reverse biased and off as initially assumed.

The current I_1 is found from the voltage drop across R_1 as

$$I_1 = \frac{V_{CC} - V_1}{R_1}$$

$$= \frac{5.0 - 2.2}{4.0 \times 10^3} \tag{6.4}$$

$$= 700 \ \mu A$$

Since the input diodes are off, the entire current I_1 is diverted through D_1 and D_2 forcing the diode current to be 700 μA.

The current I_5 can be calculated from the known voltage drop across R_B as

$$I_5 = \frac{V_B}{R_B}$$

$$= \frac{0.80}{10 \times 10^3} \tag{6.5}$$

$$= 80 \ \mu A$$

The base current into the output transistor is then

$$I_B = I_4 = I_3 - I_5$$

$$= 700 \times 10^{-6} - 80 \times 10^{-6} \tag{6.6}$$

$$= 620 \ \mu A$$

The circuit is designed to saturate Q, so that $V_{OUT} = V_{CE}(\text{sat}) = 0.10$ V and the collector current (with the output terminal not loaded) is

$$I_C = I_2 = \frac{V_{CC} - V_{CE}(\text{sat})}{R_C}$$

$$= \frac{5.0 - 0.10}{4.0 \times 10^3} \tag{6.7}$$

$$= 1.23 \ \text{mA}$$

6.1.3 **Minimum β**

In the previous calculations, we assumed that the output transistor Q was driven into saturation with a base current of 0.62 mA and a collector current of 1.23 mA. To ensure that the transistor is in saturation, the current gain β must be at least as large as the ratio of collector current to base current. For this case,

$$\beta_{min} = \frac{I_C}{I_B}$$

$$= \frac{1.23 \times 10^{-3}}{620 \times 10^{-6}} \tag{6.8}$$

$$= 1.98$$

If the forward active current gain of the transistor is less than β_{min}, Q will not be driven into saturation and the currents and voltages in the circuit will have to be recalculated. A current gain greater than 1.98 ensures that Q is driven into saturation for the given circuit parameters. The forward active current gain of the DTL transistor is usually in excess of 20, ensuring a well-saturated transistor under the current conditions described above.

6.1.4 **The NAND Operation**

So far, both input voltages have either been at a low or a high logic state. Assume now that $V_x = 0$ V and $V_y = 5.0$ V, causing diode D_x to be forward biased with $V_1 = 0.70$ V and diode D_y to be reverse biased. Since $V_1 = 0.70$ V, diodes D_1 and D_2 and the output transistor Q are off so that $V_{OUT} = V_{CC} = 5.0$ V. In this case, the entire current I_1 passes through diode D_x and the source V_x.

Similarly, if $V_x = 5.0$ V and $V_y = 0$ V, diode D_y will be forward biased and D_x will be reverse biased. Again, $V_1 = 0.70$ V and $V_{OUT} = 5.0$ V. This circuit performs the NAND operation summarized in Table 6.2 for the two input circuit. The DTL NAND logic circuit is not limited to two inputs but may have three or more inputs. The number of inputs to a logic gate is referred to as the *fan-in* of the circuit. Thus, the circuit of Figure 6.1 has a fan-in of 2, while a three-input NAND gate has a fan-in of 3.

6.1.5 **Voltage Transfer Characteristic**

The voltage transfer characteristic curve V_{OUT} versus V_{IN} is useful in determining how the output changes as the input varies from one logic state to another. The curve is obtained by connecting all of the inputs to a variable

Table 6.2. Truth Table for a Two-Input NAND Gate

V_x (V)	V_y (V)	V_{OUT} (V)
0	0	5.0
5.0	0	5.0
0	5.0	5.0
5.0	5.0	$0.10 = V_{CE}(\text{sat})$

dc voltage source and plotting the output voltage as the input voltage is slowly varied between its maximum limits.

The changes in V_B and V_1 as the input voltage varies from 0 to 5.0 V are also useful in analyzing the circuit transfer characteristic. Table 6.3 shows the results in four regions of interest for the circuit of Figure 6.1. In region 1, the diodes D_1 and D_2 and the transistor are off. The voltage V_1 is not sufficient to turn on these devices. In region 2, V_{IN} and V_1 have increased sufficiently so that diodes D_1 and D_2 begin to conduct. However, the voltage V_B is still not sufficient to turn the transistor on. In region 3, the output transistor is in its active state. Here, the collector current is directly proportional to the base drive current, causing the output voltage to drop as the voltage across R_C increases with increasing I_2. In region 4, the output transistor has reached saturation and the output voltage drops to its minimum value. In this region, the input diodes are reverse biased since V_1 is clamped to 2.2 V. The voltage transfer characteristic for this particular DTL circuit is shown in Figure 6.2, with the various regions indicated on the figure.

The function of diodes D_1 and D_2 in the DTL circuit is to provide a voltage offset. This effect is seen in the voltage transfer curve of Figure 6.2. If D_1 and D_2 were not in the circuit, the voltage V_1 would be connected directly to the base of Q. For $V_{IN} = 0$ V, $V_1 = V_B = 0.70$ V, which is right at the forward active bias of the transistor. Then, from Figure 6.2, point A would be at $V_{IN} = 0$ V and point B would be at $V_{IN} = 0.10$ V. If only one diode were in the circuit, the breakpoints A and B would move to $V_{IN} = 0.65$ and 0.80 V, respectively.

6.1.6 Pull-Down Resistor

In the basic DTL NAND circuit shown in Figure 6.1, a resistor R_B is connected between the base of the output transistor and ground. This resistor is referred to as a pull-down resistor and is necessary to decrease the switching time of the output transistor as it goes from saturation to cutoff. As noted in Chapter 4, it is necessary to remove excess minority carriers from the base before a transistor can be cut off. This removal of base charge produces a current out of the base terminal until the transistor is cut off. Without the pull-down resistor, this reverse base current would be limited to the reverse bias leakage current of D_1 and D_2, establishing a relatively long turn-off time. The

Table 6.3. Regions of the Transfer Characteristic

Region	V_{IN} (V)	V_1 (V)	V_B (V)	V_{OUT} (V)
1	0–0.60	0.70–1.30	0	5.0
2	0.60–1.35	1.30–2.05	0–0.65	5.0
3	1.35–1.55	2.05–2.20	0.65–0.80	5.0–0.10
4	1.55–5.0	2.20	0.80	0.10

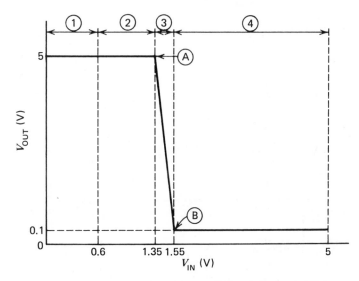

Figure 6.2. Voltage transfer characteristic of the basic DTL NAND circuit showing the four regions of operation described in Table 6.3.

pull-down resistor provides a path for the reverse base current. The excess minority carriers in the base can be removed more rapidly if R_B is reduced in value. The larger the reverse base current, the shorter the turn-off time of the output transistor. However, a tradeoff must be made in choosing the value of R_B. A small value of R_B provides faster switching but lowers the base current to the transistor in the on-state by diverting some drive current to ground. A lower base drive reduces the drive capability of the circuit or fan-out, as is discussed in the next section.

In some DTL circuits, the pull-down resistor is connected to a negative power supply voltage instead of ground. This results in a faster switching logic circuit, but requires a more complicated power supply arrangement. A typical value for this voltage is -2.0 V.

6.1.7 **Fan-Out**

The term *fan-out* is defined as the maximum number of similar-type logic circuits that may be connected to a single output and still maintain proper circuit operation. Proper circuit operation implies that the output transistor remains in saturation when the input voltage is in the logic high state. Fan-out also implies that capacitive loading is not sufficient to seriously degrade the transition times as the output waveform changes from low to high or from high to low. In the previous analysis of the basic DTL circuit, the output terminal was assumed to be unconnected to any load gate. Normally, in digital systems,

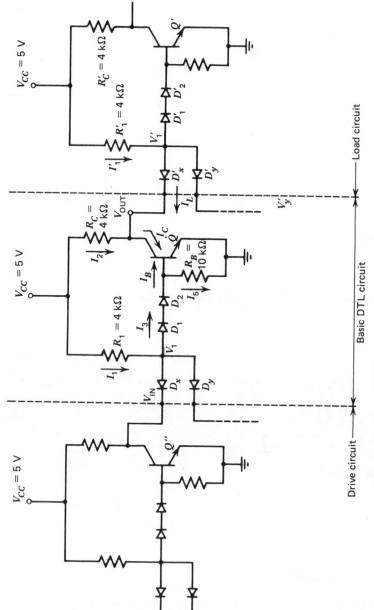

Figure 6.3. A basic DTL logic circuit with drive and load circuits.

a logic circuit will drive other similar type logic circuits and this can present limitations on the circuit.

Figure 6.3 shows a basic DTL NAND circuit with similar DTL circuits connected at both its input and output. If the inputs to both diodes D_x and D_y are in the logic high state, these diodes are off. The voltage V_1 is again clamped at 2.2 V, Q is driven into saturation, and V_{OUT} is decreased to $V_{CE}(\text{sat})$. These currents were calculated previously as

$$I_1 = I_3 = 700 \ \mu A$$

$$I_5 = 80 \ \mu A$$

$$I_B = 620 \ \mu A$$

$$I_2 = 1.23 \ \text{mA}$$

With $V_{OUT} = V_{CE}(\text{sat}) = 0.10$ V, the diode D_x' is forward biased and

$$V_1' = V_{OUT} + V_D \qquad (6.9)$$

$$= 0.80 \ \text{V}$$

If V_y' is in the high state, then $I_1' = I_L$ and

$$I_1' = \frac{V_{CC} - V_1'}{R_1'}$$

$$= \frac{5.0 - 0.80}{4.0 \times 10^3} \qquad (6.10)$$

$$= 1.05 \ \text{mA} = I_L$$

This load current now enters the collector of the output transistor, forcing the circuit to receive current from the load instead of supplying current to the load. This is called *current sinking*. The total collector current in the output transistor Q increased when the load was connected and is now

$$I_C = I_2 + I_L$$

$$= 1.23 \times 10^{-3} + 1.05^{-3} \qquad (6.11)$$

$$= 2.28 \ \text{mA}$$

Since the output transistor Q must remain in saturation, a new minimum forward active current gain is required. Then,

$$\beta_{\min} = \frac{I_C}{I_B} = \frac{2.28 \times 10^{-3}}{620 \times 10^{-6}}$$

$$= 3.68$$

whereas in the no-load situation the $\beta_{\min}$ was 1.98.

The effect of connecting one DTL load circuit to the output of a DTL driver circuit is to require current sinking and to increase the $\beta_{\min}$ of the output transistor. If we assume the output transistor has a fixed current gain, then the number of identical logic circuits that may be connected to the output without changing the circuit characteristics is limited.

The base drive current to the output transistor Q is independent of the load or collector current in this simplified piecewise linear model of the transistor. Hence, the base current, $I_B = 620\ \mu A$, is independent of the number of load circuits connected. Figure 6.4 shows N identical circuits connected to the output terminal. The currents from the individual loads, I_L, I_L, . . . , and I_L^N, are the same and given by $I_L' = 1.05$ mA. The total load current in the collector of Q is then $N \times I_L'$, or $N \times 1.05$ mA. The total collector current in Q is $I_2 + NI_L'$. The maximum collector current without pulling Q out of saturation is then

$$I_C(\max) = \beta I_B = I_2 + NI_L' \tag{6.12}$$

where β is the forward active current gain of the output transistor Q. If we assume that $\beta = 20$, then

$$20 \times 620 \times 10^{-6} = 1.23 \times 10^{-3} + N \times 1.05 \times 10^{-3}$$

The solution for N is 10.64, but N must be rounded off to the nearest lower integer, for a fan-out of 10. An implicit assumption in this calculation is that the output transistor is capable of handling this amount of current. If the maximum collector current rating for the output transistor is less than the 12.4 mA calculated, the fan-out will be limited by the maximum collector current rating.

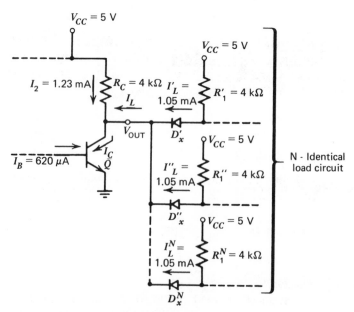

Figure 6.4. The DTL output transistor connected to N-identical DTL input diodes.

The input diodes of the load circuits becomes reverse biased when the driver output voltage goes to its high logic state. In this case, the only current supplied by the basic driver DTL circuit is the sum of the reverse bias diode leakage currents. These leakage currents are normally so small in DTL circuits that the fan-out for the output high state is not a limitation. The dominant fan-out limitation in the static case illustrated here is the requirement of sinking load current in the output low-voltage state.

6.1.8 **Noise Margin**

Noise margin in digital circuits provides a measure of the safety in assuring that transistors are either in cutoff or saturation. Extraneous noise signals may be picked up from a number of sources on the terminal connections of an integrated circuit. A noise signal on an input terminal should not produce a change in the output voltage, and it is thus desirable to have a margin of safety built into the circuit.

For the simplified piecewise linear model, noise margin is defined as the maximum value that the input voltage may change either its high or low state without a change in the output voltage. Noise margin is most easily determined from the voltage transfer characteristic. Figure 6.5 shows the piecewise linear voltage transfer curve of the DTL logic circuit shown in Figure 6.1.

The input voltage to the DTL circuit is derived from a similar type of

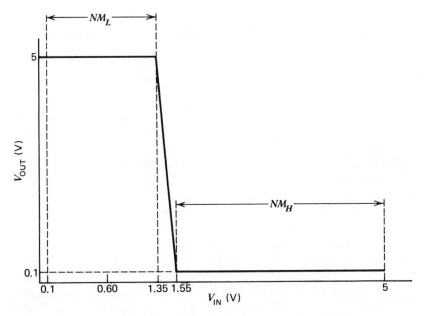

Figure 6.5. Voltage transfer characteristic showing noise margin definition.

DTL circuit giving a minimum input voltage of $V_{CE}(\text{sat}) = 0.10$ V. The low-state noise margin for this piecewise linear model is then

$$NM_L = 1.35 - 0.10 = 1.25 \text{ V}$$

while the high-state noise margin is defined as the value

$$NM_H = 5.0 - 1.55 = 3.45 \text{ V}$$

Diodes D_1 and D_2 are necessary to provide a voltage offset that increases the low-state noise margin. The NM_L will decrease with less than two diodes and will increase with more than two diodes. The incorporation of two diodes is a compromise between the number of circuit components required and a reasonable value for both the high and low noise margins.

6.1.9 Propagation Delay Time

The finite switching speed of the transistors, as discussed in Chapter 4, leads to propagation delay times throughout digital logic circuits. A change in the input signal from one logic state to another causes the transistors within the circuit to switch, and, a finite time later, the output signal will respond to the change.

The propagation delay time is defined with the aid of Figure 6.6, where the input and output waveforms are assumed to be similar in shape. The parameter t_{PHL} is defined as the propagation delay time between input and output when the output goes from the high to the low state. Similarly, t_{PLH} is the propagation delay time between input and output when the output goes from the low to the high state. The overall propagation delay time is defined as the average of these two values, or

$$t_{PD} = \frac{t_{PHL} + t_{PLH}}{2} \tag{6.13}$$

The propagation delay times are measured at a defined level, V_1, for the input and output voltage. Typically, V_1 is defined as 50% of the high-logic state value, or at some fixed voltage such as 1.3 V.

The propagation delays in DTL circuits are typically between 30 and 60 ns. In general, the time to switch a transistor from saturation to cutoff will be greater than the time required from cutoff to saturation. For this reason, t_{PHL} is about 25% smaller than t_{PLH}. A major contributor to propagation delay is the storage time delay associated with the removal of excess minority charge from the base region of the saturated output transistor.

A pulse will also impart delay to a digital system if the rise and fall times of the output waveforms are excessive. The rise time t_r is defined as the time required for a signal to increase from 10% to 90% of its final value, while the fall time is defined as the time for a signal to decrease from 90% to 10% of its final value. Figure 6.6b illustrates these definitions for an output waveform. The rise and fall times for a logic gate are primarily influenced by the capacitive

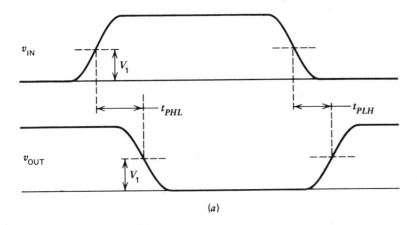

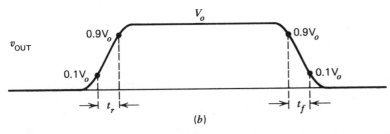

Figure 6.6. (*a*) Propagation delay time definition (*b*) Rise and fall time definition.

loading, the β values of the transistors, the base drive, and the signal amplitude. A small capacitive loading and signal amplitude will reduce rise and fall times. A similar effect is observed for increased base drive and β values. Designers continually balance the tradeoffs involved with optimization, since increased base drive means higher power dissipation while a lower signal voltage swing reduces the noise margins. The capacitance at the input terminal of a DTL logic gate is approximately 5.5 pF.

6.2 TRANSISTOR–TRANSISTOR LOGIC (TTL)

Transistor–transistor logic (TTL) evolved directly from DTL and provided reduced propagation delay times. Figure 6.7*a* shows a basic DTL circuit with one input diode D_x and one offset diode D_1. The structure of these two back-to-back diodes is the same as the *npn* transistor indicated in Figure 6.7*b*. The base–emitter junction of Q_1 corresponds to the input diode D_x, and the base–collector junction corresponds to the offset diode D_1.

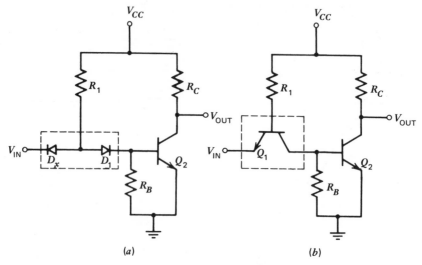

Figure 6.7. (a) A basic DTL circuit. (b) A primitive TTL circuit.

In isoplanar integrated circuit technology, the emitter of a bipolar transistor is normally fabricated within the base region. Therefore, the addition of more emitters within the same base region to form a multiemitter, multiinput device is readily accomplished. Figure 6.8a shows the cross section of a three emitter transistor used as the input device in a TTL circuit. Figure 6.8b shows a circuit schematic of the basic TTL circuit with the multiemitter input transistor. This circuit performs the same NAND operation as its DTL counterpart. The multiemitter transistor actually reduces silicon area compared to the DTL input diodes and it increases the switching speed of the circuit. The transistor action of Q_1 assists in the exit of Q_2 from saturation during a low to high transition of the output. The pull-down resistor R_B in Figure 6.7b is no longer necessary, since the excess minority carriers in the base of Q_2 use the Q_1 transistor as a path to ground.

6.2.1 The Input Transistor

The operation of the input transistor Q_1 in Figure 6.8b is somewhat unconventional. If any one of the three inputs to Q_1 are in a low state (approximately 0 V), the base–emitter junction is forward biased through R_1 and V_{CC}. The base current enters Q_1, and the emitter current exits the particular emitter that is connected to the low input. Transistor action forces collector current into Q_1, but the only steady-state collector current in this direction is a reverse bias saturation current out of the base of Q_2. The steady-state collector current of Q_1 is usually much smaller than the base current, implying that Q_1 is in saturation.

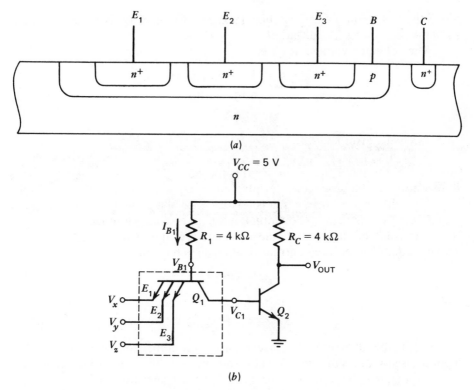

Figure 6.8. (a) Cross section of a multiemitter bipolar transistor. (b) A simple TTL NAND logic circuit with a multiemitter input transistor.

The condition that at least one input is in a low state ($V_x = 0.10$ V) sets the voltage V_{B1} as

$$V_{B1} = V_x + V_{BE1}(\text{sat}) \tag{6.14}$$
$$= 0.10 + 0.80 = 0.90 \text{ V}$$

Then,

$$I_{B1} = \frac{V_{CC} - V_{B1}}{R_1}$$

$$= \frac{5.0 - 0.90}{4.0 \times 10^3} \tag{6.15}$$

$$= 1.03 \text{ mA}$$

and, with a forward current gain β_F of 20, Q_1 will be in saturation if

$$I_{C1} < \beta_F I_{B1} = 20 \times 1.03 \times 10^{-3}$$

$$= 20.6 \text{ mA}$$

Since the base–emitter reverse bias leakage current of Q_2 is much less than 20.6 mA, Q_1 will be in saturation.

The collector voltage of Q_1 is

$$V_{C1} = V_x + V_{CE1}(\text{sat})$$
$$= 0.10 + 0.10 \tag{6.16}$$
$$= 0.20 \text{ V}$$

This voltage is insufficient to turn on Q_2, so that V_{OUT} is 5.0 V.

If all inputs are high, $V_x = V_y = V_z = 5.0$ V, then all three base–emitter junctions of the input transistor are reverse biased. Then, V_{B1} is large enough to forward bias the base–collector junction of Q_1 and drive Q_2 into saturation. Since the base–collector junction is forward biased and the base–emitter junction is reverse biased, Q_1 is operating in its inverse active mode.

The roles of the emitter and collector are interchanged in the inverse active mode. The terminal current relationships become

$$-I_E = \beta_R I_B \tag{6.17}$$

and

$$I_C = (1 + \beta_R) I_B \tag{6.18}$$

where β_R is the inverse mode current gain of the transistor. In Figure 6.8, the input transistor has a fan-in of 3 that results in a modification of Equation 6.17. Transistor Q_1 may be considered as three separate transistors having their bases and collectors connected. When all inputs are high, the current I_{B1} splits approximately equally for the three transistors, making

$$I_{Ex} = I_{Ey} = I_{Ez} = \frac{1}{3}\beta_R I_{B1} \tag{6.19}$$

The collector current of Q_1 is the sum of the individual collector current contributions, giving

$$I_{C1} = (1 + \beta_R) I_{B1} \tag{6.20}$$

If the inverse mode current gain β_R of Q_1 in Figure 6.8 is 0.10, then the emitter and collectors currents are found from

$$V_{B1} = V_{BE2}(\text{sat}) + V_{BC1}$$
$$= 0.80 + 0.70 \tag{6.21}$$
$$= 1.5 \text{ V}$$

$$I_{B1} = (V_{CC} - V_{B1})R_1$$
$$= \frac{5.0 - 1.5}{4.0 \times 10^3} \tag{6.22}$$
$$= 8.75 \ \mu\text{A}$$

$$I_{Ex} = I_{Ey} = I_{Ez} = \frac{1}{3}\beta_R I_{B1}$$

$$= \frac{1}{3}(0.10)(875 \times 10^{-6}) \tag{6.19}$$

$$= 29.2 \ \mu A$$

and

$$I_{C1} = I_{B2} = (1 + \beta_R)I_{B1}$$

$$= (1 + 0.10)(875 \times 10^{-6}) \tag{6.20}$$

$$= 963 \ \mu A$$

The output transistor Q_2 is designed to be driven into saturation with this base current. The inverse mode emitter currents in Q_1 are not desirable, since they draw current from preceding driver gates in the high state. These currents tend to lower the high-state voltage and reduce the noise margin (NM_H). It is for this reason that β_R is intentionally made small, and in TTL has typical values less than 0.1.

The major advantage of TTL over DTL circuits lies in switching the output transistor from saturation to cutoff. If all inputs are initially high and at least one input is switched to a logic-low state at 0.10 V, the base–emitter junction of Q_1 is forward biased dropping V_{B1} to approximately $0.10 + 0.70 = 0.80$ V. The voltage V_{C1} is held at 0.80 V as long as Q_2 remains in saturation. At this instant, Q_1 is biased in its forward active region with $V_{E1} = 0.10$ V, $V_{B1} = 0.80$ V, and $V_{C1} = 0.80$ V. With Q_1 in its active region, a large collector current may exist, drawing a large current out of the base of Q_2. A large base current drawn from an *npn* transistor in saturation very quickly brings that transistor out of saturation. When the excess minority carriers are drawn out of the base of Q_2, the collector current in Q_1 drops to a value of a reverse saturation current from Q_2. Transistor Q_1 then enters the saturation region of operation.

The input transistor of the TTL circuit reduces the propagation delay time compared to a DTL circuit by actively driving excess minority charge from the base of the saturated output transistor. The reduction in propagation delay from a DTL to a TTL NAND gate is typically from 40 ns to 10 ns.

6.2.2 A Basic TTL NAND Circuit

The fan-out capability and voltage transfer characteristic of the basic TTL circuit of Figure 6.8*b* may be improved by adding a second transistor stage. The second stage adds current gain to increase fan-out capability and provides another diode offset voltage to improve the noise margin.

CURRENT–VOLTAGE ANALYSIS. The TTL NAND circuit in Figure 6.9 uses the base–emitter junction of Q_2 to provide the additional diode offset

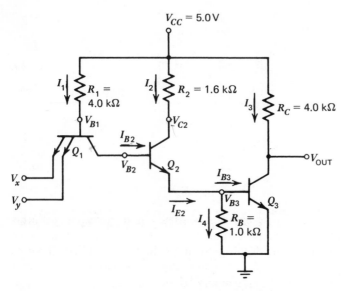

Figure 6.9. A basic TTL NAND logic circuit.

voltage. In this circuit, both Q_2 and Q_3 are driven into saturation, but during the switching process, Q_1 brings Q_2 and Q_3 out of saturation very quickly. A pull-down resistor R_B must be added to provide a charge path for the remaining excess minority carriers to exit the base of Q_3 when Q_2 is cut off.

Assume that $V_x = V_y = 0.10$ V. Then

$$V_{B1} = V_x + V_{BE}(\text{sat})$$
$$= 0.10 + 0.80 \tag{6.23}$$
$$= 0.90 \text{ V}$$

and

$$I_1 = \frac{V_{CC} - V_{B1}}{R_1}$$
$$= \frac{5.0 - 0.90}{4.0 \times 10^3} \tag{6.24}$$
$$= 1.03 \text{ mA}$$

Both Q_2 and Q_3, are off for $V_{B1} = 0.90$ V, forcing I_1 into the emitter terminals of Q_1 and the voltage sources V_x and V_y.

If the input voltages are in the high-logic state of $V_x = V_y = 5.0$ V, then Q_1 goes into the inverse active mode and Q_2 and Q_3 are driven into saturation where

$$V_{B1} = V_{BE3}(\text{sat}) + V_{BE2}(\text{sat}) + V_{BC1}(\text{act})$$
$$= 0.80 + 0.80 + 0.70 \qquad (6.25)$$
$$= 2.3 \text{ V}$$

and

$$I_1 = \frac{V_{CC} - V_{B1}}{R_1}$$
$$= \frac{5.0 - 2.3}{4.0 \times 10^3} \qquad (6.26)$$
$$= 675 \ \mu A \simeq I_{B2}$$

Setting $I_1 \simeq I_{B2}$ assumes that the inverse current gain β_R of Q_1 is negligibly small. The collector current of Q_2 can be calculated from the collector voltage:

$$V_{C2} = V_{BE3}(\text{sat}) + V_{CE2}(\text{sat})$$
$$= 0.80 + 0.10 \qquad (6.27)$$
$$= 0.9 \text{ V}$$

Then,

$$I_2 = \frac{V_{CC} - V_{C2}}{R_2}$$
$$= \frac{5.0 - 0.9}{1.6 \times 10^3} \qquad (6.28)$$
$$= 2.56 \text{ mA}$$

and

$$I_{E2} = I_{B2} + I_2$$
$$= 675 \times 10^{-6} + 2.56 \times 10^{-3} \qquad (6.29)$$
$$= 3.24 \text{ mA}$$

The pull-down resistor current is

$$I_4 = \frac{V_{BE3}(\text{sat})}{R_B}$$
$$= \frac{0.80}{1.0 \times 10^3} \qquad (6.30)$$
$$= 800 \ \mu A$$

and the base drive current I_{B3} is given by

$$I_{B3} = I_{E2} - I_4$$
$$= 3.24 \times 10^{-3} - 800 \times 10^{-6} \qquad (6.31)$$
$$= 2.44 \text{ mA}$$

In TTL circuits, the output stage is designed so that transistor Q_3 is saturated in the output-low state so that $V_{\text{OUT}} = V_{CE}(\text{sat}) \simeq 0.10$ V.

VOLTAGE TRANSFER CHARACTERISTIC. The voltage transfer characteristic of the TTL circuit in Figure 6.9 can be analyzed with the aid of Table 6.4. This table shows the significant voltage ranges and conducting states of the three transistors in the circuit. The two input leads are tied together so that $V_{\text{in}} = V_x = V_y$. Because of the complex nature of the input transistor, the transfer characteristic is somewhat more difficult to calculate than for the DTL circuit. Thus, we initially consider the transfer characteristic from V_{B2} to V_{OUT} and then discuss the influence of Q_1 on this analysis.

Region 1. When $V_{\text{IN}} = 0$ V, Q_1 is in saturation so that $V_{B2} = 0.10$ V. This voltage ensures that both Q_2 and Q_3 are off. There is a negligible emitter current in Q_1, and V_{B3} is held at ground potential. When Q_3 is off, there is no collector current, and V_{OUT} increases to $V_{CC} = 5.0$ V.

Region 2. When $V_{B2} = V_{BE}(\text{cut-in}) = 0.65$ V, Q_2 enters its active region, and there is an emitter current. Most of this emitter current is initially in R_B because R_B is much less than the base–emitter junction resistance of Q_3 at cut-in. Then,

$$V_{B3} \simeq I_{E2}R_B \qquad (6.32)$$

When Q_3 is at the edge of cutoff, $V_{B3} = 0.65$ V, so

$$I_{E2} \simeq \frac{V_{BE3}(\text{cut-in})}{R_B} \qquad (6.33)$$

For a large value of β_F, $I_{C2} = I_2 \simeq I_{E2}$. Then,

$$
\begin{aligned}
V_{C2} &= V_{CC} - I_2 R_2 \\
&= 5.0 - (700 \times 10^{-6})(1.6 \times 10^3) \qquad (6.34) \\
&= 3.88 \text{ V}
\end{aligned}
$$

Since $V_{CE2} \gg V_{CE}(\text{sat})$, Q_2 is still in its active region; and when Q_3 is at the edge of cutoff,

Table 6.4. Circuit Conditions for the Regions of the Transfer Characteristic of a TTL Circuit

Region	V_{IN} (V)	V_{B1} (V)	Q_1	V_{B2} (V)	V_{B3} (V)	Q_2	Q_3	V_{OUT} (V)
1	0–0.55	0.80–1.35	sat	0.10–0.65	0	off	off	5.0
2	0.55–1.25	1.35–2.05	sat	0.65–1.35	0–0.65	act	off	5.0
3	1.25–1.4	2.15–2.2	sat	1.35–1.5	0.65–0.80	act	act	5.0–0.1
4	1.5–1.6	2.2–2.3	sat	1.5–1.6	0.80	act	sat	0.1
5	1.6–5.0	2.3	inv act	1.6	0.80	sat	sat	0.1

$$V_{B2} = V_{BE3}(\text{cut-in}) + V_{BE2}(\text{act})$$

$$= 0.65 + 0.70 \tag{6.35}$$

$$= 1.35 \text{ V}$$

Region 3. As V_{B3} increases from 0.65 V to 0.80 V, Q_3 traverses its active region. The output voltage decreases from 5.0 V to 0.10 V as Q_3 reaches saturation, and

$$I_{C3}(\text{sat}) = I_3 = \frac{V_{CC} - V_{CE3}(\text{sat})}{R_C}$$

$$= \frac{5.0 - 0.10}{4.0 \times 10^3} \tag{6.36}$$

$$= 1.23 \text{ mA}$$

At the onset of saturation

$$I_{B3} = \frac{I_C(\text{sat})}{\beta} \tag{6.37}$$

If $\beta = 20$, then $I_{B3} = 1.23 \times 10^{-3}/20 = 62$ μA and $I_{E2} = 800 \times 10^{-6} + 62 \times 10^{-6} = 862$ μA. If Q_2 is still active, $I_2 \simeq I_{E2}$, and

$$V_{C2} = V_{CC} - I_2 R_2$$

$$= 5.0 - (862 \times 10^{-6})(1.6 \times 10^3) \tag{6.38}$$

$$= 3.62 \text{ V}$$

This shows that indeed Q_2 is still in its active region during this period.

Region 4. The output transistor is now saturated so that V_{B3} is clamped at 0.80 V. Transistor Q_2 goes from its active region into saturation so that V_{B2} ranges from 1.5 to 1.6 V while the output voltage remains at 0.10 V. When Q_2 goes into saturation, I_2 and the base current I_{B3} increase, driving Q_3 further into saturation.

Region 5. In this region, both Q_3 and Q_2 are in saturation so that V_{B2} is clamped at 1.6 V.

Figure 6.10 shows the transfer characteristics of V_{OUT} as a function of V_{B2} and the associated regions of interest for the circuit of Figure 6.9. When all inputs to Q_1 are in the high logic state, Q_1 is in the inverse active mode, Q_3 and Q_2 are in saturation, and V_{B1} is clamped at

$$V_{B1} = V_{B2} + V_{BC1}(\text{act})$$

$$= 1.6 + 0.70 \tag{6.39}$$

$$= 2.3 \text{ V}$$

The base current of Q_1 supplies the majority of the base current to Q_2, since the inverse current gain β_R of Q_1 is small by design. This situation is maintained

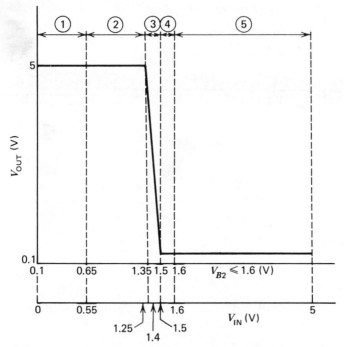

Figure 6.10. Voltage transfer characteristic of the basic TTL NAND logic circuit illustrating five regions of operation. The abcissa for V_{B2} does not exceed 1.6 V.

as long as the input voltage V_{IN} is positive with respect to V_{B2}, holding Q_1 in the inverse active region.

As V_{IN} decreases to $V_{IN} \simeq V_{B2}$, the current I_1 begins to divert from the base–collector junction of Q_1 into the base–emitter junction. At $V_{IN} = V_{B2}$, the collector–emitter voltage of Q_1 is zero, and the emitter current is approximately zero. The current I_{B2} is reduced by a small amount. However, this small drop in I_{B2} normally will not cause Q_2 to come out of saturation.

As V_{IN} decreases slightly below V_{B2}, more current is diverted from the base–collector junction so that Q_2 and Q_3 are brought out of saturation and eventually are turned off. The change in the collector–emitter voltage to completely divert the current from the base–collector to the base–emitter junction is only about 0.10 V. Therefore, in Table 6.4, the values of V_{IN} in regions 2, 3, and 4 differ from those of V_{B2} by no more than 0.10 V.

The plot of V_{OUT} versus V_{IN} for the TTL circuit is of more significance but is very similar to the plot of V_{OUT} versus V_{B2} in regions 1 to 4, as shown in Figure 6.10. When V_{IN} is less than V_{B1}, Q_1 is in saturation and $V_{B2} \simeq V_{IN}$ + 0.10 V. As V_{IN} exceeds V_{B1}, Q_1 enters the inverse active mode and V_{B1} clamps at 1.6 V. Therefore, for $V_{IN} < 1.6$ V, V_{B2} and V_{IN} differ by approximately

0.1 V, while for $V_{IN} > 1.6$ V, V_{B2} clamps at 1.6 V. Figure 6.10 shows separate abscissas for V_{B2} and V_{IN} to illustrate this relation.

OUTPUT-LOW FAN-OUT. The calculation of the static fan-out capability of the TTL circuit is similar to that for the DTL circuits in the previous section. Figure 6.11 shows a TTL circuit driving N TTL inverter load circuits. The inverter represents the worst case load for fan-out calculations.

If $V_x = V_y = 5.0$ V, then Q_2 and Q_3 are driven into saturation, and $V_{OUT} = V_{CE3}(\text{sat}) = 0.10$ V. In this condition, the individual load currents are calculated as

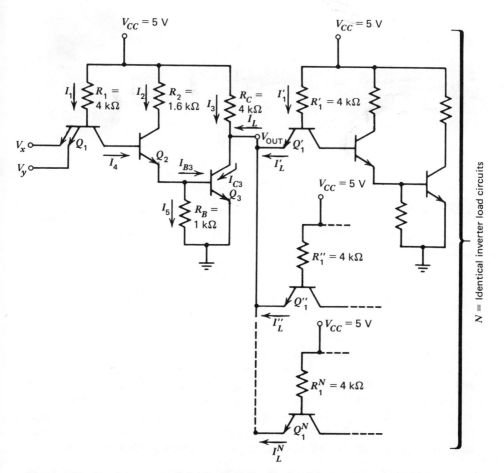

Figure 6.11. Load currents of the basic TTL circuit with the output in the low-voltage state.

$$I'_L = I''_L = I^N_L = \frac{V_{CC} - [V_{OUT} + V'_{BE1} \text{ (sat)}]}{R'_1}$$

$$= \frac{5.0 - (0.10 + 0.80)}{4.0 \times 10^3} \tag{6.40}$$

$$= 1.03 \text{ mA}$$

Then,

$$I_L = NI'_L = N \times 1.03 \text{ mA}$$

The current in R_C is

$$I_3 = \frac{V_{CC} - V_{CE3}(\text{sat})}{R_C}$$

$$= \frac{5.0 - 0.10}{4.0 \times 10^3} \tag{6.41}$$

$$= 1.23 \text{ mA}$$

The total collector current of the output transistor Q_3 is $I_{C3} = I_3 + I_L$, and the total base drive current was calculated as $I_{B3} = 2.44$ mA. If the β of Q_3 is approximately 20, then the maximum collector current that still maintains Q_3 in saturation is

$$I_{C3}(\text{max}) = \beta I_{B3} = I_3 + NI'_L$$

$$= 20 \times 2.44 \times 10^{-3} \tag{6.42}$$

$$= 48.8 \text{ mA}$$

In this equation, the value of N is the only unknown. Then,

$$48.8 \times 10^{-3} = 1.23 \times 10^{-3} + N \times 1.03 \times 10^{-3}$$

or

$$N = 46.2$$

for a fan-out of

$$N = 46$$

Therefore, with the inputs in the high-voltage state, a maximum of 46 similar circuits can be driven and still maintain the output transistor in saturation. It should be noted that $N = 46$ represents an ideal calculation that did not take into account the variation of $V_{CE}(\text{sat})$ from 0.10 to 0.40 V as Q_3 entered the forward active state. Manufacturers specify a more conservative value of fan-out that also considers the degradation in performance with increased capacitive loading. The fan-out for an industrial version of Figure 6.11 is 10.

OUTPUT-HIGH FAN-OUT. The output transistor Q_3 is off when at least one of the input voltages to the driver circuit is in the low-voltage state. This

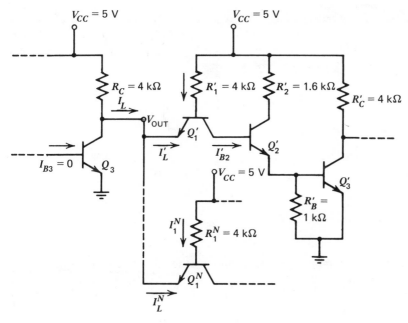

Figure 6.12. Load currents of the basic TTL circuit with the output in the high-voltage state.

situation is shown in Figure 6.12 with $I_{B3} = 0$. In this case, V_{OUT} goes high, driving Q_1', into its inverse active mode. The current I_1' was previously calculated as

$$I_1' = \frac{V_{CC} - [V_{BE3}' \text{ (sat)} + V_{BE2}' \text{ (sat)} + V_{BC1}' \text{ (act)}]}{R_1'}$$

$$= \frac{5.0 - (0.80 + 0.80 + 0.70)}{4.0 \times 10^3} \qquad (6.43)$$

$$= 675 \ \mu\text{A}$$

The inverse mode emitter current I_L' is a function of the base current I_1' and of the inverse mode current gain β_R of the transistor. Then,

$$I_L' = \beta_R I_1' \qquad (6.44)$$

If the inverse current gain is 0.1,

$$I_L' = 0.1 \times 675 \times 10^{-6}$$

$$= 67.5 \ \mu\text{A}$$

and

$$I_L = N \times 67.5 \ \mu\text{A}$$

The limiting factor in calculating static fan-out in this output-high mode is the decrease in output voltage that can be tolerated. If the minimum output voltage corresponding to a *high* output is defined to be 2.4 V, then the maximum load current I_L results in a 2.6-V drop across R_C. Therefore,

$$I_L(\text{max}) = NI'_L = \frac{2.6}{4.0 \times 10^3} \tag{6.45}$$

or

$$N \times 67.5 \times 10^{-6} = \frac{2.6}{4.0 \times 10^3}$$
$$N = 9.63$$

for a fan-out of

$$N = 9$$

In this case, the static fan-out is limited to 9 by the conditions when the output voltage is in the high state. The low value of the static fan-out in this mode can be drastically increased by modifying the output stage of the TTL circuit, as shown in the next section.

6.2.3 TTL Output Stages

The fan-out capability in the output-high state as well as the propagation delay time are improved by replacing the collector resistor of the output transistor by a current source. It was shown in the last section that fan-out for the high-output state was limited by the ability of the circuit to supply current. When the output changes from its low state to its high state, the capacitance associated with the load circuits must be charged through the collector pull-up resistor. The load capacitance is composed of the input capacitance of the load circuits as well as that of the connecting wires. The associated RC time constant for a load capacitance of 15 pF and a collector resistor of 4.0 kΩ is 60 ns, which is large when compared to a commercial TTL circuit.

THE TOTEM-POLE OUTPUT STAGE. In Figure 6.13, the combination of Q_4, D_1, and Q_3 form an output stage called a "totem-pole." The transistor Q_2 forms a phase splitter, since the collector and emitter voltages are 180° out of phase. To analyze this circuit, consider Figure 6.14 with N similar circuits connected as loads and $V_x = V_y = 5.0$ V. Then,

$$
\begin{aligned}
I_1 &= \frac{V_{CC} - [2V_{BE}(\text{sat}) + V_{BC}(\text{act})]}{R_1} \\
&= \frac{5.0 - (2 \times 0.80 + 0.70)}{4.0 \times 10^3} \\
&= 675 \ \mu\text{A}
\end{aligned}
\tag{6.46}
$$

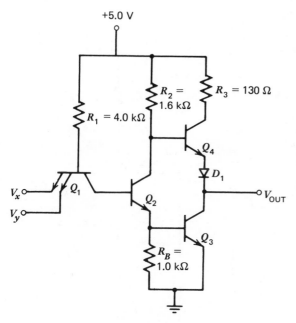

+5.0 V

$R_2 = 1.6\ \text{k}\Omega$

$R_3 = 130\ \Omega$

$R_1 = 4.0\ \text{k}\Omega$

Q_4

D_1

V_x

Q_1

V_y

Q_2

V_{OUT}

Q_3

$R_B = 1.0\ \text{k}\Omega$

Figure 6.13. The TTL NAND logic circuit with a totem-pole output stage.

and, for a reverse current gain β_R of 0.10 in the input transistor Q_1,

$$I_{B2} = I_1 + \beta_R I_1$$

$$= 675 \times 10^{-6} + (0.10 \times 675 \times 10^{-6}) \qquad (6.47)$$

$$= 743\ \mu\text{A}$$

The transistor Q_2 is driven into saturation so that

$$I_{C2} = \frac{V_{CC} - [V_{BE3}(\text{sat}) + V_{CE2}(\text{sat})]}{R_2}$$

$$= \frac{5.0 - (0.80 + 0.10)}{1.6 \times 10^3} \qquad (6.48)$$

$$= 2.56\ \text{mA}$$

Then,

$$I_{E2} = I_{B2} + I_{C2}$$

$$= 743 \times 10^{-6} + 2.56 \times 10^{-3} \qquad (6.49)$$

$$= 3.30\ \text{mA}$$

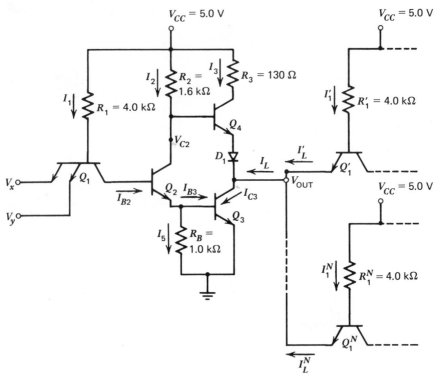

Figure 6.14. TTL gate load currents with the output in the low-voltage state.

and

$$I_5 = \frac{V_{BE3}(\text{sat})}{R_B}$$

$$= \frac{0.80}{1.0 \times 10^3} = 800 \ \mu\text{A}$$

(6.50)

resulting in

$$I_{B3} = I_{E2} - I_5$$

$$= 3.30 \times 10^{-3} - 0.80 \times 10^{-3}$$

$$= 2.50 \ \text{mA}$$

(6.51)

The output transistor Q_3 is driven into saturation with $V_{OUT} = 0.10$ V. Note that the voltage difference between V_{C2} and V_{OUT} is now

$$V_{C2} - V_{OUT} = 0.90 - 0.10$$

$$= 0.80 \ \text{V}$$

This voltage is across the base–emitter of Q_4 and the diode D_1 and is insufficient to turn on both devices. If the diode were not included, the 0.80-V drop of $V_{C2} - V_{OUT}$ would be entirely across the base–emitter of Q_4, turning this transistor on. However, with Q_3 in saturation and the output voltage in its low-voltage state, transistor Q_4 will be off.

The fan-out capability of the circuit in this state is calculated as before. The output transistor Q_3 sinks the load current, given by

$$I_{C3} = I_L = I'_L + I''_L + I'''_L + \ldots + I^N_L \qquad (6.52)$$

Since Q_4 and D_1 are off in this state, there is no equivalent current in the collector resistor. The fan-out is again determined by the condition that the output transistor remain in saturation and that, under worst-case conditions, $I'_I = I'_L$.

To determine the output-high fan-out capability, consider the condition that at least one of the inputs, V_x or V_y, goes to its low-voltage state. In Figure 6.15, Q_2 and Q_3 are then turned off, causing V_{C2} to increase and turn on Q_4 and D_1. The input transistors of the load circuits go into the inverse active mode as V_{OUT} rises. The input transistors of the load gates are considered here as one-input NAND (inverter) gates to illustrate the maximum current under input-high conditions ($I'_L = \beta_R I_{B1}$). The load current I_L is now in the opposite direction from the previous case, and

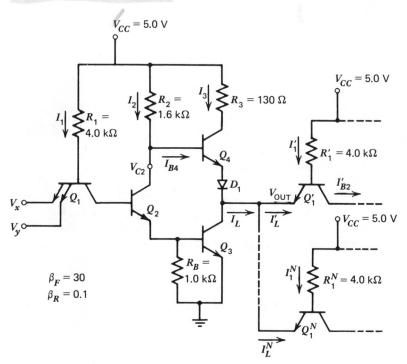

Figure 6.15. TTL gate load currents with the output in the high-voltage state.

$$I_L = N_1 I'_L \tag{6.53}$$

The current I'_L is determined by the β'_R of the input load transistor, while N_1 is the fan-out for V_{OUT} in its high-logic state. With Q'_1 in the inverse active mode, Q'_2 and Q'_3 are driven into saturation so that

$$
\begin{aligned}
I'_1 &= \frac{V_{CC} - [2V_{BE}(\text{sat}) + V_{BC}(\text{act})]}{R'_1} \\[2mm]
&= \frac{5.0 - (2 \times 0.80 + 0.70)}{4.0 \times 10^3} \\[2mm]
&= 675 \ \mu A
\end{aligned}
\tag{6.54}
$$

Assuming a $\beta_R = 0.10$,

$$
\begin{aligned}
I'_L &= \beta_R I'_1 \\
&= 0.10 \times 675 \times 10^{-6} \\
&= 67.5 \ \mu A
\end{aligned}
\tag{6.55}
$$

and

$$I_L = N_1 \times 67.5 \times 10^{-6} \tag{6.56}$$

With Q_4 in its active region,

$$I_2 = \frac{I_L}{1 + \beta_4} \tag{6.57}$$

where β_4 is the forward current gain of Q_4. The fan-out capability in the output-high mode is again determined by the allowable drop in the output voltage.

The ideal case occurs if $I'_L \approx 0$ and $I_L \approx 0$. If no current existed, then the output voltage would be $V_{CE3} = 5.0$ V. However, I_{E4} is always finite, with or without output high load currents. If no loads are driven, then the current I_{E4} is equal to the reverse bias saturation current arising from the reverse bias base–collector junction in Q_3. Even though this current may be on the order of picoamperes, it is sufficient to generate a few hundred millivolts across the base–emitter of Q_4 and D_1. Under no load conditions, TTL output high voltages are slightly above 4.0 V. A realistic situation has load gates that draw tens or hundreds of microamperes through Q_4 and D_1. If the voltage drop across R_2 is neglected, then

$$
\begin{aligned}
V_{OUT} &\approx V_{CC} - V_{BE4}(\text{act}) - V_{D1} \\
&= 5.0 - 0.70 - 0.70 \\
&= 3.6 \text{ V}
\end{aligned}
\tag{6.58}
$$

If, due to loading, the output voltage is allowed to drop from 3.6 V to only 3.5 V, then

$$V_{C2} = V_{OUT} + V_{D1} + V_{BE4}(\text{act})$$
$$= 3.5 + 0.70 + 0.70 \qquad (6.59)$$
$$= 4.9 \text{ V}$$

This means that the maximum allowed voltage drop across R_2 is 0.10 V. Then,

$$V_{CC} - V_{C2} = 0.10 = I_2 R_2 = \frac{I_L R_2}{1 + \beta_4} \qquad (6.60)$$

$$0.10 = \frac{N_1 \times 67.5 \times 10^{-6} \times 1.6 \times 10^3}{1 + \beta_4}$$

If a nominal value of $\beta_4 = 30$ is assumed,

$$N_1 = \frac{0.10 \times 31}{67.5 \times 10^{-6} \times 1.6 \times 10^3} = 28.7$$

which rounds off to

$$N_1 = 28 \text{ gates}$$

In addition to the higher fan-out capability in the output-high condition, the totem-pole output circuit also provides a much lower output impedance. This lowers the output time constant and decreases the switching time of the circuit. These advantages are compromised by a lower output voltage swing. A collector resistor allowed an ideal output voltage swing between 0.10 and 5.0 V. With the totem-pole output circuit, the logic swing is from 0.10 to approximately 3.6 V.

MODIFIED TOTEM-POLE OUTPUT STAGE. A modified totem-pole output stage is shown in Figure 6.16. The addition of the transistor Q_5 in place of a diode has several advantages. First, the transistor pair of Q_4 and Q_5 provides greater current gain so that the fan-out capability of this circuit in its high state is further increased. Secondly, the output impedance in the high state is reduced over that of the single transistor, decreasing the switching time even more. The base–emitter junction of Q_5 fulfills the function of diode D_1 in the totem-pole output stage, so that the diode is no longer required to provide a voltage offset. The fabrication of transistors in integrated circuits is no more complex than the fabrication of diodes.

A pair of transistors connected like Q_4 and Q_5 in Figure 6.16 is called a Darlington pair. The collectors are tied together, and the emitter of one transistor is connected to the base of the other. The Darlington pair provides very high current gain, which makes it useful to source load currents in the output-high condition. The resistor R_4 is included to increase the current in Q_4 during its conducting states. The emitter current in Q_4 in the output-high state is $I_{E4} = V_{B5}/R_4$, or $I_{E4} \simeq 4.3/4.0 \times 10^3 = 1.08$ mA. Without R_4, this current

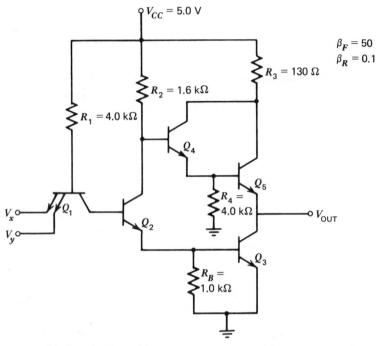

Figure 6.16. TTL gate with modified totem pole-output state.

would be $I_{E4} = I_{B5} = I_L/(1 + \beta_F)$. If $\beta_F = 50$ and $I_L = 100 \ \mu A$, then $I_{E4} = 100 \times 10^{-6}/51 = 1.96 \ \mu A$. This current is at a sufficiently low level that recombination losses in the base would dominate. The β_F in this case would be quite low.

6.2.4 **Tristate Output**

The output impedance of the TTL logic circuits with totem-pole output stages considered so far is extremely low when the output voltage is either in the high or low state. Situations arise, however, in memory circuit applications in which the outputs of many TTL circuits need to be connected together to form a single output. This creates a serious loading situation demanding that all other TTL outputs be disabled or put into a high-impedance state. This is shown symbolically in Figure 6.17 where, G1 and G3 are disconnected from the output so that the output voltage V_{OUT} measures only the output of G2.

The TTL circuit shown in Figure 6.18 may be used to put the logic output into a high-impedance state. If the voltage at $\overline{D} = 5.0$ V, the state of the input transistor Q_1 is controlled by the inputs V_x and V_y. Under these circumstances, the diode D_2 is always reverse biased, no current flows in this diode, and the circuit function is unchanged. Now consider the case when $\overline{D}$ is driven to a low-voltage state of 0.10 V from a similar type of TTL circuit. A low voltage

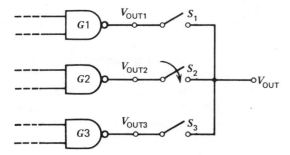

Figure 6.17. Connection of NAND gate outputs with disabling switches to prevent impedance loading of unused gates.

to the emitter of Q_1 ensures that both Q_2 and Q_3 are turned off and that diode D_2 is forward biased with

$$V_{C2} = 0.10 + 0.70 = 0.80 \text{ V}$$

This voltage is sufficient to turn on Q_4 but prevents Q_5 from turning on. Transistor Q_4 turns on in its active region so that

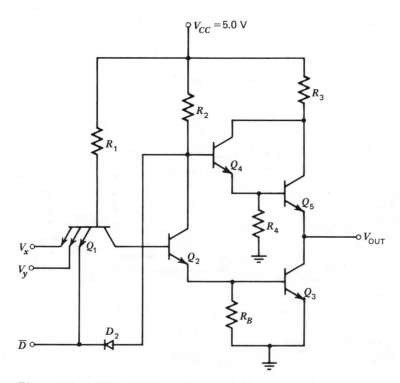

Figure 6.18. A TTL NAND logic circuit with tristate output stage.

$$V_{E4} = V_{C2} - V_{BE}(\text{act}) = 0.80 - 0.70$$
$$= 0.10 \text{ V}$$

This low voltage assures that Q_5 remains off and that the impedance into the emitter of Q_5 is then very high (nominally megohms). Therefore, with $\overline{D} =$ logic 0, the impedance at the output terminal looking back into the TTL circuit is extremely high. When TTL circuits are paralleled to increase the capability of a digital system, the tristate output stage is either enabled or disabled via the $\overline{D}$ select line. Only the output stage on any one TTL circuit may be enabled at any one time.

6.3 SCHOTTKY TRANSISTOR–TRANSISTOR LOGIC (S-TTL)

The TTL circuits considered thus far drive the output and phase splitter transistors between cutoff in the output-high state and saturation in the output-low state. The input transistor is also driven into saturation. Since the propagation delay time through a TTL gate is a strong function of the storage time of the saturated transistors, a nonsaturating logic circuit would be an advantage. The Schottky clamped transistor does prevent the transistor from being driven into deep saturation and has a storage time of about 50 ps. The Schottky clamped transistors switch faster than saturated transistors and have been incorporated in many of the recent bipolar transistor logic families.

6.3.1 The Schottky Clamped Transistor

The symbol for the Schottky clamped transistor is shown in Figure 6.19a, and its equivalent configuration is given in Figure 6.19b. The Schottky diode was discussed in Chapter 3. In the Schottky clamped transistor, a Schottky barrier diode with a low forward voltage drop (approximately 0.3 V) is connected between the base and the collector. When the transistor is in its active region of operation, the base–collector junction is reverse biased. The Schottky diode is also reverse biased and effectively out of the circuit. In this case, the Schottky clamped transistor behaves as a normal npn bipolar transistor. As the Schottky clamped transistor goes into saturation, the base–collector junction becomes forward biased, and the base–collector voltage is effectively clamped at the 0.3-V Schottky diode voltage. The excess base current is shunted through the diode and the transistor is prevented from going heavily into saturation.

Figure 6.20 shows the equivalent configuration of the Schottky clamped transistor with designated current and voltage parameters. If this transistor is the output transistor of a TTL circuit, the current I_L would correspond to the load current from the TTL load circuits.

The three defining equations for this Schottky clamped transistor are

$$I_C = I_D + I_L \tag{6.61}$$

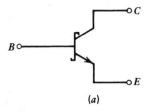

(a)

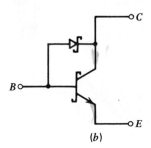

(b)

Figure 6.19. (a) Symbol of Schottky-clamped transistor. (b) Equivalent circuit of a Schottky-clamped transistor.

$$I_{IN} = I_B + I_D \tag{6.62}$$

and

$$I_C = \beta I_B \tag{6.63}$$

The third equation is appropriate since the transistor is clamped at the edge of saturation. The second equation can be written as

$$I_D = I_{IN} - I_B = I_{IN} - \frac{I_C}{\beta} \tag{6.64}$$

Substituting this into Equation 6.61 gives

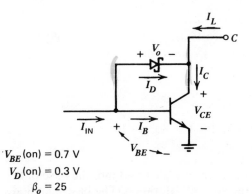

$V_{BE}(\text{on}) = 0.7 \text{ V}$
$V_D(\text{on}) = 0.3 \text{ V}$
$\beta_o = 25$

Figure 6.20. Equivalent circuit of Schottky-clamped transistor with currents and voltages.

$$I_C = I_{IN} - \frac{I_C}{\beta} + I_L \tag{6.65}$$

yielding

$$I_C = \frac{I_{IN} + I_L}{1 + (1/\beta)} \tag{6.66}$$

Assume that the input current I_{IN} is 1.0 mA, the load current I_L is 2.0 mA, and the current gain β is 25. This gives

$$I_C = \frac{1.0 \times 10^{-3} + 2.0 \times 10^{-3}}{1 + (1/25)}$$

$$= 2.89 \text{ mA}$$

$$I_B = \frac{2.89 \times 10^{-3}}{25}$$

$$= 115 \ \mu\text{A}$$

and

$$I_D = 1.0 \times 10^{-3} - 115 \times 10^{-6}$$

$$= 885 \ \mu\text{A}$$

These calculations show that a large part of the input current is shunted through the Schottky diode. For a relatively small load current of 2.0 mA, only a small base current is needed to hold the transistor at the edge of saturation.

Suppose the load current is increased to $I_L = 20$ mA, while maintaining an input current of $I_{IN} = 1.0$ mA. Then,

$$I_C = \frac{I_{IN} + I_L}{1 + (1/\beta)}$$

$$= \frac{1.0 \times 10^{-3} + 20 \times 10^{-3}}{1 + (1/25)}$$

$$= 20.2 \text{ mA}$$

$$I_B = \frac{I_C}{\beta} = \frac{20.2 \times 10^{-3}}{25}$$

$$= 808 \ \mu\text{A}$$

and

$$I_D = I_{IN} - I_B = 1.0 \times 10^{-3} - 808 \times 10^{-6}$$

$$= 192 \ \mu\text{A}$$

In this case, the major part of the input current is diverted into the base of the transistor, keeping the transistor at the edge of saturation. The base and diode currents change with the load conditions while the transistor remains at the edge of saturation. The Schottky transistor is defined as coming out of the low-voltage state when the current in the Schottky diode becomes zero. The Schottky clamped transistor recovers very quickly when the input base drive is cut off, since the Schottky barrier diode has no minority carrier charge storage and the transistor is never fully saturated.

6.3.2 A Schottky TTL NAND Circuit

Figure 6.21 shows a Schottky TTL NAND circuit in which all of the transistors except Q_5 are Schottky clamped transistors. The connection of Q_4 across the base–collector junction of Q_5 prevents this junction from becoming forward biased, ensuring that Q_5 never goes into saturation. Another difference between this circuit and the standard TTL circuit of Figure 6.16 is that the pull-down resistor of the output transistor Q_3 has been replaced by Q_6 and two resistors. This arrangement is referred to as a squaring network, since it squares or sharpens the voltage transfer characteristics of the circuit. The conduction of Q_2 is prevented until the input voltage is large enough to turn on both Q_2 and Q_3 simultaneously. Recall that a passive pull-down resistor was needed in

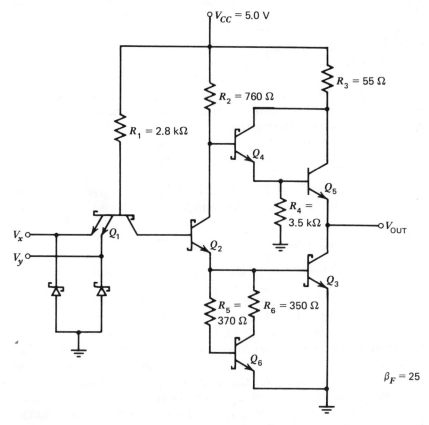

Figure 6.21. A Schottky TTL NAND logic circuit.

the TTL circuit to allow a pathway for removal of stored charge in the base of the output transistor when that transistor turned off from the saturated state. During the cut-in of Q_3, the passive pull-down resistor drained base current from Q_3, causing an appreciable slope in the voltage transfer curve. The active pull-down circuit presents a high resistive load during the cut-in phase diverting most of the phase-splitter emitter current into the base of the output transistor. This provides a sharper voltage transfer curve and enhances the noise margin. The squaring network also improves the propagation delay time by providing a low-resistance path to discharge the base–emitter capacitance of Q_3 during turn-off.

The two Schottky diodes from the input terminals to ground act as clamps to suppress any ringing that might occur from voltage transitions. The input diodes clamp negative undershoots at approximately -0.3 V.

The dc current–voltage analysis of the Schottky TTL circuit in Figure 6.21 is similar to that for the standard TTL circuit in Figure 6.16. The major difference between the two circuits is the quantity of excess minority carrier storage in the transistors when they are driven into or near saturation. The Schottky transistor is held at the edge of saturation by the Schottky clamping diode, lowering the propagation delay time of a Schottky logic circuit to approximately 2 to 5 ns. This compares to a nominal 10 to 15 ns propagation delay time in standard TTL logic circuits.

Another slight differences between Schottky TTL and standard TTL is the value of the output voltage in the logic 0 state. The output voltage of a standard TTL circuit drops to $V_{CE}(\text{sat})$ values in the range of 0.1 to 0.2 V, while the Schottky output transistor is driven only to the edge of saturation. The nominal voltage on the output transistor can be found from Figure 6.20. When the output transistor is driven on, $V_{BE}(\text{act}) = 0.7$ V, the diode drop across the Schottky diode is clamped at 0.3 V, and $V_{CE} = 0.4$ V. Thus, the output voltage of a Schottky TTL circuit, in its logic 0 state, tends to be slightly higher than in the conventional TTL circuits. The output voltage in the logic 1 state is essentially the same in both conventional and Schottky TTL circuits.

6.3.3 Low-Power Schottky TTL (LS-TTL)

The Schottky TTL circuit in Figure 6.21 and the standard TTL circuit dissipate approximately the same power, since the voltages and resistance values in the two circuits are similar. The advantage of the Schottky TTL circuit over the conventional TTL circuit, however, is the reduction in propagation delay time by a factor of 3 to 10.

Propagation delay times depend not only on the type of transistors (Schottky clamped or regular) used in the circuit, but also on the current levels in the circuit. The storage time of a regular transistor is a function of the reverse base current that pulls the transistor out of saturation. The turn-on time of a transistor depends on the current level charging the base–emitter junction capacitance. A desirable tradeoff can be made between current levels (power dissipation) and propagation delay time. Lower current levels lead to smaller power dissipation in the circuit at the expense of increased propagation delay times. This tradeoff has been successful in commercial applications since very short propagation delay times are not always necessary, but reduced power supply requirements are always an advantage.

A low-power Schottky TTL NAND circuit is shown in Figure 6.22. With few exceptions, these LS-TTL circuits do not use the multiemitter input transistor that originally gave TTL its name. Most LS-TTL circuits use a DTL type of input circuit with Schottky diodes performing the AND function. This circuit is faster than the classical multiemitter input transistor circuit, and the input breakdown voltage is also increased. It should be noted that the elimination of a transistor input stage in Figure 6.22 decreases the noise margin to positive input noise (NM_H) by one diode drop since the base–collector junction

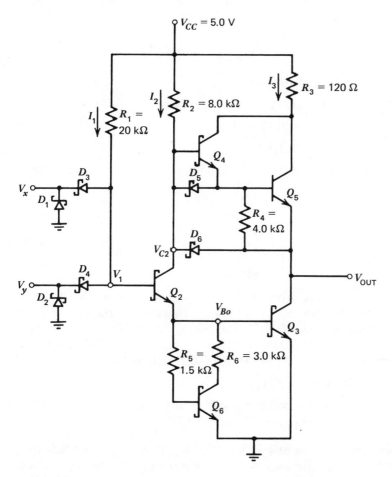

Figure 6.22. A low-power Schottky TTL NAND logic circuit.

of an input stage transistor was not replaced. Even though the LS-TTL circuit has lost its classical TTL input structure, the TTL name is still used because of the established reputation of the TTL line of logic gates.

The dc analysis of the LS-TTL circuit in Figure 6.22 shows the drastic reduction in power dissipation compared to the conventional or S-TTL circuits. When $V_x = V_y = $ logic $1 = 3.6$ V, the input diodes D_3 and D_4 are off, and the voltage V_1 is clamped at

$$
\begin{aligned}
V_1 &= V_{BE3}(\text{act}) + V_{BE2}(\text{act}) \\
&= 0.70 + 0.70 \\
&= 1.4 \text{ V}
\end{aligned}
\tag{6.67}
$$

Then,

$$
\begin{aligned}
I_1 &= \frac{V_{CC} - V_1}{R_1} \\
&= \frac{5.0 - 1.4}{20 \times 10^3} \\
&= 180 \ \mu\text{A}
\end{aligned}
\tag{6.68}
$$

The collector voltage of Q_2 is found from

$$
\begin{aligned}
V_{C2} &= V_{BE3}(\text{act}) + V_{CE2}(\text{sat}) \\
&= 0.70 + 0.40 \\
&= 1.1 \text{ V}
\end{aligned}
\tag{6.69}
$$

using the higher value of $V_{CE}(\text{sat})$ for the Schottky clamped transistors. Then,

$$
\begin{aligned}
I_2 &= \frac{V_{CC} - V_{C2}}{R_2} \\
&= \frac{5.0 - 1.1}{8.0 \times 10^3} \\
&= 488 \ \mu\text{A}
\end{aligned}
\tag{6.70}
$$

In this mode, Q_3 and Q_2 are driven to the edge of saturation so that $V_{\text{OUT}} = 0.40$ V. The difference between V_{C2} and V_{OUT} appears across the base–emitter of Q_4 and the resistor R_4 and is 0.70 V. This puts Q_4 at the edge of cutoff, with negligible emitter current. The transistor Q_3 is off, and $I_3 = 0$. The total power dissipated in this LS-TTL circuit in the input-high state is then

$$
\begin{aligned}
P &= (I_1 + I_2)V_{CC} \\
&= (180 \times 10^{-6} + 488 \times 10^{-6})5.0 \\
&= 3.34 \text{ mW}
\end{aligned}
\tag{6.71}
$$

This power dissipation is almost a factor of 10 less than that in the Schottky TTL circuit of Figure 6.21. Because of the lower current levels, the propagation delay time of the LS-TTL circuit is larger than that of the Schottky TTL. The

LS-TTL circuit shown in Figure 6.22 has a propagation delay time of approximately 10 ns. This compares closely with propagation delay time for a standard TTL circuit but has a considerably reduced power dissipation.

Two speed-up diodes, D_5 and D_6, are included to reduce the switching time of this circuit. With at least one input in the low state, the output is in the logic 1 state, and Q_4 and Q_5 are on. When both inputs are switched to their logic 1 state, Q_2 turns on and V_{C2} decreases, turning on D_5 and D_6. The diode D_5 helps to discharge the capacitance at the base of Q_5, turning the transistor off more rapidly. Diode D_6 helps discharge the load capacitance and bring V_{OUT} to its logic 0 state more rapidly.

6.4 ADVANCED SCHOTTKY TTL LOGIC CIRCUITS

By the late 1970s, continued innovations in integrated circuit technology had created even better performance in TTL circuits. The performance of the Advanced Schottky TTL exceeded that of its TTL predecessors in decreased power and average propagation delay. Table 6.5 compares the performance of the TTL series.

The advanced low-power Schottky circuit possesses the lowest speed–power product with a propagation delay short enough to accommodate a large number of digital applications. The choice of Advanced Low-Power Schottky or Advanced Schottky lies in the requirement for low propagation delay or low power. The Texas Instrument Advanced Low-Power Schottky (ALS) inverter (74ALS04) is shown in Figure 6.23. The major modification in this circuit over the Low-Power Schottky circuit lies in the design of the input circuitry. The phase-splitter, totem-pole, squaring and output transistors are similar in Low-Power Schottky and Advanced Low-Power Schottky circuits.

The input circuitry of Figure 6.23 contains a *pnp* transistor (Q_1), a current amplification transistor (Q_2), and a Schottky diode (D_2) from the base of Q_3 to the input. The diode D_2 provides a low-impedance path to ground when the

Table 6.5 Typical Performance Characteristics by 74/TTL Series

Series	Propagation delay (ns)	Power dissipation (mW)	Speed–power product (pJ)
Advanced Low-Power Schottky	4	1	4
Low-Power Schottky	9.5	2	19
Advanced Schottky	1.5	20	30
Low-Power Standard	33	1	33
Schottky	3	19	57
Standard	10	10	100

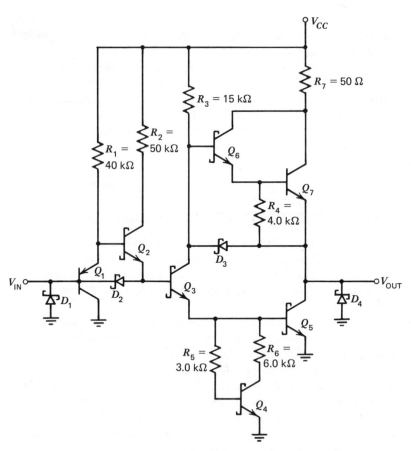

Figure 6.23. Advanced Low Power Schottky (ALS) Inverter Gate (74ALSO4) by Texas Instruments, Inc.

input makes a high-to-low transition. This enhances the switching time of the inverter. The current driver transistor Q_1 provide a faster transition when the input goes from low to high than if a Schottky diode input stage is used. Transistor Q_1 provides the switch element that steers current from R_1 to either Q_2 or to the input source. In some versions of Advanced Schottky TTL circuits, this *pnp* transistor is replaced by a diode.

Most of the Advanced Schottky TTL circuits have transistor rather than diode input stages. The transistor input stage of Q_1 and Q_2 uses slightly more power than a diode stage but increases the noise immunity to positive noise voltages (NM_H). This is important since the threshold for switching a TTL circuit is skewed toward the direction of low-input voltages, as shown in Figure 6.24.

The phase splitter transistor Q_3 controls the on and off states of Q_4, Q_5,

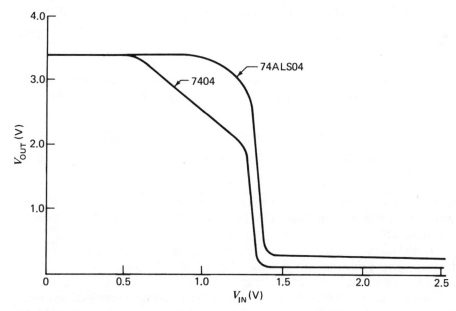

Figure 6.24. Voltage transfer characteristics for the standard 7404 and its ALS counterpart (74ALS04).

Q_6, and Q_7. Transistor Q_4 actively pulls down Q_5 during the output low-to-high transition. As discussed before, Q_4 also "squares" the transfer curve of Figure 6.24 by diverting most of the emitter current of Q_3 into the base of Q_5 during the initial cut-in of Q_5. The Darlington pair, Q_6 and Q_7, is the part of the output stage providing increased current drive during output low-to-high transitions. The output transistor Q_5 provides an output-low voltage of approximately 0.40 V because of the Schottky nature of the transistor.

The Schottky diode D_3 is connected from the output to the collector of the phase splitter transistor Q_3. This diode assists the active pull-down of the output during high-to-low transitions. Charge from the load capacitance is partially eliminated through D_3 and Q_3 to the base of Q_5. This increases the base drive to Q_5 and more rapidly lowers the output voltage. The position of D_3 also assists in a more rapid turn-off of Q_6 and Q_7.

The static voltage transfer curve for the Advanced Low-Power Schottky inverter is shown in Figure 6.24. The output voltage begins to drop when the base of Q_3 acquires sufficient voltage to turn on Q_4 and Q_5. This occurs at approximately

$$
\begin{aligned}
V_{IN} &= V_{BE3}(\text{cut-in}) + V_{BE5}(\text{cut-in}) - V_{D2} \\
&= 0.65 + 0.65 - 0.30 \\
&= 1.00 \text{ V}
\end{aligned}
\tag{6.72}
$$

A second breakpoint in the transfer curve occurs when

$$V_{IN} = V_{BE2}(\text{cut-in}) + V_{BE3}(\text{cut-in}) + V_{BE5}(\text{cut-in}) - V_{EB1}(\text{act})$$
$$= 0.65 + 0.65 + 0.65 - 0.70 \tag{6.73}$$
$$= 1.25 \text{ V}$$

The output voltage reaches its minimum level when the voltage at the base of Q_2 is sufficient to fully drive Q_2, Q_3, and Q_5 to the edge of saturation. This occurs at

$$V_{IN} = V_{BE2}(\text{act}) + V_{BE3}(\text{act}) + V_{BE5}(\text{act}) - V_{EB1}(\text{cut-in})$$
$$= 0.70 + 0.70 + 0.70 - 0.65 \tag{6.74}$$
$$= 1.45 \text{ V}$$

During the output high-to-low transition, transistors Q_2, Q_3, Q_4, and Q_5 are turning on with increasing base drive current, while transistors Q_1, Q_6, and Q_7 are turning off.

Logic circuit designers must be aware of different problems that arise when using very fast circuits such as the Advanced Schottky configuration. The inductance, the capacitance, and signal delay introduce problems that require a transmission line theory approach. Most manufacturers of fast logic circuits provide helpful literature in addressing the interconnection problem, but the designer must draw upon the concepts of transmission line theory.

6.5 WIRED-AND LOGIC CONFIGURATIONS

The AND operation can be performed by two methods in TTL or DTL. An easy way to perform the AND function on a single gate is to include another inverter stage following the phase-splitter transistor. Figure 6.25 shows a TTL AND gate with Q_3 inserted to convert the NAND to an AND operation. This circuit draws slightly more current than a comparable TTL NAND gate (Figure 6.9) and undergoes a slight degradation in propagation delay due to the addition of Q_3.

Another useful method of performing the AND function in TTL or DTL logic is accomplished by directly connecting the collectors of individual gate output transistors. Figure 6.26 shows an example of a two-input NAND gate connected to a three-input NAND gate at their output collectors. The output Y_{OUT} will be in a high-logic state only if both Q_3 and Q_3', are off. If either or both Q_3 and Q_3' are saturated, then Y_{OUT} is in the low-voltage state. The two gates in this configuration then perform the logic function

$$Y_{OUT} = \overline{AB} \cdot \overline{CDE}$$

This design tool for performing the AND function in TTL and DTL is called *wired-AND*. Note that to accomplish this function without wired-AND would require an inverter gate following both Y_1 and Y_2 and that the inverter outputs would then feed a two-input AND gate.

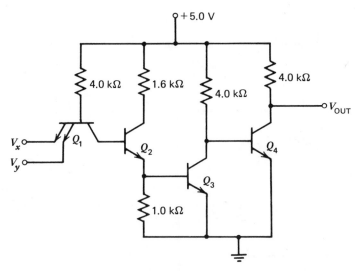

Figure 6.25. TTL AND gate.

A potential problem exists in the wired-AND of TTL and DTL gates. Figure 6.26 illustrates circuits with passive pull-up resistors R_C and R_3' and represents a correct configuration for the wired-AND. However, if totem-pole output stages were used, an intolerable short-circuit condition arises when one of the output transistors is on and the other is off. Figure 6.27 shows the consequences if totem-pole output stages are connected as wired-AND. If Q_3' is saturated and Q_3 is off, then Q_4 and Q_5 drive excessive current into Q_3'. This current is

$$I_1 = I_{B4} + I_C$$

This can be verified as $I_{B4} = 2.13$ mA and $I_C = 40$ mA, or $I_1 = 42.1$ mA. The total current into Q_3' becomes

$$I_{C3}' = I_1 + I_L$$
$$= 42.1 \times 10^{-3} + 1.03 \times 10^{-3}$$
$$= 43.1 \text{ mA}$$

The current may be of sufficient amplitude to destroy Q_3'. If more than two NAND gates are connected as wired-AND, the problem becomes more severe.

6.6 **SUMMARY**

Bipolar saturating transistor logic is a major contributor to digital logic systems. The most popular of these circuits come from the TTL families. Conventional TTL is an effective logic system but consumes power and has a relatively long

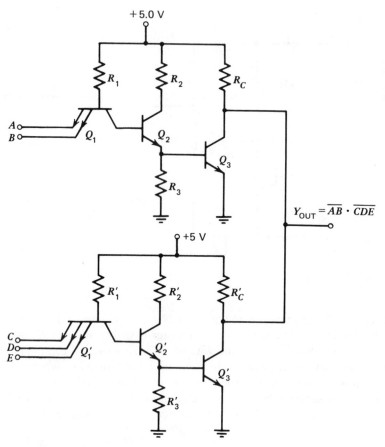

Figure 6.26. A wired-AND connection of a two-input NAND gate and a three-input NAND gate.

propagation delay due to saturation of the transistors in the circuits. Power dissipation in these circuits can be reduced by increasing the resistance values in the circuits, but the effect is an increase in the propagation delay time. A significant reduction in the propagation delay time can be achieved by incorporating Schottky barrier diodes across the base–collector junctions of the transistors in these circuits that normally would be driven into saturation. The result of these circuit improvements is a variety of high-speed and low-power TTL circuits to satisfy most of the needs of the digital circuit designer. Of particular importance are the low-power Schottky (LS) and advanced low-power Schottky (ALS) families, which provide electrical characteristics that approach those of the competing technologies described in Chapters 7 and 8.

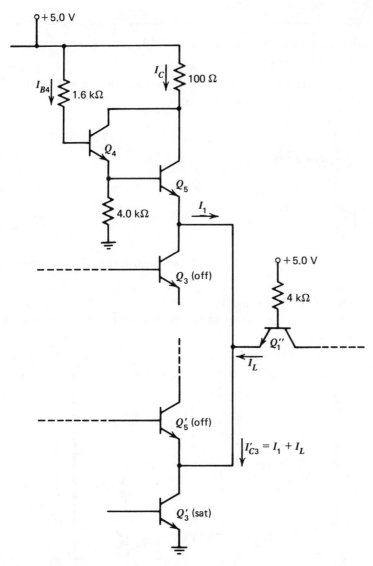

Figure 6.27. Two totem pole-output stages present an undesirable wired-AND condition.

PROBLEMS

Note: The piecewise linear parameters listed in Table 6.1 will be used on all problems unless otherwise noted.

6.1 The bipolar inverter circuit shown in Figure 6.28 has $\beta = 35$. Find I, I_B, and V_{OUT} for (a) $V_{IN} = 0.10$ V and (b) $V_{IN} = 5.0$ V.

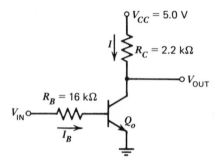

Figure 6.28. Circuit for Problem 6.1.

6.2. The bipolar inverter shown in Figure 6.29 feeds eight identical inverters. (a) If $V_{IN} = 5.0$ V, find I and V_{OUT}. (b) If $V_{IN} = 0.10$ V, calculate I and V_{OUT} assuming that the transistors in the eight inverters are driven into saturation.

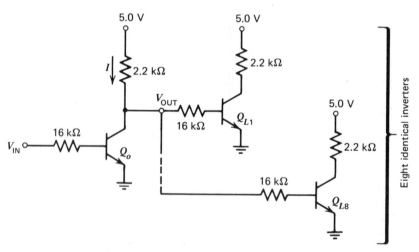

Figure 6.29. Figure for Problem 6.2.

6.3. The transistor in the circuit in Figure 6.30 has $\beta = 25$. Find the currents and voltages listed in the table for the three sets of input conditions.

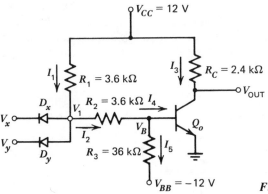

Figure 6.30. Figure for Problem 6.3.

	I_1 (mA)	I_2 (mA)	I_3 (mA)	I_4 (mA)	I_5 (mA)	V_1 (V)	V_B (V)	V_{OUT} (v)
Input								
$V_x = V_y = 0.10$ V								
$V_x = 12$ V, $V_y = 0.10$ V								
$V_x = V_y = 12$ V								

6.4. The transistors in Figure 6.31 have $\beta = 20$. (*a*) Find the currents and voltages I_1, I_3, I_4, and V_1 for the input conditions (1) $V_x = V_y = 0.10$ V and (2) $V_x = V_y = 5.0$ V. (*b*) Calculate the fan-out for the output-low condition. (*c*) If the maximum collector current of Q_1 is limited to 5.0 mA, find the fan-out.

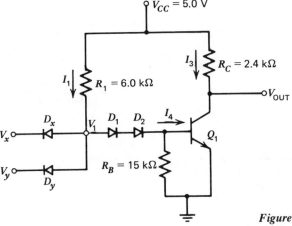

Figure 6.31. Figure for Problem 6.4.

6.5. The DTL inverter shown in Figure 6.32*a* has an applied signal of V_x = 5.0 V. Draw load lines in Figure 6.32*b* for the following three conditions at the output: (*a*) no load, (*b*) one load, and (*c*) two loads. Mark the approximate operating point as the intersection of the load line with the base current of Q_o and comment on the trend as more loads are added.

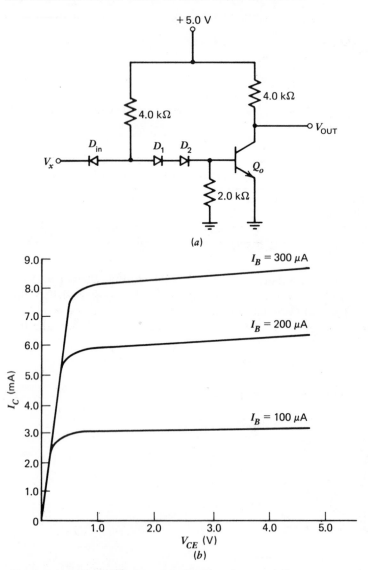

(*a*)

(*b*)

Figure 6.32. (*a*) A DTL inverter. (*b*) The characteristic curves (I_C vs. V_{CE}) of the output transistor Q_O.

6.6. Figure 6.33 shows an improved version of a DTL circuit. An offset diode is replaced by a transistor Q_1 providing increased current drive to Q_2 as well as maintaining the same noise margin. (*a*) Let $\beta = 20$ for Q_2 and $V_x = V_y = 5.0$ V, and assume that both transistors are driven into saturation. Calculate I_{B2} and I_{CC}. Compare I_{B2} with that obtained for the DTL circuit of Figure 6.1. (*b*) Calculate the fan-out for the output-low condition. Compare this with the fan-out determined for the circuit of Figure 6.3.

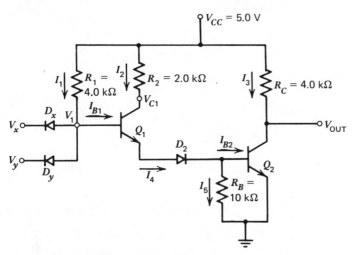

Figure 6.33. An improved DTL NAND logic circuit.

6.7. The transistor in the circuit of Figure 6.34 has $\beta = 25$. (*a*) Find the currents I_1, I_2, I_3, I_4, and I_5 and the voltages V_1, V_B, and V_{OUT} for the following input conditions: (1) $V_x = V_y = 0.10$ V and (2) $V_x = V_y = 5.0$ V. (*b*) Calculate the fan-out for the output-low condition.

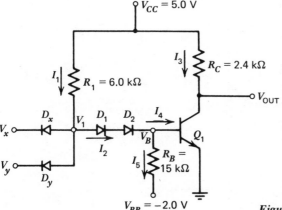

Figure 6.34. Circuit for Problem 6.7

6.8. The transistors in the DTL circuit shown in Figure 6.35 have $\beta = 25$. (*a*) Find the currents I_1, I_2, I_3, I_4, and I_5 and the voltages V_1, V_B, and V_{OUT} for the following input conditions: (1) $V_x = V_y = 0.10$ V and (2) $V_x = V_y = 5.0$ V. (*b*) Find the power dissipated in the circuit for the conditions in (*a*). (*c*) Calculate the fan-out for the output-low condition.

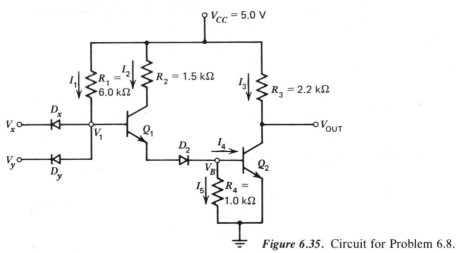

Figure 6.35. Circuit for Problem 6.8.

6.9. The transistors in the DTL circuit in Figure 6.36 have $\beta = 30$. (*a*) Find the currents I_1, I_2, I_3, and I_4 for the following input conditions: (1) $V_x = V_y = 0.10$ V and (2) $V_x = V_y = 5.0$ V. (*Hint:* Transistor Q_1 does not enter saturation. Transistors Q_2 and Q_3 are either off or saturated.) (*b*) Find the power dissipated in the circuit for the input conditions in (*a*). (*c*) Calculate the fan-out for the output-low condition.

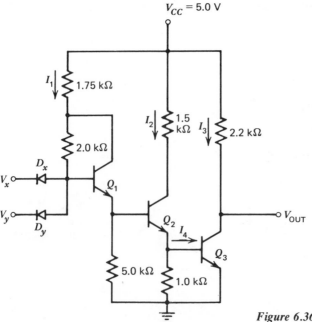

Figure 6.36. Circuit for Problem 6.9.

6.10. The transistors in the TTL circuit shown in Figure 6.37 have $\beta_F = 20$ and $\beta_R = 0$. (a) Determine the currents I_1, I_2, I_3, I_4, I_{B2}, and I_{B3} for the input following conditions: (1) $V_x = V_y = 0.10$ V and (2) $V_x = V_y = 5.0$ V. (b) Show that, for $V_x = V_y = 5.0$ V, the transistors Q_2 and Q_3 are in saturation.

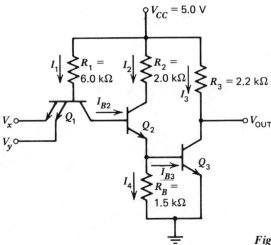

Figure 6.37. Circuit for Problem 6.10.

6.11. The transistors in the TTL circuit of Figure 6.38 have $\beta_F = 20$ and $\beta_R = 0.10$. For the diode, $V_D = 0.70$ V. (a) Calculate the fan-out for $V_x = V_y = 5.0$ V. (b) Calculate the fan-out for $V_x = V_y = 0.10$ V. (Assume that V_{OUT} is allowed to decrease by 0.10 V.)

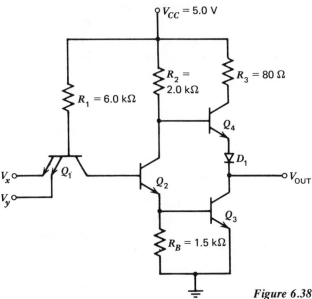

Figure 6.38. Circuit for Problem 6.11.

6.12. A two-input, low-power TTL NAND circuit is shown in Figure 6.39. The piecewise linear parameters are $V_{BE}(\text{act}) = V_D = 0.50$ V, $V_{BE}(\text{sat}) = 0.70$ V, $V_{CE}(\text{sat}) = 0.10$ V, $\beta_F = 20$, and $\beta_R = 0.5$. Determine the power dissipation of the circuit when (1) $V_x = V_y = 0.10$ V and (2) $V_x = V_y = 4.0$ V. Assume a fan-out of 2 for these calculations.

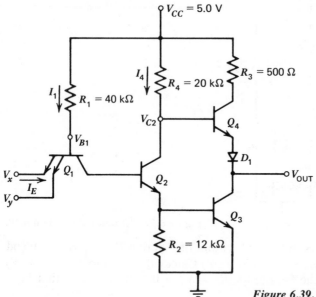

Figure 6.39. Circuit for Problem 6.12.

6.13. Consider the Schottky clamped transistor shown in Figure 6.40. Assume that $\beta_F = 10$, $V_{BE} = 0.75$ V, and $V_D = 0.30$ V. (*a*) For no load, $I_L = 0$. Find the diode current I_D. (*b*) Determine the maximum load current that the transistor can sink and still remain at the edge of saturation.

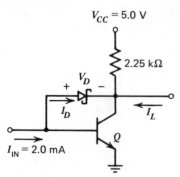

Figure 6.40. Circuit for Problem 6.13.

6.14. A modified low-power Schottky logic circuit is shown in Figure 6.41. Assume that $\beta_F = 15$, $V_{BE}(\text{act}) = 0.70$ V, $V_{CE}(\text{sat}) = 0.40$ V, $V_D(\text{Schottky}) = 0.30$ V, and $V_D(pn$ junction$) = 0.70$ V. Calculate the fan-out for the case when $V_x = V_y = 3.6$ V.

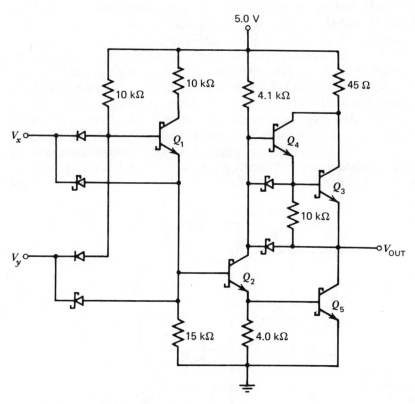

Figure 6.41. Circuit for Problem 6.14.

6.15. Figure 6.42 shows a modified low-power Schottky TTL NAND circuit that has three gain stages instead of the usual two in the standard TTL circuits. The addition of this third stage raises the input threshold voltage, provides a higher noise margin in the input-low state, and increases the output drive. Diodes D_7 and D_8 are speed-up diodes and perform the same function as the diodes D_5 and D_6 in Figure 6.22 for the LS-TTL circuit. Diode D_7 helps turn off transistor Q_5 when the output switches from its high to low state; the D_8 helps discharge the load capacitance. The Schottky diode D_5 acts as a low-resistance path to discharge the capacitances (both parasitic capacitance and junction capacitance) connected to the base of Q_2 when V_x makes a high-to-low transition. Diode D_6 performs the same function when V_y makes a high-to-low transition. The propagation delay time through this circuit is nominally 4.0 ns. Assume that $V_{BE}(\text{sat}) = 0.70$ V and $V_{CE}(\text{sat}) = 0.40$ V. (*a*) Compute the power dissipated in the circuit of Figure 6.42 when all inputs are low. (*b*) Repeat this calculation when all inputs are high. (*c*) Sketch a voltage transfer curve for V_{OUT}/V_x when $V_y = 5.0$ V. Label the voltage values at the significant break points.

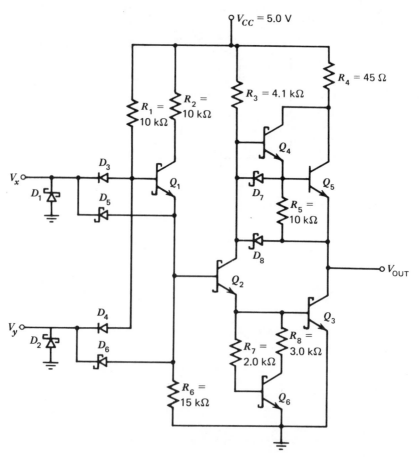

Figure 6.42. A modified low-power Schottky TTL NAND logic circuit for Problem 6.15.

6.16. (*a*) Calculate the power dissipation in the circuit of Figure 6.23 when $V_{IN} = 0$. (*b*) Repeat this calculation for $V_{IN} = 3.6$ V. (*c*) Calculate the output short-circuit current. Assume that $V_{CC} = 5.0$ V.

6.17. In Figure 6.23, the output-high state current and voltage are 5.0 mA and 3.0 V, respectively. Is Q_6 at the edge of saturation or well into the active region?

chapter 7

BIPOLAR CURRENT MODE DIGITAL LOGIC CIRCUITS

The most widely used bipolar logic family, transistor–transistor logic (TTL), was considered in the previous chapter. In this chapter, we introduce two other important bipolar digital logic technologies, integrated injection logic (I^2L) [also called merged transistor logic (MTL)] and emitter-coupled logic (ECL). Each of these technologies has particular advantages that must be considered by a digital systems designer. Both I^2L and ECL circuits switch a constant current from one part of the circuit to another to achieve the two distinct logic states. For this reason, I^2L and ECL are referred to as current mode logic. ECL circuitry is noted for its very low propagation delay time, while I^2L possesses a high component density and a low power–delay product.

7.1 CONVENTIONAL INTEGRATED INJECTION LOGIC

Both high packing density and low power–delay product make I^2L a viable bipolar technology for large-scale integrated circuit (LSI) applications. I^2L can be used over a wide dynamic range of operating power level, has bipolar current drive capability, and is compatible with existing bipolar technologies. Schottky transistors and/or Schottky diodes are commonly used with I^2L to reduce propagation delay times.

The logic voltage swing in I^2L circuits is significantly less than in TTL circuits; and, consequently, the noise margins of I^2L circuits are much less than in TTL circuits. The problem of lower noise margins has been effectively eliminated by using I^2L within integrated circuits to perform the various logic functions, while using TTL for the input and output circuitry. In the interior of the integrated circuit, I^2L has low power dissipation and high packing density, while the TTL at the input and output terminals gives a high noise immunity. Interface circuits couple the TTL output circuits and the internal I^2L circuitry.

7.1.1 The Basic Inverter

The basic I^2L inverter is shown in Figure 7.1. The constant-current sources I_1, I_2, and I_3 have a magnitude I_o, and Q_o is the output transistor of a previous stage. If Q_o is driven into saturation, V_x drops to a low voltage (approximately 0.10 V), forcing the switching transistor Q_1 into cutoff. The current I_1 becomes the collector current of Q_o, and I_{B1} is zero. Since Q_1 is cut off, the collector current I_{C1} is zero and the constant current I_2 is forced into the base of Q_2, driving it into saturation. The voltage V_{o1}, then equals $V_{BE}(sat)$ of Q_2 (approximately 0.70 V). With Q_2 in saturation, $V_{o2} = V_{CE}(sat) \approx 0.10$ V, which forces Q_3 into cutoff. The current I_3 is then the collector current of Q_2, so that

$$I_{C2} = I_3 = I_o \tag{7.1}$$

If we apply the piecewise linear parameters listed in Table 7.1 to the I^2L circuits, the logic 1 and the logic 0 states correspond to 0.70 and 0.10 V, respectively. Each succeeding voltage at the circuit nodes of Figure 7.1 is the complement

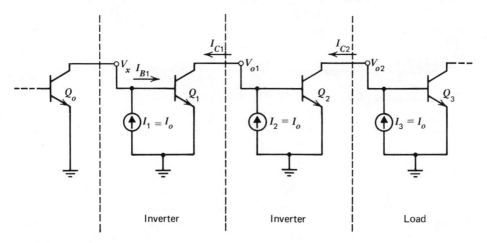

Figure 7.1. Basic I^2L circuit showing inverters in series.

Table 7.1. Piecewise Linear Parameters
for I²L Transistors shown in Figure 7.2.

Parameter	Source Transistor Q_3	Inverter Transistor Q_1
V_{BE}(cut-in) (V)	− 0.70	0.60
V_{BE}(sat) (V)	− 0.80	0.70
V_{CE}(sat) (V)	− 0.10	0.10

of the previous value. The combination of a constant-current source and a switching transistor (for example, I_1 and Q_1) defines the basic I²L inverter.

The switching transistors are designed to be either in cutoff or saturation. If, in Figure 7.1, Q_o is assumed to be in saturation, then Q_1 is cut off, Q_2 turns on with a base current I_o, and the collector current of Q_2 becomes I_o. The minimum current gain β_{min} of these switching transistors must be unity to ensure that they are driven into saturation.

In the case where transistor Q_o is cut off, the constant current I_1 is forced into the base of Q_1, driving that transistor into saturation and $V_x = V_{BE1}$(sat) = 0.70 V. The Q_1 output voltage becomes $V_{o1} = 0.10$ V and $I_{C1} = I_o$. Here, Q_2 is cut off and I_3 is forced into the base of Q_3, driving it into saturation. The voltage V_{o2} is then equal to 0.70 V. In this circuit, the second inverter, Q_2, is the load circuit for the first inverter, while Q_3 provides the load for the second inverter.

CONSTANT-CURRENT SOURCE. In the I²L inverter circuits shown in Figure 7.1, a constant-current source is included between the base and ground terminals of each switching transistor. One way to achieve a constant-current source is with a common-base *pnp* transistor. Figure 7.2 shows a basic inverter Q_1 and load Q_2 with injector current transistors Q_3 and Q_4. The base–emitter junctions of the current source transistors are forward biased to give an emitter current I_E. The collector currents of Q_3 and Q_4 are the constant currents I_o. The collector voltages on the current source transistors are always positive at 0.70 or 0.10 V, ensuring that the collector–base junctions of Q_3 and Q_4 are always forward biased and operating in or near saturation.

The common-base *pnp* current source transistor provides an output current I_o that is related to the driving current I_E by

$$I_o \simeq \alpha_F I_E \tag{7.2}$$

where α_F is the forward active common-base current gain. A detailed analysis shows that Equation 7.2 is valid even though the current source transistor is operating in saturation and the base–collector junction voltage changes with the two bias states. Therefore, when the emitter voltage and α_F are constant, the common-base *pnp* transistor provides a constant current I_o independent of the logic state of a particular switching transistor.

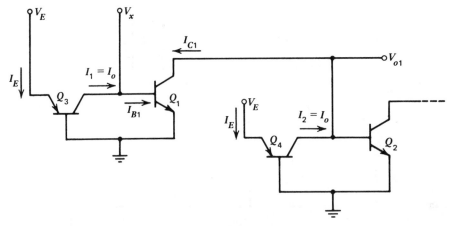

Figure 7.2. I²L logic circuit with *pnp* current source transistors Q_3 and Q_4, and inverter transistors Q_1 and Q_2.

INTEGRATED INJECTION LOGIC STRUCTURE. The high packing density of I²L is achieved by integrating or merging the source transistor with the switching transistor. A pictorial view of the I²L circuit of Figure 7.2 is shown in Figure 7.3. The emitter of the *npn* switching transistor, Q_1, is the n_2 epitaxial region as well as the heavily doped n_3^+-substrate. The base of the current source transistor Q_3 is also *n*-type and electrically connected to the switching transistor emitter. The same *n*-type substrate, therefore, serves as the switching transistor emitter and the source transistor base. The substrate has a single external terminal (E_1, B_3) connected to the ground potential. Since the substrate is common to all switching transistor emitters, each I²L inverter does not have to be isolated from the others. Elimination of the isolation requirement between transistors greatly reduces the area of a single inverter. The packing density of the I²L circuits thus becomes suitable for LSI applications.

A second integration occurs in the p_2-region between the source transistor and the switching transistor. This region serves as the base of the switching transistor and the collector of the source transistor. With this geometry, the source transistor is a lateral *pnp* device having a relatively low forward current gain. However, a forward common-base current gain α_F of 0.8 is sufficient for this current source application.

The switching transistor is a vertical *npn* structure but, due to the geometry shown in Figure 7.3, operates in the inverse mode. In practice, the heavily doped n_3^+-substrate gives a high injection efficiency so that the common-emitter forward current gain β_F is typically 5. This current gain is sufficient for most I²L applications.

The p_1^+-region of Figure 7.3 serves as the emitter for both current sources. Several current sources may be served by one p^+-emitter strip, since it can

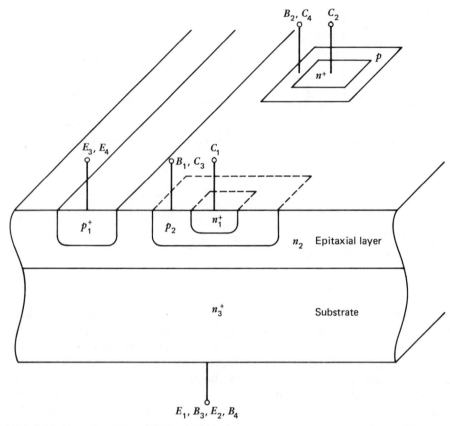

Figure 7.3. Cross section of I²L circuit structure from the circuit of Figure 7.2.

run the length of several switching transistors. The merged base and collector of Q_2 and Q_4 are shown in Figure 7.3, as well as the collector output of Q_2. The emitter of Q_2 and the base of Q_4 are merged with the ground points of the other transistors through the n_3^+-substrate. In addition, multiple collectors may be fabricated in the p_2-base region in much the same way as the multiemitter TTL device. As will be seen, multicollectors can be built into the switching transistor providing a multioutput I²L device.

POWER DISSIPATION. All of the current source transistors can be biased with a single external resistor and supply voltage V_{CC}, as shown in Figure 7.4. This capability allows a bias resistor to be selected that minimizes power consumption in a particular I²L circuit.

Example. Calculate the total current and external resistance required to bias an I²L circuit having 1000 current sources similar to those in Figure

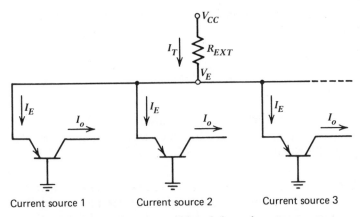

Current source 1   Current source 2   Current source 3

Figure 7.4. I²L current sources biased through a common external resistor R_{EXT}.

7.4. Assume that for the current source transistor, $V_{EB}(\text{sat}) = 0.80$ V, $\alpha_F = 0.80$, $I_o = 50$ μA, and the supply voltage $V_{CC} = 5.0$ V.

Solution. The individual current source emitter currents I_E may be calculated from Equation 7.2 as

$$I_E = \frac{I_o}{\alpha_F}$$

$$= \frac{50 \times 10^{-6}}{0.80} \tag{7.3}$$

$$= 62.5 \ \mu A$$

The total current I_T is given by

$$I_T = \frac{V_{CC} - V_{EB}(\text{sat})}{R_{EXT}} = N I_E$$

$$= 1000 \times 62.5 \times 10^{-6} \tag{7.4}$$

$$= 62.5 \ \text{mA}$$

Then,

$$R_{EXT} = \frac{5.0 - 0.80}{62.5 \times 10^{-3}}$$

$$= 67.2 \ \Omega$$

The injector current value I_o can easily be varied over a wide range by changing the external resistance value.

In this example, the total power dissipated on the chip is the product of the total current I_T and the voltage $V_{EB}(\text{sat})$ across the emitter–base junction of the current source transistors. The total power dissipated in the above example is

$$P_T = I_T V_{EB}(\text{sat})$$
$$= (62.5 \times 10^{-3})(0.80) \tag{7.5}$$
$$= 50 \text{ mW}$$

For 1000 I^2L gate on this chip, the average power dissipated per gate is 50 μW. This value is approximately 300 times less than the power dissipated per gate of a standard TTL circuit.

POWER–DELAY PRODUCT. The propagation delay time is primarily a function of the time needed to charge or discharge transistor junction capacitance. The removal of excess minority carriers from the base of the switching transistor is less important at the low levels of collector current used in I^2L. For a fixed value of capacitance, the propagation delay time is decreased by increasing the current drive. In I^2L circuits, the propagation delay time is inversely proportional to the injector current over a wide range of current levels. Figure 7.5 shows a typical plot of propagation delay time as a function of injector current. As the injector current level increases, the transistors are driven further into saturation. The propagation delay time is eventually dominated by the saturation storage time. At high injector current levels, the change in the propagation delay time reverses and begins to increase because of the increased charge storage time.

A useful figure of merit for logic gates is the product of power dissipation and propagation delay time, referred to as the *power–delay product,* or *speed–power product.* The product of power and time is energy, and the power–delay product is usually expressed in picojoules (pJ). The power dissipated per gate of an I^2L circuit is approximately proportional to the injector current, since $V_{EB}(\text{sat})$ changes very little over a wide range of current. As shown in Figure 7.5, the propagation delay time is inversely proportional to the injector current so that the power–delay product is a constant for lower current values. A low value of the power–delay product implies that the gate can operate at high speed and low power. Typical values of the power–delay product for I^2L range from 0.1 to 4 pJ, compared to approximately 100 pJ for a standard TTL circuit.

Since the injector current level can be varied over a wide range by changing the value of the external bias resistor or the voltage supply, the propagation delay time and power dissipation of a gate can also be controlled over a wide range. The supply current may be programmed to increase for faster operation at certain desired times and then reduced to allow low power dissipation during noncritical periods of time.

VOLTAGE TRANSFER CHARACTERISTIC. Consider the I^2L inverter shown in Figure 7.2. The voltage transfer characteristic, V_{o1} versus V_x, can be

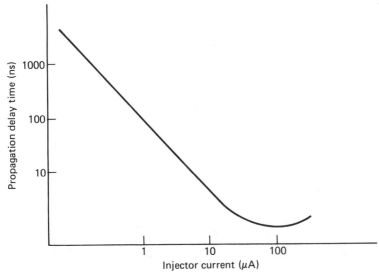

Figure 7.5. Propagation delay vs. injector current level for an I²L inverter.

analyzed using the piecewise linear model of the transistor from Table 7.1. The resultant piecewise linear transfer characteristic is shown in Figure 7.6.

For $V_x < 0.60$ V, the inverter transistor is cut off so that V_{o1} is equal to the $V_{BE}(\text{sat})$ of the load transistor Q_2. For 0.60 V $< V_x <$ 0.70 V, Q_1 is in its active region causing the output voltage to change directly as the input voltage varies. When $V_x = 0.70$ V, Q_1 is in saturation and $V_{o1} = 0.10$ V. The maximum input voltage is $V_x = 0.70$ V and is the result of the current I_1 driving Q_1 into saturation.

NOISE MARGIN. The noise margin of the I²L inverter can be determined qualitatively from the transfer characteristics shown in Figure 7.6. The low-state noise margin is given by

$$NM_L = 0.60 - 0.10 = 0.50 \text{ V}$$

where the 0.60-V level is the maximum input voltage before the output begins to change, and the 0.10-V level is the minimum value of the input voltage.

The high-state noise margin is more ambiguous. From the definition used to this point, we would conclude that $NM_H = 0$ V, since the maximum input voltage is 0.70 V and any input voltage less than this value causes a change in the output voltage. Fortunately, there are no discrete resistors between I²L logic gates, and therefore a noise voltage cannot be produced between the output of one gate and input of the next gate. In addition, I²L circuits are normally used within an integrated circuit and are not connected directly to input and output terminals where noise margins are more critical. TTL circuits

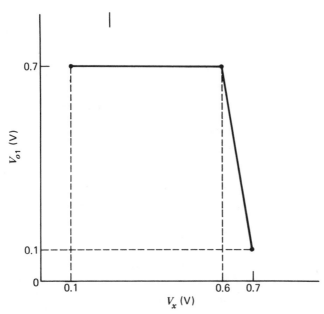

Figure 7.6. Piecewise linear transistor curve for V_{O1} vs. V_x from the I^2L circuit in Figure 7.2.

with high margins may be used at the input and output terminals and with interface circuits between the TTL and the I^2L. The noise margin factors within the integrated circuit are not as critical as they are at the terminals. In summary, the noise margins in an I^2L logic gate are low, but the circuitry and integrated circuit layout minimize noise errors.

7.1.2 I²L Logic Circuits

THE INTEGRATED INJECTION LOGIC NOR CIRCUIT. The circuit shown in Figure 7.7 is an I^2L NOR gate. The injector currents are supplied by the common-base lateral *pnp* transistors. If both the V_x and V_y inputs are at logic 0 (= 0.10 V), then both Q_1 and Q_2 are cut off. The injector current from Q_6 forces Q_3 into saturation so that the output voltage $V_{OUT} = V_{BE}(sat) =$ logic 1 ≈ 0.70 V.

When V_x = logic 1 and V_y = logic 0, Q_1 goes into saturation and Q_2 remains in cutoff. In this case, V_{OUT} goes to logic 0, Q_3 is in cutoff, and $I_{C1} = I_o$. If V_x and V_y are at logic 1 levels, then Q_1 and Q_2 go into saturation, and V_{OUT} is again at a logic 0 state. In this case, the injector current I_o from Q_6 splits between Q_1 and Q_2, so that,

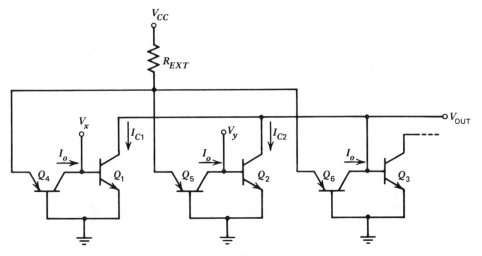

Figure 7.7. I²L NOR gate circuit.

$$I_{C1} = I_{C2} = \frac{I_o}{2} \tag{7.6}$$

This circuit performs the NOR function.

AN INTEGRATED INJECTION LOGIC EXAMPLE. The I²L circuit in Figure 7.8 contains multicollector switching transistors. Multicollectors allow decoupling between signals, providing the same function as the isolating diodes used in other types of logic circuits. For simplicity, each injector current is represented schematically by a constant-current source, although, in the actual circuit, each source would be the common-base *pnp* connected to an external resistor as shown in Figure 7.7.

If both inputs V_x and V_y are at logic 0, then Q_1 and Q_2 are cut off, and all of the collector currents into Q_1 and Q_2 are zero. The outputs are

$$V_1 = V_2 = V_4 = V_5 = 0.70 \text{ V} = \text{logic 1}$$

and

$$V_3 = 0.10 \text{ V} = \text{logic 0} \tag{7.7}$$

Now consider that $V_x = 0.70$ V and $V_y = 0.10$ V holding Q_2 in cutoff. All collector currents into Q_2 are still zero, but the collector currents into Q_1 are given by

$$I_{11} = I_o$$
$$I_{12} = I_3 = I_o \tag{7.8}$$
$$I_{13} = I_4 = I_o$$

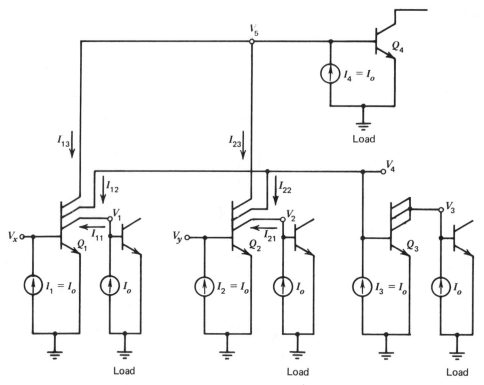

Figure 7.8. I²L circuit that performs multiple logic functions.

The output voltages are

$$V_1 = V_4 = V_5 = 0.10 \text{ V} = \text{logic } 0 \qquad (7.9)$$
$$V_2 = V_3 = 0.70 \text{ V} = \text{logic } 1$$

Since Q_1 is in saturation, all of the collector voltages of Q_1 are at the same 0.10-V potential. However, the collectors of Q_2 are not all at the same potential, since different collector potentials may exist on a cutoff transistor.

Now let $V_x = V_y = \text{logic } 1 = 0.70 \text{ V}$, where both Q_1 and Q_2 are in saturation. If Q_1 and Q_2 are identical transistors, the collector currents are

$$I_{11} = I_o$$

$$I_{12} = \frac{I_3}{2} = \frac{I_o}{2}$$

$$I_{13} = \frac{I_4}{2} = \frac{I_o}{2}$$

$$I_{21} = I_o$$

(7.10)

$$I_{22} = \frac{I_3}{2} = \frac{I_o}{2}$$

$$I_{23} = \frac{I_3}{2} = \frac{I_o}{2}$$

The output voltages are

$$V_1 = V_2 = V_4 = V_5 = 0.10 \text{ V} = \text{logic } 0 \qquad (7.11)$$

$$V_3 = 0.70 \text{ V} = \text{logic } 1$$

The total collector current into Q_1 has changed from $3I_o$ in the previous case to $2I_o$ in this case. The total collector current depends upon the condition at the other inputs.

Multiple-logic functions are performed by this circuit. These functions are

$$V_1 = \overline{V}_x \qquad (7.12a)$$

$$V_2 = \overline{V}_y \qquad (7.12b)$$

$$V_3 = \text{OR} \qquad (7.12c)$$

$$V_4 = V_5 = \text{NOR} \qquad (7.12d)$$

An advantage of I²L is the ability to implement complex logic functions with circuits that are relatively simple compared to TTL.

7.1.3 Integrated Injection Logic Fan-Out

The I²1 circuit of Figure 7.8 illustrates multicollector inverters that must sink as much as three injector currents ($3I_o$). The forward current gain β of the inverter transistor must be large enough to sink these load currents without allowing the transistor to leave saturation. Since the base current has a value equal to one injector current level, the minimum β of the transitor must be equal to the number of outputs on the collector. For example, the minimum β of Q_1 and Q_2 in Figure 7.8 is $\beta_{\min} = 3$. This is critical in I²L circuits since the vertical *npn* transistors are operated in an inverse mode physical layout and consequently have low values of β. The βs of these transistors may typically range from 5 to 8.

7.2 SCHOTTKY INTEGRATED INJECTION LOGIC

The I²L is an extremely useful technology because of its high density and low power–delay product. An additional improvement may be realized by incorporating Schottky diodes into the basic I²L circuit. The Schottky diodes decrease propagation delay by reducing both charge storage effects and the output voltage swing. Two basic approaches of Schottky I²L are (1) Schottky-coupled transistor logic (SCTL), and (2) Schottky transistor logic (STL).

7.2.1 **Schottky-Coupled Transistor Logic (SCTL)**

In Figure 7.9a, the multicollectors are replaced by a single collector and multiple Schottky diodes to provide the multiple output function. The single

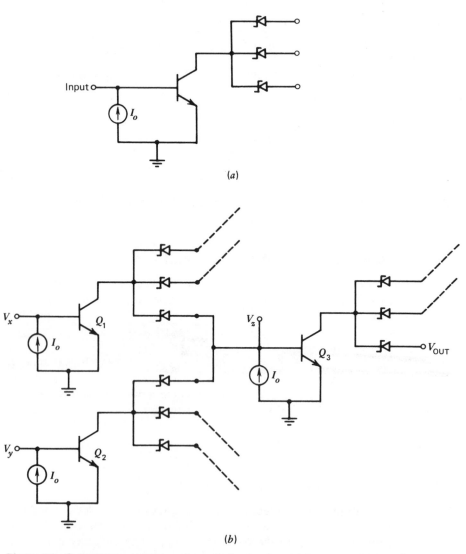

(a)

(b)

Figure 7.9. Schottky-coupled transistor logic circuits. (a) Inverter with multiple Schottky diodes on collector. (b) OR and NOR logic by SCTL circuit.

collector allows a better design for the *npn* transistor, and the Schottky diodes decouple the outputs. Figure 7.9*b* shows how Schottky diode outputs can be used to perform an OR and NOR operation. The OR logic is between inputs V_x and V_y and the output V_{OUT}. The NOR logic is between inputs V_x and V_y and the output V_z.

An output diode in series with a load switching transistor reduces the output logic swing. The reduction in output logic swing means a reduced voltage swing at the base–emitter input of the following circuit. The reduced base–emitter input voltage swing shortens the propagation delay since the input capacitance is not charged or discharged over as wide a voltage range.

The reduced voltage swing is observed in the piecewise linear model for the voltage transfer characteristic, V_z versus V_x with $V_y = 0$, of the circuit in Figure 7.9*b*. Transistor Q_1 is cut off for $V_x < 0.60$ V, but the output voltage $V_z = V_{BE}(\text{sat}) = 0.70$ V. For $0.60 < V_x < 0.70$ V, Q_1 goes through its active region reaching saturation at $V_x = 0.70$ V. The output voltage for Q_1 in saturation is given by

$$V_z = V_{CE}(\text{sat}) + V_D$$
$$= 0.10 + 0.30 \tag{7.13}$$
$$= 0.40 \text{ V}$$

where V_D is the turn-on voltage of the Schottky diode and is assumed to be 0.30 V. Figure 7.10 shows the resulting piecewise linear voltage transfer char-

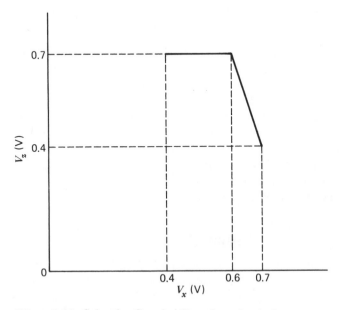

Figure 7.10. Schottky-Coupled Transistor Logic Inverter circuit voltage transfer function.

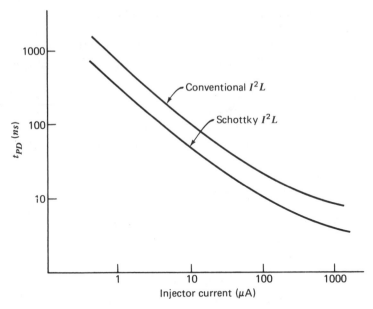

Figure 7.11. The propagation delay t_{PD} vs. injector current for conventional and SCTL gates.

acteristic. The V_z output voltage swing is

$$\Delta V_z = 0.70 - 0.40$$
$$= 0.30 \text{ V}$$

This compares to 0.60 V for the conventional I²L inverter. Since V_x is the output of a SCTL circuit, V_x is shown in Figure 7.10 to change within the range of 0.40 to 0.70 V.

The propagation delay time of the SCTL circuit is reduced by approximately a factor of 2 over that of the conventional I²L circuit, as shown in Figure 7.11. This reduction is a direct result of the smaller voltage swing on the base of the switching transistor. The use of the Schottky diode for I²L is different from that for TTL. In TTL, the Schottky diode is used to hold transistors out of saturation, while in SCTL, the diodes improve performance by substituting for the multiple collector transistor.

7.2.2 Schottky Transistor Logic (STL)

Figure 7.12 shows the STL approach to incorporating Schottky junctions into the I²L circuit. The switching transistor may have a Schottky collector region consisting entirely of metal. Unlike in Schottky clamped transistors, the Schottky diode replaces the collector–base junction in this structure. Therefore, additional Schottky diodes in the collector cannot be fabricated. Instead, Schottky

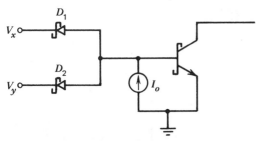

Figure 7.12. A Schottky Transistor Logic (STL) two-input NAND gate.

diodes are placed in the base of the switching transistor and the NAND operation is performed by the input diodes D_1 and D_2.

In the configuration of Figure 7.12, the Schottky diodes need to be fabricated on the p-type base material. However, Schottky diodes are much easier to fabricate on low-doped n-type material. Figure 7.13 shows the complementary structure built on low-doped n-type material with a *pnp* switching transistor and *npn* current source transistor. The bias voltage as well as the input/output voltages are now negative with respect to ground.

As with the SCTL structure, the voltage swing at the base of the switching transistor is smaller than the conventional I²L. With an input voltage of -0.10 V (logic 1), the input diodes are forward biased, and the base voltage of the switching transistor is

$$V_B = -0.10 - 0.30$$

$$= -0.40 \text{ V}$$

As the input voltage decreases, V_B decreases until $V_B = -0.70$ V, at which time the switching transistor is fully on. V_B is then clamped at this voltage and does not decrease further. The voltage swing at the base is 0.30 V, and the propagation delay time is approximately one half that of the conventional I²L circuit.

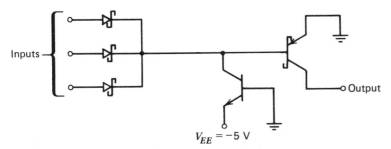

Figure 7.13. A STL three-input NAND gate using a *pnp* switching transistor and an *npn* current source transistor.

7.3 **EMITTER-COUPLED LOGIC (ECL)**

The parameter most affecting propagation delay times of the logic circuits considered thus far is the storage time of the transistors that are driven into saturation. One method of decreasing the propagation delay time is to prevent the transistor from entering saturation. Schottky TTL uses this approach by clamping the base–collector voltage at the edge of the saturation condition. This significantly reduces the propagation delay, but even the slight saturation of the Schottky transistor adds to the switching time. The transistors in emitter-coupled logic circuits are never driven into saturation. This unique nonsaturated logic family is described below.

7.3.1 **The Differential Amplifier**

The ECL circuit is based on the differential amplifier shown in Figure 7.14. Analysis shows that if the input voltages V_x and V_y differ by no more than 100 mV, the output voltage V_{o2} is directly proportional to the difference $V_x - V_y$. Similarly, the output V_{o1} is directly proportional to the difference $V_y - V_x$. This direct proportionality assumes that transistors Q_1 and Q_2 were initially biased in their active regions. This same configuration exhibits digital properties. If V_x exceeds V_y by approximately 120 mV, then, effectively, Q_1 is on and Q_2 is off. This digital property may be demonstrated by using a modeling equation that uses the diode equation (Equation 3.6) to approximate the collector current as a function of the base–emitter voltage. For Q_1 and Q_2 in Figure 7.14, this modeling equation can be written as

$$I_{C1} \simeq I_s \exp \frac{qV_{BE1}}{kT} \tag{7.14}$$

$$I_{C2} \simeq I_s \exp \frac{qV_{BE2}}{kT} \tag{7.15}$$

where I_s is the reverse bias saturation current of the base–emitter junction. It is the same value in both equations, since Q_1 and Q_2 are assumed to be identical transistors.

If $V_{BE1} = V_{BE2} + 0.12$, then the ratio of I_{C1} to I_{C2} is

$$\frac{I_{C1}}{I_{C2}} = \exp \left[\frac{q(V_{BE2} + 0.12)}{kT} - \frac{qV_{BE2}}{kT} \right]$$

$$= \exp \left(\frac{0.12}{0.026} \right) \tag{7.16}$$

$$= 101$$

where $kT/q = 0.026$ V at room temperature. This calculation shows that when the base–emitter voltage of Q_1 is 120 mV greater than the base–emitter voltage of Q_2, the collector current of Q_1 is 100 times that of Q_2. Since the emitters of Q_1 and Q_2 are tied together and the emitter voltage changes by at most 120

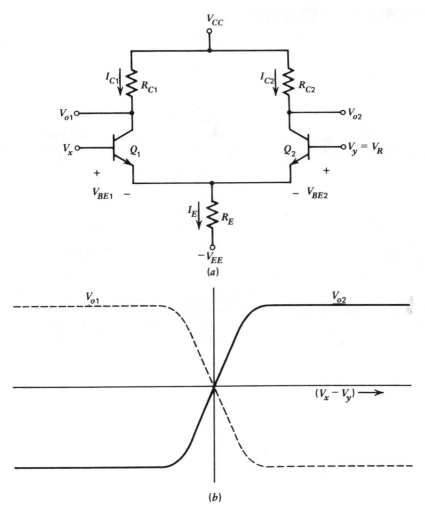

Figure 7.14. (*a*) Differential amplifier. (*b*) Voltage transfer curve for differential amplifier.

mV, the total current through R_E remains essentially constant, and virtually all of the current is now in Q_1 and very little is in Q_2.

Conversely, if V_x is less than V_y by approximately 120 mV, then Q_1 is off and Q_2 is on. This difference amplifier, when operated as a digital-type device, is basically a current switch. It switches an approximately constant current through R_E from Q_2 to Q_1 as V_x changes from less than V_y to greater than V_y.

The approximate transfer characteristics for output voltages V_{o1} and V_{o2} are shown in Figure 7.14*b*. As Q_1 turns on, all of the current in R_E is directed through Q_1 and the output voltage V_{o1} decreases because of the $I_{C1}R_{C1}$ voltage drop. At the same time, I_{C2} decreases and the output voltage V_{o2} increases because of the reduced $I_{C2}R_{C2}$ voltage drop.

7.3.2 **The Basic ECL Gate**

A two-input OR/NOR circuit is shown in Figure 7.15 with the two input transistors Q_1 and Q_3 connected in parallel. If both V_x and V_y are less than V_R, then both Q_1 and Q_3 are off while Q_2 is on. In this condition, the output voltage V_{o1} is greater than V_{o2}. If either V_x or V_y becomes greater than V_R Q_2 turns off causing V_{o2} to become greater than V_{o1}. The OR logic is at the V_{o2} output, and the NOR logic is at the V_{o1} output. The availability of complementary outputs is an advantage of ECL gates and avoids the necessity of including separate inverters to provide these complementary outputs.

One problem with the OR/NOR circuit in Figure 7.15 is that the output voltage levels differ from the required input voltage level. This voltage mismatch arises since the ECL transistors operate between their cutoff and their active regions, requiring that the base–collector junctions be reverse biased. Note that a high-output voltage V_{oH} for the circuit of Figure 7.15 has a value of V_{CC}. If this voltage level is applied to either the V_x or V_y input, Q_3 or Q_1 will turn on and the collector voltage V_{o1} will drop. The base–collector voltage will then become positive for the conducting transistor and saturation occurs. Emitter follower circuits are added to provide outputs that are compatible with the inputs of similar-type gates.

The ECL circuit shown in Figure 7.16 has emitter followers added to the

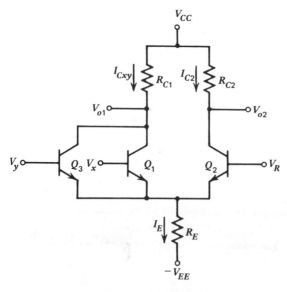

Figure 7.15. A two-input OR/NOR ECL gate.

OR/NOR outputs and a supply voltage V_{CC} that is zero. The ground and power supply voltages are reversed in ECL circuits because analysis shows that placing the ground nearer the collectors results in less noise sensitivity. If the forward active current gain of the transistors is on the order or 100, then the dc base currents can be neglected compared to the collector and emitter currents, with little error in the calculations.

If either V_x or V_y is a logic 1 level, Q_2 is off and $I_{C2} = 0$. Very little current then exists in R_{C2}, making $V_{o2} \simeq 0$. The output V_{OR} is a base–emitter voltage drop lower than V_{o2}, or $V_{OR} \simeq -0.70$ V = logic 1. If V_x = logic 1 = -0.70 V, then Q_1 is on, and

$$V_E = V_x - V_{BE1}$$
$$= -0.70 - 0.70$$
$$= -1.40 \text{ V}$$

Then,

$$I_E = \frac{V_E - (-V_{EE})}{R_E}$$
$$= \frac{-1.40 + 5.20}{1.18 \times 10^3} \tag{7.17}$$
$$= 3.22 \text{ mA}$$

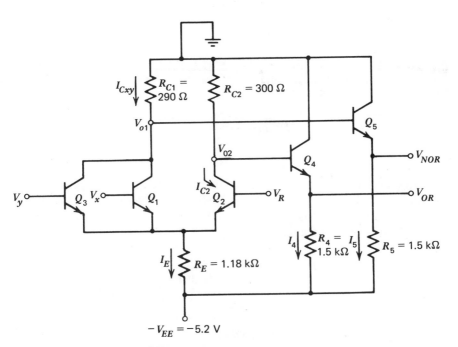

Figure 7.16. A two-input OR/NOR ECL gate with emitter-follower output stages.

The current $I_{Cxy} = I_E$, since Q_1 is on and Q_2 is off. In this case,

$$V_{o1} = 0 - I_{C1}R_{C1}$$
$$= 0 - (3.22 \times 0.29) \tag{7.18}$$
$$= -0.93 \text{ V}$$

and

$$V_{NOR} = -0.93 - 0.70 = -1.63 \text{ V} = \text{logic } 0$$

The current I_4 is

$$I_4 = \frac{V_{E4} - (-V_{EE})}{R_4}$$
$$= \frac{-0.70 - (-5.2)}{1.5 \times 10^3} \tag{7.19}$$
$$= 3.0 \text{ mA}$$

and

$$I_5 = \frac{V_{E5} - (-V_{EE})}{R_5}$$
$$= \frac{-1.63 - (-5.2)}{1.5 \times 10^3} \tag{7.20}$$
$$= 2.38 \text{ mA}$$

The input voltages V_x and V_y are greater than V_R when in a logic 1 state and less than V_R when in a logic 0 state. If V_R is set at the midpoint between the logic 0 and logic 1 levels, then

$$V_R = \frac{-0.70 - 1.63}{2}$$
$$= -1.17 \text{ V}$$

Another circuit is required to provide this reference voltage. A complete two-input OR/NOR ECL logic circuit is shown in Figure 7.17. The reference circuit consists of the resistors R_1, R_2, and R_3, diodes D_1 and D_2, and transistor Q_6. Neglecting base currents,

$$I_1 = I_2 = \frac{(0 - 2V_D) - (-V_{EE})}{R_1 + R_2}$$
$$= \frac{-1.40 + 5.20}{300 + (2.14 \times 10^3)} \tag{7.21}$$
$$= 1.56 \text{ mA}$$

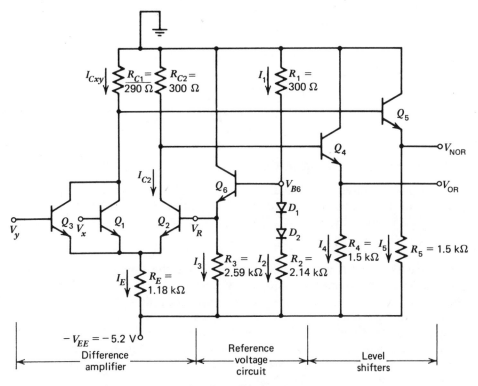

Figure 7.17. A typical ECL gate in IC applications.

Then,

$$V_{B6} = -I_1 R_1$$
$$= -1.56 \times 10^{-3} \times 300 \qquad (7.22)$$
$$= -0.47 \text{ V}$$

and

$$V_R = V_{B6} - V_{BE6}(\text{act})$$
$$= -0.47 - 0.70 \qquad (7.23)$$
$$= -1.17 \text{ V}$$

giving the required reference voltage. The diodes D_1 and D_2 provide temperature compensation for the base–emitter junction of Q_6. Good temperature compensation requires that the current through the emitter of Q_6 be the same as the current through diodes D_1 and D_2. The resistor value R_3 is found from the relation

$$I_3 = I_2 = 1.56 \text{ mA} = \frac{V_R - (-V_{EE})}{R_3} \qquad (7.24)$$

or

$$R_3 = \frac{-1.17 + 5.20}{1.56 \times 10^{-3}}$$

$$= 2.59 \text{ k}\Omega$$

7.3.3 **Power Dissipation**

An important characteristic of a logic circuit is the power dissipated in the circuit. The power dissipated in the unloaded ECL OR/NOR circuit of Figure 7.17 is given by

$$P_D = (I_E + I_2 + I_3 + I_4 + I_5)V_{EE} \qquad (7.25)$$

The currents I_2 and I_3 in the reference circuit and the supply voltage V_{EE} are independent of the input logic states of V_x and V_y. The sum of the currents in the output emitter followers, $I_4 + I_5$, changes very little as the input voltages go from one logic state to the other. When the input voltages are a logic 1, $I_{Cxy} = I_E$ and $I_{C2} = 0$, while, in the opposite state, $I_{C2} = I_E$ and $I_{Cxy} = 0$. The current I_E changes very little as the input voltages change from one state to another, and therefore little power dissipation differences exist between the high and low state of the ECL circuit.

When $V_x = V_y = $ logic 1, the reference transistor Q_2 is off; and, as previously calculated,

$$I_E = 3.22 \text{ mA}$$

Since Q_2 is off, $I_{C2} = 0$, and the power dissipation in the ECL circuit is

$$P_D = (3.22 + 1.56 + 1.56 + 3.0 + 2.38) \times 10^{-3} \times 5.2$$

$$= 60.9 \text{ mW}$$

When the input voltages change to a logic 0 state, the power dissipated in the ECL circuit is 59.1 mW due to the slight decrease in I_E. This power level is considerably higher than the 2.0 mW dissipation of the advanced Schottky TTL circuits or the microwatt levels of the I²L gates.

7.3.4 **Transfer Characteristics**

A good approximation to the voltage transfer characteristics is obtained from the piecewise linear model for the two input transistors and the reference transistor. The result is given in Figure 7.18. If the inputs V_x and V_y are a logic 0, or -1.63 V, then Q_1 and Q_3 are off and $V_{NOR} \simeq -0.70$ V. The reference transistor Q_2 is on, and

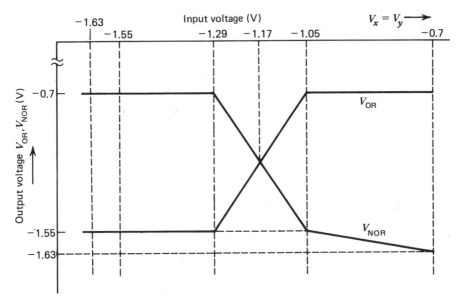

Figure 7.18. Voltage transfer curves for the OR and NOR outputs of the ECL logic gate of Figure 7.17

$$I_E = \frac{V_R - V_{BE}(\text{act}) - (-V_{EE})}{R_E}$$

$$= \frac{-1.17 - 0.70 + 5.20}{1.18 \times 10^3} \qquad (7.26)$$

$$= 2.83 \text{ mA} = I_{C2}$$

then,

$$V_{B4} = -I_{C2}R_{C2}$$

$$= -2.83 \times 10^3 \times 300 \qquad (7.27)$$

$$= -0.85 \text{ V}$$

and

$$V_{OR} = V_{B4} - V_{BE}(\text{act})$$

$$= -0.85 - 0.70 \qquad (7.28)$$

$$= -1.55 \text{ V}$$

If V_y remains in a logic 0 state, the outputs do not change as long as the V_x input is less than $V_R - 0.12 = -1.29$ V. When V_x becomes greater than -1.29 V, V_x is within 120 mV of V_R, causing Q_1 to begin conduction. During the interval when V_x is within 120 mV of V_R, the output voltage levels are changing.

When $V_x = V_R + 0.12 = -1.05$ V, Q_1 is in the active region of operation, and Q_2 is off. At this point,

$$
\begin{aligned}
I_E &= \frac{V_x - V_{BE}(\text{act}) - (-V_{EE})}{R_E} \\
&= \frac{-1.05 - 0.70 + 5.20}{1.18 \times 10^3} \\
&= 2.93 \text{ mA} = I_{C1}
\end{aligned}
\tag{7.29}
$$

and

$$
\begin{aligned}
V_{B5} &= -I_{C1}R_{C1} \\
&= -2.93 \times 10^3 \times 290 \\
&= -0.85 \text{ V}
\end{aligned}
\tag{7.30}
$$

Then,

$$
\begin{aligned}
V_{NOR} &= -0.85 - 0.70 \\
&= -1.55 \text{ V}
\end{aligned}
\tag{7.31}
$$

and

$$
V_{OR} \simeq -0.70 \text{ V}
\tag{7.32}
$$

As V_x increases to the logic 1 level of -0.70 V, the voltage V_E and the current I_E change slightly so that V_{NOR} also changes slightly. However, since Q_2 remains off, V_{OR} remains constant at -0.70 V. As shown previously, when $V_x = -0.70$ V, $V_{NOR} = -1.63$ V. The resultant transfer curves are given in Figure 7.18.

7.3.5 Noise Margin

A simplified idea of the noise margin can be determined from these transfer characteristics. The low-state noise margin of the V_{OR} output is determined as

$$
\begin{aligned}
NM_L &= -1.29 - (-1.55) \\
&= 0.26 \text{ V}
\end{aligned}
\tag{7.33}
$$

where the -1.55 V is the input logic state of a low output OR terminal and the -1.29 V is the input voltage at which the output begins to change. Similarly, the high-state noise margin is given by

$$
\begin{aligned}
NM_H &= -0.70 - (-1.05) \\
&= 0.35 \text{ V}
\end{aligned}
\tag{7.34}
$$

These noise margin values for the ECL circuit are considerably less than those calculated for the TTL circuits in the previous chapter.

7.3.6 **Propagation Delay Time**

The major advantage of ECL circuits over saturating transistor logic circuits is reduced propagation delay time. The transistors are not driven into saturation, thus eliminating the charge storage effect that dominates the switching time of TTL and I^2L circuits. The elimination of the charge storage term gives a much reduced propagation delay time. Nominally, the ECL OR/NOR circuits have propagation delay times on the order of 1 ns. However, the tradeoff for the reduced propagation delay time is higher power dissipation and reduced noise margin.

The ECL circuits are very fast and, as with advanced Schottky TTL, require special attention to the problems of transmission line effects. Carelessly designed ECL circuit boards lead to ringing or oscillations in the waveforms. These problems have less to do with the ECL circuits than with the interconnections between circuits. Care must be taken to properly terminate signal lines. Manufacturers usually supply applications literature to assist the designer with these problems.

7.3.7 **Fan-Out**

Figure 7.19 shows the emitter follower output stage of an ECL circuit driving a difference amplifier input stage of an ECL load circuit. When V_{OR} is at a logic 0 level, the input load transistor Q_1' is cut off, eliminating a load current from the driver output stage. With V_{OR} at a logic 1 level, the input load

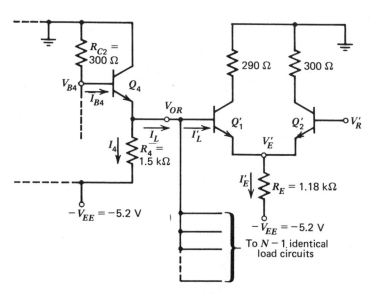

Figure 7.19. Circuit illustrating coupling between ECL gate driver transistor and ECL gate load circuits.

transistor Q_1' is on, and input base currents I_L' will exist. The load current I_L will decrease the output voltage, so that the fan-out is then determined by the maximum amplitude that the output voltage is allowed to change.

To calculate fan-out, suppose that V_{OR} is allowed to decrease, at most, by 50 mV from -0.70 V to -0.75 V. Assume that the β of all transistors is 50. Then,

$$I_E' = \frac{V_{OR} - V_{BE}(\text{act}) - (-V_{EE})}{R_E}$$

$$= \frac{-0.75 - 0.70 + 5.20}{1.18 \times 10^3} \tag{7.35}$$

$$= 3.18 \text{ mA}$$

The input base current is

$$I_L' = \frac{I_E'}{1 + \beta}$$

$$= \frac{3.18 \times 10^{-3}}{51} \tag{7.36}$$

$$= 62.3 \ \mu A$$

and the load current is

$$I_L = N I_L' \tag{7.37}$$

The base current I_{B4} required to achieve both the load current I_L and the current I_4 is

$$I_{B4} = \frac{I_4 + I_L}{1 + \beta} = \frac{0 - V_{B4}}{R_{C2}}$$

$$= \frac{0 - [V_{OR} + V_{BE}(\text{act})]}{R_{C2}} \tag{7.38}$$

or

$$\frac{I_4 + N I_L'}{1 + \beta} = \frac{N(62.3 \times 10^{-6})}{51}$$

$$= \frac{0.75 - 0.70}{300} \tag{7.39}$$

and

$$I_4 = \frac{V_{OR} - (-V_{EE})}{R_4}$$

$$= \frac{5.2 - 0.75}{1.5 \times 10^3}$$

$$= 2.97 \text{ mA}$$

resulting in

$$N = 88$$

Fan-out, then, is not a problem if the only limitation is the availability of drive current. The primary advantage of ECL circuits is short propagation delay, which is adversely affected by a large fan-out, since each load circuit increases the load capacitance by approximately 3 pF. An "ac fan-out," of about 15, is usually recommended to keep the propagation delay time within specified limits.

7.3.8 The Negative Supply Voltage

A common practice in ECL circuits is to ground the positive terminal of the supply voltage, resulting in reduced noise signals at the output terminal. An emitter follower output stage is shown in Figure 7.20 with the supply voltage V_{CC} in series with a "noise" source V_n. The voltage V_o is taken as the output voltage if the positive terminal of V_{CC} is grounded. If the negative terminal of V_{CC} were at ground, then V'_o would be the output voltage. To calculate the effect of the "noise" voltage on the output, assume that Q_2 is cut off, $h_{ie4} = 1.3$ kΩ and $h_{fe} = 150$. Then,

$$V_n = 300I_{b4} + h_{ie4}I_{b4} + 1.5 \times 10^3 \, I_4 \tag{7.40}$$

and

$$I_4 = (1 + h_{fe})I_{b4} \tag{7.41}$$

Solving for I_{B4} yields

$$I_{b4} = \frac{V_n}{300 + h_{ie4} + 151 \times 1.5 \times 10^3} \tag{7.42}$$

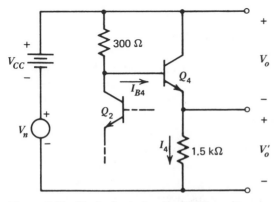

Figure 7.20. Equivalent circuit of ECL emitter-follower output stage and noise generator V_n.

Since

$$V_o = 300 I_{b4} + h_{ie4} I_{b4} \tag{7.43}$$

then

$$V_o \simeq 0.007 V_n \tag{7.44}$$

The magnitude of V_o' can be found by first solving for I_4 in Equation 7.40 as

$$I_4 = \frac{V_n}{\dfrac{300 + h_{ie4}}{151} + 1.5 \times 10^3} \tag{7.45}$$

Since

$$V_o' = I_4 (1.5 \times 10^3) \tag{7.46}$$

then,

$$V_o' = 0.98 V_n \tag{7.47}$$

Equations 7.44 and 7.47 show that the effect of V_n on the output voltage V_o is much less than the effect of V_n on the output voltage V_o'. The advantage is to use V_o, which implies that the positive terminal of V_{CC} is grounded. The noise insensitivity gained with a negative power supply is critical in a logic circuit having a noise margin as low as that in the ECL circuit.

7.4 MODIFIED ECL CIRCUIT CONFIGURATION

The circuit shown in Figure 7.17 is representative of a commercial ECL OR/NOR gate. Several modifications to this gate exist that are designed to improve stability and allow more efficient interface connections. In addition, different internal rearrangements may be made in the ECL gate to provide a diversity in logic performance other than the standard OR/NOR outputs.

Figure 7.21 shows a simplified ECL OR/NOR gate that has been modified by a current source stage Q_4, open emitter outputs, and base leak resistors R_P. The current source stabilizes the output voltage against changes in the power supply, and the open emitter output configuration provides flexibility to the designer when driving transmission lines directly. The internal pull-down resistors normally present in Q_5 and Q_6 increase the power dissipation in the circuit and may cause ringing when very high-speed ECL circuits drive a transmission line. The base leak resistors are about 50 kΩ and serve to drain off base charge from the input transistors. This is critical if there are unused input transistors on the ECL gate and charge builds on the base from the reverse bias saturation current of the base–collector junction. The leak resistors eliminate the necessity to tie unused inputs to the power supply, $-V_{EE}$.

Transistor Q_4 in Figure 7.21 operates in the forward active mode using a bias voltage $-V_{BB}$ that is stabilized with respect to temperature. The major

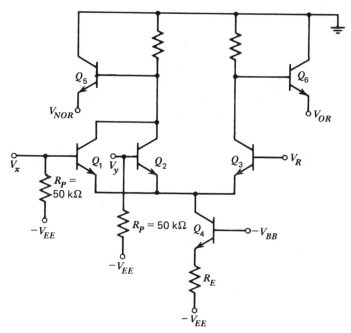

Figure 7.21. Simplified ECL OR/NOR gate with current
source Q_3, open emitter outputs, and base leak resistors R_P.

advantage of Q_4 over a single resistor lies in the stability imparted to I_{C4} and
the output voltage against fluctuations in $-V_{EE}$. The bias voltage $-V_{BB}$ is
derived from a series network of resistors and diodes similar to the reference
voltage subcircuit of Figure 7.17. Consequently, for rises or drops in $-V_{BB}$,
the voltage drop across V_{BE4} and R_E does not change appreciably, and then
neither does I_{C4}. This regulation of I_{C4} has allowed the low-output voltages in
commercial circuits to vary by as little as 20 mV for 1-V changes in $-V_{EE}$.
The standard ECL OR/NOR logic circuit of Figure 7.21 may be modified to
accomplish other logic functions. There are three configurations, called series
gating, collector dotting, and wired-OR, that provide NAND, AND, and OR
functions, respectively.

Figure 7.22 shows a simplified ECL circuit that uses series gating collector
dotting to achieve the NAND and AND functions. Series gating refers to a
configuration that stacks transistors such as Q_1 and Q_3 in Figure 7.22 in a
current switch tree network. The NAND function is generated at the collector
of Q_1. Only if both A and B are high does current exist in R_1 and thereby drop
the collector voltage to a low state.

Collector dotting is an ECL logic technique that refers to the connection
of collector nodes separated in a vertical stack. Transistors Q_2 and Q_4 are
collector dotted and perform the AND function on inputs A and B. Transistors
Q_2 and Q_4 share the collector resistor R_{C2}. This situation means that their

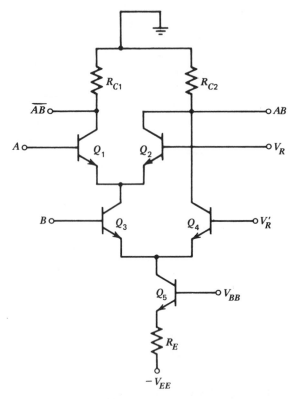

Figure 7.22. Simplified ECL gate illustrating series gating (Q_1 and Q_3) and collector dotting (Q_2 and Q_4) to achieve the NAND and AND functions.

common collector node can only be high if both Q_2 and Q_4 are off. This situation occurs only when both Q_1 and Q_3 are high, and for no other condition.

Figure 7.23 illustrates the complete circuit of a Motorola ECL gate that uses series gating to perform multiple NAND operations. The current source transistor is Q_1, and the bias voltage to this transistor is generated by Q_8 and the series resistor diode path connected to the base of Q_8. The input voltage level to the Q_2/Q_3 switch must be lowered in order to adequately control the logic current in Q_2 and Q_3. Transistor Q_0 shifts the logic voltage level through its base–emitter voltage drop, a diode, and a resistor. The emitter follower buffer transistors are open emitters. The reference voltage to Q_5 and Q_7 is obtained from Q_9 and the resistor diode network connected to the Q_9 base.

Figure 7.24 shows the complete circuit of a Motorola ECL gate that uses collector dotting to perform the AND function on pairs of two inputs. The collector dotting occurs between transistors Q_1 and Q_2 and provides the AND function of $(A + B) \cdot (C + D)$. The OR function is accomplished by the parallel

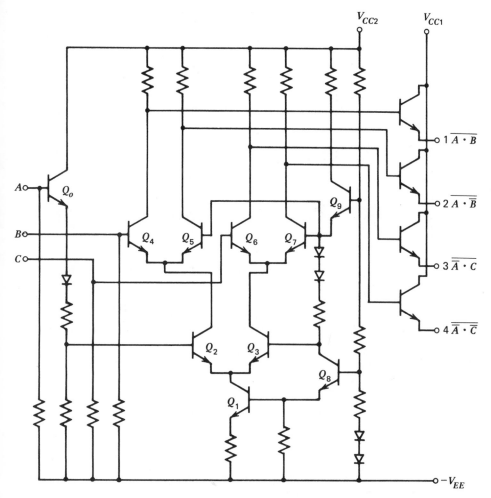

Figure 7.23. An ECL gate that uses the series gating technique to perform NAND functions on the three inputs A, B, and C.

input transistors whose bases are connected to A and B, and C and D. Bias resistors at the emitters of Q_1 and Q_2 provide the current to the switches. The reference circuit and base leak resistors have been described previously.

Figure 7.25 shows the output stages of two ECL circuits connected at their output transistor emitters. This common connection is tied to $-V_{EE}$ through a resistor R_{OR}. This use of the output nodes is referred to as wired-OR, since an OR-ing of the outputs occurs. The output voltage is high if either or both Q_3 or Q_3' are at a logic high. If Q_3 is high and Q_3' is low, then the output voltage is high and Q_3' is turned off. The output level is low only when both Q_3 and

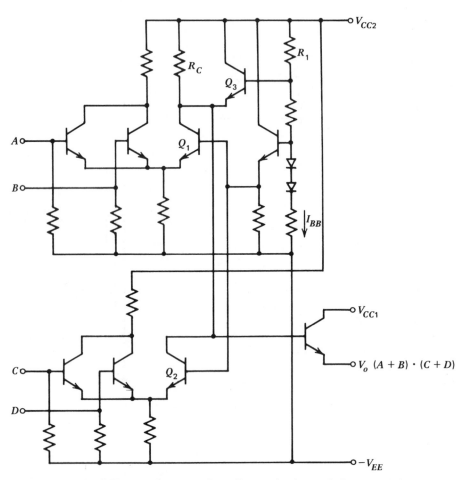

Figure 7.24. An ECL gate that uses the collector dotting technique to perform AND functions on the multiple inputs.

Q_3' are low. The wired-OR configuration is immensely useful to a designer because of the simplicity in implementing the OR function.

7.5 **SUMMARY**

Two bipolar logic technologies were analyzed in this chapter. The I²L and ECL circuits, along with TTL, comprise the three bipolar transistor approaches to logic circuit construction. The I²L technique is distinctive for high packing density on an LSI chip, low power dissipation per gate, and an ability to trade off the propagation delay and power dissipation of the gate. Among the three bipolar logic families, it is the most suitable for LSI.

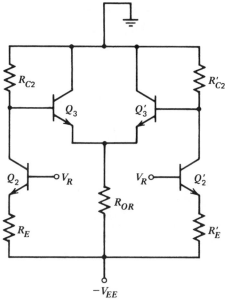

Figure 7.25. The outputs of two ECL gates connected in a wired-OR configuration.

ECL is the fastest of the bipolar logic circuits but dissipates the most power. The ECL technology finds primary applications in those areas of a computer that require many repetitive operations. The low noise margins of ECL demand close attention to noise protection.

The TTL technology discussed in Chapter 6 finds the largest use in SSI and MSI applications. The TTL combines good noise margins with high current sourcing and sink capabilities. The propagation delay and power dissipations have improved such that in SSI and MSI applications TTL would be preferred over ECL or I²L. The TTL circuits will probably not compete in LSI applications because of the large area required on a chip to implement a logic function and the relatively high power dissipation. The TTL circuits however do appear as interface circuits in LSI circuits where TTL noise margins and load interface capability are superior.

REFERENCES

1. *MECL Systems Design Handbook* (Phoenix, Ariz.: Motorola Semiconductor Products, 1980).

2. *ECL Data Book* (Mountain View, Calif.: Fairchild Semiconductor, 1977).

PROBLEMS

7.1. Calculate I_{CQ} and V_{CEQ} for the *pnp* transistor amplifier of Figure 7.26.

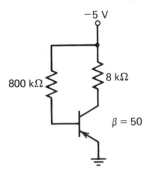

Figure 7.26. Circuit for Problem 7.1.

7.2. Calculate I_{CQ} and V_{CEQ} for the transistor amplifier of Figure 7.27.

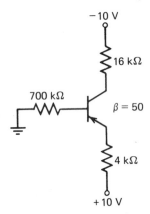

Figure 7.27. Circuit for Problem 7.2.

7.3. In Figure 7.28, calculate a value for R_E so that $I_C = 100\ \mu A$. Calculate V_{CE}.

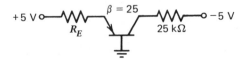

Figure 7.28. Circuit for Problem 7.3.

7.4. Show why the minimum current gain β_{min} of an I²L switching transistor must be unity to ensure that the transistor is driven into saturation.

7.5. What logic function does the I²L circuit in Figure 7.29 perform on inputs *A* and *B*? Choose an external resistor such that the current sources deliver an injector current of 200 μA ($\alpha_F = 0.80$).

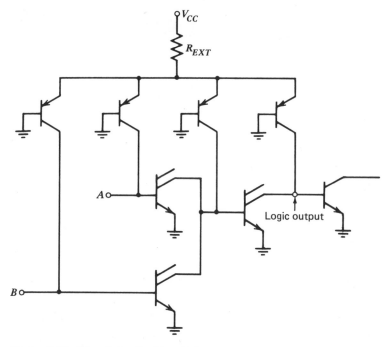

Figure 7.29. Circuit for Problem 7.5.

7.6. An I²L network consists of 1500 inverter gates as shown in Figure 7.30, with each gate having an injector current I_o of 100 nA. The collector current is 75% of the emitter current in the source transistors and V_{EB} of each source transistor is 0.80 V. Each gate has a power–delay product of 0.50 pJ. (*a*) Calculate the required external resistance R_{ext}, the total power dissipation in all 1500 gates, and the propagation delay time of a gate. (*b*) For $R_{ext} = 80 \, \Omega$, calculate the injector current I_o of a gate, the total power dissipation in all 1500 gates, and the propagation delay time of a gate.

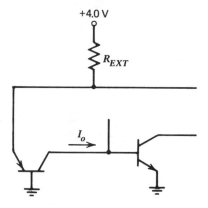

Figure 7.30. Circuit for Problem 7.6.

7.7. Calculate I_E, V_{o1}, and V_{o2} in the differential amplifier of Figure 7.31 for (a) $V_{IN} = -2.7$ V and (b) $V_{IN} = -1.3$ V. Let $V_{BE}(\text{act}) = 0.70$ V and $\beta = 150$.

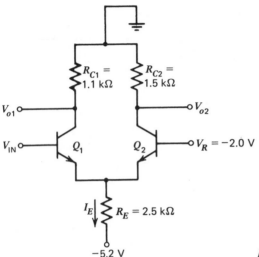

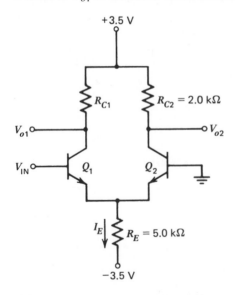

Figure 7.31. Circuit for Problem 7.7.

7.8. For the differential amplifier of Figure 7.32, let $V_{BE}(\text{act}) = 0.70$ V and neglect the base currents. (a) For $V_{IN} = -1.5$ V, calculate I_E, V_{o1}, and V_{o2}. (b) For $V_{IN} = 1.0$ V, calculate I_E and V_{o2}. Determine R_{C1} so that the logic 0 level at V_{o1} is the same value as the logic 0 value at V_{o2}.

Figure 7.32. Circuit for Problem 7.8.

7.9. Calculate the power dissipated for both logic states in the basic differential amplifier for (a) the circuit in Problem 7.7 and (b) the circuit in Problem 7.8.

7.10. Plot the voltage transfer characteristic V_{o1} versus V_{IN} and V_{o2} versus V_{IN} for $-2.7 \text{ V} \le V_{IN} \le -1.3 \text{ V}$ for the differential amplifier in problem 7.7.

7.11. Consider the ECL circuit in Figure 7.33. Assume that $V_{BE} = V_D = 0.70 \text{ V}$ and neglect the base currents. (a) Determine the reference voltage V_R and (b) determine the logic 0 and logic 1 voltage values at each output V_{o1} and V_{o2}. Assume that the inputs V_x and V_y are the same values as the logic levels at V_{o1} and V_{o2}.

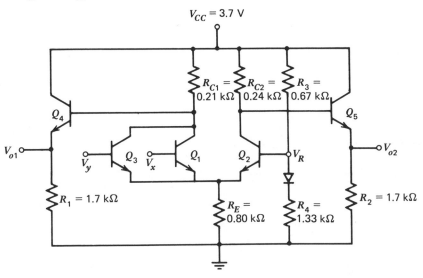

Figure 7.33. Circuit for Problem 7.11.

7.12. For the circuit in Problem 7.11, assume that the input voltages have the following values: logic 1 = 3.0 V and logic 0 = 2.4 V. Calculate the power dissipated in the circuit for (a) $V_x = V_y =$ logic 0 and (b) $V_x = V_y =$ logic 1.

7.13. For the ECL circuit in Figure 7.34, the logic swing at the output is symmetric about the ground potential. The input voltages V_x and V_y have the same logic values as V_{OUT}. The emitter current of Q_4 is 1.25 mA when V_{OUT} is a logic 1. (a) Calculate the values for R_1 and R_2, and (b) calculate the logic 1 and logic 0 values.

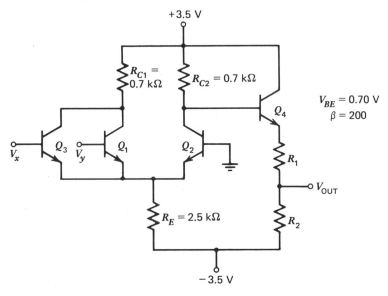

Figure 7.34. Circuit for Problem 7.13.

7.14. Consider the ECL circuit in Figure 7.35. Let $V_{BE}(\text{act}) = 0.70$ V and let the base currents be negligible. Calculate all resistor values so that the following specifications are satisfied: (1) logic $1 = 1.0$ V and logic $1 = 0.0$ V; (2) V_R is to be the average of logical 1 and 0; (3) $I_E = 1.0$ mA when Q_2 is on; (4) $I_1 = I_2 = 1.0$ mA; (5) when $V_{OR} = $ logic 1, $I_4 = 3.0$ mA; and (6) when $V_{NOR} = $ logic 0, $I_5 = 3.0$ mA.

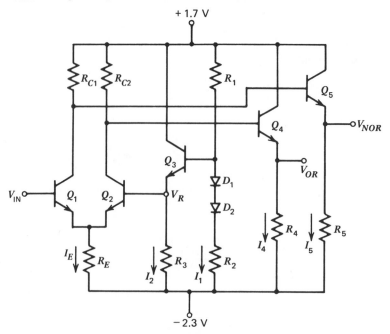

Figure 7.35. Circuit for Problem 7.14.

7.15. In the ECL circuit of Figure 7.36, the outputs have a logic swing of 0.60 V (logic 1 − logic 0 = 0.60 V) symmetric about the reference voltage. The maximum emitter current for all transistors is 5.0 mA. Assume the logic voltages at V_{IN} are compatible with the output voltage levels. Calculate the resistance values for R_{C1}, R_{C2}, R_E, R_4, and R_5.

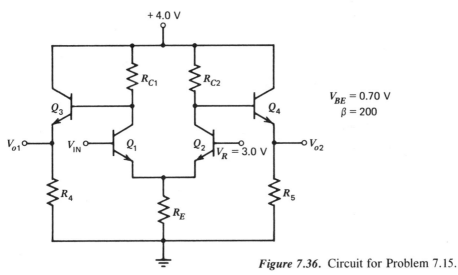

Figure 7.36. Circuit for Problem 7.15.

7.16. In the ECL circuit of Problem 7.15, let R_E = 1.2 kΩ and R_4 and R_5 = 1.5 kΩ, and the logic levels are unchanged. (*a*) Calculate the resistance values for R_{C1} and R_{C2}, and (*b*) calculate the power dissipation for V_{IN} = logic 1 and V_{IN} = logic 0.

chapter 8

FET DIGITAL LOGIC CIRCUITS

The metal oxide–semiconductor field effect transistor (MOSFET) is simpler to fabricate and occupies less space in integrated form than the bipolar transistor. The packing density (number of devices per unit area) of MOSFET devices is extremely high, and MOSFET circuits can be fabricated for LSI and VLSI applications. In addition, a MOSFET can be connected to act as a resistive load. This is an advantage in LSI circuits since transistors occupy much less area than the diffused resistors in bipolar integrated circuits. Three general types of MOSFET circuits are used. The first of these uses n-channel MOSFETs exclusively and is called NMOS. The second uses only p-channel MOSFETs and is called PMOS. The third type uses both kinds of devices and is called complementary MOS or CMOS. In this chapter, we consider the analysis of NMOS and CMOS. The analysis of PMOS is identical with that of NMOS except for voltage polarity and current direction. NMOS dominates over PMOS in commercial applications at the present time.

It is also possible to use junction FETs and metal–Schottky barrier FETs (MESFETs) in digital integrated circuits. MESFETs are usually fabricated in GaAs, where the electron mobility is higher than that in silicon, resulting in faster devices. MESFET logic circuits are fabricated for their outstanding speed capabilities.

8.1 NMOS CIRCUITS

NMOS circuits use n-channel transistors to perform essentially all of the circuit functions. In many cases, all the transistors in the circuits are enhancement mode devices, which simplifies the fabrication procedure for integrated circuits. However, advantages in circuit speed may be realized if both enhancement mode and depletion mode devices are used in the same circuit. These two types of devices can be fabricated in the same circuit using ion implantation techniques.

8.1.1 NMOS Inverters

INVERTER WITH A RESISTIVE LOAD. The basis of NMOS digital circuits is the inverter. An NMOS inverter with a load resistor is shown in Figure 8.1a. The substrate is connected to the source terminal at ground potential. The input voltage is on the gate terminal, and the output voltage is at the drain terminal.

Figure 8.1b shows the current–voltage characteristics of an n-channel MOSFET. The parametric curve dividing the saturation and nonsaturation regions of the transistor characteristics is given by

$$V_{DS}(\text{sat}) = V_{GS} - V_T \tag{8.1}$$

where $V_{DS}(\text{sat})$ is the drain-to-source saturation voltage, V_{GS} is the gate-to-source voltage, and V_T is the threshold voltage. In the saturation region, the drain current is given by

$$I_D = k(V_{GS} - V_T)^2 \tag{8.2}$$

and in the nonsaturation region, it is given by

$$I_D = k[2(V_{GS} - V_T)V_{DS} - V_{DS}^2] \tag{8.3}$$

where k is the conduction parameter of the transistor.

The load line of the inverter circuit is given by

$$V_{DS} = V_{DD} - I_D R_D \tag{8.4}$$

In this case, R_D results in a linear load line superimposed on the device characteristics shown in Figure 8.1b.

The voltage characteristic of the inverter is determined from Equations 8.2, 8.3, and 8.4. Assume that the n-channel MOSFET in Figure 8.1 is an enhancement mode device with a threshold voltage $V_T = 1.0$ V. Also assume that $k = 1.0$ mA/V^2, $R_D = 2.0$ kΩ, and $V_{DD} = 5.0$ V. For $V_x \le V_T = 1.0$ V, the transistor is cut off so that $V_{\text{OUT}} = V_{DS} = 5.0$ V. As V_x becomes slightly greater than V_T, the transistor begins to conduct and is operating in the saturation region. As in the case of the bipolar transistor, the n-channel MOSFET always operates on the load line. Combining Equations 8.2 and 8.4 gives

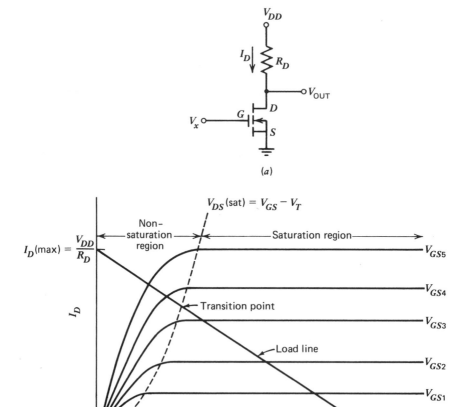

Figure 8.1. (a) NMOS inverter with a passive load resistor. (b) Circuit load line superimposed on I_D-vs.-V_{DS} family of curves.

$$V_{DS} = V_{DD} - k(V_{GS} - V_T)^2 R_D \qquad (8.5)$$

or

$$V_{OUT} = 5.0 - (1.0 \times 10^{-3})(V_x - 1.0)^2(2.0 \times 10^3) \qquad (8.6)$$

Equation 8.6 describes the voltage transfer characteristic as long as the transistor remains in the saturation region.

The transition point between the saturation and nonsaturation regions is found from Equation 8.1 and may be rewritten as

$$V_{OUT} = V_x - 1.0 \tag{8.7}$$

Equating 8.7 and 8.6 gives

$$V_x - 1.0 = 5.0 - 2.0(V_x - 1.0)^2 \tag{8.8}$$

whose solution is $V_x = 2.35$ V. Substituting this value of V_x into Equation 8.7 gives $V_{OUT} = 1.35$ V. These values of V_x and V_{OUT} are the input and output voltages at the transition point. For $V_x > 2.35$ V, the n-channel MOSFET is operating in the nonsaturation region. Combining Equations 8.4 and 8.3 yields

$$V_{OUT} = 5.0 - 2.0[2(V_x - 1.0)V_{OUT} - V_{OUT}^2] \tag{8.9}$$

This equation describes the voltage transfer characteristic for $V_x > 2.35$ V. The resulting voltage transfer characteristic is given in Figure 8.2a, and the drain current I_D versus input voltage V_x is shown in Figure 8.2b.

INVERTER WITH SATURATED LOAD. In integrated circuit fabrication, resistors consume large areas on a chip compared to transistors. For this reason, it is desirable to replace the resistor R_D in the inverter circuit of Figure 8.1 by a transistor that acts like a resistive load. Consider the enhancement mode n-channel transistor in Figure 8.3a in which the gate is connected to the drain terminal. The drain current will be zero as long as

$$V_{GS} = V_{DS} \le V_T \tag{8.10}$$

For the condition $V_{GS} = V_{DS} > V_T$, drain current will exist. Note that the following condition is always satisfied:

$$V_{DS} > (V_{GS} - V_T) = (V_{DS} - V_T) = V_{DS}(\text{sat}) \tag{8.11}$$

For this condition, the enhancement mode transistor in Figure 8.3a is always operating in the saturation region. The drain current-versus-drain voltage characteristic is then given by

$$I_D = k(V_{DS} - V_T)^2 \tag{8.12}$$

This plot is shown in Figure 8.3b and indicates that, for $V_{DS} \ge V_T$, the enhancement mode transistor with this connection acts like a nonlinear resistor.

Now consider the NMOS inverter in Figure 8.4a with the saturated transistor as a load. In integrated circuit fabrication, all substrates of the n-channel transistors are electrically connected, usually at ground potential. This makes the threshold voltage of the load transistor nonconstant, as discussed in Chapter 5. The variable threshold voltage is due to the body effect since the source voltage of the load transistor varies. However, for simplicity, we assume that the threshold voltage of the load transistor is a constant. This makes the analysis much easier without seriously affecting the results.

The current–voltage characteristics of the driver transistor are shown in Figure 8.4b with the load characteristic superimposed. For $V_x \le V_{TD}$, where V_{TD} is the threshold voltage of the driver transistor, the driver transistor is cut

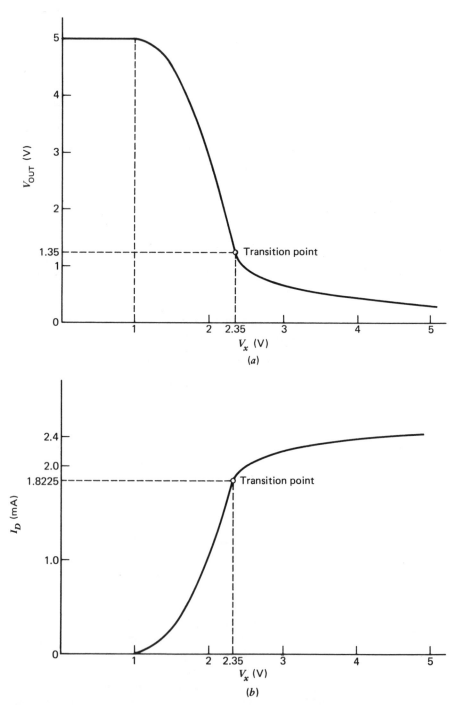

Figure 8.2. Transfer characteristics for NMOS inverter of Figure 8.1a. (a) V_{OUT} vs. V_x. (b) I_D vs. V_x.

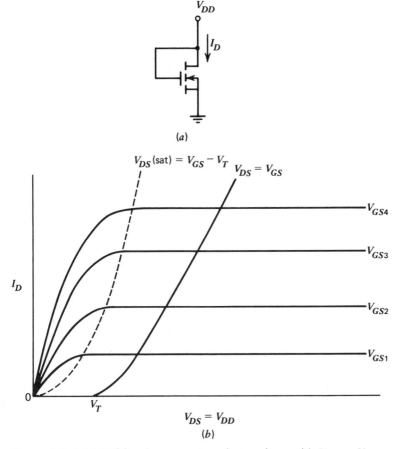

Figure 8.3. (a) NMOS enhancement mode transistor with $V_{GS} = V_{DS}$. (b) I_D vs. $V_{DS} = V_{GS}$ for the NMOS enhancement mode transistor of (a).

off so that the drain current through the two transistors must be zero. For the saturated load,

$$I_{DL} = 0 = k_L(V_{GSL} - V_{TL})^2 \tag{8.13}$$

where k_L is the conduction parameter for the load, V_{GSL} is the gate-to-source voltage of the load transistor, and V_{TL} is the threshold voltage of the load transistor. We then have

$$V_{GSL} = V_{DD} - V_{OUT} \tag{8.14}$$

and Equation 8.13 becomes

$$0 = k_L(V_{DD} - V_{OUT} - V_{TL})^2 \tag{8.15}$$

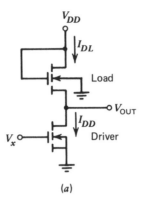

(a)

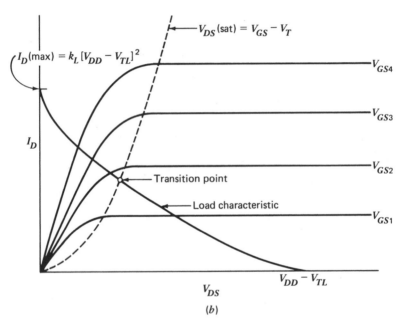

(b)

Figure 8.4. (a) NMOS inverter with saturated enhancement mode NMOS load. (b) The I_D-vs.-V_{DS} properties of the load transistor with superimposed load line.

This gives

$$V_{OUT} = V_{DD} - V_{TL} \tag{8.16}$$

for the case when $V_x \leq V_{TD}$. When a saturated load is used, the maximum output voltage is less than V_{DD} by the threshold value of the load transistor.

When V_x becomes slightly greater than V_{TD}, the driver transistor turns on and is in its saturation region of operation. The steady-state analysis of the circuit is accomplished by equating the drain currents in the two series transistors. Since any load circuit connected to V_{OUT} will be purely capacitive, there will be no steady-state output current. Then,

$$I_{DL} = I_{DD} \tag{8.17}$$

so that

$$k_L(V_{GSL} - V_{TL})^2 = k_D(V_{GSD} - V_{TD})^2 \tag{8.18}$$

Substituting the values for the corresponding gate-to-source voltages gives

$$k_L(V_{DD} - V_{OUT} - V_{TL})^2 = k_D(V_x - V_{TD})^2 \tag{8.19}$$

or

$$V_{OUT} = V_{DD} - V_{TL} - \left(\frac{k_D}{k_L}\right)^{1/2} (V_x - V_{TD}) \tag{8.20}$$

for $V_x \geq V_{TD}$. When both transistors are in saturation, the output voltage is a linear function of the input voltage.

The transition point of the driver transistor is found from

$$V_{DSD}(\text{sat}) = V_{GSD} - V_{TD} \tag{8.21}$$

where $V_{DSD}(\text{sat})$ is the drain-to-source saturation voltage of the driver transistor. Equation 8.21 many be rewritten in terms of the input and output voltages to give

$$V_{OUT} = V_x - V_{TD} \tag{8.22}$$

Equating 8.22 with 8.20 gives the input voltage at the transition point:

$$V_x = \frac{V_{DD} - V_{TL} + V_{TD}[1 + (k_D/k_L)^{1/2}]}{1 + (k_D/k_L)^{1/2}} \tag{8.23}$$

When the input voltage becomes greater than this transition point value, the driver goes into its nonsaturation region of operation so that, using Equation 8.17, we get

$$k_L(V_{GSL} - V_{TL})^2 = k_D[2(V_{GSD} - V_{TD})V_{DSD} - V_{DSD}^2] \tag{8.24}$$

Substituting the values for the corresponding load and driver terminal voltages, we obtain

$$k_L(V_{DD} - V_{OUT} - V_{TL})^2 = k_D[2(V_x - V_{TD})V_{OUT} - V_{OUT}^2] \tag{8.25}$$

which relates the input and output voltages over the remaining input voltage range.

Assuming that $V_{DD} = 5.0$ V and $V_{TD} = V_{TL} = 1.0$ V, Figure 8.5 shows plots of the voltage and current transfer characteristic of an NMOS inverter with a saturated load for various values of k_D/k_L. The curves show that, as the ratio of k_D to k_L becomes larger, a steeper transfer characteristic is obtained. This effect is usually desirable in digital logic circuits. In addition, for larger

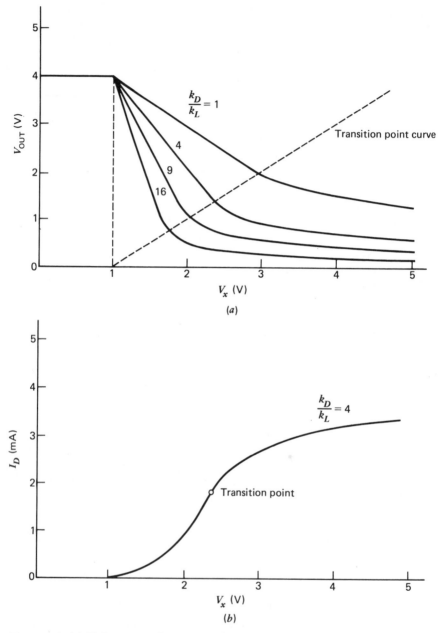

Figure 8.5. (a) Voltage transfer curves of an NMOS inverter with a saturated NMOS inverter for various k_D/k_L ratios. (b) Drain current vs input voltage for the driver transistor for $k_D/k_L = 4$.

values of k_D/k_L, the minimum output voltage that occurs at $V_x = V_{DD} = 5.0$ V becomes smaller. Note that for $k_D/k_L = 1$, $V_{OUT} = 1.17$ at $V_x = 5.0$ V. If the output of this circuit were connected to the input of a similar type of inverter, the driver device of the succeeding inverter would never be cut off, since V_{OUT} never falls below the threshold voltage. In digital logic circuits, this is unacceptable.

INVERTER WITH DEPLETION LOAD. Another popular type of NMOS inverter circuit makes use of a depletion mode transistor as the load device. Consider the depletion mode MOSFET ($V_T < 0$) in Figure 8.6(a) in which the gate is connected to the source. This connection makes $V_{GS} = 0$, and the corresponding current–voltage characteristic for the depletion mode transistor is shown in Figure 8.6b.

An NMOS inverter with depletion mode load device and an enhancement mode driver transistor is shown in Figure 8.7a. The substrate of the load transistor is again electrically connected to ground potential, which implies that

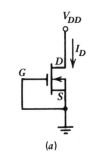

(a)

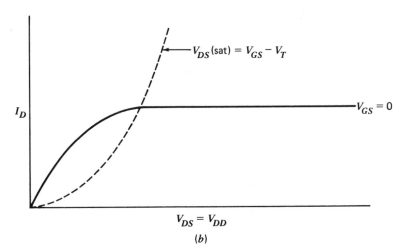

(b)

Figure 8.6. (a) Depletion mode NMOS transistor with $V_{GS} = 0$. (b) I_D-vs.-V_{DS} transfer curve for $V_{GS} = 0$.

the body effect makes the threshold voltage of the load transistor nonconstant. However, for simplicity, we assume a constant threshold voltage for the load transistor in the analysis. For $V_x \leq V_{TD}$, the driver transistor is cut off, and no drain current exists. Since the depletion load device always operates on the curve in Figure 8.6b, the drain-to-source voltage of this load transistor must be zero to give zero drain current. Hence, when $V_x \leq V_{TD}$,

$$V_{OUT} = V_{DD} \qquad (8.26)$$

This result is a major difference between the saturated load inverter and the depletion load inverter. For the depletion load inverter, there is no threshold voltage drop between V_{DD} and V_{OUT} when V_{OUT} is a logic 1.

When V_x becomes slightly larger than V_{TD}, the drain current begins. The driver is again in the saturation region, and the load is in the nonsaturation region as determined from Equation 8.1. The drain currents in the load and driver transistors are again equal, so that

$$I_{DL} = I_{DD} \qquad (8.17)$$

or

$$k_L[2(V_{GSL} - V_{TL})V_{DSL} - V_{DSL}^2] = k_D(V_{GSD} - V_{TD})^2 \qquad (8.27)$$

Substituting the corresponding terminal voltage symbols, we obtain

$$k_L[2(-V_{TL})(V_{DD} - V_{OUT}) - (V_{DD} - V_{OUT})^2] = k_D(V_x - V_{TD})^2 \quad (8.28)$$

This equation relates the input and output voltages as long as the driver is in the saturation region and the load is in the nonsaturation region. There are two transition points for the inverter with a depletion load: one for the driver and one for the load. The transition point from the nonsaturated to the saturated region for the load is given by Equation 8.1 as

$$V_{DSL} = V_{DD} - V_{OUT} = V_{GSL} - V_{TL} = -V_{TL} \qquad (8.29)$$

or

$$V_{OUT} = V_{DD} + V_{TL} \qquad (8.30)$$

The transition point for the driver from the saturation to the nonsaturation region is given by

$$V_{DSD} = V_{OUT} = V_{GSD} - V_{TD} = V_x - V_{TD} \qquad (8.31)$$

Figure 8.8 shows the threshold point locus for both the load and driver transistors assuming that $V_{DD} = 5.0$ V, $V_{TD} = 1.0$ V, and $V_{TL} = -2.0$ V. This figure will aid in determining which transistor reaches its transition point first.

In the region where both the driver and load transistors are in the saturation region, we have

$$\begin{aligned} I_{DL} &= k_L(V_{GSL} - V_{TL})^2 \\ &= k_L[0.0 - (-2.0)]^2 = 4.0k_L \end{aligned} \qquad (8.32)$$

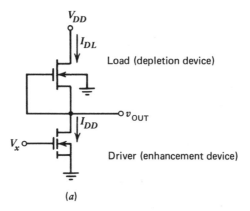

(a)

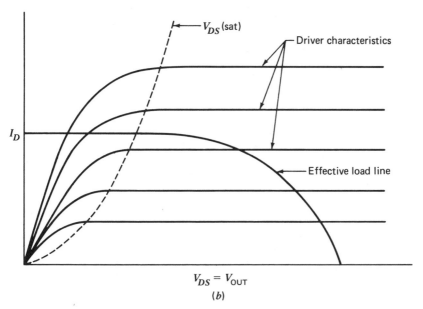

$V_{DS} = V_{OUT}$

(b)

Figure 8.7. (*a*) NMOS inverter with depletion mode load transistor. (*b*) I_D-vs.-V_{DS} transfer curve of load transistor.

When the load device is in the saturation region, the depletion mode device provides a constant current through the inverter. When both devices are in saturation we have

$$I_{DL} = I_{DD} \tag{8.17}$$

or

$$4.0k_L = k_D(V_{GSD} - V_{TD})^2 \tag{8.33}$$

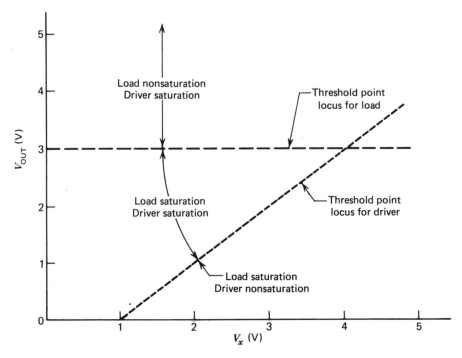

Figure 8.8. Threshold point locus for both driver and load transistors of Figure 8.7a.

which becomes

$$4.0k_L = k_D(V_x - V_{TD})^2 \tag{8.34}$$

This equation gives the input voltage when both transistors are in the saturation region.

Finally, the driver transistor goes into its nonsaturation region, and the input and output voltages are again related through Equation 8.17 as

$$4.0k_L = k_D[2(V_x - V_{TD})V_{OUT} - V_{OUT}^2] \tag{8.35}$$

The voltage transfer characteristics are plotted in Figure 8.9a for several values of the ratio k_D/k_L using the same transistor constants as in Figure 8.8. The voltage transfer curves show a progressively narrower transition region as the ratio of k_D/k_L becomes larger. This result is similar to the case of the saturated load. The dependence of the voltage transfer curve on the ratio of k_D/k_L has led to the term *ratio inverter* for the circuits of Figures 8.4 and 8.7. Figure 8.9b is a plot of the drain current versus input voltage for $k_D/k_L = 4$. As seen from the curve, the depletion load provides a constant current once the depletion mode transistor is in the saturation region.

Figure 8.10 shows the load lines associated with the three basic types of

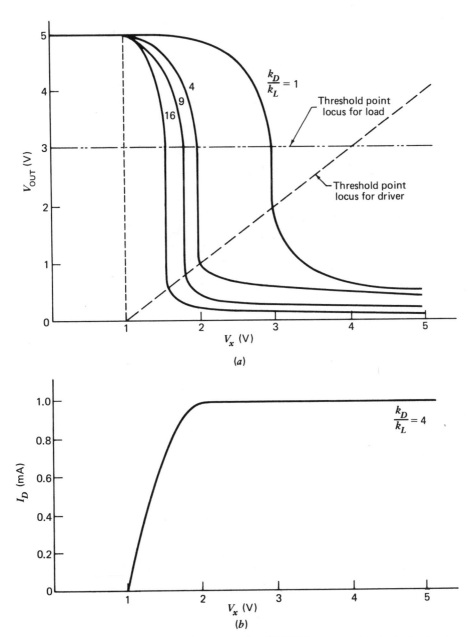

Figure 8.9. (a) V_{OUT}-vs.-V_x transfer curves for NMOS inverter with depletion load for various k_D/k_L ratios. (b) I_D vs. V_x for $k_D/k_L = 4$.

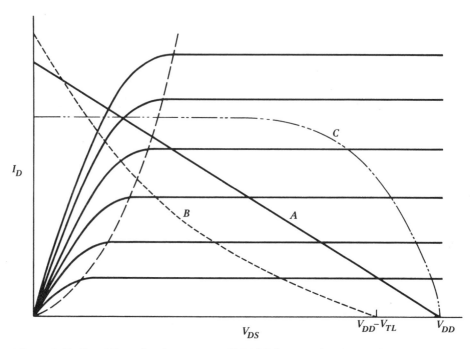

I_D

V_{DS}

$V_{DD}\text{-}V_{TL}$ V_{DD}

Figure 8.10. Load lines for three types of NMOS inverter loads superimposed on characteristic curves of NMOS inverter-driver transistor. *A*, resistor load. *B*, saturated load. *C*, depletion load.

load components superimposed on the drain characteristics of a driver transistor. The constant current over a wide range of V_{DS} provided by the depletion load implies that this type of inverter will switch a capacitive load more rapidly than the other two inverter configurations. These curves are the result of assuming constant threshold voltages for the load transistors. The load lines for the saturated load and depletion load inverters change somewhat in actual inverters fabricated as integrated circuits. This change is due to the body effect, since the source terminals of the load devices are all connected to the substrate. The body effect decreases the current drive for the depletion mode load device at high voltages.

8.1.2 **NMOS Transmission Gate**

Another versatile type of circuit frequently used in digital circuits is the NMOS transmission gate. The *n*-channel enhancement mode transistor shown in Figure 8.11 is a transmission gate connected to an effective load capacitance C_L. In this circuit, the *n*-channel transistor is assumed to be completely bilateral. The substrate voltage V_{SUB} is connected to the most negative potential in the circuit rather than to the source terminal. The two terminals are considered to

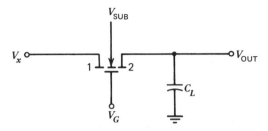

Figure 8.11. NMOS transistor transmission gate.

be completely equivalent. Assume that V_G and V_x operate between 0.0 and 5.0 V and that the threshold voltage V_T is constant at 1.0 V. Several cases are of interest. For this example, we assume that $V_{SUB} = 0.0\ V$.

Case I. Let $V_G = 0.0$ V. For this case, the gate is never positive with respect to either terminal so that the transistor is always cut off. The input and output terminals are then isolated from each other.

Case II. Let $V_G = 5.0$ V and $V_x = 0.0$ V. The voltage from the gate to terminal 1 is 5.0 V, so that the transistor is turned on. No current will exist in the steady state because of the capacitor. The only way there can be zero current in this transistor is to have a zero drain-to-source voltage. Therefore, in this case, V_{OUT} will be 0.0 V.

Case III. Let $V_G = 5.0$ V, $V_x = 5.0$ V, and, initially, $V_{OUT} = 0.0$ V. In this case, terminal 1 acts as the drain, terminal 2 acts as the source, and the current direction is from terminal 1 to terminal 2. The current charges the load capacitance until the gate-to-terminal 2 voltage is equal to the threshold voltage. That is,

$$V_G - V_{OUT} = V_T$$

or

$$5.0 - V_{OUT} = 1.0$$

and

$$V_{OUT} = 4.0\ V$$

Thus, under these conditions, the maximum value V_{OUT} will reach is 4.0 V. When V_G goes to zero, the transistor is cut off, and the input and output terminals are isolated. The input voltage V_x may then take on any value without affecting the output voltage.

Case IV. Let $V_G = 5.0$ V, $V_x = 0.0$ V, and, initially, $V_{OUT} = 4.0$ V. In this case, terminal 2 acts as the drain, terminal 1 acts as the source, and the current direction is now from terminal 2 to terminal 1. The current discharges the capacitance. The transistor will always be on so that in the steady state, V_{OUT} will discharge to $V_{OUT} = 0.0$ V.

The transmission gate has three major applications. First, it will perform an analog switching function. The input voltage of Figure 8.11 may have only segments of its waveform passed to the output. The voltage V_G determines by its pulse width and time of occurrence what segment of the input voltage is passed to the output. For example, a gated sine wave burst may be desired to drive an ultrasonic crystal. The number of sine wave periods and the period between bursts is controlled by the pulse width and period of V_G.

A second major application of transmission gates lies in the sample and hold circuits necessary to convert analog signals to their digital equivalents. An analog-to-digital (A/D) converter requires a small amount of time to "read" the analog voltage, and the analog voltage must be held constant during the read function. The transmission gate accomplishes this function by sampling the input analog wave in a series of short, gated time periods through V_G. Each

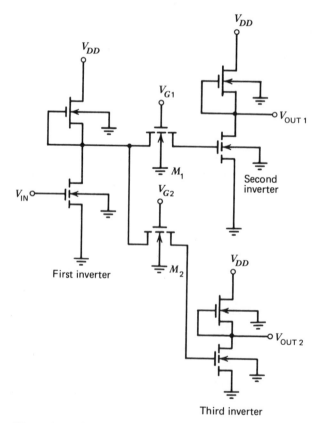

Figure 8.12. Steering logic using transistors M_1 and M_2 to control logic between inverter stages.

individual analog sample is held briefly at the output when V_G goes to zero. This analog sample at the output of the transmission gate is then read by the A/D converter.

A third application of the transmission gate lies in the implementation of steering logic using pass transistors. Figure 8.12 shows depletion load inverter stages interconnected by transmission gates M_1 and M_2. The transmission gates are called pass transistors in these logic functions. The logic output of the first inverter is steered by the voltage levels on the gates of M_1 and M_2. If $V_{G1} = V_{DD}$, the transistor M_1 is on and provides a conducting path to the second inverter. If $V_{G2} = V_{DD}$, the transistor M_2 is on and provides a conducting path to the third inverter. Logic-low signals applied to the gates of M_1 and M_2 turn off the pass transistors and inhibit the flow of information between the inverters. The pass transistors in this case control the flow of data in alternate paths, performing the OR function.

The pass transistors in Figure 8.12 suffer one disadvantage in that the voltage level at the output is less than the input voltage level by one threshold voltage. Consider the output of the first inverter at a high logic state of V_{DD} volts. If V_{G1} is switched from a low to a high state, then M_1 forms a conducting channel, and charge is conducted from the output of the first inverter to the gate of the second inverter. As the gate voltage rises on the second inverter input, the V_{GS1} value decreases. When V_{GS1} decreases to $V_{GS1} = V_T$, then M_1 cuts off, leaving a voltage at the second inverter input of $V_{DD} - V_T$. The second inverter is now driven by a "weak" or reduced value for a high state input. This weak input causes the output low voltage of the second inverter to be slightly higher than that obtained for a strong or full voltage, V_{DD}, applied to its input gate. The second inverter is usually adjusted to have a higher k_D/k_L ratio in order to have a logic-low voltage compatible with the logic-low values in the rest of the circuit.

8.1.3. NMOS Digital Logic Circuits

STATIC NOR AND NAND GATES. The NMOS inverter is the basis of the NMOS digital logic circuits. Consider the circuit in Figure 8.13 in which two parallel enhancement mode driver transistors are in series with a depletion mode load. This circuit performs the NOR logic function. The circuit is similar to that of the depletion load inverter.

If both input voltages V_x and V_y are less than the driver transistor threshold voltage, then $V_{OUT} = V_{DD} = $ logic 1 level. If $V_x = V_{DD} = $ logic 1, then M_1 turns on, and the output voltage drops to some low value corresponding to a logic 0 state. If both V_x and V_y are equal to the logic 1 level, then both M_1 and M_2 turn on, and the output voltage is again at a low voltage level. The exact value of the low output voltage depends on the values of the conduction parameters of the three transistors. The logic 0 value will also change slightly depending upon whether only one of the driver transistors is turned on or both

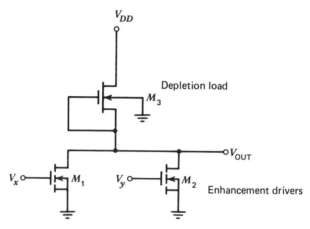

Figure 8.13. Two-input NOR logic gate with depletion load transistor.

driver transistors are turned on. The same logic function would be obtained if the depletion load in Figure 8.13 were replaced with a saturated enhancement mode load transistor. However, in this case, the logic 1 state would be $V_{DD} - V_{T3}$ rather than V_{DD}.

In Figure 8.14, a two-input NMOS NAND logic circuit with a depletion load is shown. If both input voltages V_x and V_y are less than the driver threshold voltages, then $V_{OUT} = V_{DD} = $ logic 1 level. If $V_x = V_{DD}$ and V_y is still less than V_{T2} (the threshold voltage of M_2), then M_2 is still in cutoff and $V_{OUT} = V_{DD}$. Only when both V_x and V_y are a logic 1 will both M_1 and M_2 turn on. The output voltage then drops to a low voltage level corresponding to a logic 0

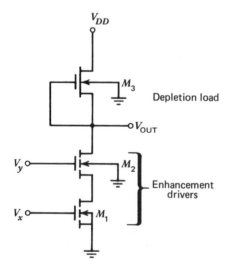

Figure 8.14. Two-input NAND gate with depletion load transistor.

state. This low voltage level is a function of the conduction parameters of the load and driver transistors. As in the previous case, the same logic function could be obtained with a saturated load transistor, but with a reduced output in the high-logic state.

More complex logic functions may be performed in an NMOS circuit by appropriately stacking NMOS driver transistors. The circuit in Figure 8.15*a* is one example. This circuit also has a depletion load and enhancement mode driver transistors. The equivalent logic circuit is shown in Figure 8.15*b*. By stacking the NMOS transistors and using only one load transistor, a great savings in chip area may be realized as well as reducing propagation delay times.

DYNAMIC NMOS GATES. The NMOS transmission gate is used in NMOS dynamic logic and memory circuits. One example is that shown in Figure 8.16.

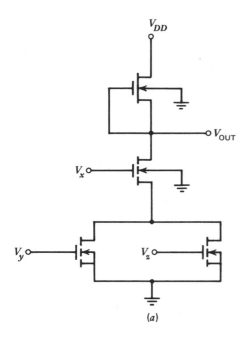

(a)

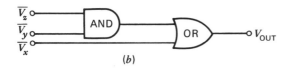

(b)

Figure 8.15. (*a*) NMOS logic circuit with depletion load. (*b*) Equivalent logic diagram for circuit in (*a*).

A saturated load device is shown, but a depletion mode device could be used in its place. This circuit may be used to temporarily store a bit of information and is one stage of a dynamic shift register.

If the gate voltage V_{G1} goes to a logic 1 level, then M_1 turns on. If V_x is zero, the gate voltage to M_2 is zero and $V_{OUT} = V_{DD} - V_{T3}$ since M_2 is a saturated load transistor. If $V_x = V_{DD}$, the gate voltage of M_2 becomes $V_{G2} = V_{DD} - V_{T1}$ so that V_{OUT} drops to a low voltage level. When V_{G1} returns to zero volts, M_1 turns off and the state of the inverter remains unchanged for a period of time. If the gate voltage of M_2 is high, and M_1 is off, the M_2 gate voltage may discharge through the reverse-biased terminal 2 to a substrate *pn* junction. The equivalent RC time constant including the high impedance of the reverse-biased junction may be in the tens of millisecond range. The inverter may then store a bit of information for a period of time.

A series of the inverters shown in Figure 8.16 may be constructed with two clock signals to form a simple dynamic shift register. This type of circuit, along with the necessary clock pulses, is shown in Figure 8.17. At time t_1, the V_x data are transmitted to the gate of M_2 so that $V_{o1} = \overline{V}_x$. From t_1 to t_2, M_4 is off and the V_{o1} data are not transmitted any further. At time t_3, M_4 turns on and the V_{o1} data are transmitted through the second inverter so that $V_{o2} = V'_{o1} = V_x$. Meanwhile, M_1 is off so that no new data enter the first inverter during the time the information is being transmitted to the second inverter. Therefore, information is consistently shifted through the series of inverters that form the shift register.

A shift register is essential in moving words of digital information. For example, a digital word may be retrieved from memory by first shifting the word to a storage register. The word may then be transferred on command to another digital element such as a computer arithmetic logic unit.

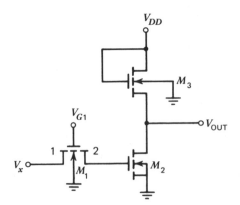

Figure 8.16. Single-stage dynamic shift register.

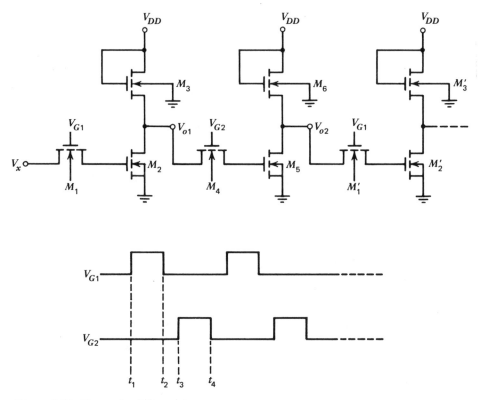

Figure 8.17. Dynamic shift register.

8.2 **CMOS CIRCUITS**

Complementary or CMOS circuits use both n-channel enhancement mode and p-channel enhancement mode transistors in a special configuration in the same circuit. CMOS circuits have unique advantages that make them one of the most competitive technological families.

8.2.1 **The CMOS Inverter**

Figure 8.18a shows a CMOS inverter with each substrate terminal tied to its source. The transistor M_N is the n-channel device and M_P is the p-channel device. The input signal is connected to both transistor gates, while the output node is at the point where both drains are connected. As the input voltage changes from zero to a maximum value of V_{DD} volts, the output voltage responds as an inverter, as shown in Figure 8.18b.

At $V_{IN} = V_{GSN} = 0$, the n-channel MOSFET, M_N, is off, while the p-channel MOSFET, M_P, has $V_{SGP} = V_{DD}$. This places M_P in its nonsaturation

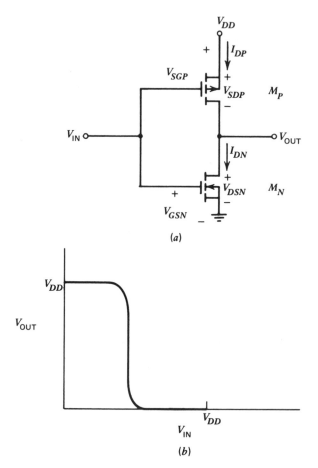

Figure 8.18. (*a*) CMOS inverter. (*b*) CMOS inverter transfer curve.

region of operation. It is significant that virtually no current exists when the inverter is in its output-high state, nor, as we shall see, when it is in its output-low state. As V_{IN} increases to the threshold voltage V_{TN} of M_N, the n-channel transistor starts to conduct. The conduction increases as V_{IN} increases beyond V_{TN}, causing $V_{OUT} = V_{DSN}$ to drop. At the midpoint in the transition, both transistors are in their saturation region. With further increase in V_{IN}, the source-to-gate voltage of M_P drops below the threshold voltage V_{TP} of the p-channel MOSFET cutting off M_P. The inverter is now in its output-low state with virtually no current existing in the transistors.

Some major advantages of CMOS are evident in this description. First, negligible current is required when the CMOS inverter is in either its low- or high-output state. The CMOS gate is said to power down in its static condition. The CMOS configuration also allows the full power supply voltage V_{DD} to be present at the output-high condition. The power supply voltage can have any value between 3 and 15 V without changing the logic operation of the circuit. The higher power supply voltages provide larger noise margins, while the lower power supply voltages result in lower power dissipation during the transition time. A CMOS circuit operated at 3 V has the lowest power dissipation of the commercial logic technology families. Recent technology and circuit developments have dropped the propagation delay times of CMOS gates to the nanosecond level, making CMOS competitive in market areas formerly held by TTL and ECL.

We first examine the static transfer characteristic of the CMOS inverter shown in Figure 8.18a, using the analytical expressions of the MOSFET in the saturated and nonsaturated regions. Figure 8.19 illustrates a CMOS static transfer curve in more detail than Figure 8.18b. Let us track the output voltage V_{OUT} as the input voltage V_{IN} increases from zero to V_{DD} volts. For V_{IN} less than the NMOS threshold voltage, M_N is off, and no current exists through both

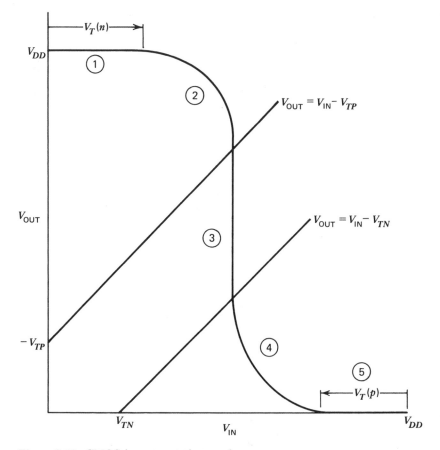

Figure 8.19. CMOS inverter static transfer curve.

transistors. At this low level of V_{IN}, the gate–source voltage of M_P is large and negative, driving M_P into its nonsaturated state. The channel resistance of the off-transistor M_N is very much greater than that of the on-transistor M_P. Consequently, M_N and M_P form a voltage divider and V_{OUT} will have a value of V_{DD} as shown in region 1 of Figure 8.19. When V_{IN} increases to the NMOS threshold voltage, M_N will start to conduct. The transistor M_N will be in the saturated mode for the conditions

$$V_{IN} = V_{GSN} > V_{TN}$$

and

$$V_{IN} = V_{GSN} < V_{DSN} + V_{TN}$$

Note that $V_{DSN} = V_{OUT}$ and $V_{SDP} = V_{DD} - V_{OUT}$. In region 2 of Figure 8.19, M_N is in the saturated state and M_P is in the nonsaturated state. Since

$$I_{DN} = I_{DP}$$

and

$$k_N(V_{GSN} - V_{TN})^2 = k_P[2(V_{SGP} + V_{TP})V_{SDP} - V_{SDP}^2]$$

then we may write

$$k_N(V_{IN} - V_{TN})^2 = k_P[2(V_{DD} - V_{IN} + V_{TP})(V_{DD} - V_{OUT}) - (V_{DD} - V_{OUT})^2] \tag{8.36}$$

Recall that for enhancement mode devices, $V_{TN} > 0$ and $V_{TP} < 0$. Equation 8.36 may be used to plot values of V_{OUT} versus V_{IN} in region 2.

As V_{IN} increases further, V_{GSN} increases and V_{SGP} decreases. At two specific points in the transfer curve, M_P makes the transition from the non-saturated to the saturated state and M_N goes from the saturated to the non-saturated state. Transistor M_N enters the nonsaturated state when

$$V_{GSN} = V_{DSN} + V_{TN}$$

or

$$V_{IN} = V_{OUT} + V_{TN} \tag{8.37}$$

Transistor M_P enters the saturated state when

$$V_{SGP} = V_{SDP} - V_{TP}$$

or

$$V_{DD} - V_{IN} = V_{DD} - V_{OUT} - V_{TP}$$

resulting in

$$V_{IN} = V_{OUT} + V_{TP} \tag{8.38}$$

The transition points for change of state as given in Equations 8.37 and 8.38 may be solved respectively for V_{OUT} as

$$V_{OUT} = V_{IN} - V_{TN} \tag{8.39}$$

and

$$V_{OUT} = V_{IN} - V_{TP} \tag{8.40}$$

Equation 8.39 is plotted in Figure 8.19 as a line intercepting the V_{IN} axis at V_{TN}. Equation 8.40 intercepts the V_{OUT} axis at $-V_{TP}$ (a positive quantity). The intersections of these two lines with the transfer function curve denote the demarcation between transistor states. Thus, in region 3, both M_N and M_P are saturated.

The equation describing region 3 is found by equating the drain current in both saturated transistors. This yields

$$k_N(V_{GSN} - V_{TN})^2 = k_P(V_{SGP} + V_{TP})^2$$

or

$$k_N(V_{IN} - V_{TN})^2 = k_P(V_{DD} - V_{IN} + V_{TP})^2 \tag{8.41}$$

The solution for V_{IN} in this region is

$$V_{IN}(\text{region 3}) = \frac{V_{DD} + V_{TP} + (k_N/k_P)^{1/2}V_{TN}}{1 + (k_N/k_P)^{1/2}} \tag{8.42}$$

Note that this solution for V_{IN} does not possess a term for V_{OUT} but predicts the very sharp transition region for the CMOS inverter. It is desirable that this sharp transition occur at a value for $V_{IN} = V_{DD}/2$ to maximize the noise immunity of the circuit. This situation occurs when $|V_{TN}| = |V_{TP}|$ and $k_P = k_N$. If $|V_{TN}|$ is not equal to $|V_{TP}|$, k_P and k_N can be adjusted through their transistor geometries to achieve a value of $V_{IN}(\text{region 3}) \simeq V_{DD}/2$.

It is of interest to observe that both transistors are in their saturated or linear amplification mode in region 3. A CMOS inverter may provide linear amplification if biased at the midpoint of region 3. This property has been used in analog design but is not the primary discussion of this chapter.

In region 4, M_N is not saturated and M_P is saturated. Equating the drain currents gives

$$I_{DN} = I_{DP}$$

or

$$k_N[2(V_{GSN} - V_{TN})V_{DSN} - V_{DSN}^2] = k_P(V_{SGP} + V_{TP})^2$$

This can be written as

$$k_N[2(V_{IN} - V_{TN})V_{OUT} - V_{OUT}^2] = k_P(V_{DD} - V_{IN} + V_{TP})^2 \tag{8.43}$$

Region 5 is a state with virtually no current through either transistor since the gate–source voltage of M_P is below the threshold voltage and M_P is cut off. M_N is in the nonsaturation state. The analysis of the transfer curve seems tedious but is quite important to the integrated circuit designer. The shape of the curve can be manipulated by the physical properties of the transistors. It should be emphasized that this analysis pertains to a static measurement situation and does not include the important capacitive effects inherent in MOS circuitry.

8.2.2 The CMOS Transmission Gate

A CMOS transmission gate performs the same type of switching function as the NMOS transmission gate but with an advantage. Figure 8.20 shows a CMOS transmission gate connected to a load capacitance. Both transistors are assumed to be completely bilateral. The substrate of the n-channel device is connected to the most negative potential (assumed to be ground potential), and the substrate of the p-channel transistor is connected to the most positive potential in the circuit. The two terminals (source and drain) of each transistor are considered to be completely equivalent. The input gate voltage to the p-channel transistor is the complement of the gate voltage to the n-channel device. Assume that V_x and V_G operate between 0.0 and 5.0 V, $V_{DD} = 5.0$ V, $V_{TN} = 1.0$ V, and $V_{TP} = -1.0$ V. As with the NMOS transmission gate, several cases are of interest.

Case I. Let $V_G = 0.0$ V, $\overline{V}_G = 5.0$ V, and $V_x = 0.0$ V. The gate voltage of M_N is never positive with respect to either terminal, so that M_N is always cut off. Similarly, the gate voltage of M_P is never negative with respect to either terminal, so that M_P is also always cut off. The input and output terminals are then isolated from each other.

Case II. Let $V_G = 5.0$ V, $\overline{V}_G = 0.0$ V, and $V_x = 0.0$ V. The gate-to-terminal 1 voltage of M_N is 5.0 V, so that M_N is turned on. In the steady state, zero current will exist because of the capacitor. The only way that zero current can exist is to have zero drain-to-source voltage on M_N. Therefore, in this case,

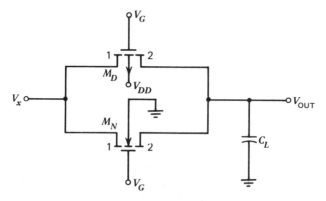

Figure 8.20. CMOS transmission gate with load capacitance C_L.

V_{OUT} will be 0.0 V. With $\overline{V}_G = 0.0$ V and with $V_x = V_{OUT} = 0.0$ V, M_P will be cut off.

Case III. Let $V_G = 5.0$ V, $\overline{V}_G = 0.0$ V, $V_x = 5.0$ V, and initially $V_{OUT} = 0.0$ V. In this case terminal 1 of M_N acts as the drain, terminal 2 of M_N acts as the source, and the current direction is from terminal 1 to terminal 2. The current charges the load capacitance until the current goes to zero. When $V_{OUT} = 4.0$ V, the gate-to-source voltage of M_N is 1.0 V $= V_{TN}$, and M_N turns off. However, with $\overline{V}_G = 0.0$ V on the gate of M_P, the p-channel transistor is still turned on. In this case, the current will go to zero only when the drain-to-source voltage on M_P is zero. Then, V_{OUT} will charge all the way up to $V_{OUT} = 5.0$ V. This is the advantage of a CMOS transmission gate. The output will charge to the full value of V_x without a threshold voltage drop as occurred in the NMOS transmission gate.

Case IV. Let $V_G = 5.0$ V, $\overline{V}_G = 0.0$ V, $V_x = 0.0$ V, and initially $V_{OUT} = 5.0$ V. In this case, terminal 2 acts as the drain of M_N and the source of M_P, so the current direction is from terminal 2 to terminal 1. The current now discharges the capacitance. The n-channel transistor will always be turned on, so that C_L will discharge completely to $V_{OUT} = 0.0$ V.

The CMOS transmission gate has the advantage that V_{OUT} will always equal V_x when the transmission gate is conducting, but the CMOS gate requires both a gate voltage V_G and its complement for successful operation.

8.2.3 CMOS Digital Logic Circuits

The CMOS inverter is the basis of the CMOS digital logic circuits. Consider the circuit in Figure 8.21 in which two parallel enhancement mode n-channel transistors are in series with two enhancement mode p-channel transistors. This circuit is a two-input NOR logic circuit. The substrates of the n-channel devices are all connected to the most negative potential, while the substrates of the p-channel devices are all connected to the most positive potential.

If both input voltages V_x and V_y are less than the transistor threshold voltage, then M_{N1} and M_{N2} are cut off. At the same time, the p-channel transistors M_{P1} and M_{P2} are turned on. As with the simple CMOS inverter, the output voltage goes to a high state, or $V_{OUT} = V_{DD}$.

Consider the case when V_x is a logic 1 or V_{DD}. Then, the M_{N1} transistor turns on and M_{P1} turns off. In this case, the impedance from V_{OUT} to ground is very small compared to the impedance from V_{OUT} to V_{DD}. The output voltage then goes to a low state, or 0.0 V. The same result exists if V_y is a logic 1 and V_x is a logic 0. If both V_x and V_y are logic 1, then both M_{N1} and M_{N2} are turned on, and both M_{P1} and M_{P2} are in cutoff. Again, V_{OUT} will be a logic 0 level. This circuit, then, performs the NOR logic level.

Consider the current which exists in the circuit. For $V_x = V_y = $ logic 0, both M_{N1} and M_{N2} are cut off, so the current is zero. If one or both of the

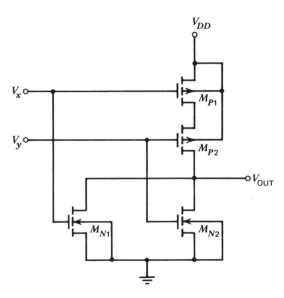

Figure 8.21. Two-input NOR CMOS gate.

inputs is a logic 1, at least one of the p-channel transistors is in cutoff, so that the current will again be zero. Since the current in the static situation is always zero (except for small reverse-biased drain-to-substrate leakage currents), the power dissipation is essentially zero. This is the most attractive feature of CMOS technology. Only as the circuit goes through a transition from one state to another will current exist and power be dissipated.

Figure 8.22 shows a two-input CMOS NAND logic circuit. In this case the p-channel devices are in parallel and the n-channel devices are series. If the inputs V_x and V_y are both logic 0, then M_{N1} and M_{N2} are in cutoff, and M_{P1} and M_{P2} are turned on. The output voltage is then $V_{OUT} = V_{DD} = $ logic 1. If, for example, $V_y = $ logic 1 and $V_x = $ logic 0, then M_{N1} turns on and M_{P1} turns off. However, M_{N2} remains cut off and M_{P2} remains on. The impedance from V_{OUT} to ground is still very high, while the impedance through M_{P2} to V_{DD} is very low. Thus, the output remains in a logic 1 state. Only when both V_x and V_y are a logic 1 so that both M_{N1} and M_{N2} turn on and both M_{P1} and M_{P2} turn off will the output change state to a logic 0.

8.3 LIMITATIONS OF SPEED, POWER, AND SIZE IN MOSFETS

The MOSFET logic circuits are competitively separated into PMOS, NMOS, CMOS, and mixed MOS transistor lines. The pure PMOS logic circuit was quickly overtaken by the faster NMOS circuits after the processing problems were eliminated in NMOS transistors. The NMOS and CMOS circuits competed

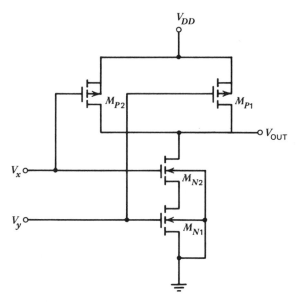

Figure 8.22. Two-input CMOS NAND gate.

during the 1970s for market shares. CMOS provided very low power but was not as dense as the NMOS circuits. A pure CMOS circuit contains equal numbers of p- and n-channel FETs, even though the logic information is contained in a little more than half of these transistors. The NMOS logic circuit performs equivalent logic functions with fewer transistors but, in contrast to CMOS, dissipates considerable power during the nontransitional periods of the circuit. In very large-scale integrated circuits (VLSI), NMOS circuits may become thermally limited, in that the power dissipated by thousands of transistors exceeds the ability of the chip to remove the heat.

The CMOS circuit draws a spike of current during the level transitions that reaches a maximum when both the p- and n-channel transistors are in their saturated states. The current spike is brief if the CMOS gate is driven by a fast rise time input pulse but may reach currents on the order of 1 mA.

The industrial competitive drive to overcome the inherent weaknesses of CMOS and NMOS has led to both processing improvements and novel configurations that use combination of n- and p-channel transistors but minimize the number of p-channel transistors. The MOS circuits have experienced greater speeds by a scaling down of the dimensions of the devices. Line widths have been reduced from 7.0 μm, which was standard in 1980, to less than 2.0 μm, with continuing demand for smaller dimensions. Let us follow these modern trends by observing the problems of the standard CMOS inverter during voltage transitions. Figure 8.23 shows two CMOS inverters with two load capacitances, C_P and C_N, representing the input capacitances of the p- and n-channel transistors of the driven gate. During a high-to-low transition at the output of the

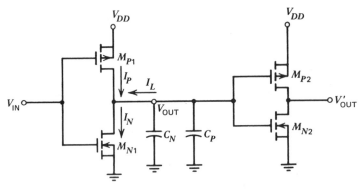

Figure 8.23. Two CMOS inverters showing load capacitance of M_{P2} and M_{N2} as C_N and C_P.

first gate, the current direction is as shown in Figure 8.23. The capacitances must be discharged as rapidly as possible to allow the second gate to respond to its logic input. Two degradations are apparent for rapid discharge in a CMOS circuit. First, observe that the PMOS transistor (M_{P1}) drives current in such a direction to charge the output capacitors rather than discharge them. Similarly, when the output is driven from low to high, the NMOS transistor drains current from the PMOS transistor that is needed to rapidly charge the load capacitance.

Secondly, the PMOS transistors are constructed to be approximately twice as large as the NMOS transistors. This is done to compensate for the reduced hole mobility in the PMOS, allowing equal current drives in the NMOS and PMOS transistors. Equal current drive will give approximately equal rise and fall times to the CMOS gate. However, the disadvantage of a large PMOS transistor is that the input capacitance is more than twice as large as an NMOS transistor. Thus, the PMOS transistor contributes an inordinate amount of input capacitance and delays the pull-down of the output voltage during a high-to-low transition. In summary, the two central problems with CMOS logic are the large area and large capacitance of each PMOS transistor and the counter-productive action of each complementary transistor as the output capacitance tries to charge or discharge.

Different circuit configurations have evolved to reduce both the numbers of p-channel transistors required in a CMOS gate and the magnitude of the current spike during the transition. One such logic gate was developed by Bell Laboratories in 1981 and is called a CMOS domino circuit. This type of logic circuit is shown in Figure 8.24. It represents the AND/OR INVERT (AOI) function for one three-input AND and one two-input AND. The circuit has seven NMOS and two PMOS transistors. The basic logic unit consists of the NMOS transistors M_{N1} to M_{N5}, while M_{N6} and M_{P1} control the charging of current to the logic unit. The CMOS inverter (M_{N7}, M_{P2}) performs a voltage buffering function between the logic unit and any subsequent driven logic units.

When the clock pulse goes from high to low, M_{N6} turns off and M_{P1} turns

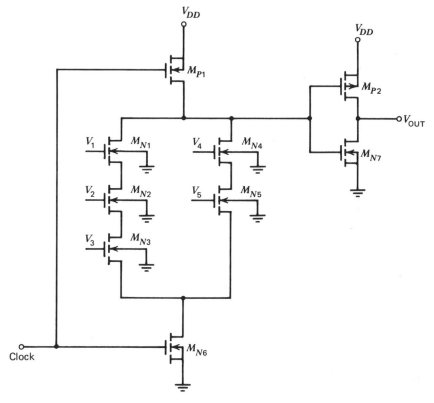

Figure 8.24. A domino CMOS logic circuit performing the 3-2 AND-OR-Invert function.

on. The output buffer then receives a high input and the buffer output goes low. This means that the inputs to all gates in the total circuit, including the AOI circuit inputs, receive a low input voltage. When the clock is low, all gate inputs to logic transistors are low. During this low clock pulse period, the capacitance at the logic unit output node (the drain of M_{P1}) is charged. This period is called the precharge phase. Note that, since M_{N6} is off, there is no large current spike that occurs during the clock transition. This is a significant reduction over the current spike occurring in a pure CMOS circuit during voltage level changes.

When the clock signal goes from low to high, M_{P1} cuts off and M_{N6} turns on, discharging the drain node of M_{N6}. Momentarily, there is no input signal on the majority of NMOS transistors in the logic units of a complex circuit since they are held low by a low output from the output buffer CMOS inverters that drive them. However, the external logic signals that drive the complex circuit are active (either high or low) and initiate a logic response from the first logic unit of the complex circuit. The output buffer of that first logic stage now responds and drives the next set of logic units. The process of rapid sequential activation of logic units within a complex circuit gives rise to the name domino CMOS.

The other problem of CMOS logic is the high input capacitance of the PMOS transistor. This has been virtually eliminated in the domino CMOS circuit by constructing the logic unit transistors out of NMOS transistors. The logic unit then has the advantages of smaller NMOS transistors with the added advantage of lower input capacitance. The nearly zero power dissipation that occurs in a CMOS circuit is also found in the domino CMOS circuit. This is accomplished through M_{N6} and M_{P1} that never allow the power supply V_{DD} to see a highly conducting pathway to ground.

8.4 SUMMARY

NMOS and CMOS circuits as applied to digital circuits have been analyzed in this chapter. The NMOS inverter is the basis of NMOS digital circuits, although the NMOS transmission gate is extremely useful in steering–logic functions. The use of a depletion load transistor allows the logic 1 level to be the same as the supply voltage and also gives the shortest NMOS switching times.

Innovations such as the domino CMOS have advanced CMOS logic as a leading contender in the VLSI market. The important factor is not only the speed and density of the circuit but the low power dissipation of CMOS. There is a thermal limit to the packing density of transistors that can only be advanced by reducing the power dissipation per transistor.

REFERENCES

1. M. Elmasry, *Digital MOS Integrated Circuits* (New York: Institute of Electrical and Electronic Engineers, 1981).
2. C. Mead and L. Conway, *Introduction to VLSI Systems* (Reading, Mass.: Addison-Wesley, 1980).
3. J. G. Posa, *Electronics* 54(21), 103 (1981).
4. R. H. Krambeck et al., *IEEE J. Solid State Circuits* SC-17(3), 614 (1982).

PROBLEMS

8.1. Consider the n-channel enhancement mode MOSFET inverter shown in Figure 8.1. Let $V_{DD} = 5.0$ V, $R_D = 5.0$ kΩ, and $V_T = 1.0$ V. Find the transition point and find V_{OUT} when $V_x = 5.0$ V for (a) $k = 100$ μA/V^2 and (b) $k = 500$ μA/V^2. Sketch the voltage transfer characteristic for each case.

8.2. For the n-MOS inverter with resistive load shown in Figure 8.1, let $V_{DD} = 5.0$ V, $R_D = 8.0$ kΩ, and $V_T = 0.80$ V. Find the conduction parameter k so that $V_{OUT} = 0.50$ V when $V_x = 5.0$ V.

8.3. An n-MOS inverter with saturated load is shown in Figure 8.4. Let $V_{DD} = 10$ V, $V_{TL} = V_{TD} = 2.0$ V, $k_D = 100$ μA/V², and $k_L = 20$ μA/V². Calculate the transition point and calculate V_{OUT} when $V_x = 10$ V. Sketch the voltage transfer characteristic.

8.4. For the NMOS inverter with saturated load given in Figure 8.4, let $V_{DD} = 5.0$ V and $V_{TL} = V_{TD} = 0.70$ V. (a) Find the ratio k_D/k_L so that $V_{OUT} = 0.25$ V when $V_x = 5.0$ V. (b) Find the ratio k_D/k_L so that $V_{OUT} = 0.25$ V when $V_x = 4.3$ V.

8.5. For the depletion load n-MOS inverter circuit of Figure 8.7, assume that $V_{DD} = 5.0$ V, $k_L = 100$ μA/V², $V_{TL} = -2.0$ V, $k_D = 1.6$ mA/V², and $V_{TD} = +1.0$ V. (a) Find the transition points for the load and driver transistors. (b) Calculate the value of V_{OUT} for $V_x = 5.0$ V. (c) Calculate I_D when $V_x = 5.0$ V. (d) Sketch the voltage transfer characteristic for $0 < V_x < 5.0$ V.

8.6. In the depletion load n-MOS inverter circuit shown in Figure 8.7, let $V_{DD} = 5.0$ V, $k_L = 50$ μA/V², $k_D = 500$ μA/V², and $V_{TD} = +1.0$ V. Calculate the value of V_{TL} so that $V_{OUT} = 0.10$ V when $V_x = 5.0$ V.

8.7. Calculate the power dissipated in the inverter of (a) Problem 8.1 for $V_x = 0.20$ V and $V_x = 5.0$ V, (b) Problem 8.3 for $V_x = 0.75$ V and $V_x = 8.0$ V, and (c) Problem 8.5 for $V_x = 0.50$ V and $V_x = 5.0$ V.

8.8. In the NMOS transmission gate shown in Figure 8.11, let $C_L = 4.0$ pF, $k = 50$ μA/V², and $V_T = 0.75$ V. (a) If $V_G = 5.0$ V, what is the steady-state value of V_{OUT} for (1) $V_x = 0.0$ V, (2) $V_x = 5.0$ V, and (3) $V_x = 4.0$ V. (b) Let V_G switch to 0.0 V when $V_{OUT} = 4.25$ V. Assume that the transmission gate can be modeled as a reverse biased pn junction with a reverse saturation current of 1.2 μA. How long will it take the output to discharge to 1.0 V?

8.9. Consider the NMOS logic circuit shown in Figure 8.25. Determine the output voltage V_{OUT} for each combination of input voltage and derive the equivalent logic circuit.

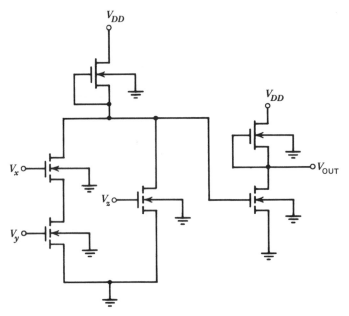

Figure 8.25. Circuit for Problem 8.9.

8.10. Determine the state of each transistor (on or off) and the voltage outputs at Q and $\overline{Q}$ for the circuit in Figure 8.26 for the input conditions listed. Assume the input conditions are sequential in time from state 1 to state 4. This circuit is an RS flip-flop.

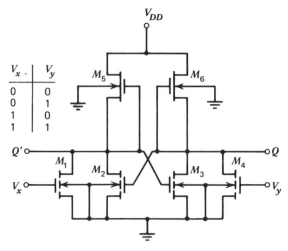

V_x .	V_y
0	0
0	1
1	0
1	1

Figure 8.26. Circuit for Problem 8.10.

8.11. Consider the dynamic circuit shown in Figure 8.27. Assume the transistor parameters are as follows: $k_1 = k_2 = 1.0$ mA/V^2, $V_{T1} = V_{T2} = 1.0$ V, $k_3 = 100$ μA/V^2, and $V_{T3} = -2.0$ V. (a) Calculate V_{OUT} for $V_x = V_G = 5.0$ V. (b) Sketch the voltage transfer characteristic V_{OUT} versus V_x for $0 < V_x < 5.0$ V and for $V_G = 5.0$ V. Indicate the various transition points.

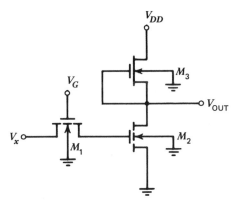

Figure 8.27. Circuit for Problem 8.11.

8.12. Given the dynamic circuit in Figure 8.16, let $V_{T1} = V_{T2} = V_{T3} = 0.80$ V and $V_{DD} = 5.0$ V. For $V_x = V_{G1} = 5.0$ V, find the ratio k_3/k_2 so that $V_{OUT} = 0.25$ V.

8.13. Consider the CMOS inverter given in Figure 8.18. Let $k_P = k_N$, $V_{TN} = 1.0$ V, $V_{TP} = -1.0$ V, and $V_{DD} = 5.0$ V. (a) Find the transition points for the p-channel and n-channel transistors. (b) Sketch the voltage transfer characteristic including the appropriate voltage values at the transition points. (c) Find V_{OUT} for $V_{IN} = 2.0$ V and V_{OUT} for $V_{IN} = 3.0$ V.

8.14. For the CMOS inverter in Figure 8.18, let $V_{TN} = +1.5$V, $V_{TP} = -1.5$ V, $k_N = 100$ μA/V^2, $k_P = 50$ μA/V^2, and $V_{DD} = 10$ V. (a) Find the transition points for the p-channel and n-channel transistors. (b) Sketch the voltage transfer characteristic including the appropriate voltage values at the transition points. (c) Find V_{IN} for $V_{OUT} = 9.0$ V and V_{IN} for $V_{OUT} = 1.0$ V.

8.15. In the CMOS transmission gate circuit of Figure 8.20, let $V_G = V_{DD} = 5.0$ V. Sketch the dc voltage transfer characteristic, V_{OUT} versus V_x, for $0 < V_x < 5.0$ V. Indicate the regions over which the n-channel and p-channel devices are either turned on or in cutoff.

8.16. Consider the CMOS circuit in Figure 8.28. For the input conditions listed, find the output voltage, and state whether the transistors M_{N1}, M_{N2}, M_{P1}, and M_{P2} are either on or cut off.

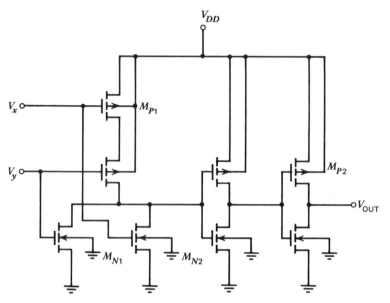

Figure 8.28. Circuit for Problem 8.16.

8.17. Figure 8.29 shows one-stage of a CMOS dynamic shift register with its timing diagram. Determine the voltages V_{o1}, V_{o2}, V_{o3}, and V_{OUT} at times t_1, t_2, t_3, and t_4 for (*a*) $V_x = 0.0$ V and (*b*) $V_x = V_{DD}$.

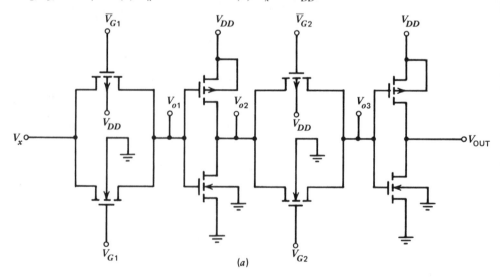

(*a*)

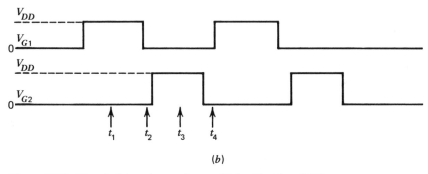

Figure 8.29. Circuit (*a*) and waveforms (*b*) for Problem 8.17.

8.18. Figure 8.30 shows a CMOS RS flip-flop. Determine the state of each transistor (on or off) and the voltage outputs at Q and $\overline{Q}$ for the input conditions listed. Assume the input conditions are sequential in time from state 1 to state 4. Assume that V_{TN} (each transistor) = 0.50 V and V_{TP} (each transistor) = −0.50 V.

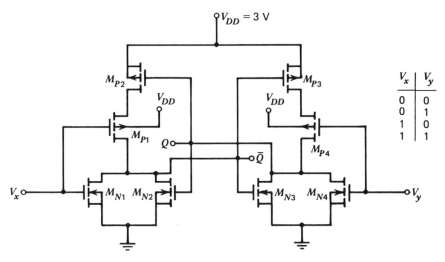

Figure 8.30. (*a*) Circuit and (*b*) Input listing for Problem 8.18.

chapter 9
MEMORY
CIRCUITS

In Chapters 6 through 8, various logic circuits were considered. Combinations of gates can be used to perform logic functions such as addition, multiplication, and multiplexing. In addition to these combinatorial logic functions, digital systems usually require some way of storing information. Semiconductor circuits form one type of memory and define a class of digital electronic circuits that are just as important as the logic circuits.

A memory cell is a circuit, or in some cases just a single device, that can store one bit of information. A systematic arrangement of memory cells constitutes a memory. The memory must also include peripheral circuits to address and write data into the cells as well as detect data that are stored in the cells.

Three basic types of semiconductor memory will be considered. The first is the *random access memory* (RAM), in which each individual cell can be addressed at any particular time. The access time to each cell is virtually the same. Implicit in the definition of the RAM is that both the read and write operations are permissible in each cell. A second class of semiconductor memory is the *read-only memory* (ROM), which has a set of data that cannot be altered. The data were fixed either by the manufacturer during fabrication or by the user. If the memory is fixed by the manufacturer, it is referred to as a mask-programmed ROM. If the user can program the memory once and only once, it is referred to as a *programmable read-only memory* (PROM). In either case, the data content cannot be altered once it is placed in memory. The third class of semiconductor memories is the erasable-programmable read-only memory (EPROM), or electrically erasable PROM (EEPROM). This memory is

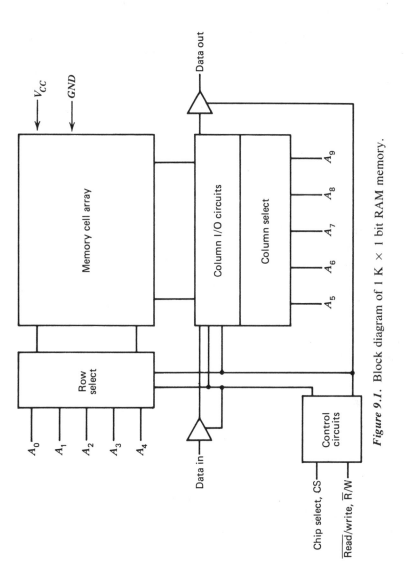

Figure 9.1. Block diagram of 1 K × 1 bit RAM memory.

281

programmed by the user, and once the memory is programmed, the data can remain fixed for a period of years. However, the data in the memory can be altered or erased by the user, but the time required for this operation is considerably longer than for the RAM cell.

A volatile memory is one that loses its data when power is removed from the circuit, while a nonvolatile memory retains its data when power is removed. As we shall see in the following sections, a random access memory is, in general, a volatile memory, while ROMs and EPROMs are nonvolatile. A RAM can be made nonvolatile by adding circuitry that switches to a battery when the power is interrupted.

Figure 9.1 shows block diagram of a 1 K $\times$ 1 bit memory. Note that an upper case K is used to represent 1024 in memory applications. There are four basic parts to the memory: (1) the memory cell array in which the data are stored, (2) the row- and column-select decodes, (3) the input/output circuits that sense the data stored or are used to write new data, and (4) the control circuits that determine if data are to be read or new data are to be stored. The memory cells can be arranged in a square matrix that, for the example of a 1 K $\times$ 1 bit memory in Figure 9.2, results in 32 rows and 32 columns. The additional lines needed to detect the data stored or to write in new data are not shown. In order to address a cell, 32 row-select lines and 32 column-select lines are required for this particular cell array.

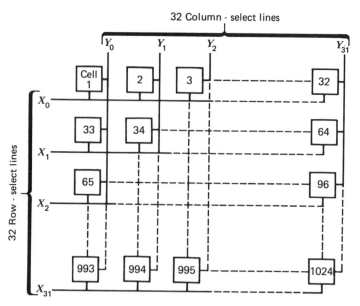

Figure 9.2. Memory cell layout of a 1 K $\times$ 1 bit memory.

9.1 ADDRESS DECODERS

The 1024 bits of the 1 K memory given in the example of Figures 9.1 and 9.2 must be addressed. Word decoders can reduce the number of input address terminals to 10 and provide 2^{10}, or 1024 combinations. A five-input word is used to address the 32 rows, and a second five-input word is used to address the 32 columns.

Figure 9.3 shows a simple decoder with a two-bit input. The decoder shown uses NAND logic circuits. The input word goes through input buffers that generate the complements as well as the signal. The same type of decoder could also be implemented with NOR logic gates.

Figure 9.4*a* shows a pair of input buffer–inverters, and Figure 9.4*b* shows a five-input NOR logic address decoder circuit using *n*-channel enhancement mode MOSFETs. A pair of input buffer–inverters is required for each input address line. The input is then required to drive only an inverter, while the buffer–inverter pair can be designed to drive the remainder of the logic circuits. The NOR gate shown in Figure 9.4*b* would decode the address word and select the sixth row for a read or write operation.

The NOR type decoder can be implemented in an array form. Figure 9.5 shows a two-input NOR word decoder with depletion loads. The array can easily be expanded to accommodate a three-input or more word. The substrate connection in each of the transistors is not shown but would, in the actual integrated circuit, be connected to ground potential.

As the size of the memory increases, the length of the address word must increase. For example, a 16 K × 1 bit memory must have a seven-input word for the row decoder and a seven-input word for the column decoder. A seven-input NOR logic gate to implement a NOR decoder becomes rather complex, and the number of transistors and power dissipation become large. In addition, the drain-to-substrate capacitance becomes large so that the propagation delay times may become significant.

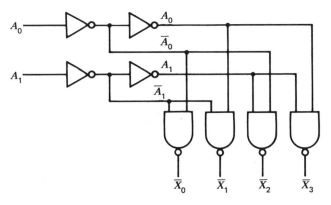

Figure 9.3. Simplified decoder with two-bit input.

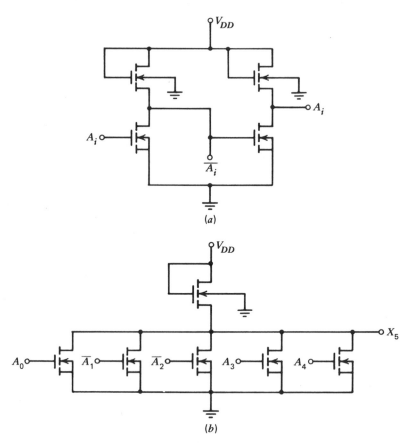

Figure 9.4. (a) An input buffer–inverter pair. (b) A five-input NOR logic address decoder.

A two-stage decoder using both NOR gates and NAND gates will reduce the number of transistors required to design a decoder. For example, Figure 9.6 shows a four-input word feeding a set of NOR gates with the NOR outputs driving a set of AND gates (NAND plus inverter). As the input word becomes longer, the savings in number of devices becomes progressively greater.

9.2 **CONTROL CIRCUITS**

Control circuits are used to enable or select a particular memory array and also to select whether data are to be read from or written into the memory cell. Memory chips are designed to be paralleled so that the memory capacity can be increased. The additional lines needed to address parallel chips are called the *chip select* signals.

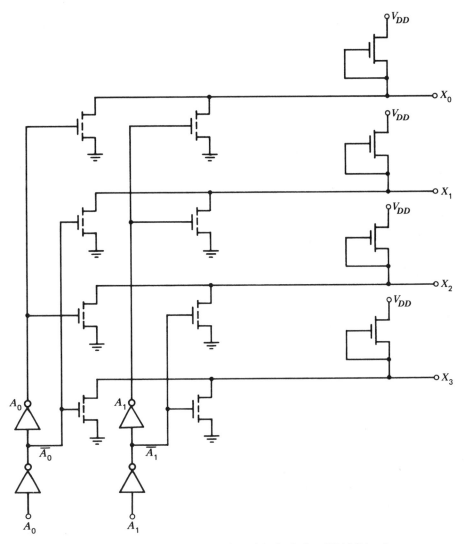

Figure 9.5. A two-input NOR word decoder with depletion NMOS loads.

Figure 9.7 shows a simple example of a chip select. The $\overline{CS}$ signal goes into the NOR gates of the two-stage NOR/NAND word decoder, as was shown in Figure 9.6. If $CS = 0$, the chip is not selected and $\overline{CS} = 1$. Then, all outputs, W_0, W_1, W_2, . . ., are zero, so that all select line signals are also zero. In this case, no memory cell is addressed. At the same time, the chip select signal controls the tristate output of the data-in and data-out buffers. In this way, the data-in and data-out lines to and from several memory chips may be connected together without interfering with each other.

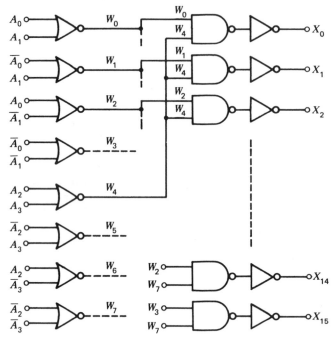

Figure 9.6. A four-input two stage decoder using NOR and NAND logic.

An example of a bipolar read/write control circuit and write amplifier is shown in Figure 9.8. If the data in the memory cell are to be read, then $R = 1$ or $\overline{R} = 0$. Under these circumstances, at least one input to both NAND gates will be a zero and the outputs will be a logic 1. The logic 1 level input to the bases of Q_1 and Q_2 will drive both Q_1 and Q_2 into saturation so the diodes D_1 and D_2 will both be turned off. In this condition, the DATA and $\overline{\text{DATA}}$ lines are disconnected from the write amplifier. The data that exist on the DATA and $\overline{\text{DATA}}$ lines can then be detected by the sense amplifiers that are discussed in the next section.

If new data are to be written or stored in a memory cell, $W = 1$ in Figure 9.8. Suppose that a logic 1 is to be stored so that DATA IN $= 1$. In this condition, both inputs to the $N2$ gate are 1's so that the output is logic 0. Then, Q_2 is turned off. The collector voltage of Q_2 rises turning on D_2, and the DATA line is brought high. The resistance values R_{C2} and R_{D2} must be such that the voltage on the DATA line does in fact represent a logic 1. If a logic 0 were to be stored, then the output of the $N1$ gate would turn Q_1 off and the $\overline{\text{DATA}}$ line would be brought high.

Figure 9.9 is an example of a read/write control circuit using NMOS. If $R = 1$ so that $\overline{R} = 0$, then the M_1 n-channel transmission gate is turned off

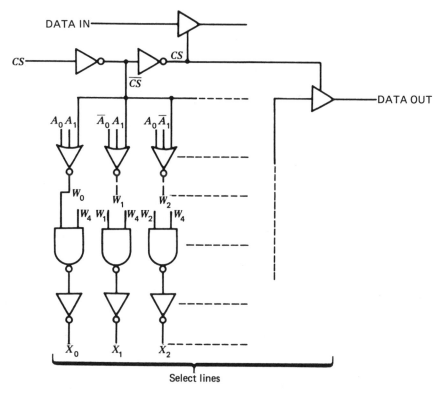

Figure 9.7. Chip select decoder using NOR/NAND logic decoders and tristate input and output buffers.

and the DATA OUT transmission gate M_2 is turned on. Then, if, for example, the Y_1 column is addressed, data will be read through the M_6 transmission gate. If new data are to be written into a memory cell, then $W = 1$, so that the M_1 gate becomes conducting and the M_2 gate is turned off. If a logic 1 is to be stored, then the DATA line will be brought to logic 1.

9.3 STATIC RANDOM ACCESS MEMORY (RAM) CELLS

Several examples of RAM cells and the sense amplifiers used to detect the logic state of these cells are considered in this section. The RAM cells considered here are examples that illustrate the basic principles involved the design of a RAM cell but do not represent a complete listing of possible RAM cell designs. The data stored in a static RAM cell are held as long as power is applied to the circuit, since the cell is constructed by cross-coupling the inputs and outputs of two inverters.

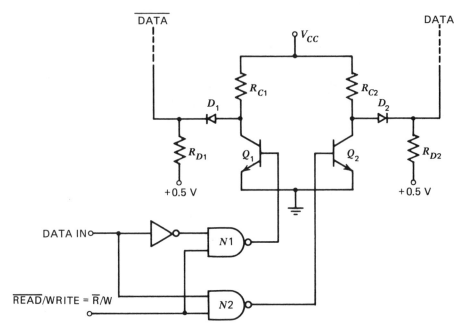

Figure 9.8. A NAND logic read/write control circuit (N_1 and N_2) and a write-amplifier using bipolar transistors Q_1 and Q_2.

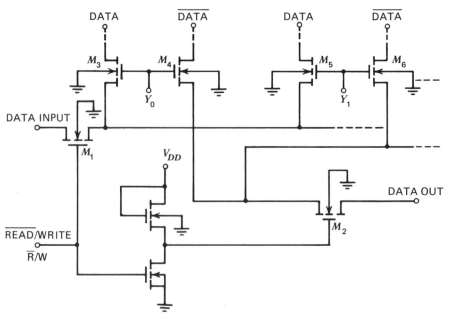

Figure 9.9. Read/write control circuit using NMOS.

9.3.1 **TTL Static RAM Cell**

A TTL static RAM cell is shown in Figure 9.10, in which the collector voltage of each transistor is cross-coupled back to the base of the opposite transistor. In the static condition, one transistor is in saturation while the other transistor is in cutoff. The two emitters E_{x1} and E_{x2} are connected to the row-select line, and the E_{y1} and E_{y2} emitters are connected to the column-select

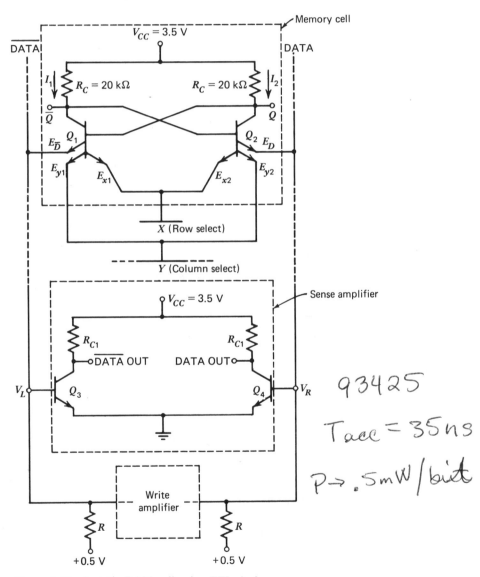

Figure 9.10. A static RAM cell using TTL design.

lines. In the standby condition, the row- and column-select lines will be 0.3 V. Also in standby, the E_x and E_y emitters are at a lower potential (by 0.2 V) than the E_D or $E_{\overline{D}}$ emitters so that the E_D and $E_{\overline{D}}$ emitters will be turned off.

Assume that the piecewise linear parameters $V_{BE}(\text{sat})$ and $V_{CE}(\text{sat})$ are 0.8 and 0.2 V, respectively, and that the state of the circuit is such that Q_1 is in saturation and Q_2 is in cutoff. Then, in standby, $V_{B1} = X + V_{BE}(\text{sat}) = 0.3 + 0.8 = 1.1$ V and $\overline{Q} = X + V_{CE}(\text{sat}) = 0.3 + 0.2 = 0.5$ V. The transistor Q_2 is then in cutoff, with $\overline{Q} = 0.5$ V and $X = Y = 0.3$ V. The currents are then

$$
\begin{aligned}
I_1 &= \frac{V_{CC} - \overline{Q}}{R_C} \\[6pt]
&= \frac{3.5 - 0.5}{20 \times 10^3} \\[6pt]
&= 0.15 \text{ mA}
\end{aligned}
\tag{9.1}
$$

$$
\begin{aligned}
I_2 &= \frac{V_{CC} - V_{B1}}{R_C} \\[6pt]
&= \frac{3.5 - 1.1}{20 \times 10^3} \\[6pt]
&= 0.12 \text{ mA}
\end{aligned}
\tag{9.2}
$$

The current I_1 is the collector current of Q_1, and I_2 is the base current of Q_1. The two currents add to give the emitter current in Q_1 that enters the X- and Y-select lines. In this standby situation, no current leaves the $E_{\overline{D}}$ emitter terminal.

To address the cell, both the X- and Y-select lines are raised to approximately 3.0 V. In this condition, both of the base–emitter junctions associated with the E_x and E_y emitters are turned off; and with Q_1 in saturation, the $E_{\overline{D}}$ base–emitter junction turns on. In this situation, the current flows in the DATA' line through the left sense resistor R_{SL} and into the 0.5-V source. The voltage V_L then increases compared to the right data line voltage V_R. The increase in V_L turns Q_3 on, driving it into saturation, while transistor Q_4 remains off (considering the piecewise linear model). The outputs then show $\overline{\text{DATA OUT}}$ = logic 0 and DATA OUT = logic 1.

If, for example, the X is selected ($X = 3.0$ V) but the Y is still in standby ($Y = 0.3$ V), then the E_x emitters are reverse biased but the E_y emitters are still on. If Q_1 is in saturation, then charge flows out the E_{y1} emitter terminal into the column-select line. It is only when both the X and Y are selected that the cell is addressed and the state of the circuit determined by sensing the current in the left or right data line.

To write data into the cell, the cell is again addressed ($X = Y = 3.0$ V). If we assume that Q_1 is initially in saturation and Q_2 is in cutoff (logic 1) and we wish to write in a logic 0, then the $\overline{\text{DATA}}$ line voltage is raised to 3.0 V through the WRITE amplifier. Transistor Q_1 is turned off with all three emitters at 3.0 V. The voltage $\overline{Q}$ begins to rise, which turns Q_2 on, and the state of the circuit has then been changed and a logic 0 is stored. If Q_1 were initially in cutoff, raising the $\overline{DATA}$ line voltage will not change the state of the circuit.

An important characteristic of a memory cell is its power dissipation. Since a cell is in a standby mode most of the time, the standby power is of primary interest. We had previously calculated $I_1 = 0.15$ mA and $I_2 = 0.12$ mA for the TTL cell shown in Figure 9.10 for the case when Q_1 is in saturation and Q_2 is cut off. Then, in standby, the power dissipation is

$$P = (I_1 + I_2)(V_{CC} - X)$$
$$= (0.15 + 0.12)(3.5 - 0.3)$$
$$= 0.864 \text{ mW}$$

The maximum power dissipation on a chip is on the order of 500 mW, indicating that the maximum number of these particular cells on a single chip is 512. The power dissipation per cell must be reduced in order to have a larger memory on a single chip.

9.3.2 **NMOS Static RAM Cell**

An NMOS static RAM cell is shown in Figure 9.11. The substrate connections are not shown, but all the substrates are connected to ground potential. The two inverters use depletion loads and are cross-coupled as with the bipolar static RAM. The transistors M_1 through M_4 form the basic cell. If the transistor M_1 is turned on, the output Q is low, which means that the transistor M_2 is cut off. Since M_2 is cut off, the output $\overline{Q}$ is high, ensuring that M_1 is turned on. Thus, we have a static situation as long as V_{DD} is applied to the circuit.

The transistors M_5 and M_6 are n-channel transmission gates. With the row-select in standby, $X = $ logic 0, the transistors M_5 and M_6 are in cutoff, and the memory cell is isolated from the data lines. If $X = $ logic 1, M_5 and M_6 turn on, and the cell is connected to the data lines. If the column-select Y is also a logic 1, then the data may be read out through the sense amplifier. If $\overline{Q}$ is logic 1, then the $\overline{DATA}$ line goes high. This makes the output of the sense amplifier inverter (M_{11} and M_{13}) low, turning M_{13} off. With $\overline{DATA}$ high, the transistor M_{14} turns on, discharging any load capacitance. Conversely, if $\overline{DATA}$ is low, M_{14} is off and M_{13} turns on, charging any load capacitance.

To write data into the memory cell, the X- and Y-select lines are again held high. If a logic 1 is to be written into the cell, then the DATA IN will be a logic 1. This high value will make Q high, turning M_2 on and driving the output $\overline{Q}$ low to ensure the latch. The data will be written into the cell only if the cell is addressed and if the write signal $W = 1$ so that the M_9 transmission gate is conducting.

The power dissipation of the NMOS memory cell may be calculated if the transistor parameters are known. If, for example, the conduction parameters

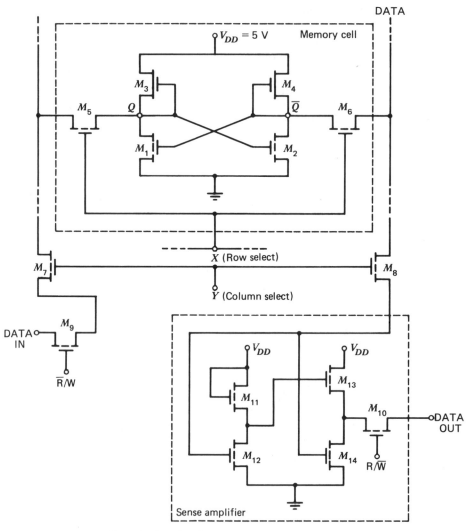

Figure 9.11. NMOS static RAM cell.

of the driver and load transistors are $k_D = 50\ \mu\text{A/V}^2$ and $k_L = 5.0\ \mu\text{A/V}^2$, and if the threshold voltages of the driver and load transistors are $V_{TD} = +1.0$ V and $V_{TL} = -2.0$ V, then the current in the on-inverter is given by

$$I_D = k_L(V_{GSL} - V_{TL})^2$$
$$= 5.0[0.0 - (-2.0)]^2$$
$$= 20\ \mu\text{A}$$

The load transistor of the on inverter is in saturation and the power dissipation in the cell is then

$$P = I_D V_{DD} = (20 \times 10^{-6})(5.0)$$
$$= 100 \; \mu\text{W}$$

This power dissipation would allow a 4 K NMOS static RAM to be fabricated on a single integrated circuit chip.

9.3.3 **CMOS Static RAM Cell**

A CMOS static RAM cell is shown in Figure 9.12 using two cross-coupled CMOS inverters to form the memory cell. If M_{N1} is on and M_{P1} off, Q is at ground potential, which means that M_{N2} is off and M_{P2} is on, which makes $\bar{Q}$ high. The advantage of CMOS is that very little current (only leakage currents) exists in the memory cell during standby. This makes the power dissipation in the cell extremely small compared to both the bipolar and NMOS static cells considered previously. The M_{N4} through M_{N7} transistors are n-channel transmission gates used to select the memory cell in the same way as in the NMOS memory circuit.

9.4 **DYNAMIC RANDOM ACCESS MEMORY (RAM) CELLS**

The number of memory cells that can be fabricated on a single monolithic chip is basically determined by the power dissipation and surface area for each cell. Both the bipolar and NMOS RAM cells dissipate power both in standby and transition; and while the CMOS static RAM dissipates very little power in standby, it requires six transistors for each cell.

A dynamic memory cell dissipates less power than most static cells and is constructed with fewer devices than the static cells. Consequently, semiconductor random access memories with more than 16,384 cells are usually dynamic. A static RAM maintains the stored data as long as power is supplied, but data stored in a dynamic RAM are in the form of charge on capacitors and consequently will leak off with time. The stored data must then be refreshed periodically to maintain the stored logic level in each cell.

Additional peripheral circuitry and timing signals need to be included in dynamic memories to provide for the refresh voltages and currents needed. This additional complication is a disadvantage of dynamic memories. However, the advantage of a 65 K × 1 (65 kilobits by one) memory on a single chip outweighs the disadvantage of the additional timing requirements. The time required to refresh an entire memory is on the order of 1 μs and must be performed every few milliseconds.

There are many configurations of a dynamic RAM using four, three, two, or one transistor. A particular example of a one-transistor dynamic RAM cell

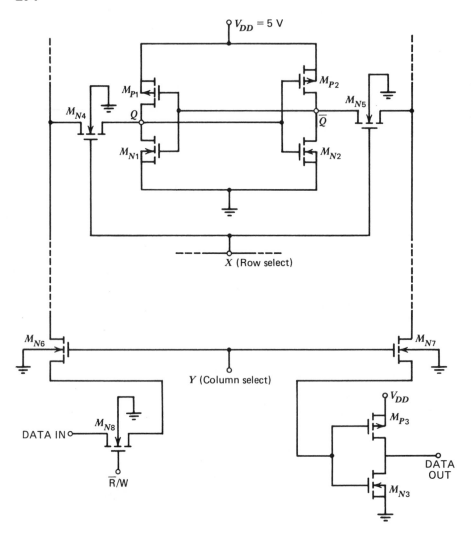

Figure 9.12. CMOS static RAM cell.

is shown in Figure 9.13. The basic memory cell consists of the capacitor C_1 and transistor M_1. Charge stored on C_1 represents a logic 1 level, and no charge on C_1 represents a logic 0 level.

There are a number of variations in the design of the single transistor/capacitor memory cell. Figure 9.14*a* shows a simplified design of a one-device cell. In general, the storage capacitance consists of an MOS capacitance in parallel with a junction capacitance with the MOS capacitance usually 5 to 10 times larger than the junction capacitance. The one-device cell shown in Figure 9.14*b* must have three interconnection lines consisting of the row-select

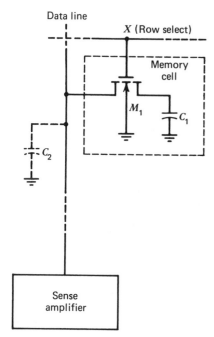

Figure 9.13. One-transistor dynamic RAM cell.

to the gate of the MOSFET, the data line to the drain of the MOSFET, and the so-called plate line to the MOS capacitor.

To read the state of the cell, the row-select voltage in Figure 9.13 is brought high, turning on transistor M_1. The data line voltage is brought either high or low depending on the data stored in C_1, and this voltage state is detected by the sense amplifier. However, this read process is complicated by the capacitance of the data line represented by C_2. The capacitor C_1 is similar geometrically to the MOS gate capacitance of the n-channel transistors, which have typical values of 0.1 pF. However, the data line connects many cells together, so that the capacitance C_2 of the data line may be 10 to 20 times that of C_1. During the read cycle, the input voltage to the sense amplifier will be a voltage division between C_1 and C_2, and because of the relative sizes of C_1 and C_2, the difference between a logic 1 and a logic 0 voltage on the data line may be as small as 100 mV.

The read process is destructive. A logic 1 level on C_1 decreases as it is read and must be written back into the cell after each read cycle. In addition, the data stored on C_1 will leak off even when the cell is in standby. If the plate and substrate connections in Figure 9.14b are at ground potential and the storage capacitor is charged so that the source potential is positive, then the source-to-substrate pn junction is reverse biased. The reverse saturation current will discharge the capacitor C_1, so the dynamic memory cell must be refreshed every few milliseconds.

The detection and refresh of a dynamic cell can be realized using a gated

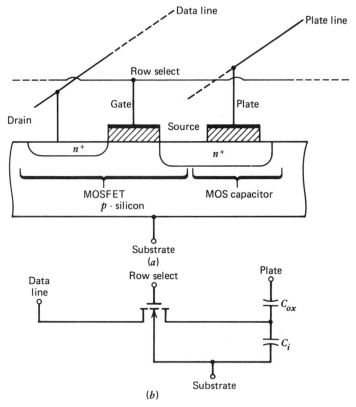

Figure 9.14. (*a*) Simplified physical layout of a one-device MOS dynamic memory cell. (*b*) Schematic of MOS dynamic memory cell.

flip-flop and a precharging technique. The precharging technique charges the data line to a voltage between the logic 1 and logic 0 levels. During the read cycle, the data line voltage will change to a value above or below the precharge level depending on the logic level stored. A two-terminal gated flip-flop is very suitable then as a sense amplifier.

Figure 9.15 shows an example of a gated flip-flop and the associated timing pulses. Initially, the memory cell is not addressed, and precharging begins by applying ϕ_2 and ϕ_3 to the flip-flop. The signal ϕ_3 turns on the cross transistor M_C as well as the dummy word line. The low conductance of M_C ensures that both the DATA line and the $\overline{\text{DATA}}$ line will be charged to a voltage midway between a logic 1 and logic 0 point. The dummy cell capacitor C_1' is also charged to this midvoltage.

The flip-flop is now turned off by switching off ϕ_2 and ϕ_3. The Row-Select and Dummy Word-Select are now activated. A voltage difference now appears at the nodes of the flip-flop since the charge stored in the dummy cell capacitor

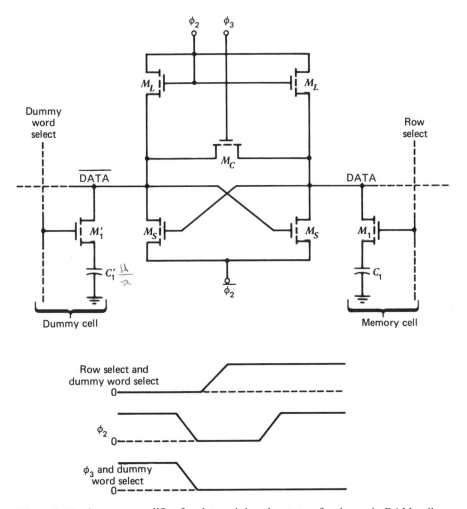

Figure 9.15. A sense amplifier for determining the state of a dynamic RAM cell.

C_1 is either greater or less than the charge in the dummy cell capacitor C_1'. This voltage difference is amplified by turning on ϕ_2, which turns on the flip-flop. If the memory cell remains turned on during this amplification, the refresh is carried out simultaneously.

9.5 READ-ONLY MEMORY (ROM) CELLS

We consider two types of ROMs. The first example is a mask-programmed ROM, in which contacts to devices are selectively included or excluded in the final manufacturing process to obtain the desired memory pattern. The second example is referred to as a PROM (programmable read-only memory), in which the data are programmed by the user. In both cases, once the data are programmed, they cannot be altered.

9.5.1 **Mask-Programmed ROM**

Figure 9.16 shows an example of a NMOS 16×1 mask-programmed ROM. Enhancement mode NMOS transistors are fabricated in each of the 16 cell positions (the grounded substrate connections are omitted for clarity). However, gate connections are fabricated only on selected transistors such as M_{00}, M_{03}, M_{11}, for example, while the gate connections are omitted on the other transistors such as M_{01}, M_{02}, M_{23}. The transistors M_1, M_2, M_3, and M_4 are column-select transistors, and M_0 is a depletion mode load device.

The inputs X_0, X_1, Y_0, and Y_1 are the row- and column-select signals. If, for example, $X_0 = \overline{X_1} = \overline{Y_0} = Y_1 = $ logic 1, then the M_{12} transistor is addressed. Transistors M_{12} and M_3 turn on with this address, forcing $V_{OUT} = $ logic 0. If the address changes to $\overline{X_0} = X_1 = \overline{Y_0} = \overline{Y_1} = $ logic 1, then the M_{23} transistor is addressed. However, M_{23} does not have a gate connection and consequently never turns on, so $V_{OUT} = $ logic 1. In this example of a ROM, a logic 1 is

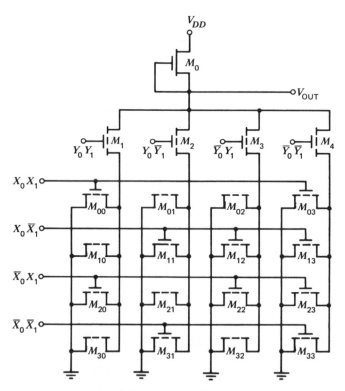

Figure 9.16. An NMOS 16×1 mask-programmable ROM.

represented by the absence of a transistor connection, and a logic 0 is represented by the presence of a complete transistor.

This example is only a 16 × 1 bit ROM, while a more useful memory would be a 16 K ROM. Memories can be organized in any desired manner such as 2048 × 8 for the 16 K memory. This ROM is a nonvolatile memory since the data stored are not lost when the power is removed.

9.5.2 **User-Programmable ROM**

A user-programmable ROM allows the data pattern to be defined after final manufacture rather than during the manufacture. One specific type is shown schematically in Figure 9.17. A small fuse is in series with each emitter,

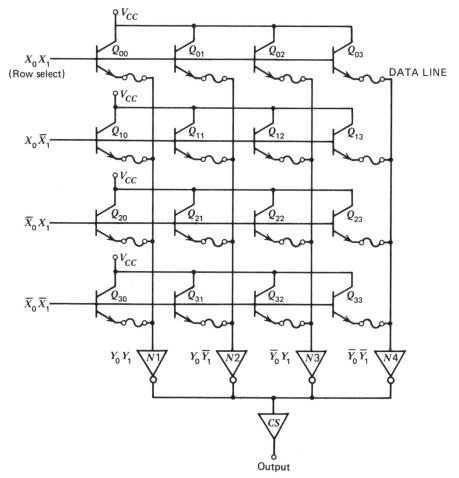

Figure 9.17. A fuse-linked user-programmable ROM (PROM).

which may be selectively "blown" or left in place by the user. If, for example, the fuse in M_{00} is left in place and $X_0 = X_1 = Y_0 = Y_1 =$ logic 1, then M_{00} turns on, raising the data line voltage at the emitter of M_{00}. The inverter $N1$ is enabled, making the output a logic 0. If the fuse is blown, then the input to the specific inverter addressed will be logic 0, so the output will be a logic 1.

An npn bipolar transistor with a polysilicon fuse is shown in Figure 9.18. The resistance of the polysilicon region is fairly low with the fuse in place and at low currents, so there is very little voltage drop across the fuse. When the current through the fuse is increased to the 20- to 30-mA range, the I^2R heating of the polysilicon fuse increases the temperature to approximately 1400°C. At this temperature, the silicon oxidizes, forming an insulator that effectively opens the path between the data line and the emitter.

The bipolar ROM circuit with the fuses either in place or "blown" form a permanent read-only memory that is not alterable and is also nonvolatile. The primary advantage is that a user may program the ROM rather than waiting for the time involved in manufacturing a mask-programmed ROM. The disadvantages of PROMs are associated with wasted area and power dissipation.

9.5.3 Electrically Erasable PROM (EEPROM)

The read-only memory cells considered in the previous section are unalterable once they are programmed. However, electrically erasable PROMs (EEPROMS) can be altered and reprogrammed. There are many possible designs, and all of these involve some method either of storing charge in the gate region of a MOSFET to permanently turn the device on, or discharging the gate region to permanently turn the device off.

Ultraviolet (UV) light-erasable, programmable read-only memories

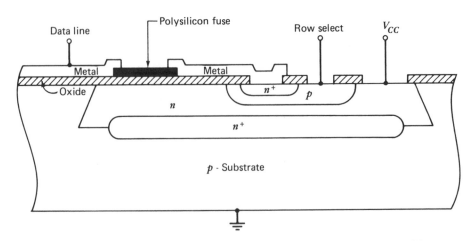

Figure 9.18. Physical cross section of an *npn* transistor with attached polysilicon fuse.

(EPROMs) have found wide application. The information stored in the UV EPROM is erased by UV light directed upon the chip through a window in the lid of the package. When the EPROM is to be erased, it must be removed from the circuit board and exposed to UV for approximately 30 minutes. The disadvantage of the memory is that the entire memory must be erased before any reprogramming can be done.

In the EEPROM, each individual cell can be erased and reprogrammed without disturbing any other cell. The most common form of EEPROM is a floating gate structure, with one such example shown in Figure 9.19. The memory transistor is similar to an n-channel MOSFET, but with a physical difference in the gate insulator region. Charge may exist on the floating gate that will alter the threshold voltage of the device. If a net positive charge exists on the floating gate, the n-channel MOSFET is turned on; whereas if zero or negative charge exists on the floating gate, the device is turned off.

The floating gate is capacitively coupled to the control gate with the tunnel oxide thickness slightly less than 0.02 μm. If 20 V is applied to the control gate while keeping $V_D = 0$, electrons tunnel from the n^+-drain region to the floating gate. This puts the MOSFET in the enhancement mode with a threshold voltage of approximately 10 V. If 0 V is applied to the control gate and 18 V is applied to the drain terminal, then electrons tunnel from the floating gate to the n^+-drain terminal. This leaves a net positive charge on the floating gate that puts the device in the depletion mode with a threshold voltage of approximately -2 V. If all voltages are kept to within 5 V during the read cycle, this structure can retain charge for a period of 10 years.

Figure 9.20 shows an EEPROM memory cell that uses two devices: a

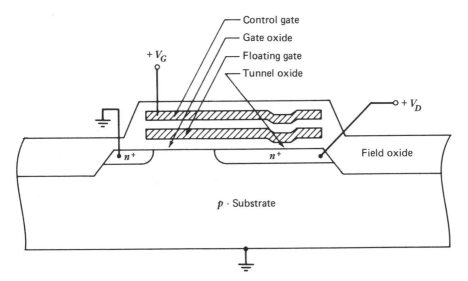

Figure 9.19. Physical cross section of a floating gate-programmable memory cell.

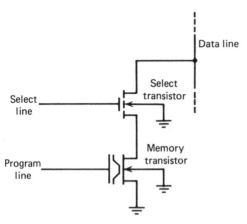

Figure 9.20. An EEPROM memory cell using a select NMOS transistor and a floating gate memory transistor.

select NMOS transistor and the floating gate memory transistor. During the read operation, 5 V is applied to the select line, program line, and data line. If the memory transistor is programmed in the depletion mode, then a drain current will exist in the memory cell as well as in the data line. The existence of the drain current is detected by a sense amplifier in the data line. The existence of drain current corresponds to a logic 0 being stored. If the memory cell is programmed in the enhancement mode with a threshold voltage of approximately 10 V, then, when the cell is selected by applying 5 V to the select line, program line, and data line, the memory cell remains off. In this case, no current exists in the data line. The nonexistence of data line current then corresponds to a logic 1 stored in the memory cell.

To write new data into the cell, the select line is brought to 20 V, which turns the select transistor on hard. If a logic 0 is to be written, the program line is held at 0 V and the data line is brought to 18 V, putting the memory transistor into the depletion mode. If a logic 1 is to be written, the program line is held at 20 V while the data line is at 0 V. The transistor is then programmed in the enhancement mode.

9.6 **SUMMARY**

Several types of memory cells and pheripheral memory circuits have been considered in this chapter. Random access memory cells are capable of being read and written with the same access time, however, the RAM cells are volatile. Static RAM cells can store data for as long as power is applied but may have significant power dissipation and many devices per cell. Dynamic

RAMs have fewer devices and dissipate less power per cell but require additional peripheral circuitry to refresh the data stored.

Read-only memory cells are nonvolatile. The mask-programmable and user-programmable ROMs are not changeable once they have been programmed. The electrically erasable PROMs can be reprogrammed, although the write time may be on the order of 10 ms whereas the read access time is in general on the order of 200 ns.

This chapter is intended only as an introduction to memories. The design of the memory cell is closely tied to the semiconductor fabrication technology so that the final design may lose some of the discrete device characteristics that we have considered. The design of dynamic memories is also very complex. In some cases, there may be as many as 23 separate clock signals involved in a dynamic memory. We have only provided an introduction to semiconductor memories in this chapter.

part III

ANALOG CIRCUITS

chapter 10 _____
TRANSISTOR
BIASING PRINCIPLES

In the digital circuits considered in the previous chapters, the transistors operated in one of two states that resulted in outputs of either a logic 1 or a logic 0. Analog circuits, however, require that transistors be biased in their active regions of operation. When no ac signal is present at the input to an amplifier, the transistors are biased at a quiescent operating point, or Q-point. Once the Q-point is established, a small-signal linear model for the transistor can be developed so that the transistor circuit is analyzed as a linear circuit.

In this chapter, we analyze the methods and circuits that achieve the Q-point for both bipolar and field effect transistors. The effects of parameter variations, such as changes in the bipolar current gain, temperature, or leakage current, on the stability of the Q-point are considered.

10.1 BIPOLAR BASE CURRENT BIAS

The method of biasing the TTL circuits in Chapter 6 provided a base current through a base resistor similar to that shown in Figure 10.1. In small-signal linear amplifiers, the transistor is biased in its active region. The ac signal may then be coupled to the circuit by a coupling capacitor such as C_C in Figure 10.1.

An approximate quiescent base current is obtained by estimating the base–emitter voltage with its forward active value. The quiescent base current

307

in Figure 10.1 is then

$$I_{BQ} = \frac{V_{CC} - V_{BE}}{R_B} \tag{10.1}$$

where V_{BE} is the forward-biased base–emitter voltage in the active region. The collector current is then

$$I_{CQ} = \beta_{dc} I_{BQ} \tag{10.2}$$

Note that the symbol β_{dc} implies the ratio of the collector current to base current, not the derivative. If the circuit has the parameters $V_{BE} = 0.70$ V, $R_B = 1.0$ MΩ, $R_C = 6.0$ kΩ, $\beta_{dc} = 100$, and $V_{CC} = 12$ V, the quiescent current and voltages are as follows:

$$I_{BQ} = \frac{V_{CC} - V_{BE}}{R_B} = \frac{12 - 0.70}{1.0 \times 10^6}$$
$$= 11.3 \ \mu A \tag{10.3a}$$

$$I_{CQ} = \beta_{dc} I_{BQ} = 100 \times 11.3 \times 10^{-6}$$
$$= 1.13 \text{ mA} \tag{10.3b}$$

$$V_{CEQ} = V_{CC} - I_{CQ} R_C = 12 - 1.13 \times 10^{-3} \times 6.0 \times 10^3$$
$$= 5.2 \text{ V} \tag{10.3c}$$

The load line superimposed on the transistor characteristics is a useful concept in visualizing and plotting the Q-point. Applying Kirchhoff's voltage law around the collector–emitter loop gives

$$V_{CE} = V_{CC} - I_C R_C \tag{10.4}$$

This equation describes the load line and is plotted in Figure 10.2 for the

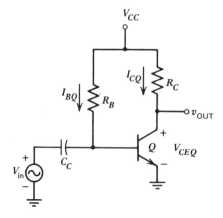

Figure 10.1. Common-emitter stage with the ac input signal V_{in} coupled to the circuit through C_C.

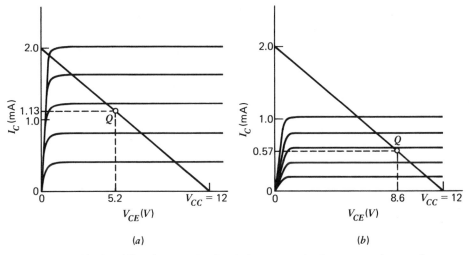

Figure 10.2. The load line for the circuit of Figure 10.1 is shown superimposed on the transistor characteristic curves for (a).

parameters given. The Q-point, calculated in Equation 10.3, is indicated in Figure 10.2 (a).

Now suppose that the transistor in Figure 10.1 is replaced by another in which the current gain β_{dc} is 50 instead of 100. The quiescent currents and voltages are now

$$I_{BQ} = 11.3 \ \mu A \ (\text{unchanged}) \tag{10.5a}$$

$$I_{CQ} = \beta_{dc}I_{BQ} = 50 \times 11.3 \times 10^{-6} \tag{10.5b}$$
$$= 0.57 \text{ mA}$$

$$V_{CEQ} = V_{CC} - I_{CQ}R_C = 12 - 0.57 \times 10^{-3} \times 6.0 \times 10^3 \tag{10.5c}$$
$$= 8.6 \text{ V}$$

The load line for the circuit in Figure 10.1 is not a function of β_{dc}. However, the quiescent operating point has changed substantially and is designated Q in Figure 10.2 (b). The small-signal linear parameters vary with the Q-point, so that the linear amplifier characteristics may change significantly. Since the β_{dc} of a discrete transistor type can vary over a 3 to 1 range (or more) due to fabrication variations, the bias scheme in Figure 10.1 has a poorly defined operating point and is unsatisfactory for linear amplifiers. Other parameter variations, including temperature, also make this biasing scheme unsatisfactory. This bias scheme is, however, adequate for TTL digital circuits, since the transistors are biased for operation either in saturation or cutoff and the β-variation can be taken into account in the bias design.

10.2   BIPOLAR COMMON-EMITTER BIASING

The circuit of Figure 10.3 is a classic example of biasing the discrete transistor. The single base resistor is replaced by a pair of resistors, R_1 and R_2, and an emitter resistor R_E and an emitter bypass capacitor C_E are added. The ac signal may still be coupled into the base via the coupling capacitor C_C. The emitter bypass capacitor shunts the ac emitter current to ground, bypassing the emitter resistor. This bias scheme is not appropriate for integrated circuits because the large bypass and coupling capacitors require too much area for use in a microcircuit.

The equivalent dc circuit is shown in Figure 10.4 in which the base circuit is the Thevenin equivalent circuit of R_1, R_2, and V_{CC}. Applying Kirchhoff's voltage law around the base loop yields

$$I_{BQ}R_B + V_{BE} + I_{EQ}R_E = V_{BB} \tag{10.6}$$

Since $I_{EQ} = I_{BQ}(1 + \beta_{dc})$, the quiescent base current is

$$I_{BQ} = \frac{V_{BB} - V_{BE}}{R_B + (1 + \beta_{dc})R_E} \tag{10.7}$$

and the quiescent collector current is

$$I_{CQ} = \frac{\beta_{dc}(V_{BB} - V_{BE})}{R_B + (1 + \beta_{dc})R_E} \tag{10.8}$$

A useful design condition is to make

$$(1 + \beta_{dc})R_E >> R_B \tag{10.9}$$

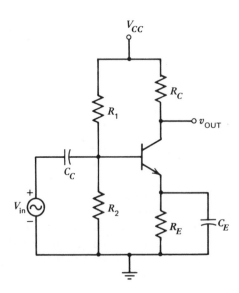

Figure 10.3. Single-stage common-emitter amplifier showing a voltage divider bias network.

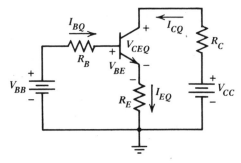

Figure 10.4. Equivalent dc circuit for the amplifier of Figure 10.3.

which makes I_{CQ} essentially independent of R_B, or

$$I_{CQ} \simeq \frac{\beta_{dc}(V_{BB} - V_{BE})}{(1 + \beta_{dc})R_E} \tag{10.10}$$

Normally, $\beta_{dc} >> 1$, so that

$$\frac{\beta_{dc}}{1 + \beta_{dc}} \simeq 1 \tag{10.11}$$

and the quiescent value of the collector current is approximately

$$I_{CQ} \simeq \frac{V_{BB} - V_{BE}}{R_E} \tag{10.12}$$

In the approximation of Equation 10.12, the quiescent collector current is a function of the dc voltage and emitter resistor and has become stabilized against current gain or β_{dc} variation. The inclusion of the emitter resistor and the magnitude relation between the emitter and base resistances in Equation 10.9 has resulted in this stabilization. This circuit has an improved stability over that in Figure 10.1 but contains more circuit components.

If Kirchhoff's voltage law is applied to the loop containing the collector–emitter terminals, then

$$I_{CQ}R_C + V_{CEQ} + I_{EQ}R_E = V_{CC} \tag{10.13}$$

The dc emitter and collector currents are related by

$$I_{EQ} = \frac{1 + \beta_{dc}}{\beta_{dc}} I_{CQ}$$

so that Equation 10.13 becomes

$$I_{CQ} = \frac{V_{CC} - V_{CEQ}}{R_C + [(1 + \beta_{dc})/\beta_{dc}]R_E} \tag{10.14}$$

which describes the dc load line for this circuit. This load line is superimposed on the transistor characteristics in Figure 10.5.

The small-signal voltage V_{in} is applied at a sufficiently high frequency that coupling and emitter bypass capacitors are assumed to be short circuits. In the small-signal analysis, we also consider that the ac signal is zero at the terminals of a dc voltage source. A dc voltage source provides a constant voltage, and hence no small-signal voltage drop appears across this device. Therefore, both terminals of this dc source are considered to be at the signal ground potential. The ac circuit configuration is shown in Figure 10.6 with ground points placed on the power supply terminals. The large coupling and bypass capacitors provide approximate short circuits over the entire operating frequency range of the input signal. Using phasor notation (upper case variables with lower case subscripts), the small-signal parameters in the collector circuit are related by the equation

$$I_c = -\frac{V_{ce}}{R_C} \qquad (10.15)$$

This equation relates the ac collector current and ac collector–emitter voltage and is called the ac load line equation. A plot of I_c versus V_{ce} is a straight line passing through the Q-point, as shown in Figure 10.5. The total instantaneous values of i_C and v_{CE} are always on the ac load line.

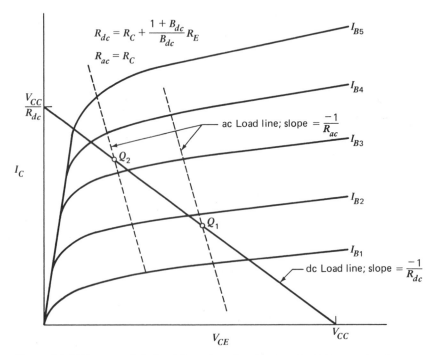

Figure 10.5 The ac and dc load line plots for the circuit of Figure 10.4 for two quiescent operating points.

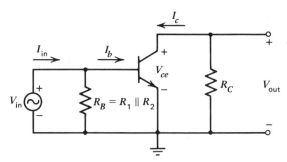

Figure 10.6. The ac equivalent circuit of the amplifier of Figure 10.3.

The stability of the Q-point is still important when considering the ac load line. With changes in current gain, the Q-point may shift from Q_1 to Q_2 as shown in Figure 10.5. The ac load line slope will remain constant, but again the small-signal transistor parameters may change substantially.

Coupling and emitter bypass capacitors do not exist in integrated circuits, since large-valued capacitors (in excess of 200 pF) cannot be fabricated with acceptable yields in integrated circuit structures. Typical coupling and bypass capacitors are more than 1000 times as large as the maximum values that can be obtained using conventional integrated circuit processing techniques. As discussed later in the chapter, it is possible to make stable dc amplifiers in microcircuits using different biasing methods, eliminating the requirements for coupling and bypass capacitors.

10.3 MOSFET COMMON-SOURCE BIASING

The n-channel MOSFET circuit in Figure 10.7 is also a classic example of discrete transistor biasing that includes coupling and source bypass capacitors. The source bypass capacitor shunts the ac source current from the source resistor to ground. The inclusion of the source resistor tends to stabilize the Q-point against changes in temperature and other parameter variations.

The quiescent gate-to-source voltage is given by

$$V_{GSQ} = \frac{R_2}{R_1 + R_2} V_{DD} - I_{DQ}R_S \qquad (10.16)$$

where I_{DQ} is the quiescent drain current. The drain current for a MOSFET in the saturation region is

$$I_D = k(V_{GS} - V_T)^2$$

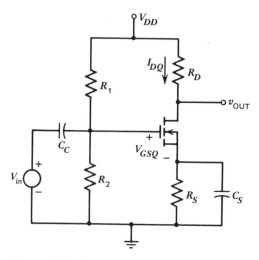

Figure 10.7. Common-source amplifier using an *n*-channel MOSFET.

However, at the Q-point, we may write

$$I_{DQ} = g_{mdc}V_{GSQ} \tag{10.17}$$

where g_{mdc} is the dc transconductance or gain. By combining Equations 10.16 and 10.17, we obtain

$$I_{DQ} = \frac{g_{mdc}R_2V_{DD}}{(1 + g_{mdc}R_S)(R_1 + R_2)} \tag{10.18}$$

If we impose the condition that

$$g_{mdc}R_S \gg 1 \tag{10.19}$$

then

$$I_{DQ} \simeq \frac{R_2V_{DD}}{R_S(R_1 + R_2)} \tag{10.20}$$

In this case, the quiescent drain current is stabilized against variations in the transistor transconductance. The result is similar to the stabilization achieved in bipolar circuits with the emitter resistor. Figure 10.8 shows the transistor characteristics with the dc load line superimposed. The Q-point should be in the active region of operation, which, for the MOSFET, is the saturation region of the transistor. Note that the ac load line has a different slope from that of the dc load line when a source bypass capacitor is present.

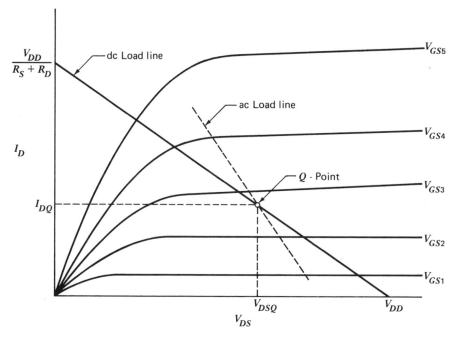

Figure 10.8. The common-source load line of Figure 10.7 superimposed on the *n*-channel MOSFET characteristic curves.

10.4 **STABILITY ANALYSIS**

In the previous sections, the stability or variation of the Q-point with dc forward current gain was discussed qualitatively. A quantitative analysis of the Q-point is desired that takes account of the other circuit variables. There is a clear need for analytical expressions to assist the designer in achieving the desired stability or sensitivity of a circuit. The measures of an integrated circuit biasing technique include both the degree of stability imparted to the circuit and the minimal area of the resistors needed to implement the bias configuration. The stability of the collector current of a bipolar transistor and the stability of the drain current of a MOSFET are analyzed in this section. However, we should keep in mind that any electrical parameter of the circuit such as the current, voltage, gain, or terminal impedance may be similarly analyzed. Any circuit, no matter the complexity, may be analyzed for its stability, especially with the aid of modern computer electronic simulation programs.

The derivation of stability equations assumes that circuit parameters are limited to small variations, allowing stability expressions to be described by linear functions. In an example here, the quiescent collector current I_{CQ} is analyzed. The circuit variables that influence I_{CQ} include β_{dc}, I_{CBo}, all of the circuit resistances and dc power supplies in the bias pathway such as R_B, R_E, and V_{BB}, and the base–emitter voltage V_{BE}. The temperature changes affect bias stability through I_{CBo} and V_{BE}. We may express these statements as

$$I_{CQ} = f(\beta_{dc}, I_{CBo}, R_B, R_E, V_{BE}, V_{BB} \ldots) \qquad (10.21)$$

from which

$$dI_{CQ} = \frac{\partial I_{CQ}}{\partial \beta_{dc}} d\beta_{dc} + \frac{\partial I_{CQ}}{\partial I_{CBo}} dI_{CBo} + \frac{\partial I_{CQ}}{\partial R_E} dR_E + \cdots \qquad (10.22)$$

The current stability factors are defined from Equation 10.22 as

$$S_\beta = \frac{\partial I_{CQ}}{\partial \beta_{dc}} \qquad S_{RE} = \frac{\partial I_{CQ}}{\partial R_E}$$

$$S_I = \frac{\partial I_{CQ}}{\partial I_{CBo}} \qquad S_V = \frac{\partial I_{CQ}}{\partial V_{BE}} \qquad (10.23)$$

$$S_{RB} = \frac{\partial I_{CQ}}{\partial R_B} \qquad S_{VBB} = \frac{\partial I_{CQ}}{\partial V_{BB}}$$

If all of the S-factors equal zero, there will be no change in I_{CQ} for changes in all the other quantities, and the operating point is stabilized. A low value for each of the S-factors is desirable. For small incremental changes, we may use the approximations

$$\frac{\partial I_{CQ}}{\partial \beta_{dc}} \simeq \frac{\Delta I_{CQ}}{\Delta \beta_{dc}}$$

$$\frac{\partial I_{CQ}}{\partial I_{CBo}} \simeq \frac{\Delta I_{CQ}}{\Delta I_{CBo}} \qquad (10.24)$$

$$\frac{\partial I_{CQ}}{\partial R_E} \simeq \frac{\Delta I_{CQ}}{\Delta R_E}$$

which allow us to write

$$\Delta I_{CQ} \simeq S_\beta \Delta \beta_{dc} + S_I \Delta I_{CBo} + S_{RB} \Delta R_B + S_{RE} \Delta R_E$$
$$+ S_V \Delta V_{BE} + S_{VBB} \Delta V_{BB} + \cdots \qquad (10.25)$$

Equation 10.25 is now applied to the common-emitter circuit of Figure 10.4. If the leakage current is included, the collector current may be written as

$$I_{CQ} = \alpha_{dc} I_{EQ} + I_{CBo} \qquad (10.26)$$

Substituting

$$\alpha_{dc} = \frac{\beta_{dc}}{1 + \beta_{dc}} \qquad (10.27)$$

into Equation 10.26 yields

$$I_{CQ} = \frac{\beta_{dc}}{1 + \beta_{dc}} I_{EQ} + I_{CBo} \qquad (10.28)$$

Then, substituting the Kirchhoff's current law expression for the transistor terminals,

$$I_{EQ} = I_{BQ} + I_{CQ} \tag{10.29}$$

into Equation 10.28 gives

$$I_{BQ} = \frac{I_{CQ}}{\beta_{dc}} - \frac{1 + \beta_{dc}}{\beta_{dc}} I_{CBo} \tag{10.30}$$

The base–emitter and collector–emitter loop equations give

$$V_{BB} = I_{BQ}R_B + V_{BE} + I_{EQ}R_E \tag{10.31}$$

$$V_{CC} = I_{CQ}R_C + V_{CEQ} + I_{EQ}R_E \tag{10.32}$$

Equations 10.29, 10.30, and 10.31 can be combined to give the quiescent collector current

$$I_{CQ} = \frac{\beta_{dc}(V_{BB} - V_{BE}) + (1 + \beta_{dc})(R_E + R_B)I_{CBo}}{R_B + (1 + \beta_{dc})R_E} \tag{10.33}$$

This general expression provides a basis for deriving the stability factors for the collector current. These stability factors can be found by taking the appropriate partial derivatives of Equation 10.33.

Example. Determine the stability factor S_β that gives the variation of I_{CQ} with changes in β_{dc}.

Solution. From Equation 10.33,

$$\begin{aligned}
S_\beta &= \frac{\partial I_{CQ}}{\partial \beta_{dc}} \\
&= \frac{(V_{BB} - V_{BE}) + I_{CBo}(R_E + R_B)}{R_B + (1 + \beta_{dc})R_E} \\
&\quad - \frac{R_E[\beta_{dc}(V_{BB} - V_{BE}) + (1 + \beta_{dc})I_{CBo}(R_E + R_B)]}{[R_B + (1 + \beta_{dc})R_E]^2}
\end{aligned} \tag{10.34}$$

Two limiting cases are significant. The first case is observed by setting $R_E = 0$. Equation 10.34 then becomes

$$S_{\beta 1} = \frac{V_{BB} - V_{BE}}{R_B} + I_{CBo} \tag{10.35}$$

The second limiting case is for $(1 + \beta_{dc})R_E \gg R_B$. In this case, Equation 10.34 becomes

$$S_{\beta 2} = \frac{V_{BB} - V_{BE}}{(1 + \beta_{dc})^2 R_E} \tag{10.36}$$

An examination of Equations 10.35 and 10.36 shows that the inclusion of R_E under the condition $(1 + \beta_{dc})R_E \gg R_B$ provides a much lower stability factor and, thus, a more stable circuit. The benefits of a stability equation are that the designer is guided to the parameters most affecting stability and the result can be quantified.

The circuit variables that influence the quiescent drain current I_{DQ} in the MOSFET circuit of Figure 10.7 include g_{mdc}, the circuit resistances, and dc power supplies. We can express this functional dependence as

$$I_{DQ} = f(g_{mdc}, R_1, R_2, R_D, R_S, V_{DD}, \ldots) \tag{10.37}$$

from which

$$dI_{DQ} = \frac{\partial I_{DQ}}{\partial g_{mdc}} dg_{mdc} + \frac{\partial I_{DQ}}{\partial R_1} dR_1 + \cdots \tag{10.38}$$

The stability factors can be defined as previously. For example

$$S_g = \frac{\partial I_{DQ}}{\partial g_{mdc}} \tag{10.39a}$$

$$S_{R1} = \frac{\partial I_{DQ}}{\partial R_1} \tag{10.39b}$$

The quiescent drain current for the circuit in Figure 10.7 was given in Equation 10.18 as

$$I_{DQ} = \frac{g_{mdc} R_2 V_{DD}}{(1 + g_{mdc} R_S)(R_1 + R_2)} \tag{10.18}$$

This equation provides the basis for deriving the stability factors for the drain current.

Example. Determine the stability factor S_g that gives the variation of I_{DQ} with changes in g_{mdc}.

Solution. From Equation 10.18,

$$S_g = \frac{\partial I_{DQ}}{\partial g_{mdc}}$$

$$= \frac{R_2 V_{DD}}{(R_1 + R_2)(1 + g_{mdc} R_S)} - \frac{g_{mdc} R_2 V_{DD} R_S}{(R_1 + R_2)(1 + g_{mdc} R_S)^2} \tag{10.40}$$

Two limiting cases are significant. The first case is for $R_S = 0$. Then, Equation 10.40 becomes

$$S_{g1} = \frac{R_2 V_{DD}}{R_1 + R_2} \tag{10.41}$$

The second limiting case is for $g_{mdc} R_S \gg 1$. Then, Equation 10.40 becomes

$$S_{g2} = \frac{R_2 V_{DD}}{(R_1 + R_2) g_{mdc} R_S} - \frac{R_2 V_{DD}}{(R_1 + R_2) g_{mdc} R_S} = 0 \tag{10.42}$$

Thus, the inclusion of R_S under the condition that $g_{mdc}R_S \gg 1$ provides an ideal stability factor.

10.5 CURRENT SOURCES

In analog circuits, the transistor must be biased in its active region to provide a linear region for the small-signal amplifier. Up to this point, biasing has been accomplished by applying a voltage to the base–emitter junction of the bipolar transistor or to the gate source of the MOSFET. Biasing may also be accomplished by using current sources that establish the quiescent collector or drain current.

10.5.1 A Simple Bipolar Current Source

A simple bipolar current source is shown in Figure 10.9. The circuit delivers a current I_C that is approximately constant, provided the base current of the transistor can be neglected. The collector voltage of Q is assumed to be sufficiently positive with respect to the emitter so that Q is biased in its active region. Kirchhoff's voltage law around the base–emitter loop gives

$$V_{BE} + I_3 R_3 = 2V_D + I_2 R_2 \tag{10.43}$$

where V_{BE} is the base–emitter voltage drop and V_D is the voltage drop across each of the diodes. If the base current is neglected, then

$$I_1 = I_2 = \frac{V_{EE} - 2V_D}{R_1 + R_2} \tag{10.44}$$

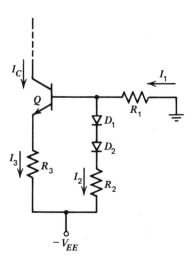

Figure 10.9. Bipolar transistor current source.

and if Equation 10.44 is substituted into 10.43, then

$$I_C \simeq I_3 = \frac{1}{R_3}\left(\frac{R_2}{R_1 + R_2}V_{EE} - V_{BE} + 2V_D\frac{R_1}{R_1 + R_2}\right) \qquad (10.45)$$

If the diodes and transistor are fabricated so that $V_D = V_{BE}$ and we choose $R_1 = R_2$, then Equation 10.45 becomes

$$I_C \simeq I_3 = \frac{V_{EE}}{2R_3} \qquad (10.46)$$

Note that the inclusion of the two diodes D_1 and D_2 has eliminated the dependence of the collector current on the temperature-sensitive base–emitter voltage. In order that $V_D = V_{BE}$, not only must the diodes and base–emitter junction be matched, but the currents I_3 and I_2 should be equal. Therefore, R_3 is chosen to obtain the desired current, and R_1 and R_2 are chosen to give $I_2 \simeq I_3$.

10.5.2 A Two-Transistor Current Source

The current source analyzed in Figure 10.9 is extremely useful but requires two diodes and three resistors in addition to the transistor. Figure 10.10 shows a two-transistor current source with Q_1 connected as a diode. The base–collector voltage of Q_1 is zero, so that the transistor acts as if it were in the forward active region of operation. If Q_1 and Q_2 are identical, then $\beta_{dc1} = \beta_{dc2}$; and since Q_1 and Q_2 have the same base–emitter voltage, $I_{B1} = I_{B2}$ and $I_{C1} = I_{C2}$. Kirchhoff's current law applied at the collector of Q_1 gives

$$I_1 = I_{C1} + I_{B1} + I_{B2} = I_{C1} + \frac{2I_{C1}}{\beta_{dc}}$$
$$= I_{C1}\left(1 + \frac{2}{\beta_{dc}}\right) \qquad (10.47)$$

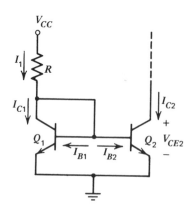

Figure 10.10. Two-transistor current source.

where $\beta_{dc} = \beta_{dc1} = \beta_{dc2}$. Then,

$$I_{C1} = I_{C2} = \frac{I_1}{1 + (2/\beta_{dc})} \qquad (10.48)$$

If β_{dc} is large, then we have

$$I_{C2} \simeq I_1 = \frac{V_{CC} - V_{BE}}{R} \qquad (10.49)$$

In the case that Q_1 and Q_2 are identical and β_{dc1} and β_{dc2} are large and equal, then the load current I_{C2} is equal to the reference current I_1.

The above analysis assumed that the dc current gain for the two transistors were identical. However, the load current I_{C2} will be a function of V_{CE2}. This effect is most easily seen by observing the common-emitter transistor characteristics (for example, Figure 10.5). The collector current will increase slowly as the collector–emitter voltage increases for a constant base current. In the current source of Figure 10.10, $V_{CE2} \geq V_{CE1}$ since $V_{CE1} = V_{BE1}$. The value of V_{CE2} depends upon the load resistance and bias voltage connected to the collector of Q_2, but normally $V_{CE2} >> V_{CE1}$. In practice, I_{C2} may be larger than I_{C1} by as much as 10%, depending upon the quiescent values of V_{CE2} and I_{C2}.

The stability factors for this circuit may be determined in a way similar to the stability factor derivations for the circuits in Figures 10.1 and 10.3. The stability factors for the variation in I_{C2} with current gain β_{dc}, base–emitter voltage V_{BE}, and collector–emitter voltage V_{CE2} are of interest. These factors are defined respectively as

$$S_\beta = \frac{\partial I_{C2}}{\partial \beta_{dc}} \qquad (10.50a)$$

$$S_{V1} = \frac{\partial I_{C2}}{\partial V_{BE}} \qquad (10.50b)$$

$$S_{V2} = \frac{\partial I_{C2}}{\partial V_{CE}} \qquad (10.50c)$$

The first two of these factors can be determined from Equations 10.48 and 10.49. From Equation 10.48,

$$S_\beta = \frac{\partial I_{C2}}{\partial \beta_{dc}} = \frac{2I_1}{(2 + \beta_{dc})^2} \qquad (10.51)$$

$$= \frac{2(V_{CC} - V_{BE})}{(2 + \beta_{dc})^2 R}$$

This stability factor is of the same order of magnitude as for the circuit in Figure 10.3, in which the emitter resistor provides the stability against Q-point variation due to changes in β_{dc}. The advantage of the two-transistor current source shown in Figure 10.10 is the reduction in the number of resistors that makes the circuit much more applicable to integrated circuits.

The second stability factor, S_{V1}, is found from Equation 10.49 as

$$S_{V1} = \frac{\partial I_{C2}}{\partial V_{BE}} = -\frac{1}{R} \tag{10.52a}$$

An increased value for R provides better stability against base–emitter voltage variation.

The third stability factor to be considered is

$$S_{V2} = \frac{\partial I_{C2}}{\partial V_{CE2}} \tag{10.52b}$$

This stability factor is the slope of the transistor characteristic, I_C versus V_{CE}, in Figure 10.5 (for example). The slope of the curve is the output admittance of the Q_2 transistor, which in this case is simply h_{oe}.

10.5.3 A Three-Transistor Current Source

To increase this Q-point stabilization, a third transistor can be added to provide more current gain, as shown in Figure 10.11. The three transistors are assumed to be identical, so that

$$\beta_{dc} = \beta_{dc1} = \beta_{dc2} = \beta_{dc3}$$

The base–emitter voltages of Q_1 and Q_2 are identical, so that $I_{B1} = I_{B2}$ and $I_{C1} = I_{C2}$. The emitter current of Q_3 is

$$I_{E3} = I_{B1} + I_{B2} = \frac{I_{C1}}{\beta_{dc}} + \frac{I_{C2}}{\beta_{dc}} = \frac{2I_{C2}}{\beta_{dc}} \tag{10.53}$$

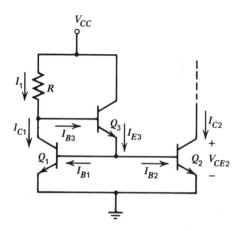

Figure 10.11. Current source using an emitter follower Q_3, to provide a more stable circuit.

The base current of Q_3 is then

$$I_{B3} = \frac{I_{E3}}{1 + \beta_{dc}} \tag{10.54}$$

$$= \frac{2I_{C2}}{\beta_{dc}(1 + \beta_{dc})}$$

Finally, the reference current is

$$I_1 = I_{C1} + I_{B3} = I_{C2} + \frac{2I_{C2}}{\beta_{dc}(1 + \beta_{dc})} \tag{10.55}$$

Solving for the output current yields

$$I_{C2} = \frac{I_1}{1 + \{2/[\beta_{dc}(1 + \beta_{dc})]\}} \tag{10.56}$$

where I_1 is given by

$$I_1 = \frac{V_{CC} - 2V_{BE}(\text{act})}{R} \tag{10.57}$$

The stability factor for current gain variation is found by differentiating Equation 10.56 with respect to β_{dc}. This gives

$$S_\beta = \frac{\partial I_{C2}}{\partial \beta_{dc}} \tag{10.58}$$

$$= \frac{2I_1}{[\beta_{dc}(1 + \beta_{dc}) + 2]^2/(1 + 2\beta_{dc})}$$

Previously, with the two transistor current source, the stability factor was approximately proportional to $1/\beta_{dc}^2$ (Equation 10.51). For the three-transistor current source, the stability factor is now nearly proportional to $1/\beta_{dc}^3$. Going from the two- to the three-transistor current source changes the stability factor by a factor of $1/\beta_{dc}$.

10.5.4 Widlar Current Source

In both the two-transistor and three-transistor current sources, the load current is approximately equal to the reference current I_1. For supply voltages of 5.5 to 10 V and load currents of a few milliamperes, the reference resistor is of the order of a few thousand ohms. However, if the load current has to be on the order of microamperes, then the reference resistor has to be in the megohm range. This large resistance value is difficult to achieve with accuracy in integrated circuits.

The load current and reference current can be made substantially different if the base–emitter voltages of Q_1 and Q_2 are made unequal. The Widlar current source shown in Figure 10.12 achieves this objective. Kirchhoff's voltage law

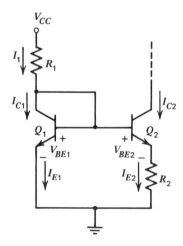

Figure 10.12. Widlar current source.

applied between the common base connection and ground results in

$$V_{BE1} = V_{BE2} + I_{E2}R_2 \tag{10.59}$$

The emitter currents of Q_1 and Q_2 can be related to the base–emitter voltages by the diode equation. If the -1 term is neglected, then

$$I_{E1} \simeq I_{s1} \exp\left(\frac{qV_{BE1}}{kT}\right) \tag{10.60a}$$

and

$$I_{E2} \simeq I_{s2} \exp\left(\frac{qV_{BE2}}{kT}\right) \tag{10.60b}$$

For identical transistors, $I_{s1} = I_{s2} \equiv I_s$. The base–emitter voltages in Equations 10.60 can be written in terms of the emitter currents as

$$V_{BE1} = \frac{kT}{q} \ln\left(\frac{I_{E1}}{I_s}\right) \tag{10.61a}$$

$$V_{BE2} = \frac{kT}{q} \ln\left(\frac{I_{E2}}{I_s}\right) \tag{10.61b}$$

Substituting these into Equation 10.59 yields

$$\frac{kT}{q} \ln\left(\frac{I_{E1}}{I_s}\right) = \frac{kT}{q} \ln\left(\frac{I_{E2}}{I_s}\right) + I_{E2}R_2 \tag{10.62}$$

which can be rearranged to give

$$\frac{kT}{q} \ln\left(\frac{I_{E1}}{I_{E2}}\right) = I_{E2}R_2 \tag{10.63}$$

If base currents are neglected, this equation can be written as

$$\frac{kT}{q} \ln \left(\frac{I_{C1}}{I_{C2}}\right) = I_{C2}R_2 \qquad (10.64)$$

and the collector current I_{C1} is approximately the same as the reference current I_1.

If, for example, we want $I_{C2} = 5.0~\mu A$, $I_{C1} \simeq I_1 = 1.0$ mA, $V_{CC} = 5.0$ V, $V_{BE1} = 0.70$ V, and, at room temperature, $kT/q = 0.026$ V, then, from Equation 10.64,

$$(0.026) \ln \left(\frac{1.0 \times 10^{-3}}{5.0 \times 10^{-6}}\right) = (5.0 \times 10^{-6})R_2 \qquad (10.65)$$

giving

$$R_2 = 27.6~k\Omega \qquad (10.66)$$

The reference resistor R_1 is found from

$$I_1 = \frac{V_{CC} - V_{BE1}}{R_1} \qquad (10.67)$$

so that, for this example,

$$R_1 = \frac{5.0 - 0.70}{1.0 \times 10^{-3}}$$
$$= 4.3~k\Omega \qquad (10.68)$$

If, on the other hand, R_1 and R_2 are given, the reference current I_1 can be determined as before from Equation 10.67, but the load current I_{C2} must be determined from Equation 10.64 by graphical techniques or by iteration.

A more generalized two-transistor current source contains resistors in the emitters of both transistors as shown in Figure 10.13. The analysis proceeds as before by writing Kirchhoff's voltage law between the common base connection and ground. If the two transistors are identical and base currents are neglected, then the load and reference currents are related by

$$\frac{kT}{q} \ln \left(\frac{I_{C1}}{I_{C2}}\right) = I_{C2}R_2 - I_{C1}R_3 \qquad (10.69)$$

and

$$I_{C1} \simeq I_1 \qquad (10.70)$$

The load current is also a function of the collector–emitter voltage of the Q_2 transistor as mentioned previously for the circuit of Figure 10.10.

The output resistance of the Widlar current source (Figure 10.12) can be determined by using the small-signal equivalent circuit shown in Figure 10.14a. Note that the base and collector of Q_1 are electrically connected. The output resistance R_o of the circuit is then found by taking the ratio of the test signals

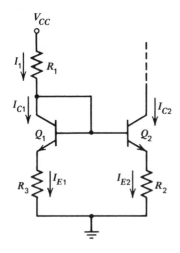

Figure 10.13. Widlar current source with two emitter resistors.

V_x and I_x, so that

$$R_o = \frac{V_x}{I_x} \tag{10.71}$$

The analysis is simplified by considering the resistance R_o', which is the resistance looking into the base of the Q_1 transistor. This resistance is defined as the ratio of V_x' to I_x' shown in Figure 10.14a. Applying Kirchhoff's current law

$$I_{b1} + h_{fe}I_{b1} + \frac{V_x'}{R_1} = I_x' \tag{10.72}$$

But we have

$$I_{b1} = \frac{V_x'}{h_{ie1}} \tag{10.73}$$

so that

$$I_x' = (1 + h_{fe})\frac{V_x'}{h_{ie1}} + \frac{V_x'}{R_1} \tag{10.74}$$

Then,

$$\frac{1}{R_o'} = \frac{I_x'}{V_x'} = \frac{1 + h_{fe}}{h_{ie1}} + \frac{1}{R_1} \tag{10.75}$$

or

$$R_o' = R_1 \| [h_{ie1}/(1 + h_{fe})] \tag{10.76}$$

In the Widlar source, $h_{ie1} < h_{ie2} \left(h_{ie} \simeq \dfrac{\beta kT}{I_{CQ}q} \right)$. In Figure 10.14a, the resistance R_o' is in series with h_{ie2}. If we assume that $h_{ie1}/(1 + h_{fe}) \ll h_{ie2}$, then $R_o' \ll$

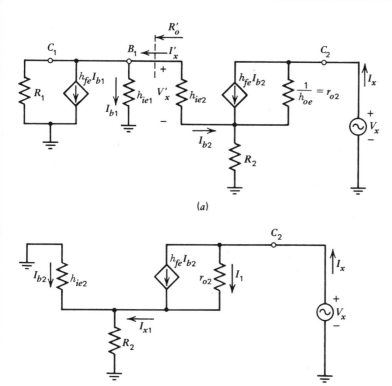

Figure 10.14. (a) The h-parameter model of the Widlar current
source of Figure 10.12. (b) The same circuit but with the effective
resistance of the Q_1 transistor neglected.

h_{ie2}, so that R_o' may be neglected compared to h_{ie2}. The small-signal equivalent
circuit can then be simplified to that of Figure 10.14b.

We observe that

$$I_1 + h_{fe}I_{b2} = I_x = I_{x1} \tag{10.77}$$

and

$$I_{b2} = \frac{-R_2 I_{x1}}{R_2 + h_{ie2}} = \frac{-R_2 I_x}{R_2 + h_{ie2}} \tag{10.78}$$

Substituting into Equation 10.77 yields

$$I_1 = I_x + \frac{R_2 I_x}{R_2 + h_{ie2}} \tag{10.79}$$

The test voltage is

$$V_x = I_1 r_{o2} - I_{b2} h_{ie2} \tag{10.80}$$

Substituting Equations 10.78 and 10.79 into Equation 10.80 results in an expression for the output resistance given by

$$R_o = \frac{V_x}{I_x} = \left(1 + \frac{h_{fe}R_2}{R_2 + h_{ie2}}\right)r_{o2} + \frac{R_2 h_{ie2}}{R_2 + h_{ie2}} \tag{10.81}$$

Note that the output resistance is substantially increased with the inclusion of the emitter resistor R_2. Analysis of the circuit in Figure 10.13 shows that including resistors in both emitters increases the output resistance even more. The output resistance may be in the megohm range, making the circuit a nearly ideal current source.

10.5.5 A Simple MOS Current Source

In analog circuits, MOS transistors are biased in their active amplifying regions, and a constant-current source may be used to achieve this result. A simple NMOS current source is shown in Figure 10.15. The transistors are assumed to be enhancement mode devices, so that M_1 and M_2 are connected as saturated transistors. The reference current I_{REF} and load current I_L are determined from the equations

$$I_{REF} = k_1(V_{GS1} - V_{T1})^2 = k_2(V_{GS2} - V_{T2})^2 \tag{10.82}$$

$$I_L = k_3(V_{GS3} - V_{T3})^2 \tag{10.83}$$

For the load current in Equation 10.83, we assume that M_3 is in its saturation region. The voltage V_{GS2}, can be found from Equation 10.82 by noting that $V_{GS1} = V_{DD} - V_{GS2}$. The result is

$$\left[1 + \left(\frac{k_1}{k_2}\right)^{1/2}\right]V_{GS2} = \left(\frac{k_1}{k_2}\right)^{1/2}(V_{DD} - V_{T1}) + V_{T2} \tag{10.84}$$

Since $V_{GS2} = V_{GS3}$, the load current I_L is then determined from Equation 10.83. If M_3 remains in the saturation region, I_L will change only slightly as the drain-to-source voltage of M_3 changes due to the output resistance of M_3.

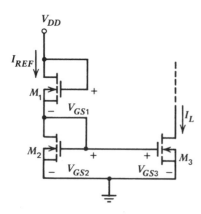

Figure 10.15. An *n*-channel MOSFET current source.

Example. In Figure 10.15, V_{DD} = 10 V, V_T = 1.0 V, and μC_{ox} = 3.0 × 10^{-5} s/V-cm². If I_L = 25 μA, I_{REF} = 100 μA, and V_{G3} is to be kept at a level of 1.5 V, find the width and length dimensions for each transistor.

Solution. Since M_3 is assumed to be in saturation, then

$$I_L = k_3(V_{GS3} - V_T)^2$$

and

$$k_3 = \frac{I_L}{(V_{GS3} - V_T)^2}$$

$$= \frac{25 \times 10^{-6}}{(1.5 - 1.0)^2}$$

$$= 1.0 \times 10^{-4} \text{ A/V}^2$$

From Equation 5.11,

$$k_3 = \frac{\mu C_{ox}}{2} \frac{W_3}{L_3}$$

then

$$\frac{W_3}{L_3} = k_3 \frac{2}{\mu C_{ox}} = (1.0 \times 10^{-4}) \frac{2}{3.0 \times 10^{-5}}$$

$$= 6.7$$

A value must now be chosen for W_3 or L_3. A reasonable low value for L_3 is 10 μm, making W_3 = 67 μm.

Transistor M_2 has I_{D2} = I_{REF} = 100 μA and V_{GS2} = V_{GS3} = 1.5 V. Since M_2 is in saturation, then

$$I_{D2} = k_2(V_{GS2} - V_T)^2$$

or

$$k_2 = \frac{I_{D2}}{(V_{GS2} - V_T)^2}$$

$$= \frac{100 \times 10^{-6}}{(1.5 - 1.0)^2}$$

$$= 4.0 \times 10^{-4} \text{ A/V}^2$$

Equation 5.11 gives

$$\frac{W_2}{L_2} = k_2 \frac{2}{\mu C_{ox}} = (4.0 \times 10^{-4}) \frac{2}{3.0 \times 10^{-5}}$$

$$= 26.7$$

If L_2 is chosen as 10 μm, then W_2 = 267 μm.

Transistor M_1 is in saturation, $I_{D1} = I_{REF} = 100 \ \mu A$, and

$$V_{GS1} = V_{DD} - V_{GS2}$$
$$= 10 - 1.5$$
$$= 8.5 \ V$$

Since

$$I_{D1} = k_1(V_{GS1} - V_T)^2$$

then

$$k_1 = \frac{I_{D1}}{(V_{GS1} - V_T)^2}$$
$$= \frac{100 \times 10^{-6}}{(8.5 - 1.0)^2}$$
$$= 1.8 \times 10^{-6} \ A/V^2$$

Equation 5.11 gives

$$\frac{W_1}{L_1} = k_1 \frac{2}{\mu C_{ox}}$$
$$= (1.8 \times 10^{-6}) \frac{2}{3.0 \times 10^{-5}}$$
$$= \frac{1}{8.3}$$

If W_1 is chosen as 10 μm, then $L_1 = 83 \ \mu m$.

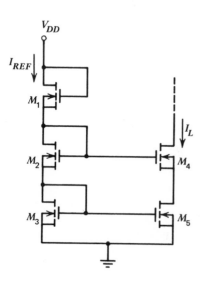

Figure 10.16. An n-channel MOSFET current source with transistor M_5 added to provide source degeneration for M_4.

The output resistance of the current source can be increased by adding additional transistors such as in Figure 10.16. A detailed analysis shows that the output resistance is given by

$$R_{\text{out}} = r_{ds2} + r_{ds4} + r_{ds2}r_{ds4}(1 + \overset{\wedge}{\Delta_4})g_{m4} \qquad (10.85)$$

where Δ is a parameter related to the body effect. This parameter is defined by the equation

$$\Delta = \frac{1}{g_m} \frac{\partial I_D}{\partial V_{BS}} \qquad (10.86)$$

For the output resistance to be maintained at this level, the transistor M_4 must remain in the saturation region.

10.6 SUMMARY

Biasing is used to establish the dc operating points of transistors in analog circuits. The power gain observed in the amplification of signals is obtained from the dc power supplies. The analysis of amplifiers is usually performed in two steps: (1) dc analysis to obtain the operating point, and (2) ac analysis to obtain the gain and impedance characteristics.

In discrete component circuits, large-valued capacitors are used to provide dc isolation between devices and to bypass ac signals in most configurations. Those large-valued capacitors are not practical in integrated circuits, and current sources using additional transistors are popular biasing schemes in both bipolar and MOS analog integrated circuits.

An important consideration in the design of bias circuitry is stability. An exhaustive study of this subject would result in the determination of a stability factor for every variable that could possibly result in a change in the quiescent operating point. A more realistic approach is to select particular variables such as temperature, current gain, or resistor values and determine their effects on the Q-point. It is important that the most significant variables be considered, since a poor selection at this point could make an otherwise good design useless.

PROBLEMS

10.1. Calculate the Q-point for the circuit in Figure 10.17 for two values of forward current gain: $\beta_{dc} = 50$ and $\beta_{dc} = 200$, letting $V_{BE} = 0.70$ V. Do this for (a) $R_B = 10$ kΩ and (b) $R_B = 100$ kΩ.

10.2. Calculate the percentage change in collector current if the β_{dc} changes from 50 to 20 in the circuit of Figure 10.18. Let $V_{BE} = 0.70$ V.

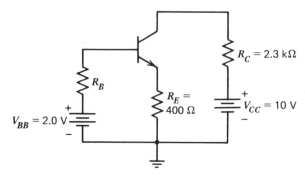

Figure 10.17. Circuit for Problem 10.1.

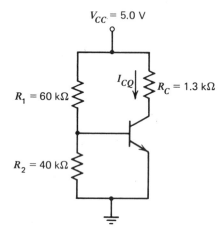

Figure 10.18. Circuit for Problem 10.2.

10.3. Calculate the quiescent values of I_C and V_{CE} in the circuit of Figure 10.19 for forward current gains of $\beta_{dc} = 100$ and $\beta_{dc} = 50$. Assume that $V_{BE} = 0.70$ V. Plot the Q-points on a dc load line.

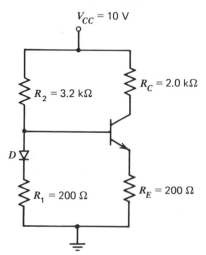

Figure 10.19. Circuit for Problem 10.3.

10.4. Find R_1, R_2, and R_C in Figure 10.20 so that $I_{CQ} = 2.0$ mA and $V_{CEQ} = 5.0$ V. Let $R_1 + R_2 = 100$ kΩ, $V_{BE} = 0.70$ V, and $\beta_{dc} = 75$.

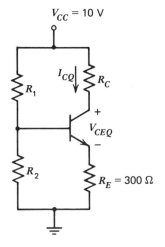

Figure 10.20. Circuit for Problem 10.4.

10.5. Calculate the stability factor S_β for the circuit in Problem 10.2 for each of the conditions listed.

10.6. Calculate the stability factor S_β for the circuit in Problem 10.4 for (a) $\beta_{dc} = 100$ and (b) $\beta_{dc} = 50$.

10.7. Find $S_I = \partial I_{CQ}/\partial I_{CBo}$ for the circuit in Figure 10.4.

10.8. An n-MOS common source circuit is shown in Figure 10.21. (a) Calculate the quiescent values for I_D and V_{DS} assuming that $k = 0.025$ mA/V² and $V_T = 0.75$ V. (b) If the value of k changes to 0.015 mA/V², find the new quiescent values.

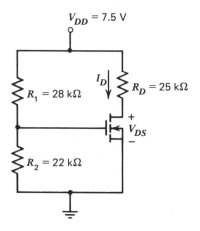

Figure 10.21. Circuit for Problem 10.8.

10.9. In the circuit of Problem 10.8, V_{DD} is changed to 10 V, but the values of R_D, k, and V_T remain constant. Calculate the values of R_1 and R_2 required to place the Q-point in the center of the load line if $k = 0.025$ mA/V². Let $R_1 + R_2 = 50$ kΩ.

10.10. For the circuit in Figure 10.22, (*a*) calculate the quiescent drain current I_{DQ} and drain to source voltage V_{DSQ} if the *n*-channel enhancement mode transistor has the properties that $k = 0.01$ mA/V² and $V_T = 1.0$ V. (*b*) If k changes to 0.05 mA/V², what are the values of I_{DQ} and V_{DSQ}?

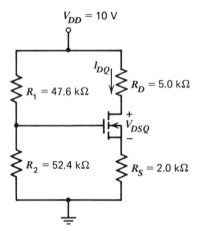

$V_{DD} = 10$ V

I_{DQ}

$R_1 = 47.6$ kΩ

$R_D = 5.0$ kΩ

$+$

V_{DSQ}

$-$

$R_2 = 52.4$ kΩ

$R_S = 2.0$ kΩ

Figure 10.22. Circuit for Problem 10.10.

10.11. Verify that the *n*-MOS transistors in Problems 10.8, 10.9, and 10.10 are biased in the saturation region.

10.12. Calculate the stability factor $S_K = \partial I_D / \partial k$ for the circuit in Problem 10.10 for (*a*) $k = 0.01$ mA/V² and (*b*) $k = 0.05$ mA/V².

10.13. Calculate the values for R_1, R_2, and R_3 for the circuit in Figure 10.23 so that $V_{CE1} = V_{CE2} = 2.0$ V. Design the current source so that $I_2 = I_3$ and that the circuit is stabilized against variations in V_{BE}. Assume that $\beta_{dc} = 200$ for each transistor, $V_{BE} = 0.70$ V, and neglect the base currents.

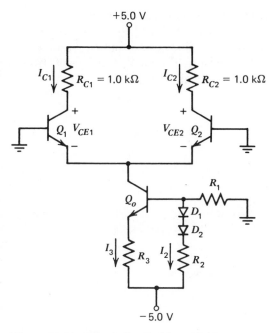

Figure 10.23. Circuit for Problem 10.13.

10.14. Calculate I_{C2} for the two transistor source shown in Figure 10.24 for $\beta_{dc} = 20$ and $V_{BE} = 0.70$ V.

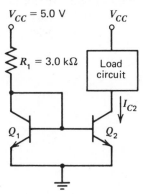

Figure 10.24. Circuit for Problem 10.14.

10.15. A two-transistor current source is shown in Figure 10.25. For V_{CE1} $= V_{CE2} = 2.0$ V and $I_{C1} = I_{C2} = 1.0$ mA (when V_1 and V_2 are at zero volts), find (a) the values of R_1 and R_2 and (b) the values of I_{C4} and I_1. Assume that $\beta_{dc} = 50$ and $V_{BE} \doteq 0.70$ V for each transistor. If both V_1 and V_2 shift from 0 to -2.0 V and $r_o = 25$ kΩ, what is the new value of I_{C4}.

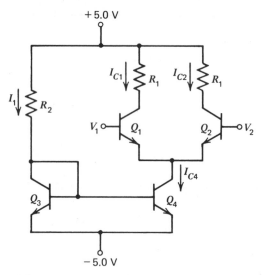

Figure 10.25. Circuit for Problem 10.15.

10.16. For the three-transistor current source shown in Figure 10.26, (a) find I_1, I_{C2}, I_{C4}, V_{CE2}, and V_{CE4}. Let $\beta_{dc} = 100$ and $V_{BE} = 0.7$ V for each transistor. (b) Find values for I_{C4}, I_{C2}, I_1, and R_1 so that $V_{CE4} = 2.5$ V.

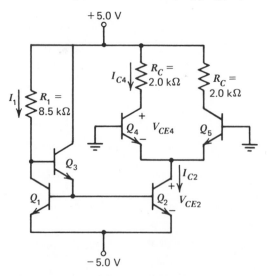

Figure 10.26. Circuit for Problem 10.16.

10.17. Find the stability factors for the following circuits: (a) Find $S_\beta = \partial I_{C4}/\partial \beta_{dc}$ in Problem 10.15. (b) Find $S_\beta = \partial I_{C2}/\partial \beta_{dc}$ in Problem 10.16 for $\beta_{dc} = 10$ and $\beta_{dc} = 50$. (c) Find $S_V = \partial I_{C2}/dV_{CE2}$ in Problem 10.16 for $r_o = 25$ kΩ.

10.18. Determine S_{V2} for the generalized two-transistor current source (see equation 10.52b).

10.19. A Widlar current source is shown in Figure 10.27 with $V_{BE1} = 0.70$ V and $\beta_{dc} = 100$. (a) Neglect base currents and calculate R_1 and R_2 so that $I_1 = 1.0$ mA and $I_{C2} = 10$ μA. (b) If $R_1 = 7.75$ kΩ and $R_2 = 5.0$ kΩ, find I_{C2}.

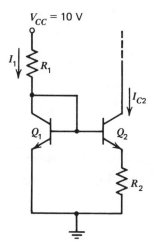

Figure 10.27. Circuit for Problem 10.19.

10.20. An n-MOS current source is shown in Figure 10.28. (a) Find I_L if all transistors have a conduction parameter of $k = 0.1$ mA/V^2 and a threshold voltage of $V_T = 1.0$ V. (b) Calculate I_{REF} and I_L for $k_1 = 0.1$ mA/V^2, $k_2 = 0.3$ mA/V^2, and $V_{T1} = V_{T2} = V_{T3} = 1.0$ V. (c) Calculate k_1, I_{REF}, V_{GS1}, V_{GS2}, and V_{GS3} for $k_2 = 0.1$ mA/V^2, $k_3 = 0.2$ mA/V^2 and $I_L = 1.0$ mA. (d) Let $V_{T1} = V_{T2} = V_{T3} = 0.5$ V, $I_L = 0.1$ mA, and $I_{REF} = 1.0$ mA. If $k_1 = 3k_3$ and $k_2 = 2k_3$ find V_{GS1}, V_{GS2}, V_{GS3}, k_1, k_2, and k_3.

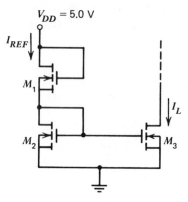

Figure 10.28. Circuit for Problem 10.20.

10.21. The n-MOS current source in Figure 10.16 has transistors with $k = 0.1$ mA/V^2 and $V_T = 1.0$ V. If $V_{DD} = 10$ V, calculate I_L.

chapter 11

SMALL-SIGNAL TRANSISTOR MODELS

In a large majority of analog circuits, analysis can be performed on the basis of "small-signal" models. The term "small signal" is a relative one and depends on the device under consideration. The basic criterion for this condition is that the excursions of the signal from the operating point are limited to magnitudes such that the device characteristic can be accurately approximated by a linear expression. For the purpose of analysis, the small-signal approximation is usually assumed, even though it may not strictly apply. This is because it greatly simplifies the analysis. Nonlinearities in device characteristics result in an output signal that is not a faithful reproduction of the input signal. Analysis of distortions requires more advanced techniques. The small-signal model is a useful tool for the analysis of analog electronic circuits.

There are two basic approaches to the development of small-signal models. One is based on the relationships between the voltages and currents at the external terminals. This is a powerful and general technique that can be applied to a variety of devices, including bipolar and field-effect transistors, and even entire integrated circuit amplifiers. The other approach is to construct a model by representing the physical processes within the device with circuit components. This technique requires a different type of model for each type of device but provides powerful insights into the device.

In this chapter, we develop small-signal models for bipolar and field effect transistors, including frequency effects.

11.1 TWO-PORT NETWORK MODELS

A device such as a bipolar or field effect transistor, biased in its amplifying region of operation, can be represented as a two-port linear network. This is indicated in Figure 11.1. For this analysis, it is assumed that the two lower terminals, *B* and *D*, are connected internally, making these terminals common to the two ports. The currents and voltages are assumed to be sinusoidal, and their amplitudes are considered to be small signals. The notation indicates the rms-valued phasors for the terminal voltages and currents.

It is customary in network theory to designate all currents as positive when they enter the network, as indicated in the Figure 11.1. There are six ways that terminal variables can be related. Three of the relationships between the terminal variables can be written in phasor notation as

Impedance parameters:

$$V_1 = z_{11}I_1 + z_{12}I_2$$
$$V_2 = z_{21}I_1 + z_{22}I_2$$

(11.1)

Admittance parameters:

$$I_1 = y_{11}V_1 + y_{12}V_2$$
$$I_2 = y_{21}V_1 + y_{22}V_2$$

(11.2)

Hybrid parameters:

$$V_1 = h_{11}I_1 + h_{12}V_2$$
$$I_2 = h_{21}I_1 + h_{22}V_2$$

(11.3)

Each pair of relationships can be represented by an equivalent circuit with two dependent sources, as shown in Figure 11.2.

An examination of the equations relating the parameters to the terminal

Figure 11.1. Two-port network with indicated current and voltage conventions.

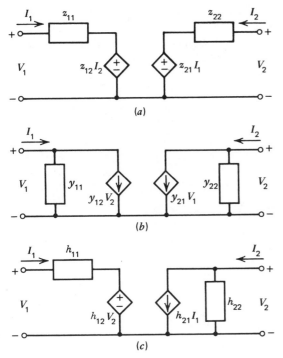

Figure 11.2. Electrical equivalent circuits of the z-, y-, and h-parameter two-port network equations.

currents and voltages indicates that the parameters can be determined by measurement. For example,

$$z_{11} = \frac{V_1}{I_1} \qquad \text{with } I_2 = 0 \qquad (11.4)$$

We must be careful in our interpretation of this equation. The statement $I_2 = 0$ means that the ac current is zero, but the dc bias current must be maintained at its previous level. It is important to note that the small-signal parameters are only meaningful when the operating point is unaffected by the measurement procedure. The ac current I_2 can be made vanishingly small by including a very large inductance in series with the current entering terminal C. In practice, this is very difficult to achieve, particularly at high frequencies, since a physical inductor has shunt capacitance associated with its windings. The admittance parameters are preferred for field effect transistors. The hybrid parameters can be readily measured at moderate frequencies, and they are the most popular of the low-frequency small-signal parameters for bipolar transistors. The admittance parameters are used for both bipolar and field effect transistors for high-frequency measurements, since they are determined by providing ac short circuits (large-valued capacitors) across the terminals.

It is important to remember that small-signal two-port parameters depend on both the operating point and the frequency. The variations of the hybrid parameters with operating point for a typical bipolar transistor are indicated in Figure 4.13.

The sets of two-port parameters are interrelated. Once a particular set of parameters has been determined (at a given frequency and operating point), the other sets of parameters can be calculated. For example, suppose that a set of admittance parameters has been measured for a two-port network and it is desired to work with the hybrid parameters for the same network under the same operating conditions. The calculation can be performed as follows:

$$I_1 = y_{11}V_1 + y_{12}V_2$$

or

$$V_1 = \frac{1}{y_{11}} I_1 - \frac{y_{12}}{y_{11}} V_2$$

and

$$I_2 = y_{21}V_1 + y_{22}V_2$$

or

$$I_2 = \frac{y_{21}}{y_{11}} I_1 - \frac{y_{21}y_{12}}{y_{11}} V_2 + y_{22}V_2$$

Then, comparing to Equation 11.3,

$$h_{11} = \frac{1}{y_{11}}$$

$$h_{12} = - \frac{y_{12}}{y_{11}}$$

$$h_{21} = \frac{y_{21}}{y_{11}}$$ (11.5)

$$h_{22} = y_{22} - \frac{y_{21}y_{12}}{y_{11}}$$

The hybrid parameters are frequently given more descriptive names such as

$$h_{11} = h_i = \text{input impedance}$$
$$h_{12} = h_r = \text{reverse voltage feedback ratio}$$
$$h_{21} = h_f = \text{forward current transfer ratio}$$
$$h_{22} = h_o = \text{output admittance}$$

This notation makes it easier to remember the units and functions of the hybrid parameters.

11.1.1 Hybrid Parameters for Bipolar Transistors

A bipolar transistor can be connected in a circuit in three ways, called the common-emitter, common-base, and common-collector connections. These configurations are shown in Figure 11.3, and the terminals marked A, B, C, and D correspond to the same terminals in Figure 11.1.

It is obvious that measurements of the hybrid parameters in the three different configurations will yield three different sets of results. The customary notation is to designate the parameters with the common terminal as a subscript. For example, one set of parameters is h_{ie}, h_{re}, h_{fe}, and h_{oe} for a common-emitter configuration, and another is h_{ib}, h_{rb}, h_{fb}, and h_{ob} for a common-base connection.

There is an interrelation between these sets of hybrid parameters. The relationship between the common-emitter hybrid parameters and the common-base hybrid parameters can be determined in the following way. For the common-emitter circuit,

$$V_{be} = h_{ie}I_b + h_{re}V_{ce} \tag{11.6}$$

$$I_c = h_{fe}I_b + h_{oe}V_{ce} \tag{11.7}$$

For the common-base circuit,

$$V_{eb} = h_{ib}I_e + h_{rb}V_{cb} \tag{11.8}$$

$$I_c = h_{fb}I_e + h_{oe}V_{cb} \tag{11.9}$$

The other important relations are

$$I_e + I_c + I_b = 0 \tag{11.10}$$

$$V_{be} = -V_{eb} \tag{11.11}$$

and

$$V_{ce} = V_{cb} + V_{be} \tag{11.12}$$

Rewriting Equation 11.8,

$$V_{be} = -h_{ib}I_e - h_{rb}V_{cb}$$

Substituting Equations 11.10 and 11.12 yields

$$V_{be} = -h_{ib}(-I_c - I_b) - h_{rb}(V_{ce} - V_{be}) \tag{11.13}$$

Similarly, Equation 11.9 can be rewritten

$$I_c = h_{fb}(-I_c - I_b) + h_{ob}(V_{ce} - V_{be})$$

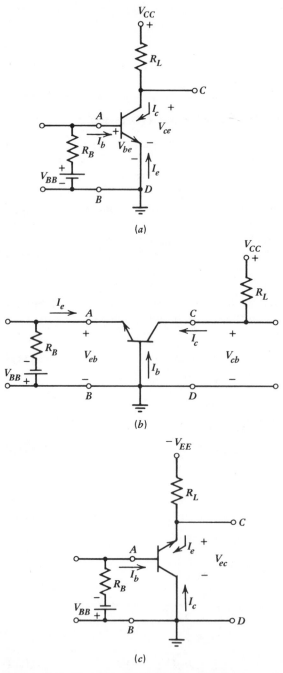

Figure 11.3. The three basic bipolar transistor configurations: (*a*) Common-emitter amplifier. (*b*) Common-base amplifier. (*c*) Common-collector amplifier.

Solving this for I_c yields

$$I_c = \frac{-h_{fb}}{1 + h_{fb}} I_b + \frac{h_{ob}}{1 + h_{fb}} (V_{ce} - V_{be}) \tag{11.14}$$

Substituting Equation 11.14 into 11.13 and rearranging provides

$$V_{be} = \frac{h_{ib}}{(1 + h_{fb})(1 - h_{rb}) + h_{ob}h_{ib}} I_b$$

$$+ \frac{h_{ob}h_{ib} - h_{rb}(1 + h_{fb})}{(1 + h_{fb})(1 - h_{rb}) + h_{ob}h_{ib}} V_{ce} \tag{11.15}$$

Similarly,

$$I_c = \frac{-h_{fb}(1 - h_{rb}) - h_{ob}h_{ib}}{(1 + h_{fb})(1 - h_{rb}) + h_{ob}h_{ib}} I_b$$

$$+ \frac{h_{ob}(1 - h_{rb})}{(1 + h_{fb})(1 - h_{rb}) + h_{ob}h_{ib}} V_{ce} \tag{11.16}$$

It is usually possible to approximate these relations by assuming that h_{rb} and h_{ob} are very small. Then, Equation 11.15 becomes

$$V_{be} \simeq \frac{h_{ib}}{1 + h_{fb}} I_b$$

$$+ \frac{h_{ib}h_{ob} - h_{rb}(1 + h_{fb})}{1 + h_{fb}} V_{ce} \tag{11.17}$$

and Equation 11.16 becomes

$$I_c \simeq \frac{-h_{fb}}{1 + h_{fb}} I_b + \frac{h_{ob}}{1 + h_{fb}} V_{ce} \tag{11.18}$$

A comparison of Equations 11.17 to 11.6 and 11.18 to 11.7 indicates the following approximate relationships between the common-emitter and common-base hybrid parameters:

$$h_{ie} = \frac{h_{ib}}{1 + h_{fb}}$$

$$h_{re} = \frac{h_{ib}h_{ob} - h_{rb}(1 + h_{fb})}{1 + h_{fb}}$$

$$h_{fe} = \frac{-h_{fb}}{1 + h_{fb}} \tag{11.19}$$

$$h_{oe} = \frac{h_{ob}}{1 + h_{fb}}$$

Example. The measured common-base hybrid parameters for a particular

2N5088 small-signal transistor at 1 kHz, with $I_{CQ} = 1$ mA and $V_{CBQ} = 10$ V, are

$$h_{ib} = 34.3 \ \Omega$$
$$h_{rb} = 3.9 \times 10^{-4}$$
$$h_{fb} = -0.997$$
$$h_{ob} = 4.4 \times 10^{-8} \ \text{S}$$

Using Equations 11.19, the common-emitter hybrid parameters are given approximately by

$$h_{ie} = \frac{34.3}{1 - 0.997} = 11.4 \ \text{k}\Omega$$

$$h_{re} = \frac{[34.3 \times 4.4 \times 10^{-8} - 3.9 \times 10^{-4}(1 - 0.997)]}{1 - 0.997}$$
$$= 1.13 \times 10^{-4}$$

$$h_{fe} = \frac{+ \ 0.997}{1 - 0.997} = 332$$

$$h_{oe} = \frac{4.4 \times 10^{-8}}{1 - 0.997} = 14.7 \times 10^{-6} \ \text{S}$$

These are typical values for hybrid parameters at low frequencies. Note that h_{re} and h_{oe} are very small quantities.

A typical circuit application is shown in Figure 11.4. The power supplies V_{CC} and V_{BB} are necessary to establish the operating point, and their effects are included in the equivalent circuit by specifying the small-signal parameters. The capacitor C_C, called a coupling capacitor, is large enough to appear as a short circuit to the signal frequency and, therefore, does not appear in the equivalent circuit. In addition, R_B is usually much larger than the input resistance looking into the base terminal and can be neglected in this equivalent circuit.

The circuit can be analyzed in the following manner. First, consider the output loop equation

$$V_{\text{out}} = - \ h_{fe} I_b (R_L \| 1/h_{oe}) \tag{11.20}$$

The input voltage can be written as

$$V_{\text{in}} = I_b (R_S + h_{ie}) + h_{re} V_{\text{out}} \tag{11.21}$$

Then,

$$I_b = \frac{V_{\text{in}} - h_{re} V_{\text{out}}}{R_S + h_{ie}} \tag{11.22}$$

and the voltage gain A_v is

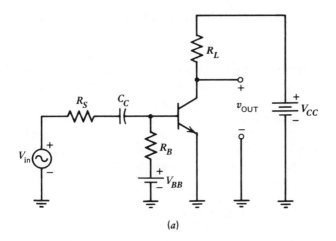

(a)

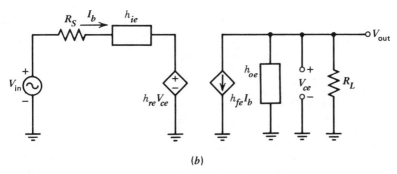

(b)

Figure 11.4. (a) Capacitively coupled common-emitter amplifier. (b) Small-signal equivalent circuit for part (a).

$$A_v = \frac{V_{\text{out}}}{V_{\text{in}}}$$

(11.23)

$$= \frac{-h_{fe}(R_L\|1/h_{oe})}{R_S + h_{ie} - h_{re}h_{fe}(R_L\|1/h_{oe})}$$

If $R_L = 5.0$ kΩ, $R_S = 1.0$ kΩ, and the same transistor parameters are used as in the previous example, then

$$A_v = \frac{-332(5.0 \times 10^3\|68 \times 10^3)}{1.0 \times 10^3 + 11.4 \times 10^3 - 1.13 \times 10^{-4} \times 332(5.0 \times 10^3\|68 \times 10^3)}$$

$$= -125$$

When h_{re} and h_{oe} are ignored in this calculation, the result is $A_v = -134$.

The solution to this problem consists of two steps. The first is the transformation from the actual circuit to the equivalent circuit. The second step is to apply the familiar techniques of circuit analysis to determine the desired quantities.

Example. Consider the common-collector circuit shown in Figure 11.5. This circuit is frequently called an emitter follower, because the output voltage at the emitter is in phase with, and approximately equal in magnitude with, the input signal between the base and ground. This circuit is often used for impedance transformation. Instead of using the set of common-collector hybrid parameters, the common-emitter equivalent circuit is used. This technique can be employed as an alternative to transforming the device parameters.

The voltage gain for this circuit can be determined in the following manner. The input loop indicates that

$$V_{in} = I_b h_{ie} + h_{re} V_{ce} + V_{out}$$

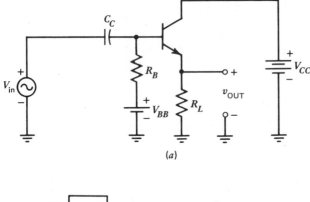

(a)

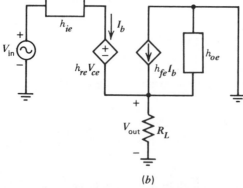

(b)

Figure 11.5. (a) Capacitively coupled common-collector amplifier. (b) Small-signal equivalent circuit for part (a).

and, since $V_{ce} = -V_{out}$

$$V_{in} = I_b h_{ie} - h_{re} V_{out} + V_{out} \tag{11.24}$$

But

$$V_{out} = (1 + h_{fe}) I_b (R_L \| 1/h_{oe}) \tag{11.25}$$

then,

$$V_{in} = -V_{out} \left[\frac{h_{ie}}{(1 + h_{fe})(R_L \| 1/h_{oe})} - h_{re} + 1 \right] \tag{11.26}$$

and

$$A_v = \frac{1}{h_{ie}/[(1 + h_{fe})(R_L \| 1/h_{oe})] - h_{re} + 1} \tag{11.27}$$

If $R_L = 1.0 \text{ k}\Omega$ and the same hybrid parameters are used as before, then

$$A_v = \frac{1}{11.4 \times 10^3/[(1 + 332)(1.0 \times 10^3 \| 68 \times 10^3)] - 1.13 \times 10^{-4} + 1}$$

$$= 0.967$$

The frequency dependence of hybrid parameters is difficult to predict. The measurements can be made over a range of frequencies and the results plotted as a function of frequency. In general, each measurement will yield complex values for impedance, admittance, or ratios. They can also be represented as a magnitude and a phase angle. High-frequency models are discussed below.

11.2 HIGH-FREQUENCY MODELS FOR BIPOLAR TRANSISTORS

11.2.1 The Hybrid-π Model

The small-signal models discussed up to this point were based on measurements of the device characteristics in network configurations. In this section, we present a different kind of model, called the *hybrid-π*. This designation came about because the components have different units and are arranged in a π-configuration. Unlike the network parameters, the hybrid-π is defined only in the common-emitter connection, although it may be used in the common-base and common-collector configurations. It is based on the concept that physical processes within the device can be accurately represented by simple electrical components. The most significant difference between the hybrid-π model and the two-port models is that the hybrid-π model can be used to predict the frequency response of the device.

The complete hybrid-π model is shown in Figure 11.6a. It is obvious that

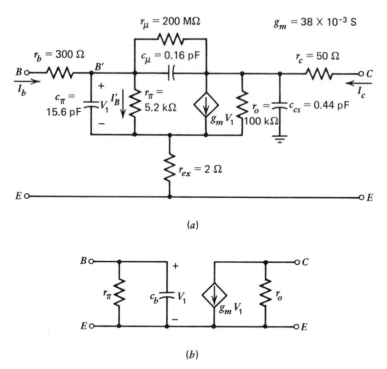

Figure 11.6. (a) Complete hybrid-π model for bipolar transistor small-signal applications. (b) Simplified hybrid-π model.

calculations involving this detailed model are cumbersome, and consequently they are usually performed on a computer. The individual parameters in the complete hybrid-π model are described in the following paragraphs.

The resistors r_b, r_c, and r_{ex} represent parasitic resistances of semiconductor material in series with each terminal due to the physical dimensions and shape of the device. Typical values for r_{ex} are from 1 to 3 Ω. This resistor can be neglected unless there is an unusually large bias current. Typical values for r_b are from 50 to 500 Ω and for r_c, from 20 to 500 Ω. The collector series resistor r_c is relatively insensitive to changes in current levels. The resistor r_b is between the base contact and that region of the base through which the carriers injected from the emitter pass.

It is useful to be able to estimate the important parameters for this model. The gain parameter is the transconductance g_m and is defined by

$$g_m = \left. \frac{\partial I_C}{\partial V_{BE}} \right|_{V_{CE} = \text{constant}} \tag{11.28}$$

Since the collector current is essentially equal to the emitter current and the emitter current is exponentially related to the base–emitter voltage by Equation

3.6, it is possible to write

$$I_C \simeq I_E = I_s \exp \left(\frac{qV_{BE}}{kT} \right) \tag{11.29}$$

Then, we may derive

$$g_m = \frac{qI_C}{kT} \tag{11.30}$$

Therefore, g_m is directly proportional to the magnitude of the collector current at the operating point.

The incremental input resistance r_π is given by

$$r_\pi = \frac{V_1}{I'_b}$$

If $R_L \ll r_o$, then

$$I_c \simeq h_{feo} I'_b$$

where h_{feo} is the low frequency common-emitter current gain. Then,

$$r_\pi \simeq \frac{h_{feo} V_1}{I_c}$$

and

$$g_m \simeq \frac{I_c}{V_1}$$

resulting in

$$r_\pi \simeq \frac{h_{feo}}{g_m} \tag{11.31}$$

An incremental change in the base–emitter voltage changes the charge stored in the transition region capacitance c_{je} and, more significantly, the charge stored in the distribution of minority carriers in the base. The charge Q_{bf} (represented as a phasor quantity) stored in the base region can be designated as if it were stored on a capacitor, c_b, such that

$$Q_{bf} = V_1 c_b$$

When the transistor is biased in the forward active region of operation,

$$I_c = \frac{Q_{bf}}{\tau_{tr}}$$

where τ_{tr} is the base transit time. Then,

$$c_b = \frac{\tau_{tr} I_c}{V_1}$$

or

$$c_b = \tau_{tr} g_m \qquad (11.32)$$

The capacitor in the model, c_π, is the parallel combination of c_{je} and c_b, or

$$c_\pi = c_{je} + c_b \qquad (11.33)$$

The depletion region capacitance is normally determined from Equation 3.12 as

$$c = \frac{c_o}{[1 - (V_D/V_o)]^{1/2}}$$

where V_o is the built-in potential of the junction, V_D is the applied junction voltage, and c_0 is the junction capacitance at zero junction voltage. However, for forward-biased junctions, this expression becomes unstable, since $1 - (V_D/V_o)$ approaches zero. The depletion region capacitance of the base–emitter junction, c_{je}, may be measured in the forward-biased mode; or in many cases, c_{je} is simply estimated by $c_{je} \simeq c_{jeo}$.

The output resistance r_o is approximately

$$r_o = \frac{\Delta V_{CE}}{\Delta I_C} \bigg|_{I_B = \text{constant}}$$

which is the inverse of the slope of the collector characteristic as measured at the operating point. Another representation of the output resistance introduces the concept of the *Early voltage*. The physical mechanism associated with the output resistance is the change in the width of that part of the base between the transition regions of the base–emitter and base–collector junctions as V_{CE} changes. Most of this change, for a device biased in the forward active region, is due to the transition region of the base–collector junction. The width of this region is a function of the base–collector voltage, and this phenomenon is called the Early effect. If the collector characteristics for a particular transistor are plotted as shown in Figure 11.7, extensions of the curves from the active region have a tendency to intersect the voltage axis at a common point. The magnitude of the voltage at which this intersection occurs is designated the Early voltage V_A. The output resistance is then approximately the inverse of the slope of the constant base curves, or

$$r_o \simeq \frac{V_A}{I_C} \qquad (11.34)$$

Typical values for the Early voltage are 50 to 100 V. Equation 11.34 predicts that the output resistance r_o increases as I_C decreases, and for small (microamp) bias currents may assume megohm values.

The components connecting the collector and base, r_μ and c_μ, provide a path for the direct interaction between the input signal and the output signal. Fortunately, r_μ is very large and c_μ is very small, making it possible to neglect

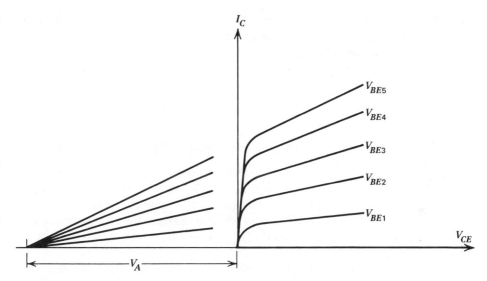

Figure 11.7. Bipolar transistor characteristic indicating the Early voltage.

them in many cases. The collector–base resistance is given by

$$r_\mu = \frac{\Delta V_{CE}}{\Delta I_{B1}} = \frac{\Delta V_{CE}}{\Delta I_C} \frac{\Delta I_C}{\Delta I_{B1}}$$

$$= \frac{r_o \Delta I_C}{\Delta I_{B1}}$$

(11.35)

where I_{B1} represents that part of the base current associated with recombination. There is a change in I_{B1} when there is a change in the width of the base. The base width decreases when the collector–base junction transition region width increases, due to an increase in the magnitude of the collector–emitter reverse bias voltage. The lower limit of r_μ is given from Equation 11.35 as

$$r_\mu = h_{feo}r_o$$

(11.36)

which assumes that the entire base current is involved in recombination. The collector–base transition region capacitance is the major contributor to c_μ. Equation 3.12 may be rewritten with the notation of the hybrid-π model as

$$c_\mu = c_{\mu o}\left(1 - \frac{V_{BC}}{V_o}\right)^{-1/n}$$

(11.37)

where V_o is the built-in voltage of the junction (approximately 0.8 V) and n is between 2 and 3.

The final component of the model is the collector-to-substrate capacitance c_{cs} if the transistor is in an integrated circuit. This capacitance can have a significant effect on the frequency response of integrated circuits if the devices are operated near the maximum frequency limit; otherwise, c_{cs} is usually neglected.

The hybrid-π model is useful and versatile. At low frequencies, the simplified model in Figure 11.6b is as easy to use as the hybrid parameter model. At higher frequencies, it is popular to use the complete model or the simplified model with the addition of r_b.

11.2.2 **Frequency-Dependent Hybrid Parameter Model**

There are several alternate ways by which frequency effects can be included in small-signal models for bipolar transistors. The hybrid-π model automatically accommodates frequency effects, but frequency trends can also be determined from measurements of two-port parameters. Regardless of the approach, it is important to recognize that there is a definite limit to the operational frequency range of the devices.

Frequency effects are often included in the hybrid parameters by assuming that one parameter, such as h_{fb}, is a single pole function with a characteristic 3 dB frequency. In this case, the frequency is called the alpha cutoff frequency and is designated f_α. Then,

$$h_{fb} = \frac{h_{fbo}}{1 + j(f/f_\alpha)} \tag{11.38}$$

where h_{fbo} is the low-frequency common-base current gain. At f_α, the magnitude of h_{fb} has been diminished to 0.707 of its low-frequency value.

For common-emitter operation, the cutoff frequency can be determined as

$$h_{fe} = -\frac{h_{fb}}{1 + h_{fb}}$$

Then,

$$h_{fe} = -\frac{h_{fbo}/[1 + j(f/f_\alpha)]}{1 + \{h_{fbo}/[1 + j(f/f_\alpha)]\}}$$

or

$$h_{fe} = -\frac{h_{fbo}}{1 + h_{fbo} + j(f/f_\alpha)}$$

The beta cutoff frequency f_β occurs when

$$1 + h_{fbo} = \frac{f}{f_\alpha}$$

or

$$f_\beta = (1 + h_{fbo})f_\alpha \tag{11.39}$$

The beta cutoff frequency is the frequency at which the common-emitter forward current gain is reduced from its dc value by 3 dB. Remember that h_{fbo} is typically $-$ 0.99, making $f_\beta \simeq f_\alpha/100$.

Another important figure of merit is the frequency at which the magnitude of h_{fe} goes to unity. This frequency, designated f_T, is often referred to as the gain–bandwidth product. It can be determined from

$$h_{fe} = \frac{h_{feo}}{1 + j(f/f_\beta)}$$

assuming that both h_{feo} and f_T/f_β are large compared to 1, so that

$$|h_{fe}| = 1 = \left| \frac{h_{feo}}{1 + j(f_T/f_\beta)} \right|$$

$$\simeq \frac{h_{feo} f_\beta}{f_T}$$

or

$$f_T \simeq h_{feo} f_\beta \qquad (11.40)$$

Equation 11.40 is shown to be the product of low-frequency current gain and the -3 dB frequency of the current gain for a particular device.

11.2.3 **Cutoff Frequency Calculations**

The dominant physical mechanism for gain degradation with increased frequency in bipolar transistors is charge storage of minority carriers in the base region. This is represented in the hybrid-π model by the capacitor c_b which is incorporated into c_π in the complete model. In the approximate model of Figure 11.6b, the effect of increasing the frequency of the input signal is a decrease in the input impedance. If the signal source is nonideal, the output signal amplitude will decrease with increasing frequency due to the increased voltage drop across the source impedance. In the complete model of Figure 11.6a, the transistor itself exhibits a cutoff frequency that is primarily due to the effects of r_b and the parallel combination of r_π and c_π. If the complete hybrid-π model is used, there will be observable frequency effects, even if the device is driven from an ideal voltage source.

Example. The beta cutoff frequency and the gain–bandwidth product for the hybrid-π model are estimated with $V_{ce} = 0$. Since c_π is almost 100 times c_μ and r_π is very small compared to r_μ, it is possible to omit r_μ and c_μ in the analysis. The simplified input circuit is shown in Figure 11.8. Using the voltage divider expression,

$$V_1 = \frac{[r_\pi \| (-j/\omega c_\pi)] V_{in}}{r_b + [r_\pi \| (-j/\omega c_\pi)]}$$

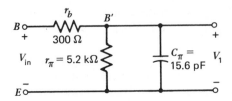

Figure 11.8. Simplified input circuit of hybrid-π model.

or

$$V_1 = \frac{r_\pi V_{in}}{(r_\pi + r_b)\left(1 + j\dfrac{r_\pi r_b \omega C_\pi}{r_\pi + r_b}\right)}$$

The beta cutoff frequency occurs when the magnitude of V_1/V_{in} is reduced to 0.707 of its low-frequency value. This occurs when

$$\frac{r_\pi r_b \omega_\beta C_\pi}{r_\pi + r_b} = 1$$

Then,

$$f_\beta = \frac{r_\pi + r_b}{2\pi r_b r_\pi C_\pi}$$

or

$$f_\beta = \frac{5.2 \times 10^3 + 300}{2\pi 5.2 \times 10^3 \times 300 \times 15.6 \times 10^{-12}}$$

$$= 35.9 \text{ MHz}$$

The gain–bandwidth product f_T is then

$$f_T \simeq h_{feo} f_\beta$$

$$= g_m r_\pi f_\beta$$

$$= 38 \times 10^{-3} \times 5.2 \times 10^3 \times 35.9 \times 10^6$$

$$= 7.1 \text{ GHz}$$

11.2.4 **Models for RF Transistors**

When transistors are used at frequencies near the physical limits of the devices, it is often useful to have more accurate models. Typical *npn* bipolar transistors have values of f_T between 100 MHz and 1.0 GHz, with 400 MHz a common value. The hybrid-π model is useful up to approximately $0.2 f_T$. For accuracy at frequencies above this, the hybrid-π model can be modified by dividing r_b into segments and replacing c_μ with several capacitors, each of which is connected between the collector and a different node in the distributed r_b. This is necessary when simple lumped models are inadequate to describe transmission line effects.

The technique of lumping all of the frequency effects into a single h-parameter is less accurate than the hybrid-π model at high frequencies. The common-base forward current ratio h_{fb} can be more accurately modeled, either as a multiple-pole function or as a single-pole function with an excess phase factor.

In most cases, better models can be obtained by direct measurement of two-port parameters as a function of frequency. Two of the h-parameters, h_{12} and h_{22} (h_{re} and h_{oe} for the common-emitter configuration), must be measured with $I_b = 0$. This could be accomplished by placing a large inductor in series with the base, but this is a cumbersome procedure. The condition for measuring the other two h-parameters is that $V_{ce} = 0$. This is easily accomplished by connecting a large-valued capacitor across the terminals from C to E. In fact, the measurements using the capacitor are so much easier than the techniques requiring inductors that it is common practice to measure the admittance (y_{ij}) parameters (all four of which are measured with capacitor short circuits in place in the circuit) rather than the hybrid parameters for frequencies above 1.0 MHz.

All four y-parameters are complex numbers ($y = g + jb$) with both real and imaginary parts that are complicated functions of frequency and operating point. Equations for the y-parameters as a function of frequency can be obtained from plots of the functions by curve fitting techniques.

For frequencies above 100 MHz, a different kind of two-port parameter is often used. The measurements are made using transmission lines with a 50-Ω characteristic impedance connected to both the input and the output ports of the transistor. The measured quantities are called *scattering* or *s* parameters. They are an indication of the mismatch between the impedances seen looking into the transistor ports and the characteristic impedance of the transmission lines. Loads that are mismatched to transmission lines result in reflections, and thus the terminology "scattering." It is common practice to plot *s*-parameters on Smith charts.

An in-depth treatment of high-frequency amplifiers is beyond the scope of this book, but the reader should be aware that accurate predictions of operation in the rf range require more sophisticated models than those for most other applications.

11.3 FET MODELS

Field effect transistors (FETs) can be used as small-signal devices in analog circuits in a manner that is analogous to bipolar transistor circuits. All three types, junction field effect transistors (JFETs), metal oxide–field effect transistors (MOSFETs), and Schottky barrier–field effect transistors (MESFETs), have similar small-signal models.

There are two approaches to modeling small-signal effects in FETs: physical models and two-port parameters. These transistors can be biased in the common-source, common-gate, and common-drain connections, but the most popular of these is the common-source connection.

Low-frequency physical models for FETs are very simple, consisting of one or two elements as shown in Figure 11.9. These models are based on the assumption that transistors are biased in the current saturation region of operation. Recall that FET terminology refers to the amplifying region of operation as the saturation region. This is shown in the static drain characteristics for all three types of field effect transistors in Figure 11.10. The negative resistance region in the $V_{GS} = 0$ curve for the GaAs MESFET is due to unusual properties of gallium arsenide associated with the formation of Gunn domains. For more information on this subject, see Reference 2. The output resistance r_o is the reciprocal of the slope of the drain characteristic at the operating point. This parameter typically ranges from 10 kΩ to infinity, with a nominal value of 100 kΩ appropriate at $I_D = 1$ mA. The value for r_o is inversely proportional to I_D.

The gain parameter g_m has the same basic form for all three types of FETs. From Chapter 5, the MOSFET transconductance is

$$g_m = \frac{kW}{L}(V_{GS} - V_T) \tag{11.41}$$

where k is the conduction constant, W and L are the width and length of the channel, V_{GS} is the gate-to-source voltage at the operating point, and V_T is the

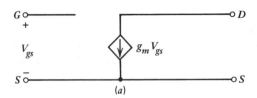

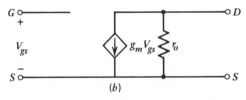

Figure 11.9. (*a*) Low-frequency two-port network electrical model of an FET. (*b*) Inclusion of a drain source resistance r_O in the model of part (*a*).

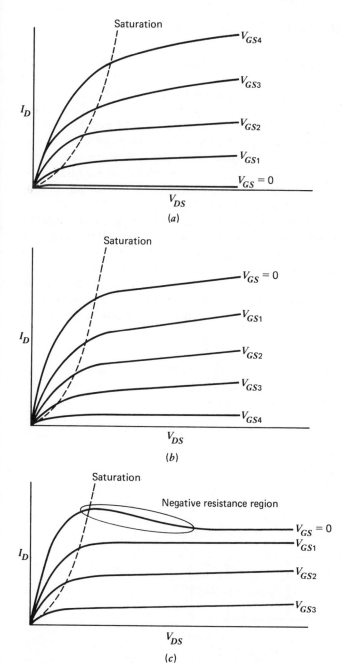

Figure 11.10. Static drain characteristics for FETs. (a) Enhancement mode MOSFET. (b) Depletion mode JFET or silicon MESFET. (c) GaAs MESFET.

threshold voltage. For the JFET and MESFET,

$$g_m = \frac{k'W}{L}(V_P - V_{GS}) \tag{11.42}$$

where k' is the conduction constant and V_P is the pinch-off voltage.

Frequency effects can be incorporated in the model by including capacitors, as shown in Figure 11.11. The construction of JFETs and MOSFETs results in parasitic resistors, r_s and r_d, in series with the source and drain. These resistors are usually in the range from from 10 to 100 Ω and are very small compared to r_o, which is typically in the range from 100 kΩ to 1.0 MΩ.

The MOSFET model in Figure 11.11b includes a fourth terminal, B, representing the substrate (sometimes called the body) contact. When discrete MOSFETs are used, the source is usually connected to the substrate. In MOS integrated circuits, however, it is common practice to use MOSFETs in a variety of configurations, including those in which the source is not connected to the substrate. Under these circumstances, a second current generator, $g_{mb}V_{bs}$, and two additional capacitors, c_{db} and c_{sb}, are included in the model. In this configuration, the device performs as if it had a second gate, represented by the substrate contact, in a common-gate connection. Variations in the source

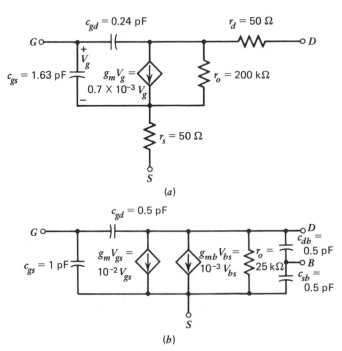

(a)

(b)

Figure 11.11. (a) Frequency-dependent FET small-signal model. (b) Frequency-dependent small-signal model for a MOSFET showing substrate capacitance elements and body effect current generator.

voltage relative to ground alter the drain current in the same manner as variations in the gate voltage relative to the source.

The two-port models for FETs are almost always common-source admittance parameters, as indicated in Figure 11.12. The admittance parameters can be related to the physical parameters in the following manner. The input admittance of the circuit in Figure 11.12b, y_{is}, is entirely capacitive and can be written

$$y_{is} = j\omega(c_{gs} + c_{gd}) \tag{11.43}$$

The forward transfer admittance is

$$y_{fs} = g_m - j\omega c_{gd} \tag{11.44}$$

The reverse transfer admittance is

$$y_{rs} = -j\omega c_{gd} \tag{11.45}$$

and the output admittance is

$$y_{os} = \frac{1}{r_o} + j\omega(c_{gd} + c_{ds}) \tag{11.46}$$

Note that all the components of the physical model can be determined from the y-parameters.

Example. For an MOS transistor, $g_m = 0.725$ mA/V, $r_o = 47$ kΩ, $c_{gs} = c_{ds} = 2.5$ pF, and $c_{gd} = 2.1$ pF. The admittance parameters at $f = 10$ kHz are

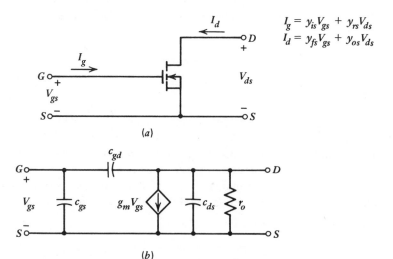

(a)

(b)

Figure 11.12. (a) Enhancement-mode MOSFET shown in two-port network configuration. (b) Small-signal model of MOSFET.

$$y_{is} = j(2\pi \times 10 \times 10^3)(2.5 + 2.1) \times 10^{-12} = j2.89 \times 10^{-7}\ S$$

$$y_{fs} = 0.725 \times 10^{-3} - j(2\pi \times 10 \times 10^3)\ (2.1 \times 10^{-12})$$

$$= 0.725 \times 10^{-3} - j1.32 \times 10^{-7}\ S$$

$$y_{rs} = +j1.32 \times 10^{-7}\ S$$

$$y_{os} = \frac{1}{47 \times 10^3} + j2.89 \times 10^{-7} = 2.13 \times 10^{-5} + j2.89 \times 10^{-7}\ S$$

11.4 COMPUTER MODELS

There are a number of computer programs that simulate the performance of electronic circuits. The most widespread of these programs was developed by the integrated circuits group of the Electronics Research Laboratory and the Department of Electrical Engineering and Computer Sciences at the University of California at Berkeley. It is called a *Simulation Program with Integrated Circuit Emphasis* (SPICE2). This is a versatile program that permits the calculation of the dc operating point, the steady-state small-signal analysis as a function of frequency, and the transient response of the circuit.

The device models in SPICE2 allow for the complete specification of bipolar and field effect transistors. Since the models must accurately describe both the small- and large-signal operation of the devices, they are much more detailed than those described in the previous sections in this chapter. The SPICE2 models contain default values for all of the parameters, and it is only necessary to insert a few pertinent values in the models to personalize them to the devices in the particular circuit under investigation. The remaining parameters take on the default values.

11.5 SUMMARY

The small-signal models for bipolar and field effect transistors can be based on two-port network theory or physical descriptions of the operation of the devices. The two-port models can be measured as a function of frequency, but the physical models permit the prediction of the frequency response. The measurement and derivation of the components of the small-signal models of transistors are an important contribution to the modeling of electronic circuits.

REFERENCES

1. I. Getreau, *Modeling the Bipolar Transistor* (Beaverton, Oreg. 97077: Tektronix, 1979).

2. H. A. Willing and P. deSantis, *IEE Electronic Letters* 13 (18), 537 (1977).
3. S. Carson, *High Frequency Amplifiers,* 2nd ed., (New York: Wiley, 1983).

PROBLEMS

11.1. Derive the z-parameters, z_{11}, z_{12}, z_{21}, and z_{22} in terms of the hybrid parameters h_{11}, h_{12}, h_{21}, and h_{22}.

11.2. The z-parameters for a particular electronic device are: $z_{11} = 1.99$ kΩ, $z_{12} = 0.1\ \Omega$, $z_{21} = -1.0 \times 10^{-5}\ \Omega$, and $z_{22} = 1.0$ kΩ. What are the h-parameters for the same device?

11.3. (*a*) Find the common-emitter h-parameters in terms of the common-collector h-parameters. (*b*) Find the common-collector h-parameters in terms of the common-emitter h-parameters.

11.4. The "dirbyh" (hybrid spelled backward) d-parameters relate I_1 and V_2 to V_1 and I_2 in the sense that the h-parameters relate V_1 and I_2 to I_1 and V_2. (*a*) Write the general circuit equations for the d-parameters. (*b*) Draw the d-parameter equivalent circuit. (*c*) Find the h-parameters in terms of the d-parameters.

11.5. The circuit in Figure 11.13 has a transistor forward current gain of $\beta = 200$. (*a*) Find I_{CQ} and V_{CEQ}. (*b*) Redraw the circuit in the small-signal form using the common-emitter h-parameters and assuming that the capacitors are ac short circuits. (*c*) If $h_{ie} = 11.4$ kΩ, $h_{re} = 1.0 \times 10^{-4}$, $h_{fe} = 200$, and $h_{oe} = 14.7 \times 10^{-6}$ S, calculate the voltage gain $A_v = V_{out}/V_{in}$ in the frequency range where the capacitors are assumed to be ac short circuits.

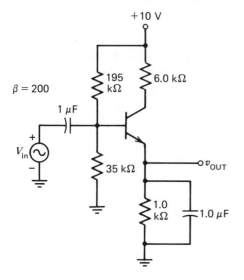

Figure 11.13. Circuit for Problem 11.5.

11.6. Calculate both the output and the input resistance for the circuit in Figure 11.14 if $h_{fe} = 150$, $h_{oe} = 25\ \mu S$, $h_{re} = 0$, and C_C is very large.

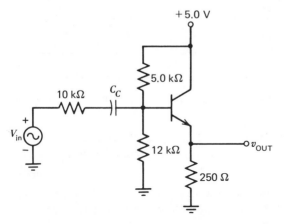

Figure 11.14. Circuit for Problem 11.6.

11.7. A hybrid-π model is shown in Figure 11.15. (*a*) Assume that $h_{re} = 0$ and determine the other *h*-parameters. (*b*) Determine the dc collector current at the quiescent operating point. (*c*) Estimate the Early voltage.

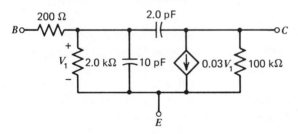

Figure 11.15. Circuit for Problem 11.7.

11.8. Calculate the resistance looking into the collector of the circuit of Figure 11.16. Assume that C_B has negligible reactance and that $h_{ie} = 1.0\ k\Omega$, $h_{re} = 0$, $h_{fe} = 100$, $h_{oe} = 25\ \mu S$, $R_1 = 40\ k\Omega$, and $R_2 = 20\ k\Omega$.

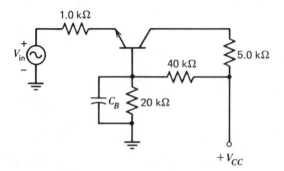

Figure 11.16. Circuit for Problem 11.8.

11.9. The voltage gain of the common-emitter amplifier shown in Figure 11.17a is $A_v = V_c/V_b = -\beta(Z_L\|1/h_{oe})/h_{ie}$. Integrated circuit designs often use a *pnp* transistor for a load in order to increase the load resistance without using large resistors. Calculate the effective load resistance in the circuit of Figure 11.17b. Assume that $h_{ie} = 10\ k\Omega$, $h_{re} = 0$, $h_{fe} = 50$, and $h_{oe} = 20\ \mu S$.

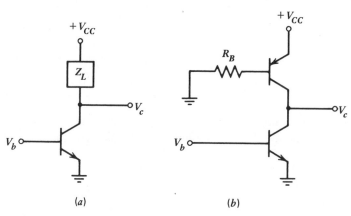

(a)   (b)

Figure 11.17. Circuit for Problem 11.9.

11.10. The circuit in Figure 11.18 has the following parameters: $V_{BE} = 0.70\ V$, $\beta_{dc} = 100$, $h_{ie} = 2.6\ k\Omega$, $1/h_{oe} = 60\ k\Omega$, $h_{re} = 0$, $h_{fe} = 100$, and the coupling capacitor is an ac short circuit. Assume that the transistor remains in the forward active region if the collector–emitter voltage remains within 1 to 9 V. (*a*) What is the maximum permissible peak-to-peak voltage of the input signal V_{in}? (*b*) Repeat part (*a*) if $R_L = 7.0\ k\Omega$.

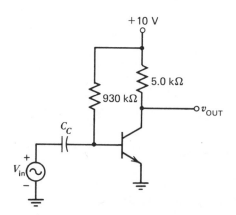

Figure 11.18. Circuit for Problem 11.10.

11.11. Calculate the input resistance of the circuit in Figure 11.19 if $h_{ie} = 3.0\ k\Omega$, $h_{re} = 0$, $h_{fe} = 100$, $h_{oe} = 25\ \mu S$, $c_{\mu o} = 1.0\ pF$, $c_{\pi} = 2.0\ pF$, $r_{\mu} = 100\ M\Omega$, and C_C is a very large capacitor. (*a*) Calculate the midband voltage gain. (*b*) Draw the hybrid-π equivalent circuit and calculate all the parameters.

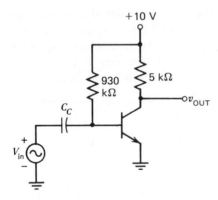

Figure 11.19. Circuit for Problem 11.11.

11.12. Use the hybrid-π model to determine a general expression for the current gain $A_i = I_{out}/I_{in}$ in the circuit of Figure 11.20. Neglect the current through R_B.

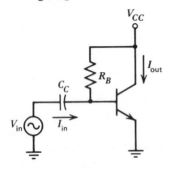

Figure 11.20. Circuit for Problem 11.12.

chapter 12
FREQUENCY RESPONSE OF TRANSISTOR CIRCUITS

The frequency response of an amplifier is generally as important a parameter as voltage gain and terminal impedances. For example, an audio amplifier whose frequency range is less than the human ear response of 15 kHz is of inferior quality. If the same amplifier had a frequency range greater than 15 kHz, unwanted noise would be amplified in the unnecessary portion of the frequency spectrum, also degrading the performance. A tuned amplifier is used at very high frequencies in radio and television to amplify only a very narrow frequency range. This narrow band capability allows the amplifier to reject unwanted signals and noise close to the frequency of the desired signal.

Another application demanding close attention to frequency is the wide-band or video amplifier. This type of amplifier with bandwidths up to 20 MHz must faithfully reproduce the many harmonics inherent in periodic pulsed waveforms. Here, the upper frequency capability of the transistor is important.

The description of an amplifier frequency response is usually given in terms of its upper and lower -3 dB frequencies. Figure 12.1 shows two different types of curves. Figure 12.1b represents the classic type of amplifier having

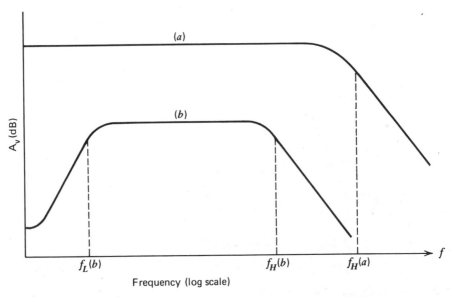

Figure 12.1. Voltage amplifier frequency response. (*a*) The dc amplifier response. (*b*) Bandpass amplifier.

frequency limitations at both the upper and lower ends of the frequency spectrum. The portion of the spectrum between f_L and f_H is referred to as the amplifier bandwidth where f_L and f_H are the -3 dB points referenced to the voltage gain level in the midband range. Most discrete amplifiers exhibit the frequency response curve shown in Figure 12.1*b*, while most analog integrated circuits and some discrete circuits exhibit the dc response of Figure 12.1*a*. The advances of integrated circuit technology allowed the achievement of dc amplification that had been a difficult problem in discrete circuit design. However, both types of amplifier designs shown in Figure 12.1 are limited at some high-frequency value. This high-frequency limitation is usually due to the internal junction capacitances of the transistor but may also be affected by other components in the circuit.

The frequency properties of the circuit in Figure 12.2*a* serve as a useful introduction to the frequency behavior of transistor circuits. If $C_p \ll C_s$, then C_s may represent a transistor circuit coupling capacitor in which C_p represents the equivalent capacitance to ground provided by the transistor *pn* junctions. The value of C_s might be fractions of a microfarad, while C_p could be on the order of hundreds of picofarads. The circuit would show a transfer frequency curve like that in Figure 12.1*b*. At low frequencies, C_s blocks signal current and reduces the voltage transfer ratio V_{out}/V_{in}. Note that phasor notation, upper case signal variables with lower case subscripts, is used throughout this chapter.

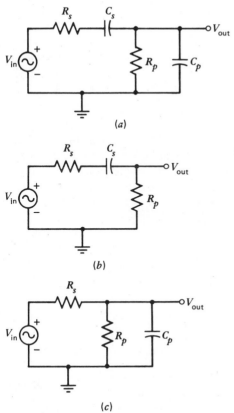

(a)

(b)

(c)

Figure 12.2. (a) Simplified equivalent circuit of a transistor amplifier. (b) Equivalent circuit at low frequencies. (c) Equivalent circuit at high frequencies.

At high frequencies, C_p provides a low-impedance shunt pathway to ground thus lowering V_{out}/V_{in}. In the midband region, the impedance of C_s becomes negligibly low while the impedance of C_p is much higher than R_p. In the midband region, then, both capacitors may be neglected.

An analysis of Figure 12.2a illustrates the complex nature of the circuit and leads to a simplified approach. A Kirchhoff's current law expression at the output node is

$$\frac{V_{out} - V_{in}}{R_s - j(1/\omega C_s)} + \frac{V_{out}}{R_p} + \frac{V_{out}}{-j(1/\omega C_p)} = 0 \tag{12.1}$$

giving

$$\frac{V_{out}}{V_{in}} = \frac{R_p}{R_s + R_p}$$

$$\times \left[\frac{1}{1 + \dfrac{R_p}{R_s + R_p} \dfrac{C_p}{C_s} + j\left(\dfrac{\omega R_p R_s C_p}{R_s + R_p} - \dfrac{1}{\omega(R_s + R_p)C_s}\right)} \right] \tag{12.2}$$

Equation 12.2 gives the exact transfer function of the circuit in Figure 12.2a
but it is awkward in its present form. The circuit may be viewed as two separate
equivalent circuits representing the low-frequency and high-frequency portions
of the spectrum. At frequencies near the lower -3 dB frequency, the circuit
of Figure 12.2b provides good representation. Since $C_p \ll C_s$ and $1/\omega C_p \gg$
R_p, C_p may be neglected. The nodal analysis of this circuit gives

$$\frac{V_{out} - V_{in}}{R_s - j(1/\omega C_s)} + \frac{V_{out}}{R_p} = 0 \tag{12.3}$$

then

$$\frac{V_{out}}{V_{in}} = \frac{R_p}{R_s + R_p} \frac{1}{1 - j\{1/[\omega(R_s + R_p)C_s]\}} \tag{12.4}$$

Equation 12.4 may be expressed in its pole-zero form as

$$\frac{V_{out}}{V_{in}} = \frac{R_p}{R_s + R_p} \frac{j\omega(R_s + R_p)C_s}{1 + j\omega(R_s + R_p)C_s} \tag{12.5}$$

Equation 12.5 describes the lower and midband portion of Figure 12.1b. The
lower -3 dB radian frequency occurs, then, at

$$\omega_L = \frac{1}{(R_s + R_p)C_s} \tag{12.6}$$

The lower -3 dB frequency is observed to be controlled by the time constant
of the series or coupling capacitor. A low-frequency design may accurately
focus on Equation 12.6 ignoring the insignificant effect of C_p.

At the higher portion of the frequency spectrum, the circuit of Figure
12.2c provides a good representation. The effects of C_p are significant at the
higher frequencies, while the low reactance of C_s allows it to be neglected in
this frequency range. The nodal equation at the output gives

$$\frac{V_{out} - V_{in}}{R_s} + \frac{V_{out}}{R_p} + \frac{V_{out}}{1/j\omega C_p} = 0 \tag{12.7}$$

giving

$$\frac{V_{out}}{V_{in}} = \frac{R_p}{R_p + R_s} \frac{1}{1 + j\omega R_p R_s C_p/(R_s + R_p)} \tag{12.8}$$

Equation 12.8 describes the upper and midband portion of Figure 12.1*b*. The upper -3 dB radian frequency occurs at the pole

$$\omega_H = \frac{1}{(R_p\|R_s)C_p} \tag{12.9}$$

The upper -3 dB frequency is controlled by the time constant associated with C_p in Figure 12.2*c*.

The circuits in Figure 12.2 and the analysis shown above can be used to describe the equivalent circuits of a transistor amplifier, except for one important feature. The midband gain or transfer function for the circuit of Figure 12.2*a* is $A_v = R_p/(R_s + R_p) < 1$. A transistor common-emitter or common-base amplifier can provide midband gain values up to several hundred.

Example. If $R_s = 50\ \Omega$, $C_s = 0.1\ \mu F$, $R_p = 10\ k\Omega$, and $C_p = 300\ pF$, find f_L, f_H, and the midband gain for the circuit of Figure 12.2*a*.

Solution. From Equation 12.6,

$$f_L = \frac{1}{2\pi} \frac{1}{(R_s + R_p)C_s}$$

$$= \frac{1}{2\pi} \frac{1}{(50 + 10 \times 10^3)0.1 \times 10^{-6}}$$

$$= 158\ Hz$$

From Equation 12.9,

$$f_H = \frac{1}{2\pi} \frac{1}{(R_s\|R_p)C_s}$$

$$= \frac{1}{2\pi} \frac{1}{(50\|10 \times 10^3)300 \times 10^{-12}}$$

$$= 10.7\ MHz$$

and

$$A_v(\text{mid}) = \frac{R_p}{R_s + R_p} = \frac{10 \times 10^3}{(10 \times 10^3) + 50}$$

$$= 0.995$$

12.1 THE COMMON-EMITTER AMPLIFIER HIGH-FREQUENCY RESPONSE

The hybrid-π model is a traditional and useful model to explore the high-frequency properties of a transistor amplifier. Figure 12.3*a* shows a common-

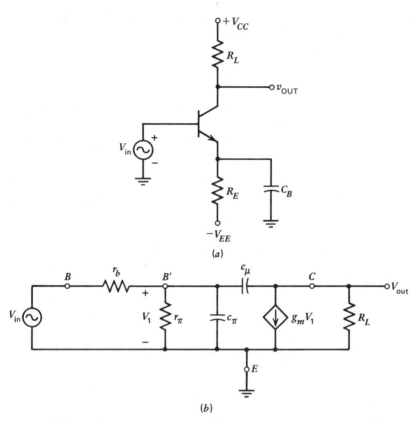

Figure 12.3. (*a*) Common-emitter amplifier with bypass capacitor and (*b*) high-frequency hybrid-π model of common-emitter amplifier.

emitter configuration. The small-signal equivalent of this circuit, using the transistor hybrid-π model, is shown in Figure 12.3*b*. The collector–base resistance r_μ is assumed to be very large and is omitted from the analysis. The output resistance r_o is incorporated in the load resistance R_L.

The junction capacitances c_π and c_μ will obviously play a strong role in the frequency limitations of the circuit. The position of c_μ provides unique and dominating properties in the circuit, even though its value is usually smaller than that of c_π. The c_μ provides a circuit path from the output collector to the input base. Since the polarity of the collector voltage is shifted 180° from the base voltage, the element c_μ is said to provide a negative feedback path in the circuit. The impedance properties imparted to an amplifying circuit from feedback elements fall in a general category called the *Miller effect*. In the common-emitter amplifier shown here, the phrase "Miller capacitance" becomes more specific. It is the Miller capacitance in the common-emitter circuit that establishes the upper frequency limits of the circuit.

Let us examine the impedance properties of the Miller capacitance by

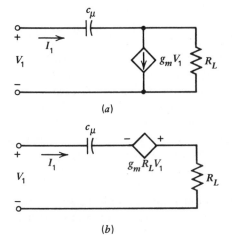

(a)

(b)

Figure 12.4. (a) Small-signal model of common-emitter amplifier as viewed to the right of B' node of Figure 12.2. (b) Series-equivalent circuit of part (a).

isolating the circuit to the right of the base node B' in Figure 12.3b. This circuit is redrawn in Figure 12.4a. The impedance is determined in this dependent source problem by applying a test voltage V_1 and then obtaining the ratio of V_1 to I_1. A Kirchhoff loop equation may be written if the current generator $g_m V_1$ and parallel resistance R_L are converted to a series resistance and voltage generator as indicated in Figure 12.4b. The loop equation becomes

$$V_1 = \frac{I_1}{j\omega c_\mu} - g_m R_L V_1 + I_1 R_L \qquad (12.10)$$

or

$$V_1 = \frac{I_1}{[1 + g_m R_L]} \left[\frac{1}{j\omega c_\mu} + R_L \right]$$

The impedance is then

$$Z = \frac{V_1}{I_1} = \frac{1}{j\omega c_\mu (1 + g_m R_L)} + \frac{R_L}{1 + g_m R_L} \qquad (12.11)$$

The value of c_μ lies in the low picofarad range, allowing the first term in the impedance to dominate at the frequencies applicable to the hybrid-π model. The effective impedance to the right of the base node is then

$$Z \simeq \frac{1}{j\omega c_\mu (1 + g_m R_L)} \qquad (12.12)$$

We observe that the capacitance c_μ is multiplied by a constant apparently increasing the capacitance seen at the base node. This is precisely what occurs. The enlarged capacitance is called the Miller capacitance and is designated

$$c_m = (1 + g_m R_L) c_\mu \qquad (12.13)$$

The hybrid-π model can be simplified by replacing the terminals of the feedback capacitor c_μ with a Thevenin equivalent circuit as viewed from the B' terminal and by a Thevenin equivalent circuit viewed from the collector terminal. The Thevenin impedance viewed to the right of the B' terminal in Figure 12.4a has already been calculated. It is approximately the Miller capacitive reactance of Equation 12.12. The Thevenin equivalent voltage to the right of terminal B' is zero, since the dependent source $g_m V_1$ is not activated when the circuit is opened at B'. The Thevenin impedance to the left of the collector terminal is closely approximated by the reactance of the c_μ capacitance, since c_μ is much smaller than c_π. The impedance presented by c_μ is also usually much larger than the small-signal resistance of the base–emitter junction, r_π. The Thevenin equivalent voltage to the left of the collector terminal is $k(\omega)V_{\text{in}}$. The results of the two Thevenin circuits are shown in Figure 12.5a. A final approximation will allow a simple view of the effect of the Miller capacitance on the amplifier frequency response. Usually, $R_L \ll 1/\omega c_\mu$, $k(\omega)V_{\text{in}} \ll V_{\text{out}}$, and little loss in accuracy results if we use the circuit of Figure 12.5b.

The output voltage is related to the voltage at B' by a constant

$$V_{\text{out}} = -g_m V_1 R_L \qquad (12.14)$$

so that the frequency response or -3 dB frequencies of V_1 and V_{out} are identical.

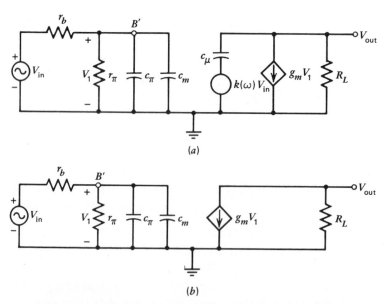

(a)

(b)

Figure 12.5. (a) Hybrid-π model of common-emitter amplifier with the c_μ element replaced by the Thevenin equivalent circuit at both nodes of c_μ. (b) Approximate circuit of (a).

The relation of V_1 to V_{in} can be written as a voltage divider expression

$$V_1 = \frac{r_\pi \| \{1/[j\omega(c_\pi + c_m)]\} V_{in}}{r_\pi \| \{1/[j\omega(c_\pi + c_m)]\} + r_b} \tag{12.15}$$

which may be reduced to a form suitable for recognition of the pole in the circuit. The result is

$$V_1 = \frac{r_\pi V_{in}}{(r_\pi + r_b)[1 + j\omega(r_\pi \| r_b)(c_\pi + c_m)]} \tag{12.16}$$

The -3 dB frequency of this circuit is then

$$\omega_{-3 \text{ dB}} = \{(r_\pi \| r_b)[c_\pi + (1 + g_m R_L)c_\mu]\}^{-1} \tag{12.17}$$

The Miller capacitance is usually the dominant factor in the frequency capability of a common-emitter circuit. In this particular example, we insert representative values to illustrate the influence of the load resistor on the Miller capacitance and the upper -3 dB frequency point.

Example. For Figure 12.5, let $r_b = 100 \ \Omega$, $r_\pi = 5.0 \ \text{k}\Omega$, $c_\pi = 100 \ \text{pF}$, $c_\mu = 5.0 \ \text{pF}$, and $g_m = 0.04 \ \text{S}$. Calculate the upper -3 dB points for (a) $R_L = 2.0 \ \text{k}\Omega$ and (b) $R_L = 20 \ \text{k}\Omega$. Substituting the appropriate values into Equation 12.17 and solving for $f_{-3 \text{ dB}}$ yields

(a)
$$f_{-3 \text{ dB}} = \frac{1}{2\pi} \{(5.0 \times 10^3 \| 100)[100 \times 10^{-12}$$
$$+ (0.04)(2.0 \times 10^3)(5 \times 10^{-12})]\}^{-1}$$
$$= 3.22 \ \text{MHz}$$

(b)
$$f_{-3 \text{ dB}} = \frac{1}{2\pi} \{(5.0 \times 10^3 \| 100)[100 \times 10^{-12}$$
$$+ (0.04)(20 \times 10^3)(5 \times 10^{-12})]\}^{-1}$$
$$= 396 \ \text{kHz}$$

The magnitude of the load resistance R_L exerts a dominant influence on the upper -3 dB frequency point through the Miller capacitance. The influence of the load resistance on the amplifier bandwidth leads to an important principle called the gain–bandwidth product.

12.1.1 Gain–Bandwidth Product

The midband voltage gain expression for the common-emitter example in Figure 12.5 is

$$A_v = \frac{V_{out}}{V_{in}} \simeq \frac{V_{out}}{V_1} = -g_m R_L \tag{12.18}$$

The load resistance R_L is in the numerator so that increases in R_L provide a direct increase in A_v. In the previous example, the absolute voltage gain will increase from 80 to 800 as R_L increases from 2.0 kΩ to 20 kΩ. The bandwidth, however, decreases by a factor slightly less than 10. This tradeoff in amplifier parameters is derived from a property known as the gain–bandwidth product (GBW). The GBW for the circuit of Figure 12.2 is the product of Equations 12.9 and 12.8 and is

$$GBW = g_m R_L \frac{1}{(r_\pi \| r_b)[c_\pi + (1 + g_m R_L)c_\mu]} \tag{12.19}$$

This figure of merit is really an approximation rather than a constant as the phrase implies. In the previous example, the GBW changed from 1.62×10^9 radians per second to 1.99×10^9 radians per second as the load resistance was changed from 2.0 kΩ to 20 kΩ. However, the concept of trading gain for bandwidth is important in the design of common-emitter amplifiers.

Our discussion looked at the effect of the interaction between the transistor internal capacitance and the circuit resistors. Although all capacitors and resistors affect the upper -3 dB frequency response, the components c_μ and R_L have, by far, the largest influence through the Miller effect.

12.2 THE COMMON-BASE AMPLIFIER

The common-base amplifier is often used for voltage amplification at high frequencies. The voltage gain of the circuit of Figure 12.6a will be analyzed to demonstrate the properties of the common-base circuit. The ac small-signal equivalent circuit is given in Figure 12.6b. The bypass capacitor C_B and the bias resistor R_B are not shown since both ends of these elements lie at ac ground. For ease of analysis, the model assumes that $r_b = 0$ and r_μ and r_o are very large. The source voltage V_{in} and its internal resistance R_i are converted to a current source V_{in}/R_i and a parallel resistance R_i. The current node equation at the emitter is

$$\frac{V_{in}}{R_i} + g_m V_1 = \left(\frac{1}{R_i} + \frac{1}{r_\pi} + j\omega c_\pi \right) V_e \tag{12.20}$$

or

$$\frac{V_{in}}{R_i} = -\left(\frac{1}{R_i} + \frac{1}{r_\pi} + g_m + j\omega c_\pi \right) V_1 \tag{12.21}$$

Writing the current node equation at the collector gives

$$-g_m V_1 = \left(\frac{1}{R_C} + j\omega c_\mu \right) V_{out} \tag{12.22}$$

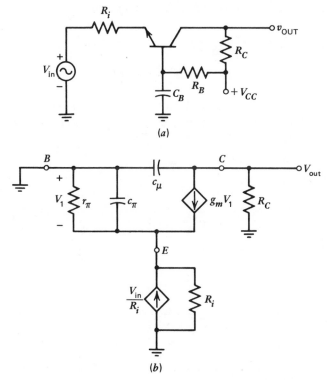

Figure 12.6. (*a*) Common-base amplifier. (*b*) The high-frequency small-signal model of the common-base amplifier.

or

$$V_1 = -\frac{1}{g_m}\left(\frac{1}{R_C} + j\omega c_\mu\right)V_{out} \qquad (12.23)$$

Substitution of Equation 12.23 into 12.21 yields

$$\frac{V_{in}}{R_i} = \left(\frac{1}{R_i} + \frac{1}{r_\pi} + g_m + j\omega c_\pi\right)\frac{1}{g_m}\left(\frac{1}{R_C} + j\omega c_\mu\right)V_{out} \qquad (12.24)$$

and

$$A_v = \frac{V_{out}}{V_{in}} = \frac{1}{R_i}\frac{g_m}{(1/R_i + 1/r_\pi + g_m + j\omega c_\pi)(1/R_C + j\omega c_\mu)} \qquad (12.25)$$

This expression for A_v may be rewritten as

$$A_v = \frac{V_{out}}{V_{in}} = \frac{g_m R_C}{R_i}(R_i\|r_\pi\|1/g_m)\frac{1}{1 + j\omega(R_i\|r_\pi\|1/g_m)c_\pi}\frac{1}{1 + j\omega R_C c_\mu} \qquad (12.26)$$

Two poles appear in the gain expression. The lower-valued pole,

$$\omega_{-3 \text{ dB}} = \frac{1}{R_C c_\mu} \tag{12.27}$$

determines the -3 dB frequency response of the amplifier.

Example. Calculate the poles and the midband gain of the common-base amplifier of Figure 12.6a. Let $R_i = r_\pi = 1.0$ kΩ, $1/g_m = 40$ Ω, $R_C = 3.0$ kΩ, $c_\pi = 30$ pF, and $c_\mu = 3.0$ pF.

$$f_2 = \frac{1}{2\pi(R_i \| r_\pi \| 1/g_m)c_\pi}$$

$$= \frac{1}{2\pi(1.0 \times 10^3 \| 1.0 \times 10^3 \| 40)30 \times 10^{-12}} \tag{12.28}$$

$$= 143 \text{ MHz}$$

$$f_1 = \frac{1}{2\pi R_C c_\mu}$$

$$= \frac{1}{2\pi \times 3.0 \times 10^3 \times 3.0 \times 10^{-12}} \tag{12.29}$$

$$= 17.7 \text{ MHz}$$

The midband gain is calculated from Equation 12.26 when the capacitive effects are negligible.

$$A_v(\text{mid}) = \frac{g_m R_C}{R_i}(R_i \| r_\pi \| 1/g_m)$$

$$= \frac{0.025 \times 3.0 \times 10^3}{1.0 \times 10^3}(1.0 \times 10^3 \| 1.0 \times 10^3 \| 40) \tag{12.30}$$

$$= 2.8$$

In this example, the midband gain is primarily determined by the ratio R_C/R_i, as seen from Equation 12.30:

$$A_v(\text{mid}) = \frac{g_m R_C}{R_i}(R_i \| r_\pi \| 1/g_m)$$

$$\approx \frac{g_m R_C}{R_i}\frac{1}{g_m} \tag{12.31}$$

$$\approx \frac{R_C}{R_i}$$

Higher gains may be achieved if R_C can be increased or R_i decreased. Note that there is no phase inversion in this circuit in the midband region.

A gain–bandwidth relation also exists for the common-base amplifier. Note that the collector or load resistor R_C is in the numerator of the gain expression (Equation 12.30) and in the denominator of the expression for the upper -3 dB cutoff frequency (Equation 12.29). An increase in R_C will increase the gain but decrease the bandwidth. For the previous example, multiplication of Equations 12.30 and 12.29 yields

$$GBW = \frac{g_m R_C}{R_i} [R_i \| r_\pi \| (1/g_m)](1/R_C c_\mu)$$

$$= \frac{g_m}{R_i c_\mu} [R_i \| r_\pi \| (1/g_m)] \qquad (12.32)$$

$$= 3.1 \times 10^9 \text{ rad/s}$$

Note that R_c does not appear in the GBW expression.

12.3 MOSFET AMPLIFIER FREQUENCY RESPONSE

The small-signal models for field effect transistors were introduced in Chapter 11. Figure 11.11b shows the MOSFET model that includes the gate–source and gate–drain capacitances c_{gs} and c_{gd}. Figure 12.7a illustrates a common-source MOSFET inverting amplifier. The load resistance R_L usually has a large value in integrated circuits and is often the equivalent resistance of a MOSFET depletion mode device. The large load resistances typically found in MOSFET designs make c_{gd} a more significant capacitance than c_{gs}. The Miller effect is present in the common-source amplifier in the same way as in the common-emitter amplifier. The main difference lies in the lower transistor transconductance found in the MOSFET than that of a bipolar transistor under similar operating conditions.

The ac equivalent circuit for the MOSFET common-source amplifier is shown in Figure 12.7b.

Example. Find (a) the Miller capacitance and (b) the -3 dB frequency for the discrete common-source amplifier of Figure 12.7 if $R_L = 50$ kΩ, $r_{ds} = 30$ kΩ, $c_{gd} = 0.4$ pF, $c_{gs} = 0.8$ pF, $g_m = 10^{-3}$ S, and $R_i = 1.0$ kΩ. In the transistor, we shall ignore the substrate capacitances and substrate source current generator $g_{mb}V_{bs}$.

Solution.

(a) The Miller capacitance can be derived as

$$c_m = [1 + g_m(r_{ds} \| R_L)]c_{gd}$$

$$= [1 + 10^{-3}(30 \times 10^3 \| 50 \times 10^3)]0.4 \times 10^{-12} \qquad (12.33)$$

$$= 7.9 \text{ pF}$$

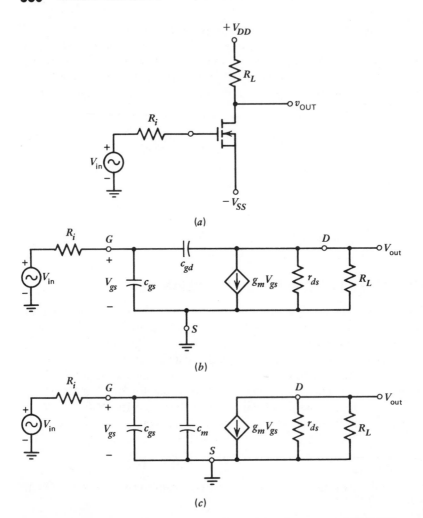

Figure 12.7. (*a*) Common-source amplifier with enhancement MOSFET. (*b*) Small-signal model of common-source amplifier. (*c*) Small-signal model with c_{gd} replaced by equivalent Thevenin circuit.

(*b*) The -3 dB frequency is obtained from expressions for V_{out} and V_{gs}:

$$V_{out} = -g_m V_{gs}(r_{ds}\|R_L) \qquad (12.34)$$

The -3 dB frequency responses of V_{out} and V_{gs} are the same since Equation 12.34 shows that these two variables are related by a constant. The -3 dB frequency of V_{gs} and V_{out} is then

$$f_{-3 \text{ dB}} = \frac{1}{2\pi} \omega_{-3 \text{ dB}}$$

$$= \frac{1}{2\pi} \frac{1}{R_i(c_{gs} + c_m)}$$ (12.35)

$$= \frac{1}{2\pi} \frac{1}{10^3(0.8 \times 10^{-12} + 7.9 \times 10^{-12})}$$

$$= 18.3 \text{ MHz}$$

An integrated circuit MOSFET might not have the source and substrate terminals connected. In this case, the full small-signal equivalent circuit of Figure 11.11*b* would be required. Analysis of the full circuit would logically be done by a computer simulation.

12.4 THE LOW-FREQUENCY RESPONSE OF AN AMPLIFIER

The lower -3 dB frequency response of an amplifier is dependent upon the presence of capacitors external to the transistors. These external capacitors, such as coupling and bypass capacitors, are needed in some amplifiers to decouple the ac and dc components of a circuit. Most integrated circuits do not use coupling or bypass capacitors, but the principle of decoupling ac and dc signals in one circuit is pervasive and merits our analysis. We shall find that the low-frequency response of a circuit is controlled not only by the value of the external capacitors but also by the magnitude of the resistive discharge path for these capacitors. In other words, the time constant associated with capacitance external to the transistor controls the lower -3 dB frequency of the amplifier.

Figure 12.8*a* shows a common-emitter amplifier with a coupling capacitor. At low frequencies, the capacitive reactance is quite large, and little signal current reaches the base of the transistor. It is then expected that the voltage gain will be low at low frequencies. As the signal frequency is increased, the capacitive reactance decreases, allowing more signal current to reach the base for amplification. At some higher frequency, the capacitive reactance is negligible compared to the resistive pathways in the circuit.

The analysis of the frequency-dependent voltage gain will use the *h*-parameter model shown in Figure 12.8*b*. The output voltage is

$$V_{\text{out}} = -h_{fe}I_b[(1/h_{oe})\|R_C]$$ (12.36)

The base current I_b is related to V_{in} by

$$I_{\text{in}} = \frac{V_{\text{in}}}{R_S + 1/j\omega C_C + (R_B\|h_{ie})}$$ (12.37)

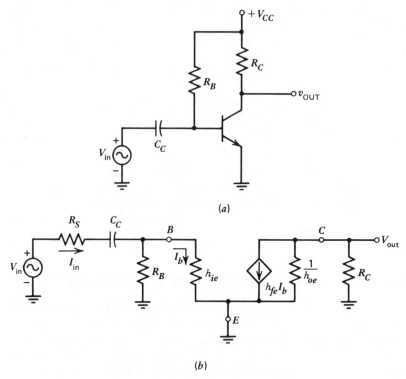

Figure 12.8. (*a*) Common-emitter amplifier with coupling capacitor C_C. (*b*) Small-signal equivalent circuit of the circuit in part (*a*).

and

$$I_b = \frac{R_B}{R_B + h_{ie}} I_{in} \tag{12.38}$$

Substitution into Equation 12.36 yields the gain expression

$$A_v = \frac{V_{out}}{V_{in}} = -h_{fe} \frac{R_B}{R_B + h_{ie}} \frac{1}{R_S + (R_B \| h_{ie}) + 1/j\omega C_C} [(1/h_{oe}) \| R_C] \tag{12.39}$$

The single pole may be observed by rewriting Equation 12.39 as

$$A_v = -h_{fe} \frac{R_B}{R_B + h_{ie}} \frac{j\omega C_C}{1 + j\omega [R_S + (R_B \| h_{ie})] C_C} [(1/h_{oe}) \| R_C] \tag{12.40}$$

The pole defining the lower -3 dB frequency is given by

$$f_1 = \frac{1}{2\pi [R_S + (R_B \| h_{ie})] C_C} \tag{12.41}$$

A very useful generality emerges from an examination of Equation 12.41. The time constant of the pole f_1 is

$$f_1 = \frac{1}{2\pi} \frac{1}{R'C_C} \tag{12.42}$$

where R' is the value of the resistive discharge path between the terminals of the capacitor.

Example. Find the lower -3 dB frequency for the circuit of Figure 12.8a if $R_S = 50$ Ω, $R_B = 68$ kΩ, $h_{ie} = 1.0$ kΩ, and $C_C = 0.1$ μF. Using Equation 12.41,

$$f_1 = \frac{1}{2\pi[50 + (68 \times 10^3 \| 1.0 \times 10^3)]0.1 \times 10^{-6}}$$

$$= 1.54 \text{ kHz}$$

Example. Analyze the circuit of Figure 12.9a. (a) Show that Equation 12.42 defines the lower -3 dB frequency of the circuit. (b) Calculate the lower -3 dB frequency. Let $R_E = 4.0$ kΩ, $R_C = R_S = 5.0$ kΩ, $C_B = 1.0$ μF, $h_{fe} = 100$, $h_{ie} = 2.5$ kΩ, and $h_{re} = h_{oe} = 0$. (c) Plot the frequency response from zero hertz to the midband gain region.

$$V_{out} = -h_{fe}I_bR_C \tag{12.43}$$

and

$$I_b = \frac{V_{in}}{R_S + h_{ie} + (1 + h_{fe})Z_B} \tag{12.44}$$

where

$$Z_B = R_E\|(1/j\omega C_B)$$

Substituting, we obtain

$$V_{out} = -h_{fe}R_CV_{in} \frac{1}{R_S + h_{ie} + (1 + h_{fe})Z_B} \tag{12.45}$$

and

$$A_v = \frac{V_{out}}{V_{in}} \tag{12.46}$$

$$= -h_{fe}R_C \frac{1}{R_S + h_{ie} + [(1 + h_{fe})R_E/(1 + j\omega R_E C_B)}$$

This expression may be rearranged to obtain

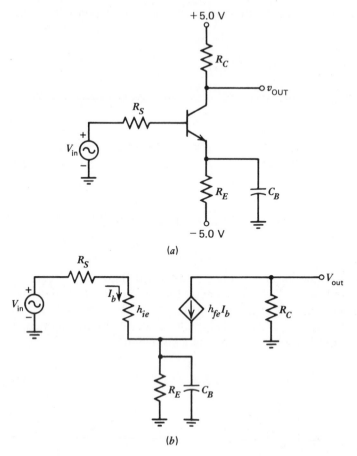

Figure 12.9. (*a*) Common-emitter amplifier with bypass capacitor C_B. (*b*) Small-signal equivalent circuit of part (*a*).

$$A_v = \frac{-h_{fe}R_C}{R_S + h_{ie} + (1 + h_{fe})R_E}$$

$$\times \frac{1 + j\omega R_E C_B}{1 + [(R_S + h_{ie})j\omega R_E C_B]/[R_S + h_{ie} + (1 + h_{fe})R_E]} \tag{12.47}$$

The lower -3 dB frequency is then

$$f_1 = \frac{1}{2\pi(R_S + h_{ie})R_E C_B]/[R_S + h_{ie} + (1 + h_{ie})R_E]} \tag{12.48}$$

The resistive discharge value of the capacitor may be more clearly seen if Equation 12.48 is rearranged to

$$f_1 = \left\{ 2\pi \frac{R_S + h_{ie}}{1 + h_{fe}} \frac{R_E C_B}{[(R_S + h_{ie})/(1 + h_{fe})] + R_E} \right\}^{-1} \quad (12.49)$$

The resistive value R' of the capacitor discharge path may then be written

$$R' = [(R_S + h_{ie})/(1 + h_{fe})] \| R_E \quad (12.50)$$

where $(R_S + h_{ie})/(1 + h_{fe})$ is the resistance seen looking into the emitter.

(b) The lower -3 dB cutoff frequency is then calculated from Equation 12.49 as

$$f_1 = 2.18 \text{ kHz}$$

(c) The plot of voltage gain versus frequency has one pole and one zero. In addition, at very low and very high frequencies, the bypass capacitor allows the assumption that R_E is either in parallel with an infinite reactance or a zero value reactance. At very low frequencies, the capacitor is neglected. A low-frequency analysis of Figure 12.9b with C_B removed is

$$V_{\text{out}} = -h_{fe} I_b R_C$$

and

$$I_b = \frac{V_{\text{in}}}{R_S + h_{ie} + (1 + h_{fe}) R_E}$$

giving

$$A_v(\text{dc}) = \frac{V_{\text{out}}}{V_{\text{in}}} = \frac{-h_{fe} R_C}{R_S + h_{ie} + (1 + h_{fe}) R_E} \quad (12.51)$$

$$= -1.23$$

A zero occurs in the gain expression of Equation 12.47 at

$$f_o = \frac{1}{2\pi R_E C_B} \quad (12.52)$$

$$= 40 \text{ Hz}$$

At f_o, the gain plot rises at a slope of 6 dB/octave until the frequency reaches the pole at f_1. At $f_1 = 4.6$ kHz, the gain plot levels off to the midband gain of

$$A_v(\text{mid}) = \frac{V_{\text{out}}}{V_{\text{in}}} = \frac{-h_{fe} R_C}{R_S + h_{ie}} \quad (12.53)$$

or

$$= \frac{-100 \times 5.0 \times 10^3}{5.0 \times 10^3 + 2.5 \times 10^3}$$

$$= -66.7$$

A plot of A_v versus frequency is shown in Figure 12.10.

Frequency analysis of circuits containing a single coupling or bypass capacitor provides a general guideline for the location of the lower -3 dB frequency point. A more complex situation arises when more than one capacitor exists in the circuit. In this case, the separate poles created by each capacitor interact, creating a complicated analysis. These types of circuits are best handled with computer analysis, keeping in mind that the principle of low-frequency analysis is the same as for single-pole circuits. In all cases, a lower -3 dB frequency response is obtained by using larger capacitors and, where possible, larger values of resistors that are in the discharge path of the capacitors.

12.5 TUNED AMPLIFIERS

An important category of frequency-sensitive amplifier uses the resonant properties of RLC networks. Figure 12.11a shows a series RLC circuit where R represents the load resistance. The impedance of the circuit is

$$Z = R + j\left(\omega L - \frac{1}{\omega C}\right) \tag{12.54}$$

At the resonant frequency ω_o, the reactive term in this expression becomes

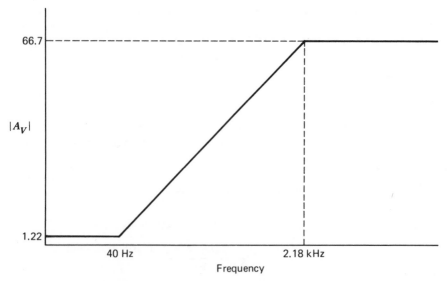

Figure 12.10. Frequency response of the common-emitter circuit of Figure 12.8.

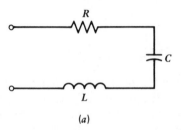

(a)

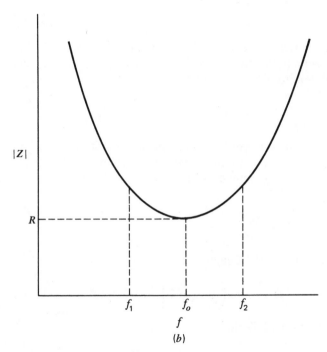

f_1 f_o f_2

f

(b)

Figure 12.11. (a) A series-resonant RLC circuit with its generalized impedance as a function of frequency shown in part (b).

zero. This frequency is determined by setting the reactive term of Equation 12.54 to zero and solving for ω:

$$\omega_o L - \frac{1}{\omega_o C} = 0 \tag{12.55}$$

or

$$\omega_o = \frac{1}{(LC)^{1/2}} \text{ rad/s} \tag{12.56}$$

and

$$f_o = \frac{1}{2\pi} \frac{1}{(LC)^{1/2}} \text{ Hz} \tag{12.57}$$

Figure 12.11*b* shows a general plot for the frequency dependence of the absolute value of Z in Equation 12.54. At the resonant frequency f_o, the impedance $Z = R$ with a zero phase angle across the resonant circuit terminals.

The -3 dB frequencies occur at the points where $Z = (2)^{1/2}R \angle \pm 45°$. The bandwidth of the resonant circuit is then

$$BW = f_2 - f_1 \tag{12.58}$$

The sharpness of the impedance curve is controlled by the values of R, L, and C. The quality factor Q defines the sharpness or selectivity of the curve, and at resonance the general definition of a circuit Q is

$$Q = 2\pi \frac{E_p}{E} \tag{12.59}$$

where E_p is the peak instantaneous energy stored in the circuit and E is the energy dissipated per cycle. For the simple resonant circuit of Figure 12.11, the quality factor is the ratio of reactance to resistance at the resonant frequency, or

$$Q = \frac{\omega_o L}{R} = \frac{1}{\omega_o RC} \tag{12.60}$$

Substitution of these Q relations into Equation 12.43 gives

$$Z = R\left[1 + jQ\left(\frac{f}{f_o} - \frac{f_o}{f}\right)\right] \tag{12.61}$$

It can be shown from Equation 12.59 that

$$Q = \frac{f_o}{f_2 - f_1} = \frac{f_o}{BW} \tag{12.62}$$

This property of circuit resonance and sharpness of the circuit impedance curve with frequency is used extensively in tuned circuit design.

The parallel RLC circuit shown in Figure 12.12*a* is equally important in tuned circuit design. The magnitude of the impedance for the parallel RLC circuit is shown in Figure 12.12*b*. The circuit admittance is

$$Y = G + j\left(\omega C - \frac{1}{\omega L}\right) \tag{12.63}$$

providing a resonant frequency at

$$f_o = \frac{1}{2\pi} \frac{1}{(LC)^{1/2}} \tag{12.64}$$

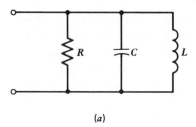

(a)

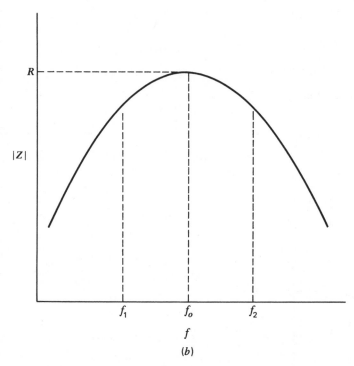

(b)

Figure 12.12. (a) Parallel RLC circuit with its generalized impedance response as a function of frequency shown in part (b).

The quality factor Q may again be derived from Equation 12.59 as

$$Q = \frac{R}{\omega_o L} = \frac{\omega_o C}{R} \qquad (12.65)$$

and

$$Q = \frac{f_o}{f_2 - f_1} = \frac{f_o}{BW} \qquad (12.66)$$

Example. The common-emitter circuit of Figure 12.13 has a tuned circuit load. Find (*a*) the load resonant frequency, (*b*) the Q of the load, (*c*) the circuit gain at resonance, (*d*) the circuit bandwidth, and (*e*) a value for R_L that reduces the circuit bandwidth to 100 kHz.

Solution.

(*a*) The load resonant frequency is

$$f_o = \frac{1}{2\pi} \frac{1}{(LC)^{1/2}}$$

$$= \frac{1}{2\pi} \frac{1}{(31.7 \times 10^{-6} \times 200 \times 10^{-12})^{1/2}}$$

$$= 2.0 \text{ MHz}$$

(*b*) The load Q is

$$Q = \frac{R_L}{\omega_o L}$$

$$= \frac{5.0 \times 10^3}{2\pi \times 2.0 \times 10^6 \times 31.7 \times 10^{-6}}$$

$$= 12.6$$

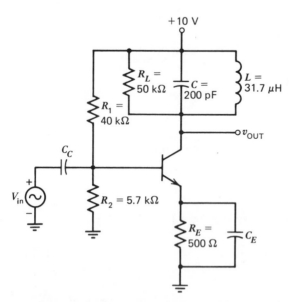

Figure 12.13. Common-emitter circuit with tuned parallel-resonant circuit.

(c) The circuit gain at resonance requires knowledge of h_{ie} that is estimated from I_{CQ}. A dc analysis of the circuit yields

$$I_{CQ} = 0.97 \text{ mA}$$

Then,

$$
\begin{aligned}
h_{ie} &\simeq \frac{h_{fe} \times 26 \times 10^{-3}}{I_{CQ}} \\
&= \frac{100 \times 26 \times 10^{-3}}{0.97 \times 10^{-3}} \\
&= 2.68 \text{ k}\Omega
\end{aligned}
$$

At resonance, the load is purely resistive since the LC parallel combination has an infinite impedance. A small-signal analysis using the h-parameter model yields

$$V_{\text{out}} = -h_{fe}I_b R_L$$

$$I_b = \frac{V_{\text{in}}}{h_{ie}}$$

and

$$
\begin{aligned}
A_v &= \frac{V_{\text{out}}}{V_{\text{in}}} = -\frac{h_{fe}R_L}{h_{ie}} \\
&= -\frac{100 \times 5.0 \times 10^3}{2.68 \times 10^3} \\
&= -187
\end{aligned}
$$

(d) The circuit bandwidth is

$$
\begin{aligned}
BW &= \frac{f_o}{Q} \\
&= \frac{2.0 \times 10^6}{12.6} \\
&= 159 \text{ kHz}
\end{aligned}
$$

(e) A decrease in bandwidth will increase the circuit Q. Since

$$
\begin{aligned}
Q &= \frac{f_o}{BW} \\
&= \frac{2.0 \times 10^6}{100 \times 10^3} \\
&= 20
\end{aligned}
$$

then

$$R_L = Q\omega_o L$$
$$= 20 \times 2\pi \times 2.0 \times 10^6 \times 31.7 \times 10^{-6}$$
$$= 8.0 \text{ k}\Omega$$

As the resonant circuit is adjusted for better selectivity (higher Q), the bandwidth is reduced and the gain is increased to $A_v = -298$.

Figure 12.14 illustrates an unusual property of the tuned circuit. The voltage at the collector, $v_C(t)$, may assume instantaneous values that are larger than the power supply voltage. The inductor (and capacitor) acquire stored energy that is in series with the power supply. A voltage up to twice the power supply value is possible before limiting occurs. Tuned amplifiers are discussed in more detail in Chapter 15.

12.6 SUMMARY

All amplifiers are subject to degradations at high frequencies. The upper limit of performance is restricted by the internal capacitances of the amplifier transistor. The hybrid-π model is a good representation of the bipolar transistor at the frequencies where gain attenuation occurs. The common-emitter and common-source amplifiers are usually described by an interaction between the load resistance and transistor capacitance called the Miller effect. The concept of

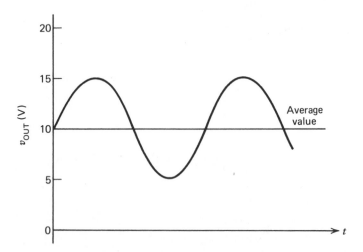

Figure 12.14. Collector voltage waveform of the common-emitter circuit of Figure 12.13.

an amplifier gain–bandwidth product shows that gain and bandwidth may be traded off for all configurations.

The low-frequency properties of an amplifier are controlled by the presence of capacitors external to the transistor. Coupling and bypass capacitors and the associated circuit time constants control the lower -3 dB frequencies of the amplifier. Finally, a tuned amplifier was presented to show the important class of amplifier that restricts the amplifier usable frequencies to a narrow band by using an RLC resonant circuit.

PROBLEMS

12.1. Determine the upper and lower -3 dB frequencies for the circuit in Figure 12.15. Use approximation techniques where the poles are considered to be far apart.

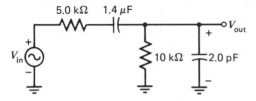

Figure 12.15. Circuit for Problem 12.1

12.2. The RC values in the circuit shown in Figure 12.16 provide poles that are not widely separated. This problem illustrates the complexity involved in manually determining the low and high -3 dB frequency points. Find the -3 dB points.

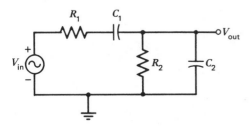

Figure 12.16. Circuit for Problem 12.2.

12.3. The transistor in the circuit of Figure 12.17 has $h_{ie} = 1.5$ kΩ, $h_{fe} = 75$, and $h_{re} = h_{oe} = 0$. (a) Find the -3 dB frequency points for the transfer function $A_v = V_{out}/V_{in}$. (b) Sketch a Bode plot for $A_i = I_1/I_{in}$. (c) For part (a), calculate the frequency at which the function A_v is within 2.0% of its maximum value (consider the magnitude only).

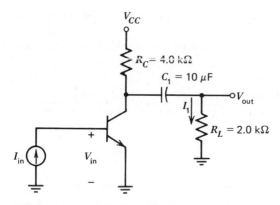

Figure 12.17. Circuit for Problem 12.3.

12.4. In the circuit of Figure 12.18, $V_{BE} = 0.70$ V and $\beta_{dc} = 50$. (a) Find R_C and R_E such that $I_{CQ} = 1.0$ mA and $V_{CEQ} = 2.5$ V. (b) What is the bandwidth of this amplifier? (c) If C_L remains fixed at 15 pF, can the bandwidth be doubled by changing R_L, R_C, and R_E? (d) If the bandwidth is doubled, what is the effect on the midband frequency small signal voltage gain?

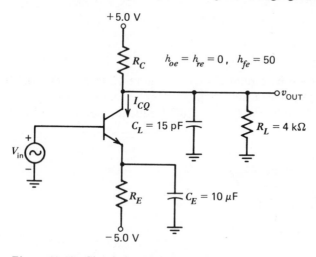

Figure 12.18. Circuit for Problem 12.4.

12.5. For the hybrid-π model shown in Figure 12.19, find (a) the Miller capacitance, (b) the midband voltage gain, and (c) the upper -3 dB frequency value.

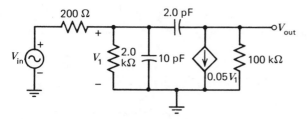

Figure 12.19. Circuit for Problem 12.5.

12.6. The transistor in the circuit of Figure 12.20 has the values $h_{fe} = 100$, $h_{re} = 0$, $h_{oe} = 25 \ \mu S$, $h_{re} = 3.0 \ k\Omega$, $c_{\mu 0} = 1.5 \ pF$, $c_{\pi 0} = 3.0 \ pF$, and $\tau_{tr} = 0.1 \ ns$. If the base–collector capacitance is due only to the depletion effect and the base–emitter capacitance is due only to the diffusion capacitance, find (*a*) the upper $-3 \ dB$ frequency response of the amplifier, (*b*) the midband frequency gain, and (*c*) the lower $-3 \ dB$ frequency response of the amplifier.

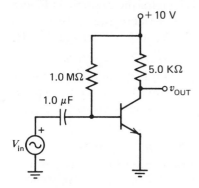

Figure 12.20. Circuit for Problem 12.6.

12.7. The gain–bandwidth properties of the common-emitter circuit in Figure 12.21*a* are to be compared with the gain–bandwidth properties of the "equivalent" common-base amplifier shown in Figure 12.21*b*. Note that the bias conditions on both transistors are identical and therefore the transistor parameters are the same. For $r_{\pi} = 1.0 \ k\Omega$, $g_m = 25 \ mS$, $R_c = 3.0 \ k\Omega$, $c_{\pi} = 30 \ pF$, $c_{\mu} = 3.0 \ pF$, and $R_B = 500 \ k\Omega$, calculate (*a*) the gain–bandwidth for the circuit in Figure 12.21*a* and (*b*) the gain–bandwidth for the circuit in Figure 12.21*b*, and (*c*) discuss the similarities or differences in your answer.

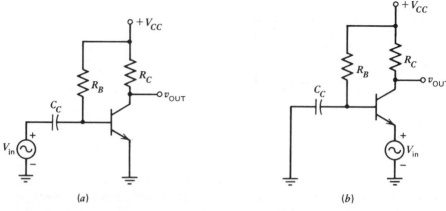

Figure 12.21. Circuits for Problem 12.7.

12.8. Use the circuit and transistor values given for the example in Figure 12.6 and find (*a*) the gain–bandwidth product if R_i is increased to 10 kΩ and (*b*) the gain–bandwidth if R_L is decreased to 1.5 kΩ and R_i remains at 1.0 kΩ.

12.9. In the circuit shown in Figure 12.7, R_L = 25 kΩ, r_{ds} = 30 kΩ, c_{gd} = 0.4 pF, c_{gs} = 0.8 pF, g_m = 1.0 mS, and R_i = 1.0 kΩ. Find (*a*) the Miller capacitance and (*b*) the -3 dB frequency point. (*c*) Discuss your solution compared to the results for the example problem that was given for Figure 12.7.

12.10. For the transistor shown in Figure 12.22, h_{ie} = 1.0 kΩ, h_{fe} = 100, and h_{re} = h_{oe} = 0. (*a*) Find C_C such that the lower -3 dB frequency is at f_1 = 10 Hz. (*b*) Sketch the asymptotic or Bode plot of $|I_c/I_{in}|$ versus frequency.

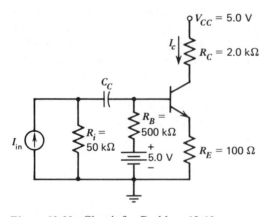

Figure 12.22. Circuit for Problem 12.10.

12.11. For the circuit shown in Figure 12.23, assume the following values: V_{BE} = 0.70 V, β_{dc} = 75, h_{fe} = 75, h_{ie} = 1.3 kΩ, and h_{re} = h_{oe} = 0. (*a*) Design the circuit so that I_{EQ} = 1.5 mA, V_{CEQ} = 5.0 V, and f_h = 1.0 kHz (where f_h

is the frequency at which A_v drops to -3 dB of its peak value). (*b*) Make a Bode plot of voltage gain versus frequency showing the appropriate breakpoint frequencies and gain values.

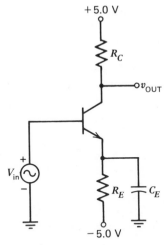

+5.0 V

R_C

v_{OUT}

V_{in}

R_E C_E

−5.0 V

Figure 12.23. Circuit for Problem 12.11.

12.12. A circuit is shown in Figure 12.24 having two resonant or tuned subcircuits. If both tuned circuits are resonant at the same frequency, (*a*) calculate the resonant frequency; (*b*) draw the small-signal equivalent circuit using hybrid parameters ignoring R_{B1} and R_{B2} (calculate h_{ie}); (*c*) calculate the voltage gain at the resonant frequency; (*d*) calculate the Q for the series resonant circuit using $Q = \omega_o L/R$, where R is the total series resistance; also calculate $Q = R/\omega_o C$. (*e*) What happens to the bandwidth when tuned circuits are cascaded?

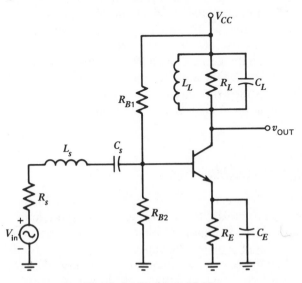

V_{CC}

L_L R_L C_L

R_{B1}

v_{OUT}

L_s C_s

R_s

R_{B2}

R_E C_E

V_{in}

Figure 12.24. Circuit for Problem 12.12.

chapter 13

MULTIPLE-TRANSISTOR CIRCUITS

The most popular analog multiple-transistor circuit is the differential amplifier. It is used as the input stage of the operational amplifier. In this type of circuit, the differential amplifier is usually followed by a high-gain stage that, in turn, is followed by an output stage. The use of multiple-transistor circuits or multistage transistor circuits may not only increase the overall small-signal voltage and/or current gain but may also change the input and output impedance levels. In this chapter, several multiple-transistor circuits and multistage transistor circuits are analyzed to determine their small-signal characteristics.

13.1 THE BASIC DIFFERENTIAL AMPLIFIER

One of the most versatile circuits in electronics is the operational amplifier, or OP AMP. A significant part of a typical OP AMP is the differential amplifier shown schematically in Figure 13.1. The output voltage v_{OUT} is proportional the difference between the two input signals. The constant K is the differential mode gain of the amplifier. Ideally, if $v_1 = v_2$, the output is identically zero. It is only when v_1 and v_2 are different that an output voltage is obtained.

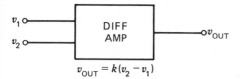

Figure 13.1. Block diagram of a differential amplifier with two inputs and one output.

13.1.1 Common-Mode and Differential-Mode Inputs

The ideal differential amplifier amplifies only the difference between the two input signals. However, in a practical device, the average value of the two input signals also has some effect on the output signal. This average value of the two input signals is called the *common-mode* signal v_c. The difference between the two input signals is called the *differential-mode* signal v_d. Thus, we have

$$v_d \equiv v_2 - v_1 \tag{13.1}$$

and

$$v_c \equiv \frac{v_1 + v_2}{2} \tag{13.2}$$

For example, if v_1 and v_2 are -10 and $+10$ μV, respectively, then the differential-mode signal v_d is 20 μV and v_c is 0.0 V. If v_1 and v_2 are changed to 90 and 110 μV, respectively, the differential-mode signal does not change, but the common-mode signal v_c is 100 μV. Although, ideally, the output voltage from a differential amplifier should be the same for these two cases, in practice the outputs will be slightly different.

13.1.2 DC Analysis

Consider the bipolar differential amplifier in Figure 13.2. The two transistors are assumed identical and biased in their active regions. The two transistors can be biased in their active regions with the input voltages v_1 and v_2 at ground potential (zero volts) by using the positive and negative supply voltages V_{CC} and $-V_{EE}$. This avoids the use of voltage divider bias resistors and coupling capacitors in the base circuits. The net result is that differences in dc voltages as well as in ac signals can be amplified. The output is the difference between v_{o1} and v_{o2}.

The resistor R_E provides an approximately constant current bias to Q_1 and Q_2. If $v_1 = v_2 = 0.0$ V and the base currents are negligibly small, then $V_E = -V_{BE}$ and

$$I_E = \frac{-V_{BE} - (-V_{EE})}{R_E} \tag{13.3}$$

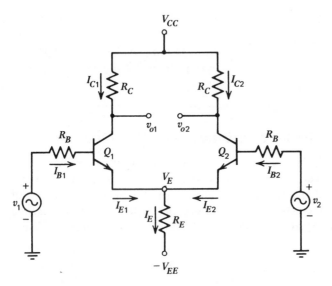

Figure 13.2. Differential amplifier with double-ended outputs v_{o1} and v_{o2}.

If Q_1 and Q_2 are matched, then I_E will split evenly, so that

$$I_{E1} = I_{E2} = \frac{I_E}{2} \tag{13.4}$$

If the current gains of the two transistors are high, then $I_{C1} = I_{C2} \approx I_{E1} = I_{E2}$, and in this case $v_{o1} = v_{o2}$.

The dc load line of Q_1 can be determined in the following manner. We have

$$
\begin{aligned}
V_{CE1} &= v_{o1} - V_E = (V_{CC} - I_{C1}R_C) - (I_E R_E - V_{EE}) \\
&= (V_{CC} + V_{EE}) - I_{C1}R_C - 2I_{C1}R_E \\
&= (V_{CC} + V_{EE}) - I_{C1}(R_C + 2R_E)
\end{aligned}
\tag{13.5}
$$

The dc load line corresponding to this equation is shown in Figure 13.3. A similar approach can be used to determine the load line of transistor Q_2.

Example. Design the circuit in Figure 13.2 to operate with $V_{CC} = V_{EE} = 6.0$ V and $I_{C1} = I_{C2} = 100 \ \mu A$ when $v_1 = v_2 = 0.0$ V.

Solution. If the current gains of Q_1 and Q_2 are high, $I_{E1} = I_{E2} \approx 100 \ \mu A$, and

$$
\begin{aligned}
I_E &= I_{E1} + I_{E2} \\
&= 200 \ \mu A
\end{aligned}
\tag{13.6}
$$

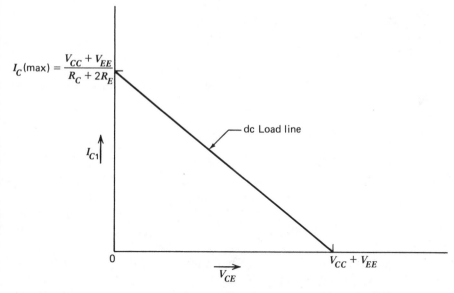

Figure 13.3. The dc load line of the differential amplifier of Figure 13.2.

Then,

$$R_E = \frac{V_{EE} - V_{BE}}{I_E}$$

$$= \frac{6.0 - 0.7}{0.2 \times 10^{-3}} \tag{13.7}$$

$$= 26.5 \text{ k}\Omega$$

Suppose we want V_{CE1} to be 6.0 V so that the Q-point is exactly in the center of the dc load line, as shown in Figure 13.4. The load line condition at $I_C(\text{max})$ is

$$I_C(\text{max}) = 200 \times 10^{-6} = \frac{V_{CC} + V_{EE}}{R_C + 2R_E} \tag{13.8}$$

$$= \frac{12.0}{R_C + 53 \times 10^3}$$

giving

$$R_C = 7.0 \text{ k}\Omega$$

This design was performed with $v_1 = v_2 = 0.0$ V. Suppose that a common-mode voltage exists, so that $v_1 = v_2 = 2.0$ V. In this case,

$$V_E = 2.0 - 0.7 = 1.3 \text{ V}$$

and

$$I_E = \frac{V_E - (-V_{EE})}{R_E}$$

$$= \frac{1.3 + 6.0}{26.5 \times 10^3} \tag{13.9}$$

$$= 275 \ \mu A$$

Then,

$$I_{C1} = I_{C2} = \frac{I_E}{2}$$

$$= \frac{275 \times 10^{-6}}{2} \tag{13.10}$$

$$= 138 \ \mu A$$

The quiescent point, or Q-point, has shifted from the center of the load line to the Q_1-point in Figure 13.4.

If another common-mode voltage exists, for $v_1 = v_2 = -2.0$ V, the Q-point becomes Q_2 in Figure 13.4. Hence, the common-mode voltage changes the Q-point on the dc load line so that the small-signal parameters

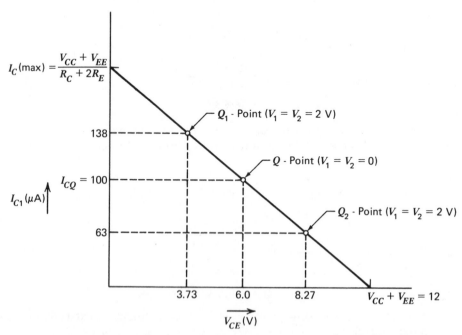

Figure 13.4. Load line with three different Q-points representing different common-mode input voltage conditions.

change. In this way, the common-mode voltage influences the differential-mode voltage gain.

An MOSFET differential amplifier is shown in Figure 13.5. The transistors M_1 and M_2 are assumed to be in the saturation region. Both positive and negative supply voltages are used so that the gates of M_1 and M_2 are at ground potential for a zero common-mode input voltage.

The resistor R_S provides an approximately constant bias current to M_1 and M_2. For $v_1 = v_2 = 0$, the current I_S is assumed to split evenly between the two matched transistors M_1 and M_2. Since $V_S = -V_{GS}$,

$$I_S = \frac{-V_{GS} - (-V_{SS})}{R_S} \tag{13.11}$$

and, for a matched pair of M_1 and M_2,

$$I_{D1} = I_{D2} = \frac{I_S}{2} \tag{13.12}$$

The dc load line of M_1 can be determined similarly to that for the bipolar circuit. We then have

$$
\begin{aligned}
V_{DS}(M_1) &= v_{o1} - V_S \\
&= (V_{DD} - I_{D1}R_D) - (I_S R_S - V_{SS}) \\
&= (V_{DD} + V_{SS}) - I_{D1}R_D - 2I_{D1}R_S \\
&= (V_{DD} + V_{SS}) - I_{D1}(R_D + 2R_S)
\end{aligned}
\tag{13.13}
$$

This dc load line is shown in Figure 13.6.

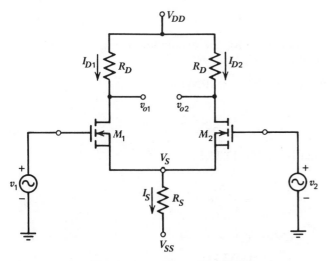

Figure 13.5. Differential amplifier with NMOS enhancement mode transistors.

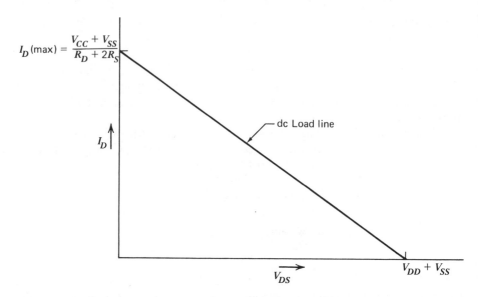

Figure 13.6. The dc load line for the MOSFET circuit of Figure 13.5.

Suppose the MOSFET circuit of Figure 13.5 is designed to operate with $V_{DD} = V_{SS} = 10$ V and $I_{D1} = I_{D2} = 0.25$ mA, when $v_1 = v_2 = 0.0$ V. Then,

$$I_S = I_{D1} + I_{D2} = 0.5 \text{ mA} \qquad (13.14)$$

The transistor characteristics of M_1 and M_2 in the saturation region are given by

$$I_D = k(V_{GS} - V_T)^2 \qquad (13.15)$$

Assuming values of $k = 0.05$ mA/V² for the conduction parameters and $V_T = 1.0$ V for the threshold voltage, then

$$0.25 \times 10^{-3} = 0.05 \times 10^{-3} (V_{GS} - 1.0)^2$$

and

$$V_{GS} = 3.24 \text{ V} \qquad (13.16)$$

The source resistance from Equation 13.11 is

$$R_S = \frac{-V_{GS} + V_{SS}}{I_S}$$

$$= \frac{-3.24 + 10}{0.5 \times 10^{-3}} \qquad (13.17)$$

$$= 13.52 \text{ k}\Omega$$

If $V_{DS}(M_1)$ is to be 10 V for maximum voltage swing, then R_D can be determined from Equation 13.13 to be

$$10 = (10 + 10) - 0.25 \times 10^{-3} (R_D + 2 \times 13.5 \times 10^3) \qquad (13.18)$$

or

$$R_D = 13.0 \text{ k}\Omega$$

As with the bipolar case, if a common mode voltage exists at the input, then the Q-point moves up or down the dc load line.

13.1.3 Bipolar Differential Amplifier Small-Signal Analysis

When the input voltages v_1 and v_2 are not equal, a difference in the output voltages v_{o1} and v_{o2} exists. To calculate the differential voltage gain, we assume that the difference between v_1 and v_2 is small enough that a linear small-signal analysis may be performed.

The small-signal equivalent circuit of the basic bipolar differential amplifier (Figure 13.2) is shown in Figure 13.7. The small-signal h-parameters are used, and it is assumed that $h_{re} = h_{oe} = 0.0$. Several methods exist to analyze the circuit. Since the small-signal circuit is linear, superposition applies, and we can write in phasor notation

$$V_{o1} = K_{11}V_1 + K_{12}V_2 \qquad (13.19)$$

and

$$V_{o2} = K_{21}V_1 + K_{22}V_2 \qquad (13.20)$$

The individual gain constants K_{ij} can be found by considering a single input and calculating a particular output.

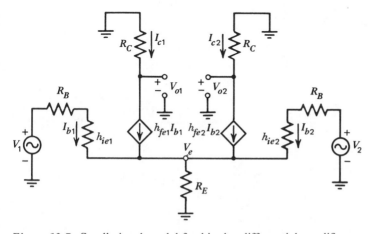

Figure 13.7. Small-signal model for bipolar differential amplifier shown in Figure 13.2.

Another approach uses nodal analysis. Kirchhoff's current law at the node common to the emitters yields

$$I_{b1} + h_{fe1}I_{b1} + h_{fe2}I_{b2} + I_{b2} = \frac{V_e}{R_E} \tag{13.21}$$

The currents I_{b1} and I_{b2} are found from

$$I_{b1} = \frac{V_1 - V_e}{R_B + h_{ie1}} \tag{13.22}$$

and

$$I_{b2} = \frac{V_2 - V_e}{R_B + h_{ie2}} \tag{13.23}$$

If the transistors are matched, then

$$h_{ie1} = h_{ie2} \equiv h_{ie} \tag{13.24}$$

and

$$h_{fe1} = h_{fe2} \equiv h_{fe} \tag{13.25}$$

The voltage V_e is found from Equations 13.21 to 13.25 to be

$$V_e = \frac{V_1 + V_2}{2 + [(R_B + h_{ie})/(1 + h_{fe})R_E]} \tag{13.26}$$

The output voltages are determined from

$$V_{o1} = -I_{c1}R_C = -h_{fe}I_{b1}R_C \tag{13.27}$$

and

$$V_{o2} = -I_{c2}R_C = -h_{fe}I_{b2}R_C \tag{13.28}$$

The output voltage V_{o1} is then

$$V_{o1} = -h_{fe}R_C I_{b1} = -\frac{h_{fe}R_C(V_1 - V_e)}{R_B + h_{ie}}$$

or

$$V_{o1} = -\frac{h_{fe}R_C}{R_B + h_{ie}}\left\{ V_1 - \frac{V_1 + V_2}{2 + [(R_B + h_{ie})/(1 + h_{fe})R_E]} \right\} \tag{13.29}$$

Rearranging, we get

$$V_{o1} = -\frac{h_{fe}R_C}{R_B + h_{ie}}\left\{ \frac{V_1\left[1 + \dfrac{R_B + h_{ie}}{(1 + h_{fe})R_E}\right]}{2 + [(R_B + h_{ie})/(1 + h_{fe})R_E]} \right.$$

$$\left. - \frac{V_2}{2 + [(R_B + h_{ie})/(1 + h_{fe})R_E]} \right\} \tag{13.30}$$

The output voltage V_{o2} is calculated in the same manner, giving

$$V_{o2} = -\frac{h_{fe}R_C}{R_B + h_{ie}} \left\{ \frac{V_2 \left[1 + \dfrac{R_B + h_{ie}}{(1 + h_{fe})R_E} \right]}{2 + [(R_B + h_{ie})/(1 + h_{fe})R_E]} - \frac{V_1}{2 + [(R_B + h_{ie})/(1 + h_{fe})R_E]} \right\} \quad (13.31)$$

If the output voltage is taken as the difference between V_{o1} and V_{o2}, then, from Equations 13.30 and 13.31, we obtain

$$V_{o1} - V_{o2} = -\frac{h_{fe}R_C}{R_B + h_{ie}}(V_1 - V_2) \quad (13.32)$$

All of the terms involving the emitter resistor R_E cancel. A differential-mode signal voltage was defined in Equation 13.1 as $V_d \equiv V_2 - V_1$ (in phasor notation), providing an output voltage from Equation 13.32 of

$$V_{o1} - V_{o2} = \frac{h_{fe}R_C}{R_B + h_{ie}} V_d \quad (13.33)$$

The factor multiplying the differential mode voltage is called the differential-mode gain. Then,

$$V_{o1} - V_{o2} = A_d V_d \quad (13.34)$$

where

$$A_d = \frac{h_{fe}R_C}{R_B + h_{ie}} \quad (13.35)$$

When the output is taken as the difference between V_{o1} and V_{o2}, only the difference between the input signals is amplified. The common-mode voltage V_c does not get amplified. This result is the ideal situation discussed from Figure 13.1. However, in most situations, a single output terminal whose voltage is measured with respect to ground is much preferred. Consider a one-sided output where V_{o2} (Equation 13.31) is taken as the output voltage measured with respect to ground.

The differential-mode and common-mode voltages were defined in Equations 13.1 and 13.2 (repeated here using phasor notation) as

$$V_d = V_2 - V_1 \quad (13.1)$$

and

$$V_c = \frac{V_1 + V_2}{2} \quad (13.2)$$

The solution for V_1 and V_2 in terms of V_d and V_c yields

$$V_1 = V_c - \frac{V_d}{2} \quad (13.36)$$

and

$$V_2 = V_c + \frac{V_d}{2} \tag{13.37}$$

The substitution of Equations 13.36 and 13.37 into Equation 13.31 results in the one-sided output voltage V_{o2} in terms of the common-mode and differential-mode voltages:

$$V_{o2} = -\frac{h_{fe}R_C}{R_B + h_{ie}} \left\{ \frac{V_d}{2} + \frac{V_c}{1 + [2(1 + h_{fe})R_E/(R_B + h_{ie})]} \right\} \tag{13.38}$$

This equation may be written as

$$V_{o2} = A_d V_d + A_c V_c \tag{13.39}$$

where A_d is the differential-mode voltage gain and A_c is the common-mode voltage gain. A comparison of Equations 13.39 and 13.38 gives

$$A_d = \frac{-h_{fe}R_C}{2(R_B + h_{ie})} \tag{13.40}$$

and

$$A_c = -\frac{h_{fe}R_C}{R_B + h_{ie}} \frac{1}{1 + [2(1 + h_{fe})R_E/(R_B + h_{ie})]} \tag{13.41}$$

Example. If $h_{fe} = 100$, $h_{ie} = 2.0 \text{ k}\Omega$, $R_C = 7.0 \text{ k}\Omega$, $R_E = 26.5 \text{ k}\Omega$, and $R_B = 0.5 \text{ k}\Omega$, the differential-mode gain is

$$A_d = -140$$

and the common-mode gain is

$$A_c = -0.13$$

13.1.4 MOSFET Differential Amplifier Small-Signal Analysis

The small-signal equivalent circuit of a MOSFET differential amplifier (Figure 13.5) is shown in Figure 13.8. The small-signal gate-to-source voltages can be written in terms of the input voltages as

$$V_{gs1} = V_1 - V_s \tag{13.42a}$$

and

$$V_{gs2} = V_2 - V_s \tag{13.42b}$$

Kirchhoff's current law at the node common to the source gives

$$g_{m1}V_{gs1} + g_{m2}V_{gs2} = \frac{V_s}{R_s} \tag{13.43}$$

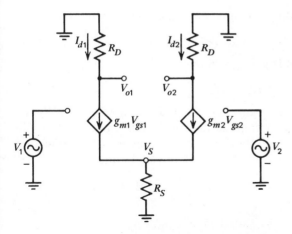

Figure 13.8. Small-signal model for the MOSFET differential amplifier shown in Figure 13.5.

If we assume that the two transistors are matched, then

$$g_{m1} = g_{m2} \equiv g_m \tag{13.44}$$

The small-signal common-source voltage V_s can be found by substituting Equations 13.42a and 13.42b into Equation 13.43:

$$V_s = \frac{V_1 + V_2}{2 + (1/g_m R_S)} \tag{13.45}$$

The small-signal output voltages are determined from

$$V_{o1} = -I_{d1} R_D = -g_m V_{gs1} R_D \tag{13.46}$$

and

$$V_{o2} = -I_{d2} R_D = -g_m V_{gs2} R_D \tag{13.47}$$

The output voltage V_{o1} is found from Equations 13.42a, 13.45, and 13.46 as

$$V_{o1} = -g_m R_D \left\{ \frac{V_1[1 + (1/g_m R_S)]}{2 + (1/g_m R_S)} - \frac{V_2}{2 + (1/g_m R_S)} \right\} \tag{13.48}$$

Similarly,

$$V_{o2} = -g_m R_D \left\{ \frac{V_2[1 + (1/g_m R_S)]}{2 + (1/g_m R_S)} - \frac{V_1}{2 + (1/g_m R_S)} \right\} \tag{13.49}$$

If the output voltage is taken as the difference between V_{o1} and V_{o2}, then, from Equations 13.48 and 13.49, we obtain

$$V_{o1} - V_{o2} = -g_m R_D (V_1 - V_2) \tag{13.50}$$

This output voltage is a pure differential-mode voltage and can be written as

$$V_{o1} - V_{o2} = g_m R_D V_d = A_d V_d \tag{13.51}$$

where A_d is the differential mode gain given by

$$A_d = g_m R_D \tag{13.52}$$

Consider the case of a one-sided output, V_{o2}. The input voltages V_1 and V_2 can be written in terms of the common-mode and differential-mode voltages in Equations 13.36 and 13.37. Substitution of these expressions into Equation 13.49 gives the one-sided output voltage V_{o2} in terms of V_c and V_d. Then,

$$V_{o2} = -g_m R_D \left[\frac{V_d}{2} + \frac{V_c}{1 + 2g_m R_S} \right] \tag{13.53}$$

A comparison of this result with Equation 13.39 shows that

$$A_d = \frac{-g_m R_D}{2} \tag{13.54}$$

and

$$A_c = \frac{-g_m R_D}{1 + 2g_m R_S} \tag{13.55}$$

13.1.5 Common-Mode Rejection Ratio

The results given by Equations 13.38 and 13.53 indicate that a one-sided output does not give an ideal differential amplifier characteristic, even if all of the components are perfectly matched. A portion of the common-mode signal is amplified, as well as the differential-mode signal. We now examine this property using a parameter called the common-mode rejection ratio (CMRR). The CMRR is a figure of merit in a differential amplifier design and is defined as the magnitude of the differential-mode gain divided by the magnitude of the common-mode gain:

$$\text{CMRR} \equiv \frac{|A_d|}{|A_c|} \tag{13.56}$$

or

$$\text{CMRR} = 20 \log \frac{|A_d|}{|A_c|} \quad \text{(in dB)}$$

The CMRR for the one-sided output of the bipolar differential amplifier is determined from Equations 13.40 and 13.41 to be

$$\text{CMRR} = \frac{1}{2} \left[1 + \frac{2(1 + h_{fe}) R_E}{R_B + h_{ie}} \right] \tag{13.57}$$

Using the parameter values from the previous section, the value of the one-sided CMRR is 1070 (60.6 dB). The common-mode gain A_c for an ideal differential amplifier with a double-sided output is zero, so that the CMRR is infinity.

The CMRR for the MOSFET differential amplifier with a one-sided output, shown in Figure 13.5, is determined from Equations 13.54 and 13.55. The result is

$$\text{CMRR} = \frac{1 + 2g_m R_S}{2} \qquad (13.58)$$

To increase the value of the CMRR, the emitter resistor R_E in the bipolar circuit and the source resistor R_S in the MOSFET circuit should be made as large as possible. However, increasing these resistor values and still maintaining the same bias current requires that the bias voltages V_{EE} and V_{SS} be increased. The values of V_{EE} and V_{SS} must be kept within reasonable values because of power considerations as well as for preventing device breakdown.

13.1.6 Input Resistance

The input resistance is an important parameter for differential amplifiers as well as for most transistor circuits. The input resistance determines the input current for a given applied voltage and becomes important when considering the type of source that will be driving the circuit. The differential amplifier requires that a differential-mode and a common-mode input resistance be defined. Both resistance concepts are helpful when discussing common-mode and differential-mode signals.

The differential-mode input resistance is defined as the ratio of the small-signal differential input voltage to the small-signal input current when a pure differential input voltage is applied. Figure 13.9a shows the small-signal equivalent circuit for the bipolar differential amplifier of Figure 13.2, when $V_1 = V_d/2$ and $V_2 = -V_d/2$, giving a pure differential input voltage of $V_d = V_1 - V_2$. The circuit may be redrawn as shown in Figure 13.9b so that the differential-mode input resistance can be determined. The symmetry of the circuit gives $I_{b2} = -I_{b1}$. Therefore,

$$V_d = I_{b1}(R_B + h_{ie1}) - I_{b2}(R_B + h_{ie2}) \qquad (13.59)$$

If $h_{ie1} = h_{ie2} \equiv h_{ie}$, then

$$R_{id} = \frac{V_d}{I_{b1}} = 2(R_B + h_{ie}) \qquad (13.60)$$

The common-mode input resistance is defined as the ratio of the small-signal common-mode input voltage V_c to the small-signal input current I_{in} when a pure common-mode input voltage is applied. Figure 13.10 shows the small-signal equivalent circuit when $V_1 = V_2 = V_c$, giving a pure common-mode input voltage. If transistors Q_1 and Q_2 are matched, then $h_{ie1} = h_{ie2} \equiv h_{ie}$ and $h_{fe1} = h_{fe2} \equiv h_{fe}$. The currents at the node common to the emitters provide

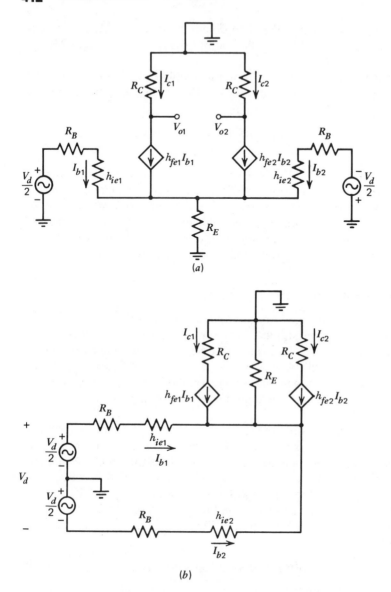

Figure 13.9. (*a*) Small-signal bipolar differential amplifier circuit showing a pure differential-mode input signal. (*b*) Circuit of part (*a*) redrawn.

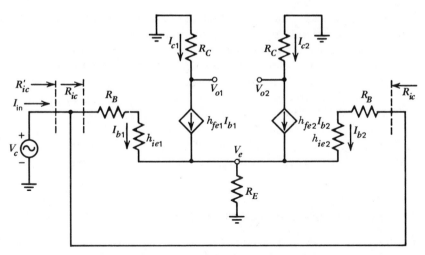

Figure 13.10. Small-signal bipolar differential amplifier circuit showing pure common-mode input signal.

$$(1 + h_{fe})I_{b1} + (1 + h_{fe})I_{b2} = \frac{V_e}{R_E} \tag{13.61}$$

and the base currents are given by

$$I_{b1} = I_{b2} = \frac{V_c - V_e}{R_B + h_{ie}} \tag{13.62}$$

The emitter voltage V_e is obtained by combining Equations 13.61 and 13.62:

$$V_e = \frac{V_c}{1 + [(R_B + h_{ie})/2(1 + h_{fe})R_E]} \tag{13.63}$$

Then, substituting Equation 13.63 into Equation 13.62 yields an expression for the two base currents:

$$I_{b1} = I_{b2} = \frac{V_c}{(R_B + h_{ie}) + 2(1 + h_{fe})R_E} \tag{13.64}$$

The input current is $I_{in} = I_{b1} + I_{b2}$ so that the common-mode input resistance R'_{ic} is

$$R'_{ic} = \frac{V_c}{I_{in}} = \tfrac{1}{2}(R_B + h_{ie}) + (1 + h_{fe})R_E \tag{13.65}$$

The common-mode input resistance R_{ic} is the resistance seen by each of the sources V_1 and V_2. This input resistance can then be calculated as

$$R_{ic} = \frac{V_c}{I_{b1}} = \frac{V_c}{I_{b2}} = (R_B + h_{ie}) + 2(1 + h_{fe})R_E \tag{13.66}$$

A comparison of the common-mode input resistance (Equation 13.66) and the differential-mode input resistance (Equation 13.60) shows that the common-mode input resistance is larger by approximately the term $2(1 + h_{fe})R_E$.

The differential-mode input resistance is the resistance parameter of most interest in dealing with the differential amplifier. The differential-mode input resistance is typically a few kilohms for the basic bipolar circuit considered previously. A high-input impedance is usually desired for a general-purpose amplifier. The MOSFET differential amplifier (Figure 13.5) has two inputs connected to the gates of M_1 and M_2. The input impedance of these gates is extremely high (megohms) so that both the differential-mode and the common-mode resistances of the MOSFET differential amplifier are very high.

13.1.7 The Differential Amplifier with Constant-Current Source

The resistors R_E and R_S in the bipolar and MOSFET circuits, respectively, should be made very large to increase the common-mode rejection ratio of the two differential amplifiers considered above. However, to maintain a given quiescent bias current through Q_1 and Q_2, the supply voltages V_{EE} and V_{SS} would also have to be increased to accommodate larger values of R_E and R_S. There is a practical limit to the extent that these supply voltages can be increased. Breakdown may become the limiting factor as the voltages increase, but the necessity of providing a high supply voltage must also be considered.

A constant-current source may be used to increase the effective value of R_E in the bipolar circuit and still maintain reasonable values of voltage. Figure 13.11 shows the basic bipolar differential amplifier with a two-transistor current source. The reference current I_1 is established by the voltage sources V_{CC} and V_{EE} and the resistor R_1. If the current gains of Q_3 and Q_4 are large, then $I_3 \approx I_1$. Also, the current I_3 divides evenly if $V_1 = V_2$ and the transistors are matched, resulting in

$$I_{C1} = I_{C2} = \frac{I_3}{2} \tag{13.67}$$

This current biases Q_1 and Q_2 in the active region.

The effective value of R_E is now the impedance seen looking into the collector of Q_3 which is $1/h_{oe}$. Values of $1/h_{oe}$ depend on the bias current and range from 20 kΩ to 1 MΩ, compared to values of 10 to 20 kΩ for R_E. For example, suppose that in Figure 13.11, $V_{CC} = V_{EE} = 5.0$ V, $R_1 = 9.35$ kΩ, $V_{BE}(\text{act}) = 0.65$ V, and $h_{fe} = 100$. Then,

$$
\begin{aligned}
I_1 &= \frac{V_{CC} + V_{EE} - V_{BE}(\text{act})}{R_1} \\
&= \frac{5.0 + 5.0 - 0.65}{9.35 \times 10^3} \\
&= 1.0 \text{ mA}
\end{aligned}
\tag{13.68}
$$

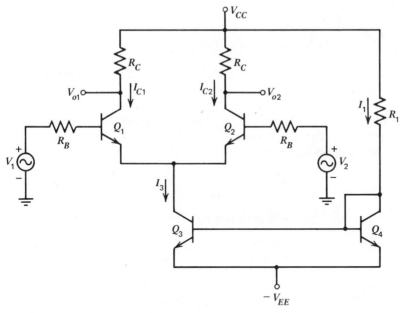

Figure 13.11. Bipolar differential amplifier biased with current mirror–current source network.

Considering the two-transistor current source

$$I_3 = \frac{I_1}{1 + (2/h_{fe})} = \frac{1.0 \times 10^{-3}}{1 + (2/100)}$$

$$\approx 1.0 \text{ mA}$$

(13.69)

Also, neglecting the base currents of Q_1 and Q_2, the quiescent collector currents are

$$I_{C1} = I_{C2} \approx \frac{I_3}{2}$$

$$= 0.5 \text{ mA}$$

(13.70)

If $R_C = 4.0 \text{ k}\Omega$, then the quiescent values of V_{o1} and V_{o2} are

$$V_{o1} = V_{o2} = V_{CC} - I_C R_C$$

$$= 5.0 - 0.5 \times 10^{-3} \times 4.0 \times 10^3$$

$$= 3.0 \text{ V}$$

(13.71)

The small-signal analysis of the differential amplifier with a constant-current source proceeds in exactly the same manner as with the resistor R_E. The one-sided output at V_{o2} is given by Equation 13.31. Replacing R_E by the $1/h_{oe}$ for the Q_3 transistor in Figure 13.11, the small-signal output voltage is

$$V_{o2} = -\frac{h_{fe}R_C}{R_B + h_{ie}} \left\{ \frac{V_2 \left[1 + \dfrac{R_B + h_{ie}}{(1 + h_{fe})(1/h_{oe})} \right]}{2 + \dfrac{R_B + h_{ie}}{(1 + h_{fe})(1/h_{oe})}} - \frac{V_1}{2 + \dfrac{R_B + h_{ie}}{(1 + h_{fe})(1/h_{oe})}} \right\}$$

(13.72)

The input voltages V_1 and V_2 may be written in terms of the common-mode and differential-mode voltages (Equations 13.36 and 13.37). Then, the differential-mode gain will be given by

$$A_d = -\frac{h_{fe}R_C}{2(R_B + h_{ie})}$$

(13.73)

and the common-mode gain will be given by

$$A_c = -\frac{h_{fe}R_C}{R_B + h_{ie}} \frac{1}{1 + \dfrac{2(1 + h_{fe})(1/h_{oe})}{R_B + h_{ie}}}$$

(13.74)

If we let $h_{fe} = 100$, $h_{ie} = 2.0$ kΩ, $R_C = 7.0$ kΩ, $R_B = 0.5$ kΩ, and $(1/h_{oe}) = 100$ kΩ, then

$$A_d = -140$$

and

$$A_c = -0.035$$

The CMRR is then

$$\text{CMRR} = |A_d/A_c| = 4040 \qquad \text{or} \qquad 72.1 \text{ dB}$$

The CMRR has been increased by approximately a factor of 4 in this example by using a constant-current source as compared to using a resistor R_E. The resistance at the collector of Q_3 can be further increased by using a Widlar current source. The effective resistance seen at the collector of the current source may then achieve values much greater than 1 megohm.

The basic MOSFET differential amplifier with a constant-current source shown in Figure 13.12 is similar to its bipolar counterpart. The resistance R_S is replaced by the output resistance $r_{o3} = 1/\lambda I_{D3}$ in the small-signal analysis. The CMRR of the MOSFET circuit for a one-sided output was given by Equation 13.58:

$$\text{CMRR} = \frac{A_d}{A_c} = \frac{1 + 2g_m R_S}{2}$$

For a value of $g_m = 5.0 \times 10^{-4}$ S and an R_S value of 13.5 kΩ, CMRR = 7.25, which is extremely small. When R_S is replaced by a constant-current source in the circuit, the effective value of R_S is r_{o3}, the output impedance of the M_3 MOSFET. Assuming a value of $r_{o3} = 200$ kΩ, the value of CMRR is 100.5. This value is nearly 14 times that obtained with the R_S resistor, but it is still

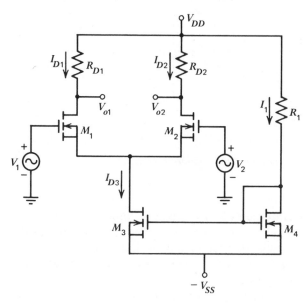

Figure 13.12. MOSFET differential amplifier biased with current mirror–current source network.

considerably smaller than for the bipolar circuit. The reason for this is that, normally, the bipolar transistor has a higher gain parameter than the MOSFET.

13.2 MULTIPLE-STAGE TRANSISTOR CIRCUITS

When the amplification of a single-stage transistor circuit is insufficient for a particular application, two or more stages may be connected together to increase the overall amplification. The input and output impedance specifications may also dictate the use of more than one transistor stage. Multistage transistor circuits are also needed if the dc level of the output voltage must be shifted to be compatible with another transistor circuit. In this section, we examine various multistage circuits and determine their characteristics.

13.2.1 Cascade Transistor Amplifiers

Transistor stages connected in cascade increase the voltage and current amplification over that of a single stage. Cascaded stages are also used to modify the input or output impedance of an amplifier. Figure 13.13 shows a cascade connection where the output voltage of the first stage becomes the input voltage of the second stage. The overall voltage amplification is now

$$A_v = \frac{V_{out2}}{V_{in1}} = \left(\frac{V_{out2}}{V_{in2}}\right)\left(\frac{V_{out1}}{V_{in1}}\right) = A_{v2}A_{v1} \qquad (13.75)$$

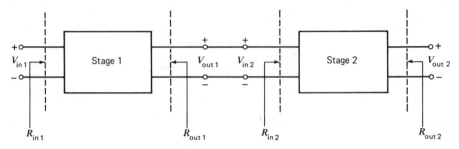

Figure 13.13. Two-stage amplifier circuit shown in block diagram form.

where $V_{out1} = V_{in2}$. The overall gain of this two-stage amplifier is then the product of the individual gains. It is important to realize that the gain of the first stage must be calculated taking into account the loading effect of the second stage. The input and output impedances of the stages also become an important characteristic of the cascade amplifier.

The two-stage bipolar amplifier in Figure 13.14 is biased so that both transistors, Q_1 and Q_2, are in their active regions. The small-signal equivalent circuit shown in Figure 13.15 applies if the circuit operates in the midband frequency range where the coupling capacitor and the emitter bypass capacitor act as virtual short circuits to the ac signal. Let the h-parameters of both transistors be

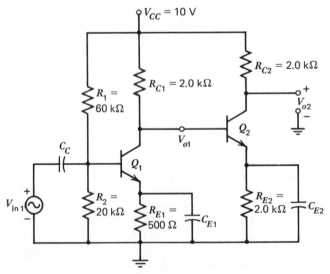

Figure 13.14. Two-stage common-emitter/common-emitter stage amplifier.

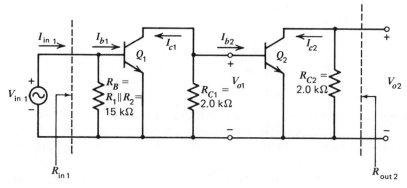

Figure 13.15. The ac equivalent circuit for the amplifier shown in Figure 13.14.

$$h_{ie} = 1.5 \text{ k}\Omega$$
$$h_{fe} = 100$$
$$h_{re} = 0.0$$
$$h_{oe} = 0.0 \text{ S}$$

It is usually advantageous to start the analysis at the output stage and work back to the input by substitution.

At the output of the circuit in Figure 13.15,

$$V_{\text{out2}} = -I_{c2}R_{C2} = -h_{fe}I_{b2}R_{C2} \tag{13.76}$$

The base current I_{b2} can be related to the collector current I_{c1} by the current divider equation, resulting in

$$I_{b2} = \frac{R_{C1}}{R_{C1} + h_{ie}}(-I_{c1}) \tag{13.77}$$

Then,

$$
\begin{aligned}
V_{\text{out2}} &= -h_{fe}R_{C2}\frac{R_{C1}}{R_{C1} + h_{ie}}(-I_{c1}) \\
&= h_{fe}R_{C2}\frac{R_{C1}}{R_{C1} + h_{ie}}(h_{fe}I_{b1}) \\
&= h_{fe}R_{C2}\frac{R_{C1}}{R_{C1} + h_{ie}}h_{fe}\frac{R_B}{R_B + h_{ie}}(I_{\text{in1}}) \\
&= h_{fe}R_{C2}\frac{R_{C1}}{R_{C1} + h_{ie}}h_{fe}\frac{R_B}{R_B + h_{ie}}\frac{V_{\text{in1}}}{R_B\|h_{ie}}
\end{aligned}
\tag{13.78}
$$

The last two expressions relate I_{b1} to I_{in} by the current divider equation and I_{in1} to V_{in1} by Ohm's law. The overall voltage gain can then be written and calculated as

$$A_v = \frac{V_{out2}}{V_{in1}}$$

$$= h_{fe}R_{C2}\,\frac{R_{C1}}{R_{C1} + h_{ie}}\,h_{fe}\,\frac{R_B}{R_B + h_{ie}}\,\frac{1}{R_B\|h_{ie}} \qquad (13.79)$$

$$= 7.6 \times 10^3$$

This voltage gain is substantially higher than that of each individual stage.

The voltage gain is also a positive quantity, which means that the output voltage is in phase with the input voltage. There is a phase reversal through each stage.

The input and output impedances of the cascade amplifier may be calculated as

$$R_{in} = R_{in1} = R_B\|h_{ie} \qquad (13.80)$$

$$= 15 \times 10^3\|1.5 \times 10^3 = 1.36 \text{ k}\Omega$$

and

$$R_{out} = R_{out2} = R_{C2} = 2.0 \text{ k}\Omega \qquad (13.81)$$

The output impedance is equal to R_{C2} because of the assumption that $h_{oe} = 0.0$.

Another example of a cascade amplifier is shown in Figure 13.16. The output stage is a common-collector or emitter–follower configuration. The equivalent small-signal circuit is shown in Figure 13.17, where the coupling and emitter bypass capacitors are again assumed to be short circuits to the ac signal.

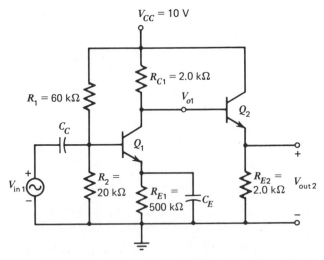

Figure 13.16. Two-stage common-emitter/common-collector amplifier.

The analysis proceeds as before, starting at the output and working back to the input:

$$V_{out2} = I_{e2}R_{E2} = (1 + h_{fe})I_{b2}R_{E2} \tag{13.82}$$

$$= (1 + h_{fe})R_{E2} \frac{R_{C1}}{R_{C1} + h_{ie} + (1 + h_{fe})R_{E2}} (-I_{c1})$$

This last expression relates I_{b2} to I_{c1} through a current divider and includes the effect of the emitter resistor. The current I_{c1} can be related to the input by

$$V_{out2} = (1 + h_{fe})R_{E2} \frac{R_{C1}}{R_{C1} + h_{ie} + (1 + h_{fe})R_{E2}} (-h_{fe}I_{b1}) \tag{13.83}$$

$$= (1 + h_{fe})R_{E2} \frac{R_{C1}}{R_{C1} + h_{ie} + (1 + h_{fe})R_{E2}} (-h_{fe}) \frac{R_B}{R_B + h_{ie}} I_{in1}$$

Finally, the voltage gain expression and calculation are (assuming the same values for the h-parameters as before)

$$A_v = \frac{V_{out2}}{V_{in1}}$$

$$= (1 + h_{fe})R_{E2} \frac{R_{C1}}{R_{C1} + h_{ie} + (1 + h_{fe})R_{E2}}$$

$$\times (-h_{fe}) \frac{R_B}{R_B + h_{ie}} \frac{1}{R_B \| h_{ie}} \tag{13.84}$$

$$= (101)(2.0 \times 10^3) \frac{2.0 \times 10^3}{[2.0 + 1.5 + (101)2.0] \times 10^3}$$

$$\times (-100) \frac{15 \times 10^3}{(15 + 1.5) \times 10^3} \frac{1}{15 \times 10^3 \| 1.5 \times 10^3}$$

$$= -131$$

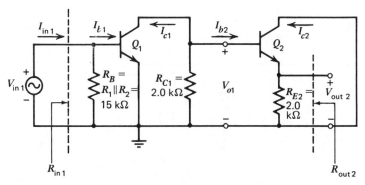

Figure 13.17. The ac equivalent circuit for the amplifier shown in Figure 13.16.

The overall voltage gain is reduced from the previous example. The reason for this is that the output stage is an emitter follower with a voltage gain slightly less than unity.

The input impedance R_{in} is, as before,

$$R_{in} = R_{in1} = R_B \| h_{ie} \tag{13.85}$$
$$= 1.36 \text{ k}\Omega$$

The output impedance R_{out} can be calculated from the equivalent circuit in Figure 13.18, where

$$R_{out} = R_{out2} = R_{E2} \| [(h_{ie} + R_{C1})/(1 + h_{fe})]$$
$$= 2.0 \times 10^3 \| [(1.5 \times 10^3 + 2.0 \times 10^3)/101] \tag{13.86}$$
$$= 34.1 \ \Omega$$

Hence, by using an emitter follower circuit as the output stage, the output impedance is reduced to a very low value. In this way, the output approaches an ideal voltage source.

MOSFET amplifier circuits may also be connected in cascade to achieve higher gain or a more desirable input or output resistance. Figure 13.19 shows a two-stage MOSFET amplifier with a source follower as the second stage. If the transistors have the parameters $V_T = 1.0$ V and $k = 0.1$ mA/V^2, the dc quiescent values are

$$I_{D1} = 0.336 \text{ mA} \qquad I_{D2} = 0.732 \text{ mA}$$
$$V_{GS1} = 2.83 \text{ V} \qquad V_{GS2} = 3.71 \text{ V}$$
$$V_{DS1} = 5.0 \text{ V} \qquad V_{DS2} = 8.54 \text{ V}$$

The small-signal transconductance at the Q-point is

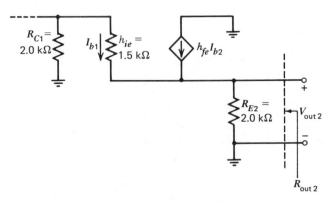

Figure 13.18. Small-signal equivalent circuit for the output stage of the amplifier shown in Figure 13.17.

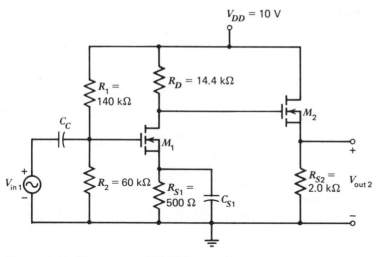

Figure 13.19. Two-stage *n*-MOSFET amplifier with a common-source/common-drain configuration.

$$g_m = \frac{\partial I_D}{\partial V_{GS}} = 2k(V_{GS} - V_T) \tag{13.87}$$

From the quiescent values,

$$g_{m1} = 0.366 \times 10^{-3} \text{ S} \qquad g_{m2} = 0.542 \times 10^{-3} \text{ S}$$

Both transistors are biased in the saturation or amplifying regions since $V_{DS} >$ $(V_{GS} - V_T)$ for each transistor.

If the circuit operates in the midband frequency range where the coupling and source bypass capacitors act as virtual short circuits to the ac signal, then the small-signal equivalent circuit becomes that shown in Figure 13.20. If the transistor small-signal output impedances $r_{d1} = r_{d2} = \infty$, the small-signal voltage

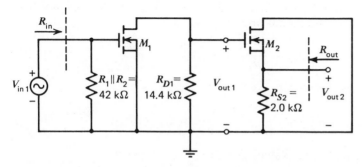

Figure 13.20. The ac equivalent circuit for the amplifier shown in Figure 13.19.

gain can be determined as follows:

$$V_{out2} = I_{d2}R_{S2} = g_{m2}V_{gs2}R_{S2} \tag{13.88}$$

and

$$V_{gs2} = V_{out1} - I_{d2}R_{S2} = V_{out1} - V_{out2} \tag{13.89}$$

Since

$$V_{out2} = \frac{g_{m2}R_{S2}V_{out1}}{1 + g_{m2}R_{S2}} \tag{13.90}$$

and

$$V_{out1} = -I_{d1}R_{D1} = -g_{m1}R_{D1}V_{gs1} \tag{13.91}$$
$$= -g_{m1}R_{D1}V_{in1}$$

Then, the overall voltage gain is

$$A_d = \frac{V_{out2}}{V_{in1}} = \frac{V_{out2}}{V_{out1}}\frac{V_{out1}}{V_{in1}} \tag{13.92}$$
$$= \frac{g_{m2}R_{S2}}{1 + g_{m2}R_{S2}}(-g_{m1}R_{D1})$$

Substituting the given values for the circuit, we obtain

$$A_d = \frac{-0.542 \times 10^{-3} \times 2.0 \times 10^3 \times 0.366 \times 10^{-3} \times 14.4 \times 10^3}{1 + (0.542 \times 10^{-3} \times 2.0 \times 10^3)}$$
$$= -2.78$$

The input impedance R_{in} as seen by the source V_{in1} is $R_1 \| R_2 = 42$ kΩ, since the input impedance at the gate of Q_1 is assumed to be infinite. The input impedance of a MOSFET amplifier is much large than the input impedance of a similar bipolar circuit.

The output impedance R_{out} can be calculated using the equivalent circuit in Figure 13.21. The independent voltage source V_{in1} is set equal to zero, causing $V_{gs1} = V_{out1} = 0$. Since $V_{gs2} = V_{out1} - V_{out2} = -V_{out2}$, then, applying Kirchhoff's current law at the output node yields

$$I_{out} = \frac{V_{out2}}{R_{S2}} + g_{m2}V_{out2} \tag{13.93}$$

so that

$$R_{out} = \frac{V_{out2}}{I_{out}} = \frac{1}{(1/R_{S2}) + g_{m2}} = R_{S2}\|(1/g_{m2}) \tag{13.94}$$

Substitution of the parameter values gives

$$R_{out} = (2.0 \times 10^3)\|(1/0.542 \times 10^{-3})$$
$$= 960 \ \Omega$$

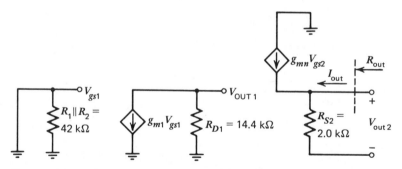

Figure 13.21. The small-signal equivalent circuit used to derive the output resistance R_{out} for the circuit shown in Figure 13.20.

13.2.2 **Darlington Pair**

The two-transistor circuit shown in Figure 13.22 is known as a Darlington pair. The emitter current gain can be found from

$$A_{ie} = \frac{I_{e1}}{I_{b2}} = \frac{I_{e1}}{I_{b1}} \frac{I_{b1}}{I_{b2}} = \frac{I_{e1}}{I_{b1}} \frac{I_{e2}}{I_{b2}}$$

$$= (1 + h_{fe1})(1 + h_{fe2}) \tag{13.95}$$

If the small-signal current gains are matched, then $h_{fe1} = h_{fe2} = h_{fe}$, and

$$A_{ie} = (1 + h_{fe})^2 \tag{13.96}$$

The overall current gain A_{ic} is given by

$$A_{ic} = \frac{I_{out}}{I_{b2}}$$

This can be found from

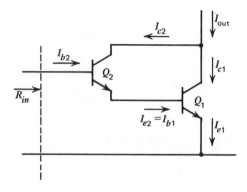

Figure 13.22. The basic configuration of the Darlington pair connection.

$$A_{ic} = \frac{I_{out}}{I_{b2}} = \frac{I_{c1} + I_{c2}}{I_{b2}}$$

$$= \frac{I_{c1}}{I_{b1}} \frac{I_{b1}}{I_{b2}} + \frac{I_{c2}}{I_{b2}}$$

which can be written as

$$A_{ic} = h_{fe1}h_{fe2} + h_{fe2} \qquad (13.97)$$

giving

$$A_{ic} \simeq (h_{fe})^2$$

Therefore, the Darlington pair can be treated as a single transistor with a very high current gain.

The small-signal input impedance can be calculated with the aid of Figure 13.23. The base–emitter impedance, h_{ie1}, of Q_1 is reflected back through the emitter of Q_2 giving

$$R_{in} = h_{ie2} + (1 + h_{fe})h_{ie1} \qquad (13.98)$$

Even though we have assumed that the current gains of the transistors are matched, the base–emitter impedances will not be equal due to the difference in the bias levels of the two devices. The base–emitter impedance h_{ie} is given by

$$h_{ie} \simeq \frac{kT}{q} \frac{1}{I_{BQ}} \qquad (13.99)$$

This gives

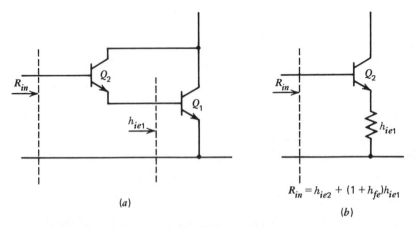

(a)

$$R_{in} = h_{ie2} + (1 + h_{fe})h_{ie1}$$

(b)

Figure 13.23. (a) Darlington pair amplifier showing the symbols for the input resistance of each transistor. (b) Equivalent circuit to calculate the input resistance R_{in} where h_{ie1} is the input resistance of Q_1.

$$h_{ie1} \simeq \frac{kT}{q} \frac{1}{I_{BQ1}}$$

(13.100)

and

$$h_{ie2} \simeq \frac{kT}{q} \frac{1}{I_{BQ2}} = \frac{kT}{q} \frac{1 + h_{fe}}{I_{BQ1}}$$
$$\simeq (1 + h_{fe})h_{ie1}$$

(13.101)

Then, the input impedance of the Darlington pair is

$$
\begin{aligned}
R_{\text{in}} &= h_{ie2} + (1 + h_{fe})h_{ie1} \\
&= (1 + h_{fe})h_{ie1} + (1 + h_{fe})h_{ie1} \\
&= 2(1 + h_{fe})h_{ie1}
\end{aligned}
$$

(13.102)

Thus, we have multiplied the impedance of Q_1 by the factor $2(1 + h_{fe})$. The input impedance of the Darlington pair is normally on the order of hundreds of kilohms.

13.2.3 The Level-Shifting Amplifier

The amplifier in Figure 13.24 is used when it is desirable to change the level of the dc voltage. The voltages V_{in} and V_{BB1} will, in general, be the small-signal output voltage and the dc output voltage, respectively, of a preceding amplifier stage. In most cases, we would like to have the dc level of the output, V_L, be zero.

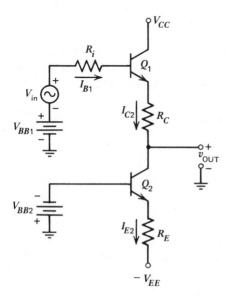

Figure 13.24. Voltage level shift circuit.

The bias V_{BB2}, the transistor Q_2, and its emitter resistor R_E form a constant-current source for the transistor Q_1. The dc analysis of the circuit proceeds as follows:

$$I_{E2} = \frac{(-V_{BB2} - V_{BE}) - (-V_{EE})}{R_E} \tag{13.103}$$

and

$$I_{C2} = \frac{\beta_0}{1 + \beta_0} I_{E2} \simeq I_{E2} \tag{13.104}$$

Kirchhoff's voltage law allows us to write

$$V_{BB1} = I_{B1}R_i + V_{BE} + I_{C2}R_C + V_{OUT} \tag{13.105}$$

The circuit is designed with Q_1 and Q_2 biased in their active regions, so that

$$I_{B1} = \frac{I_{C2}}{1 + \beta_0} \tag{13.106}$$

and

$$V_{OUT} = V_{BB1} - V_{BE} - I_{C2}R_C - \frac{I_{C2}R_i}{1 + \beta_0} \tag{13.107}$$

The value of R_C can be chosen to make the voltage V_{OUT} any particular dc level desired. In the circuit of Figure 13.24, the voltage level is shifted down from V_{BB1}. A similar circuit using *pnp* transistors would allow a positive shifting of V_{BB1}.

The small-signal ac analysis of this amplifier can be done using the small-signal equivalent circuit in Figure 13.25. The transistor Q_2 is connected in a common-base configuration. In this analysis, we assume that $h_{re} = h_{rb} = h_{oe} = 0$ for Q_1.

In this model, I_{e2} must always be zero, implying that the dependent current source $h_{fb}I_{e2}$ must also be zero. Hence, the only small-signal element remaining in Q_2 is the output admittance h_{ob}. Kirchhoff's voltage law gives

$$V_{in} = I_{b1}R_i + I_{b1}h_{ie1} + I_{c2}R_C + I_{c2}\frac{1}{h_{ob}} \tag{13.108}$$

and

$$V_{out} = I_{c2}\frac{1}{h_{ob}} \tag{13.109}$$

Substituting

$$I_{c2} = \frac{V_{out}}{1/h_{ob}} \tag{13.110}$$

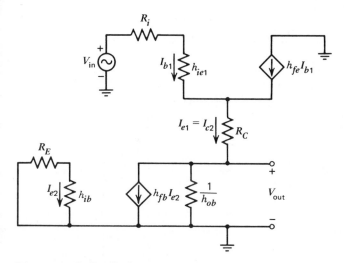

Figure 13.25. Small-signal equivalent circuit for the voltage shift circuit of Figure 13.24.

and

$$I_{b1} = \frac{I_{c2}}{1 + h_{fe1}}$$ (13.111)

into Equation 13.108 yields

$$\frac{V_{out}}{V_{in}} = \frac{1}{1 + \dfrac{[(R_i + h_{ie1})/(1 + h_{fe1})] + R_C}{1/h_{ob}}}$$ (13.112)

In the limit as $1/h_{ob}$ approaches infinity, the ratio of V_{out} to V_{in} approaches 1. The level-shifting amplifier in Figure 13.24 is then an emitter follower to the ac signal while performing a dc level shift function.

13.3 A SIMPLIFIED OPERATIONAL AMPLIFIER

An operational amplifier is composed of a differential amplifier input stage, a gain stage, and an output stage. One such example of a bipolar operational amplifier is shown in Figure 13.26. The differential amplifier stage has a constant-current source composed of Q_3 and Q_4. A one-sided output is fed to a Darlington pair gain stage. The output stage is the dc voltage level shifter with an emitter follower output.

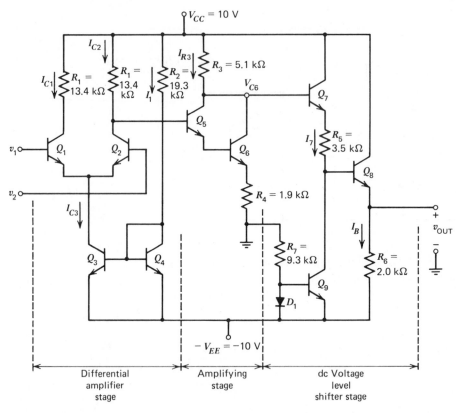

Figure 13.26. An example of a bipolar operational amplifier.

13.3.1 **DC Analysis**

The dc analysis of the circuit is simplified if all base currents are neglected and a constant $V_{BE} = 0.7$ V is assumed. This circuit is designed for the following quiescent conditions when $v_1 = v_2 = 0$:

$$I_{C1} = I_{C2} = 0.5 \text{ mA}$$
$$I_{R3} = 1.0 \text{ mA}$$
$$I_7 = 1.0 \text{ mA}$$
$$I_8 = 5.0 \text{ mA}$$
$$V_{CE1} = V_{CE2} = 4.0 \text{ V}$$
$$V_{CE6} = 3.0 \text{ V}$$
$$V_{OUT} = 0.0$$

We then have

$$I_{C3} = I_1 = I_{C1} + I_{C2}$$
$$= 1.0 \text{ mA}$$

which determines the value of R_2 as

$$R_2 = \frac{10 - 0.7 - (-10)}{1.0 \times 10^{-3}}$$

$$= 19.3 \text{ k}\Omega \tag{13.113}$$

The resistors R_1 can be determined from the equation

$$R_1 = \frac{V_{CC} - V_{CE1} - (-V_{BE})}{I_{C1}}$$

$$= \frac{10 - 4.0 + 0.7}{0.5 \times 10^{-3}} \tag{13.114}$$

$$= 13.4 \text{ k}\Omega$$

The dc voltage at the collector of Q_2 is

$$V_{C2} = V_{CC} - I_{C2}R_1 \tag{13.115}$$

$$= 3.3 \text{ V}$$

and the dc voltage at the emitter of Q_6 is

$$V_{E6} = V_{C2} - 2V_{BE} \tag{13.116}$$

$$= 1.9 \text{ V}$$

The resistor R_4 is found from

$$R_4 = \frac{V_{E6}}{I_{R3}} = \frac{1.9}{1.0 \times 10^{-3}} \tag{13.117}$$

$$= 1.9 \text{ k}\Omega$$

Assuming a value of $V_{CE6} = 3.0$ V, the dc collector voltage of Q_6 is $V_{C6} = 1.9 + 2.0 = 4.9$ V, and

$$R_3 = \frac{V_{CC} - V_{C6}}{I_{R3}} = \frac{10 - 4.9}{1 \times 10^{-3}} \tag{13.118}$$

$$= 5.1 \text{ k}\Omega$$

If the dc level of the output voltage is to be zero, then R_5 is determined from

$$V_{C6} = 2V_{BE} + I_7R_5 \tag{13.119}$$

or

$$R_5 = 3.5 \text{ k}\Omega$$

If we want the current through D_1 to be equal to the current through Q_9, then R_7 is

$$R_7 = \frac{10 - 0.7}{1.0 \times 10^{-3}}$$

$$= 9.3 \text{ k}\Omega \tag{13.120}$$

Finally, the output resistor R_6 is found from

$$R_6 = \frac{V_{EE}}{I_8} = \frac{10}{5.0 \times 10^{-3}} \tag{13.121}$$

$$= 2.0 \text{ k}\Omega$$

The circuit of Figure 13.26 now has all transistors biased in their active regions and the design specifications are satisfied.

13.3.2 Small-Signal Analysis

The small-signal analysis of the operational amplifier of Figure 13.26 can be performed by breaking the circuit into its individual stages as shown in Figure 13.27a. The equivalent cascade circuit is given in Figure 13.27b. Each stage has an input and output resistance and a voltage source that is determined from the open-circuit voltage gain of each stage. The stages are represented by their Thevenin equivalent circuits.

The open-circuit voltage gain of each stage depends on the individual

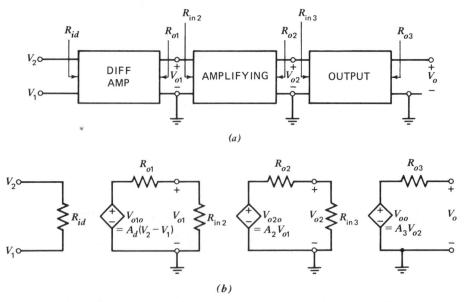

Figure 13.27. (a) Block diagram of basic sections of the operational amplifier. (b) A small-signal equivalent circuit of the amplifier sections in part (a).

transistor parameters as well as the resistor values. Assuming that h_{fe} is 100 for all transistors, the h_{ie} values for each transistor can be calculated from

$$h_{ie} = \frac{kT}{q} \frac{1}{I_B} = \frac{kT}{q} \frac{h_{fe}}{I_C}$$

We then calculate the following values:

$$h_{ie1} = h_{ie2} = 5.2 \text{ k}\Omega$$

$$h_{ie5} = 260 \text{ k}\Omega$$

$$h_{ie6} = h_{ie7} = 2.6 \text{ k}\Omega$$

$$h_{ie8} = 0.52 \text{ k}\Omega$$

The open-circuit differential voltage gain A_d of the first stage in Figure 13.26 can be determined from Equation 13.73:

$$A_d = \frac{-h_{fe}R_1}{2h_{ie1}} = -\frac{100 \times 13.4 \times 10^3}{2 \times 5.2 \times 10^3}$$

$$= -129$$

The voltage V_{o1} in Figure 13.27 forms a voltage divider with the voltage source, or

$$V_{o1} = \frac{R_{in2}}{R_{in2} + R_{o1}} V_{o1o}$$

The input resistance R_{in2} of the Darlington pair circuit is

$$R_{in2} = h_{ie5} + (1 + h_{fe})[h_{ie6} + (1 + h_{fe})R_4]$$

$$= 19.7 \text{ M}\Omega$$

The output resistance of the differential amplifier stage is

$$R_{o1} = R_1$$

$$= 13.4 \text{ k}\Omega$$

assuming that $h_{oe} = 0$ for the transistors. Then,

$$V_{o1} \simeq V_{o1o}$$

The extremely large input resistance of the Darlington circuit ensures that the differential amplifier stage is not loaded down.

The voltage gain A_2 of the amplifying stage is

$$A_2 = \frac{V_{o2o}}{V_{o1}} \cong -\frac{(h_{fe})^2 R_3}{R_{in2}}$$

$$= -2.6$$

The input resistance of the third stage is

$$R_{in3} = h_{ie7} + (1 + h_{fe})[R_5 + h_{ie8} + (1 + h_{fe})R_6]$$
$$= 20.8 \ \text{M}\Omega$$

The output resistance of the second stage is

$$R_{o2} = R_3$$
$$= 5.1 \ \text{k}\Omega$$

Then,

$$V_{o2} \simeq V_{o2o}$$

The output dc level shifter stage has a unity small-signal voltage gain given by Equation 13.112. Then, the overall small-signal voltage gain is

$$\frac{V_{\text{out}}}{V_2 - V_1} = \frac{A_3 V_{o2}}{V_2 - V_1} = \frac{A_3 V_{o2o}}{V_2 - V_1} = \frac{A_3 A_2 V_{o1}}{V_2 - V_1} = \frac{A_3 A_2 V_{o1o}}{V_2 - V_1}$$
$$= A_3 A_2 A_d$$

In this case, the overall small-signal voltage gain is the product of the gains of each individual stage. The input resistances of the second and third stages are extremely large, so loading effects are negligible. Combining the results gives

$$\frac{V_o}{V_2 - V_1} = \frac{V_o}{V_d} = A_3 A_2 A_d = 1 \times (-2.6) \times (-129)$$
$$= 335$$

13.3.3 **Bipolar Current Sources as Active Loads**

An integrated circuit version of an operational amplifier often uses *pnp* transistors as active loads in the first two stages. These active loads increase the overall gain to values in excess of 200×10^3.

Active loads are obtained with circuitry that is similar to bias circuits. In most transistor amplifiers, the small-signal voltage gain is directly proportional to the load resistance. The small-signal voltage gain may be increased by increasing this load resistance. However, this requires a larger power supply voltage and may also change the operating point of the transistor. The output resistance at the collector of a bipolar transistor may be used as a load resistance. This can result in a very large voltage gain at reasonable power supply voltages. The use of a transistor in place of a resistive load implies that the load is active.

A common-emitter amplifier with a *pnp* active load is shown in Figure 13.28. The active load circuit of Q_2 and Q_3 also provides the bias for Q_1. Assuming that both Q_1 and Q_2 are in their active regions, the collector currents can be written as

$$I_{C1} = I_{s1}\left[\exp\left(\frac{qV_{IN}}{kT}\right)\right]\left(1 + \frac{V_{CE1}}{V_{AN}}\right) \tag{13.122a}$$

$$I_{C2} = I_{s2}\left[\exp\left(\frac{qV_{EB2}}{kT}\right)\right]\left(1 + \frac{V_{EC2}}{V_{AP}}\right) \tag{13.122b}$$

where V_{AN} and V_{AP} are the Early voltages for the *npn* transistor Q_1 and the *pnp* transistor Q_2, respectively. Note from Figure 13.28 that

$$I_{C1} = I_{C2} \tag{13.123a}$$

$$V_{CE1} + V_{EC2} = V_{CC} \tag{13.123b}$$

$$V_{EC3} = V_{EB3} = V_{EB2} \tag{13.123c}$$

Assuming that Q_2 and Q_3 are identical and that base currents may be neglected, we have

$$I_1 \simeq I_{C3} = I_{s2}\left[\exp\left(\frac{qV_{EB3}}{kT}\right)\right]\left(1 + \frac{V_{EC3}}{V_{AP}}\right) \tag{13.124a}$$

or

$$I_1 = I_{s2}\left[\exp\left(\frac{qV_{EB2}}{kT}\right)\right]\left(1 + \frac{V_{EB2}}{V_{AP}}\right) \tag{13.124b}$$

Combining Equations 13.124b, 13.122a, and 13.122b, we obtain

$$I_{s1}\left[\exp\left(\frac{qV_{IN}}{kT}\right)\right]\left(1 + \frac{V_{CE1}}{V_{AN}}\right) = \frac{I_1[1 + (V_{EC2}/V_{AP})]}{1 + (V_{EB2}/V_{AP})} \tag{13.125}$$

The output voltage is $V_{OUT} = V_{CE1}$ and, from Equation 13.123b,

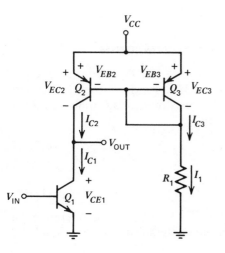

Figure 13.28. Common-emitter amplifier with *pnp* active load Q_2.

$$V_{EC2} = V_{CC} - V_{CE1} = V_{CC} - V_{OUT} \tag{13.126}$$

Then, substituting this into Equation 13.125 results in

$$I_{s1}\left[\exp\left(\frac{qV_{IN}}{kT}\right)\right] = \frac{I_1\{1 + [(V_{CC} - V_{OUT})/V_{AP}]\}}{[1 + (V_{OUT}/V_{AN})][1 + (V_{EB2}/V_{AP})]} \tag{13.127}$$

Since V_{AN} and V_{AP} are usually large values, $V_{OUT}/V_{AN} \ll 1$ and $V_{EB2} \sqrt{V_{AP}} \ll 1$, and

$$\frac{1}{[1 + (V_{OUT}/V_{AN})][1 + (V_{EB2}/V_{AP})]} \approx 1 - \frac{V_{OUT}}{V_{AN}} - \frac{V_{EB2}}{V_{AP}} \tag{13.128}$$

Then, Equation 13.127 becomes

$$I_{s1}\left[\exp\left(\frac{qV_{IN}}{kT}\right)\right] = I_1\left[1 + \frac{V_{CC}}{V_{AP}} - V_{OUT}\left(\frac{1}{V_{AP}} + \frac{1}{V_{AN}}\right) - \frac{V_{EB2}}{V_{AP}}\right] \tag{13.129}$$

Solving for the output voltage,

$$V_{OUT} = \frac{V_{AN}V_{AP}}{V_{AN} + V_{AP}}\left[1 - \frac{I_{s1}\exp\left(qV_{IN}/kT\right)}{I_1} \right.$$
$$\left. + \frac{V_{AN}}{V_{AN} + V_{AP}}(V_{CC} - V_{EB2})\right] \tag{13.130}$$

Equation 13.130 is valid as long as both Q_1 and Q_2 are in their active regions or

$$V_{CC} - V_{CE}(\text{sat})(Q_2) > V_{OUT} > V_{CE}(\text{sat})(Q_1) \tag{13.131}$$

The dc voltage transfer characteristic partially described by Equation 13.130 is shown in Figure 13.29. The parameters V_{OUTQ} and V_{INQ} represent the dc quiescent point. This point must be near the center of the active region. The small-signal voltage gain is the slope of the dc transfer curve at the quiescent operating point.

13.3.4 MOS Current Sources as Active Loads

An NMOS common-source amplifier with a PMOS current source load is shown in Figure 13.30. Transistors M_2 and M_3 form the current source load in which the current is determined by the resistor R_1. The transistor M_3 is connected as a saturated transistor so that

$$I_{REF} = k_{P3}(V_{SG3} - |V_{T3}|)^2 \tag{13.132}$$

and

$$V_{SG3} = V_{DD} - I_{REF}R_1 \tag{13.133}$$

Substituting Equation 13.133 into Equation 13.132 will give the reference current as a function of R_1 and the transistor parameters. This current establishes the voltage $V_{SG3} = V_{SG2}$.

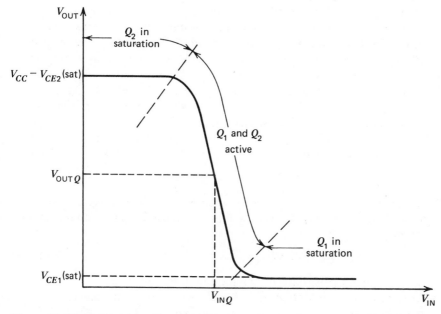

Figure 13.29. Voltage transfer curve for the active load common-emitter circuit shown in Figure 13.28.

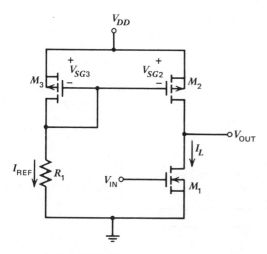

Figure 13.30. NMOS common-source amplifier with PMOS active load transistor M_2.

For $V_{IN} \leq V_{T1}$, M_1 is cut off so that there is zero current through M_2 and M_1. This makes $V_{OUT} = V_{DD}$. As V_{IN} increases above V_T, M_1 begins to conduct in its saturation region and M_2 is in its nonsaturation region, forcing the output voltage V_{OUT} to decrease. As V_{OUT} decreases, M_2 will move into its saturation region. Finally, a point will be reached where M_1 goes into its nonsaturation region. This process generates the transfer characteristic shown in Figure 13.31. This characteristic is similar to a CMOS transfer curve, except that the p-channel transistor in Figure 13.30 does not cut off.

We may assume that in the saturation region, the n-channel transistor is characterized by the equation

$$I_D = k_N(V_{GS} - V_{TN})^2 (1 + \lambda V_{DS}) \tag{13.134}$$

In the ideal device, we have assumed that $\Delta = 0$. In the MOS case, Δ is comparable to the Early voltage for the bipolar transistor. The p-channel equation is similar and is given by

$$I_D = k_P(V_{SG} - |V_{TP}|)^2 (1 + \lambda V_{SG}) \tag{13.135}$$

where I_D is defined as the current leaving the drain terminal.

If the drain currents of M_1 and M_2 of Figure 13.30 are equated for the case where both M_1 and M_2 are in their saturation regions, we have

$$k_N(V_{GS1} - V_{TN})^2 (1 + \lambda_1 V_{DS1}) = k_P(V_{SG2} - |V_{TP}|)^2 (1 + \lambda_2 V_{SD2}) \tag{13.136}$$

For this circuit, the parameters are given by $V_{GS1} = V_{IN}$, $V_{DS1} = V_{OUT}$,

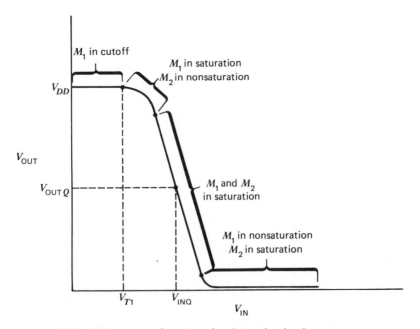

Figure 13.31. Voltage transfer curve for the active load common-source circuit shown in Figure 13.30.

$V_{SD2} = V_{DD} - V_{OUT}$, and V_{SG2} is a constant determined by the reference current in Equation 13.132. Rearranging terms, we can solve for the output voltage V_{OUT} as a function of the input voltage V_{IN}, or

$$V_{OUT} = \frac{k_P(V_{SG2} - |V_{TP}|)^2 (1 + \lambda_2 V_{DD}) - k_N(V_{IN} - V_{TN})^2}{k_P(V_{SG2} - |V_{TP}|)^2 \lambda_2 + k_N(V_{IN} - V_{TN})^2 \lambda_1} \quad (13.137)$$

The circuit in Figure 13.30 may be used as a linear amplifying stage if the quiescent point is placed near the center of the transfer characteristics as shown in Figure 13.31. The parameters, V_{OUTQ} and V_{INQ}, represent the dc quiescent point. The small signal voltage gain is the slope of the transfer curve at the quiescent point.

13.4 SUMMARY

Multiple-transistor circuits are used in a variety of electronic systems. They are particularly popular in integrated circuits. The differential amplifier makes use of matched transistor characteristics to enable it to amplify differential-mode signals while rejecting common-mode signals. The characteristics of a differential amplifier can be improved by using current source biasing.

Multistage amplifiers are characterized by large increases in voltage or current gain and changes in the impedance levels on the input or output side. Another function that can be accomplished by using multistage configurations is dc level shifting. An interesting multistage amplifier, the Darlington pair, is formed by the direct connection of transistors. This combination can be treated as a single unit with the characteristics of a very high-gain device.

A significant combination of amplifier stages is the operational amplifier, usually consisting of a differential input stage, a gain stage, and an output stage. Typical integrated circuit forms of OP AMPs have voltage gains greater than 100×10^3. One way to obtain high gain is to use active loads in the input and gain stages.

REFERENCES

1. P. R. Gray and R. G. Meyer, *The Analysis and Design of Analog Integrated Circuits* (New York: Wiley, 1977).

PROBLEMS

Note: Assume a value of $V_{BE}(\text{act}) = 0.70$ V for all bipolar transistors used in the problems below.

13.1. The differential amplifier shown in Figure 13.2 has the following circuit and transistor values: $R_C = 50$ kΩ, $R_B = 10$ kΩ, $R_E = 35$ kΩ, $\beta_1 = \beta_2$

$= 200$, $V_{CC} = 10$ V, $V_{EE} = -10$ V, and $v_1 = v_2 = 0$ V. Find (a) I_E, I_{C1}, and I_{C2}; (b) V_{CE1} and V_{CE2}; (c) If $v_1 = 1.0$ V dc, find I_{C1}, I_{C2}, and V_{CE1}.

13.2. The differential amplifier shown in Figure 13.32 has transistor h_{fe} values of $h_{fe1} = h_{fe2} = 100$. Find (a) I_{C1}, I_{C2}, and I_E; (b) V_{CE1} and V_{CE2}.

13.3. Select a value for R_E in the circuit of Figure 13.32 so that $V_{CE1} = V_{CE2} = 4.0$ V.

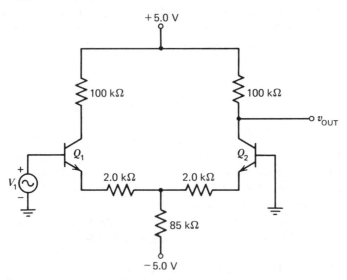

Figure 13.32. Circuit for Problems 13.2 and 13.3.

13.4. Consider the differential amplifier shown in Figure 13.33. If $h_{fe1} = h_{fe2} = 200$, (a) find R_C and R_E so that $I_{C1} = I_{C2} = 2.0$ mA and $V_{o2} = 8.0$ V when $v_1 = v_2 = 0$ V. (b) Draw the dc load line for the transistor Q_2.

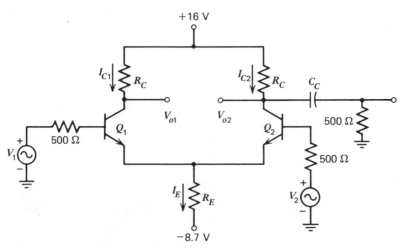

Figure 13.33. Circuit for Problem 13.4.

13.5. The n-channel MOSFET differential amplifier in Figure 13.5 has V_{DD} = 5.0 V, V_{SS} = -5.0 V, V_T = 1.0 V, R_D = 10 kΩ, and k = 0.05 μA/V². Find R_S such that I_{D1} = I_{D2} = 200 μA.

13.6. The differential amplifier shown in Figure 13.34 has h_{fe1} = h_{fe2} = 100 and h_{oe} = h_{re} = 0. Calculate R_E such that I_E = 150 μA. Then calculate A_d, A_c, CMRR (in dB), R_{ic}, and R_{id}.

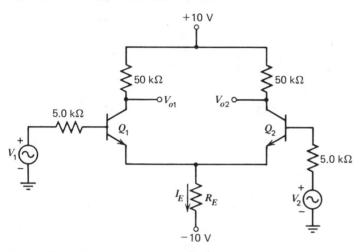

Figure 13.34. Circuit for Problem 13.6.

13.7. The transistor parameters in the circuit of Figure 13.35a are h_{fe} = 200 and h_{oe} = h_{re} = 0. (*a*) Determine the dc value of V_{OUT}. (*b*) Find the ac voltage gain V_{out}/V_{in} in general terms and then the numerical value. (*c*) Discuss the effects of the resistors labeled R_E. Could these resistors be moved to the base leads as shown in Figure 13.35b with the same effect? What value would you use for R_B in place of R_E to achieve an equal value for R_{id}?

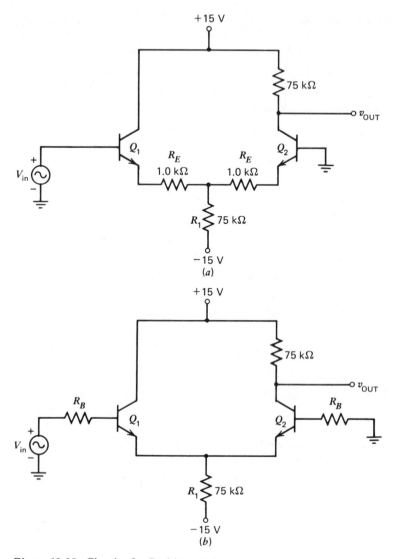

Figure 13.35. Circuits for Problem 13.7.

13.8. The MOS differential amplifier shown in Figure 13.36 has transistor values of $k = 40$ $\mu\text{A/V}^2$ and $V_T = 1.0$ V. Find A_d, A_c, and CMRR (in dB) for this amplifier.

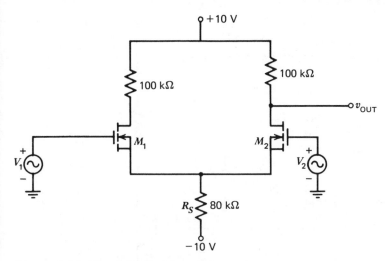

Figure 13.36. Circuit for Problem 13.8.

13.9. The transistor parameters for a circuit given in Figure 13.11 are h_{fe} = 100, h_{oe} = 8.0 μS, and h_{re} = 0. The circuit values are V_{CC} = 15 V, V_{EE} = −15 V, and R_B = 10 kΩ. (a) Design resistor values in the circuit so that I_3 = 400 μA and V_{CE1} = V_{CE2} = 10 V. (b) Calculate A_d, A_c, and CMRR (in dB) for single-ended output.

13.10. The transistor parameters for the circuit given in Figure 13.12 are V_T = 1.0 V and k = 50 μA/V^2. The power supply values are V_{DD} = 10 V and V_{SS} = 10 V. (a) Design resistor values in the circuit so that I_{DS} = 400 μA and V_{DS1} = V_{DS2} = 10 V. (b) Calculate A_d, A_c, and CMRR (in dB) if λ = 0.025/V for single ended output.

13.11. Consider the differential amplifier shown in Figure 13.37 where h_{fe} = 200 and I_0 = 2.0 mA. Assume that I_{E1} = I_{s1} exp(qV_{BE1}/kT) and I_{E2} = I_{s2} exp (qV_{BE2}/kT), where kT/q = 26 mV.

(a) If v_1 = v_2 = 0 and I_{s1} = I_{s2} = 1.0 × 10^{-23} A, find V_{o1} − V_{o2} for (1) R_{C1} = R_{C2} = 8.0 kΩ; (2) R_{C1} = 8.0 kΩ and R_{C2} = 7.9 kΩ.

(b) If v_1 = v_2 = 0, I_{s1} = 1.0 × 10^{-13} A, and I_{s2} = 1.1 × 10^{-13} A, find the dc voltage V_{o1} − V_{o2} if (1) R_{C1} = R_{C2} = 8.0 kΩ; (2) R_{C1} = 8.0 kΩ and R_{C2} = 7.9 kΩ.

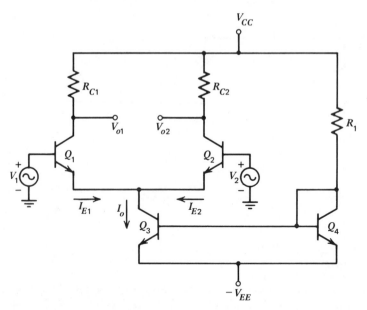

Figure 13.37. Circuit for Problem 13.11.

13.12. A cascade amplifier is shown in Figure 13.38 with the bias circuitry omitted from the first stage. The transistors parameters are $h_{fe} = 150$, $h_{oe} = 10\ \mu S$, and $h_{re} = 0$. For $I_{C2} = 400\ \mu A$, find $A_v = V_{out}/V_{in}$ and the input resistance seen by V_{in}.

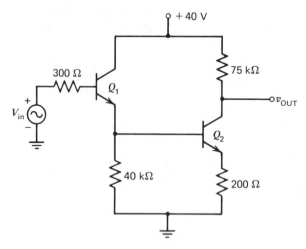

Figure 13.38. Circuit for Problem 13.12.

13.13. A Darlington pair amplifier is shown in Figure 13.39. The $h_{fe1} = h_{fe2} = 100$ and $h_{oe} = h_{re} = 0$. If $V_{CE1} = 1.0$ V, (a) find the voltage gain $A_v = V_{out}/V_{in}$; (b) find the input resistance seen by the signal generator V_{in}.

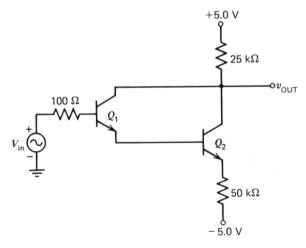

Figure 13.39. Circuit for Problem 13.13.

13.14. The level shift circuit shown in Figure 13.24 has V_{CC} = 15 V, V_{EE} = −15 V, R_i = 1.0 kΩ, and V_{BB1} = 5.0 V. If β_1 = β_2 = 100 and I_{C1} = 250 μA, (*a*) design values for V_{BB2}, R_E, and R_C so that V_{OUT} = 0.0 V dc. (*b*) If h_{oe} = 12 μS, find the voltage gain $A_v = V_1/V_{in}$.

13.15. Assume the following values for the common-emitter amplifier shown in Figure 13.40: $V_{AN} = V_{AP}$ = 100 V, V_{CC} = 5.0 V, V_{BE} = 0.65 V, I_{s1} = 1.0 × 10⁻¹² A, and V_{CE}(sat) = 0.2 V. Plot V_{OUT} versus V_{IN} for Q_1 and Q_2 in the active region for (*a*) I_{REF} = 1.0 mA and (*b*) I_{REF} = 0.1 mA.

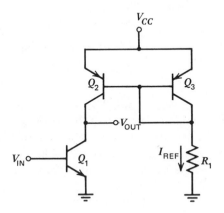

Figure 13.40. Circuit for Problem 13.15.

chapter 14
FEEDBACK

In many electronic circuits, a portion of the output signal is combined with the input signal. This "feedback" arrangement can be accidental or intentional. Negative or subtractive feedback can have beneficial effects on circuit operation. Examples of the benefits of negative feedback include:

(a) Increased stability of the operating point.
(b) Decreased distortion of the output signal.
(c) Increased bandwidth.
(d) Decreased dependence on individual transistor parameters and changes in these parameters.
(e) Control over input and output impedance.

These benefits are achieved with the sacrifice of signal gain. The reduced gain may result in a need for one or more additional amplifier stages. Under certain conditions, excessive signal phase shaft in a negative feedback configuration can appear as self-regeneration resulting in oscillations. These oscillations are certainly desirable for circuits designed as oscillators, but oscillations in signal amplifiers are an undesirable side effect.

The concept of feedback is customarily introduced in terms of block diagrams representing the forward gain and the feedback ratio. This technique of analysis emphasizes the feedback effects and provides insight to the results. It should be noted that the techniques used to analyze the electronic circuits in the preceding chapters are also applicable to the electronic feedback circuits described here. The basic approach of replacing transistors and bias circuits with small-signal models and using either nodal or loop analysis to find the voltages and currents is a legitimate and useful method for analyzing feedback circuits. The computer codes designed to aid in the analysis of electronic circuits, such as SPICE2, are also useful for dealing with the complexity of feedback circuits.

The simplest forms of feedback are associated with various connections within single-stage amplifiers. This type of feedback is sometimes called "local" feedback, in contrast to the multistage feedback found in amplifiers with more than one amplifying stage. In this chapter, we introduce the concept of feedback and analyze its general effects on amplifier performance. We then examine some specific examples of local feedback and also present the results of a computer analysis of an integrated-circuit multistage amplifier with feedback.

14.1 BASIC THEORY

In general, feedback can be described in terms of the diagram shown in Figure 14.1, although there are a variety of ways to present these concepts. We have selected an approach that stresses *negative* feedback by using a differencing node where the input signal S_{in} and the fed back portion of the output BS_{out} are compared. In some descriptions, a summing node is used so that positive feedback is represented by a positive multiplier on the feedback signal and negative feedback has a negative multiplier. The box marked A in the figure is assumed to contain both active and passive components that represent the amplifying portion of the circuit. The box marked B usually contains only passive elements so that a fraction of the output signal is presented at the differencing node.

The signals in this generalized circuit can be either voltages or currents. It is not necessary that the various signals in the system have the same units, since A and B can be dimensionless ratios, impedances, or admittances. For example, in some circuits, S_{in}, S_{error}, and BS_{out} are voltages and S_{out} is a current, making A a transfer- or trans-admittance and B a transfer- or trans-impedance.

It is important to recognize that it is not always possible to dissect the circuit into neat packages that can be identified as the differencing node, the amplifying section, and the feedback network. The partitioning of the circuit

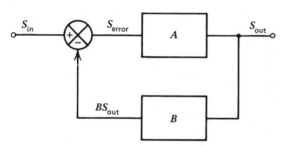

Figure 14.1. General feedback diagram.

into blocks with transfer functions A and B implies that the transfer functions do not change when the feedback is disabled. This means that the feedback components either do not load the amplifier or that the feedback must be disabled in such a way as to retain the loading effects. The amount of loading of the amplifier by the feedback network depends on the relative impedance levels in the various parts of the circuit. Consider the amplifier shown in Figure 14.2, in which resistors R_{B1}, R_{B2}, and R_E provide a stable dc operating point for the transistor. The purpose of the capacitor C_E is to provide a shunt path for the ac signal around R_E. If a large capacitance is selected, the impedance associated with C_E is negligible compared to R_E over the frequency range of interest. If, however, C_E is zero, a portion of the output voltage, $R_E V_{out}/(R_E + R_C)$, appears in series opposition with the input voltage, creating negative feedback in the circuit. For values of C_E between zero and a value large enough to completely bypass R_E, both the magnitude and the phase of the feedback voltage are functions of the frequency of the applied signal. In this example, it is not easy to visualize the circuit in the form of Figure 14.1; but, nevertheless, the feedback principles that we develop apply to both the block diagram of Figure 14.1 and the circuit of Figure 14.2.

14.1.1 **The General Feedback Expressions**

The general relations for a feedback system can be determined by analyzing Figure 14.1. At the differencing node, we have

$$S_{error} = S_{in} - B S_{out} \qquad (14.1)$$

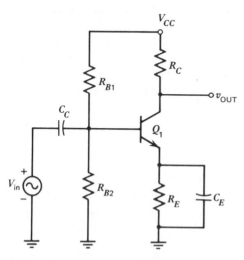

Figure 14.2. A common-emitter transistor amplifier with an emitter bypass capacitor.

Since

$$S_{out} = AS_{error} \tag{14.2}$$

then

$$S_{out} = A(S_{in} - BS_{out}) \tag{14.3}$$

or

$$\frac{S_{out}}{S_{in}} = G = \frac{A}{1 + AB} \tag{14.4}$$

Equation 14.4 is the basic equation for negative feedback, where G is called the *closed-loop gain*.

Sometimes it is useful to define another quantity, $T = AB$, the *open-loop gain* of the system. Then,

$$\frac{S_{out}}{S_{in}} = G = \frac{A}{1 + T} \tag{14.5}$$

In many cases, it is desirable for the feedback components to dominate the gain expression. If $T = AB$ is large compared to 1, then

$$G = \frac{A}{1 + AB} \simeq \frac{A}{AB} = \frac{1}{B} \tag{14.6}$$

Thus, for large values of open-loop gain, the feedback factor determines the closed-loop gain. For example, an integrated-circuit operational amplifier has a voltage gain without feedback of about 10^5, but it is common practice to use feedback to obtain closed-loop gains of 50 to 100.

Figure 14.3 illustrates how the open-loop gain function $T = AB$ might be measured in a feedback circuit. The feedback loop can be broken and the circuit driven by a test generator S_x. If the amplifier and feedback network do not unduly load each other, then $T = S_{out}/S_x$. Since T is a dimensionless quantity, S_{out} and S_x can be in voltage units. However, as we shall demonstrate, S_{out} and S_x can also be used to represent currents out of A and into B. The particular type of feedback determines whether the measured values are voltages or currents. Finally, it should be mentioned that not all feedback loops can be broken and measured as easily as indicated in Figure 14.3. However, when loop measurement is convenient, the concept if quite useful.

The derivation of Equation 14.4 was accomplished by assuming certain properties of the system in Figure 14.1. It is important to stress those assumptions, since actual electronic feedback amplifiers often differ in analysis from that expected in Equation 14.4. In Equation 14.1, we assume that the differencing or error node operates with arithmetic precision. Although this is often the case, situations may exist in electronic circuits that degrade the error node. We also assume that the feedback network B is unilateral from S_{out} through B to the error node. However, when passive resistors are used for a

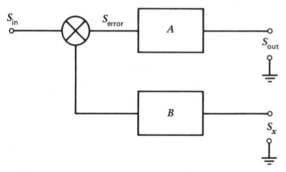

Figure 14.3. A general feedback system with the feedback disabled and sources inserted.

feedback network, signal transmission occurs in both directions. The transmission of a portion of S_{in} through the feedback network to S_{out} is called the *feed forward* effect. The feed forward signal is normally quite small but does appear in the exact analysis of feedback circuits.

In Equation 14.2, we assume that the gain of the amplifier A can be measured and that the amplifier can be inserted into the feedback configuration without alteration in the value of A. In practice, the B network may load both the input and output terminals of A, providing a different gain factor for the amplifier when it is connected to the feedback network. This loading property will provide feedback circuit analysis expressions that differ from Equation 14.4.

The origins of the statements for feedback systems contained in Equations 14.4 and 14.5 are found in control system theory that predates the application to electronic circuits. This classic equation is correct under the assumptions from which it was derived. Although the accuracy of Equations 14.4 and 14.5 suffers in many electronic configurations, our approach to feedback circuit analysis uses both of these equations for two major reasons. First, the amount of error arising in the analysis of feedback circuits using these expressions is not excessive in most applications. Second, the effects of feedback on circuit properties is best observed through the open-loop gain function $T = AB$. This is demonstrated next by observing the effect of T on amplifier bandwidth and sensitivity reduction.

14.1.2 **Bandwidth Modification**

The effect of feedback on amplifier bandwidth can be investigated by assuming that the gain of the amplifier without feedback has a frequency dependence given by

$$A(\omega) = \frac{A_0}{1 + (j\omega/\omega_u)} \qquad (14.7)$$

where ω_u is the upper cutoff frequency (-3 dB point) and A_0 is the low-frequency gain. We also assume that the amplifier and feedback network are unilateral and that loading of the amplifier by the feedback network is negligible. If B is resistive, then

$$G(\omega) = \frac{A(\omega)}{1 + A(\omega)B}$$

$$= \frac{A_0[1 + (j\omega/\omega_u)]}{1 + A_0B/[1 + (j\omega/\omega_u)]}$$

or

$$G(\omega) = \frac{A_0}{1 + (j\omega/\omega_u) + A_0B} \qquad (14.8)$$

but

$$G_0 = \frac{A_0}{1 + A_0B}$$

and

$$G(\omega) = \frac{G_0}{1 + (j\omega/\omega_u)/(1 + A_0B)}$$

This leads to the conclusion that

$$\omega_u' = \omega_u(1 + A_0B) \qquad (14.9)$$

where ω_u' is the upper cutoff frequency of the feedback amplifier. This states that the bandwidth of the amplifier is extended by the same factor that the gain is reduced, meaning that the product of the gain and the bandwidth is a constant. Resistive feedback networks are simple forms used to effect the trade between these two important amplifier characteristics.

As an example, if $A_0 = 10^5$ and $\omega_u = 50$ radians per second, using a feedback factor of $1/100$ results in a closed-loop gain of

$$G_0 = \frac{A_0}{1 + A_0B} = \frac{10^5}{1 + (10^5/100)} = 100$$

and

$$\omega_u' = \omega_u(1 + A_0B) = 50\left(1 + \frac{10^5}{100}\right)$$

$$= 50 \times 10^3 \text{ rad/s}$$

14.1.3 Gain Sensitivity

The closed-loop gain of a feedback amplifier is relatively insensitive to variations in the gain parameter A. To demonstrate this, we differentiate Equation 14.4 with respect to A:

$$\frac{dG}{dA} = \frac{1}{1 + AB} - \frac{AB}{(1 + AB)^2} \qquad (14.10)$$

$$= \frac{1}{(1 + AB)^2}$$

Then, for $A = 10^5$ and $B = \frac{1}{100}$, the variation in G is 10^{-6} times the variation in A.

A more meaningful measure of gain sensitivity is to compare dG/G to dA/A. This relationship can be found from Equation 14.10 by dividing both sides by G and rearranging the factors. The result is

$$\frac{dG}{G} = \frac{1}{1 + AB} \frac{dA}{A} \qquad (14.11)$$

Thus, the relative sensitivity to gain variations is reduced by a factor of $1/(1 + AB)$.

14.1.4 **The Effect of Feedback on Distortion**

Distortion in an amplifier is due to nonlinearities in the output-versus-input transfer characteristic. A typical nonfeedback amplifier characteristic is shown in Figure 14.4, where the change in the slope of the characteristic implies

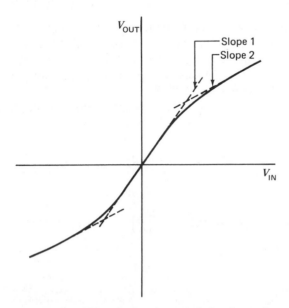

Figure 14.4. The transfer characteristic of a typical open-loop amplifier.

a change in the differential gain. This nonlinear relationship between V_{OUT} and V_{IN} is a source of distortion for this amplifier. If negative feedback is added and the closed-loop gain of the amplifier is dominated by the feedback ($AB >> 1$), then the overall gain is approximately $1/B$. The variation in gain of the basic amplifier is unimportant in the overall system gain, and the transfer plot of Figure 14.4 becomes a straight line of reduced slope.

Another viewpoint in the use of feedback to eliminate distortion is to examine the error signal. The feedback waveform has the same shape as the output waveform, only reduced in magnitude. This is subtracted from the input waveform to determine the error signal. Thus, it is the difference in waveform between the output and input (the negative of the distortion in the output) that is amplified, or exaggerated, until the error signal is reduced to a small value.

14.2 **THE BASIC FEEDBACK CONNECTIONS**

The block diagram of Figure 14.1 indicates that a portion of the output is fed back and subtracted from the input, without implying how this might be done. There are four basic configurations that can be used to implement the feedback situation. We will treat each of these configurations by analyzing the simple, one-transistor local feedback versions of these circuits as examples. In each case, the amplifier portion of the circuit is a bipolar transistor in the common-emitter configuration, and the feedback is resistive.

14.2.1 **The Transfer-Impedance Amplifier**

The closed-loop gain G_Z of a trans-impedance amplifier relates the output voltage to the input current, and it has the units of ohms. The trans-impedance amplifier is also called the parallel–parallel or shunt–shunt feedback connection, whose basic configuration is shown in Figure 14.5a. In this system, the feedback is obtained by diverting part of the output current through the feedback network and adding it to the input current. The differencing effect is due to the voltage inversion in the amplifying stage.

As an example, consider the local feedback trans-impedance amplifier circuit shown in Figure 14.5b. The biasing network is not shown in here so that we can concentrate on the feedback aspects of the circuit. The equivalent circuit is shown in Figure 14.6, using the simplified h-parameter model (with h_{re} and h_{oe} omitted). A nodal analysis at nodes x and y gives

$$\frac{V_x}{h_{ie}} + \frac{V_x - V_y}{R_F} = I_{in} \tag{14.12}$$

and

$$h_{fe}I_b + \frac{V_y - V_x}{R_F} + \frac{V_y}{R_C} = 0 \tag{14.13}$$

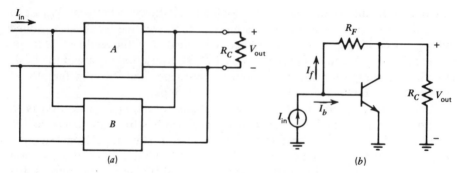

Figure 14.5. The trans-impedance or parallel–parallel feedback configuration. (*a*) The basic configuration. (*b*) The one-transistor implementation.

Also,

$$V_y = V_{\text{out}} \tag{14.14}$$

and

$$I_b = \frac{V_x}{h_{ie}} \tag{14.15}$$

Then,

$$\frac{V_x}{h_{ie}} + \frac{V_x - V_{\text{out}}}{R_F} = I_{\text{in}} \tag{14.16}$$

and

$$\frac{h_{fe}V_x}{h_{ie}} + \frac{V_{\text{out}} - V_x}{R_F} + \frac{V_{\text{out}}}{R_C} = 0 \tag{14.17}$$

Eliminating V_x from these equations and solving for $G_Z = V_{\text{out}}/I_{\text{in}}$ yields

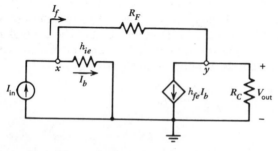

Figure 14.6. The equivalent circuit of a local feedback trans-impedance amplifier.

$$G_Z = \frac{-h_{fe}R_C}{\left(1 + \dfrac{R_C}{R_F}\right)\dfrac{(1/h_{ie}) + (1/R_F)}{(1/h_{ie}) - [1/(h_{fe}R_F)]} + h_{fe}\dfrac{R_C}{R_F}} \tag{14.18}$$

Equation 14.18 is an exact expression for the impedance transfer function G_Z of the feedback amplifier shown in Figure 14.6. However, Equation 14.18 does not match the form of the general feedback control expression given in Equation 14.4. The difference in the two equations lies in the first term of the denominator of Equation 14.18. If that term were unity, then the analysis of G_Z would align with the control theory prediction. The reason for the discrepancy lies in the fact that we used certain assumptions in the derivation of Equation 14.4, while basically no assumptions were used in the derivation of Equation 14.18.

In the derivation of the general expression, $G = A/(1 + AB)$, we assumed that both the amplifier A and the feedback network B possessed unilateral signal transmission properties. We also assumed that the impedance at the terminals of the feedback network did not load the amplifier input or output impedance. Since all of these assumptions do not apply for the feedback amplifier of Figure 14.6, it is expected that the transfer function G_Z would differ from the ideal.

If we place restrictions on the value of R_F in Figure 14.6, then we can force the amplifier to fulfill the assumptions that led to Equation 14.4. If $R_F \gg h_{ie}$ and $R_F \gg R_C$, then the terminals of the amplifier A will not be unduly loaded by the feedback network. In addition, the small component of the input signal that feeds forward to the output terminal through R_F is reduced if $R_F \gg h_{ie}$. Equation 14.18 reduces to Equation 14.19 under these assumptions, giving

$$G_Z \simeq \frac{-h_{fe}R_C}{1 + h_{fe}R_C(1/R_F)} \tag{14.19}$$

Equation 14.19 now has the same form as Equations 14.4 and 14.5 where

$$A = -h_{fe}R_C \tag{14.20}$$

$$B = -\frac{1}{R_F} \tag{14.21}$$

and

$$T = h_{fe}R_C \frac{1}{R_F} \tag{14.22}$$

The advantage of forcing Equation 14.18 into the theoretical form of Equation 14.19 is that the dominant terms in the feedback process that affect loop gain can be identified. A knowledge of $T = h_{fe}R_C(1/R_F)$ permits the designer to estimate the bandwidth increases (Equation 14.9) or the sensitivity reduction (Equation 14.11) as the loop gain is altered. We also examine below how the loop gain modifies the input and output impedances of an amplifier.

If $h_{fe}R_C$ is large compared to R_F in Equation 14.19, the gain parameter is dominated by the feedback, and

$$G_Z \simeq -R_F \qquad (14.23)$$

Example. If $R_F = 20$ kΩ, $R_C = 2.0$ kΩ, $h_{ie} = 1.0$ kΩ, and $h_{fe} = 100$, then, using Equation 14.18,

$$G_Z = \cfrac{-100 \times 2.0 \times 10^3}{\left(1 + \cfrac{2.0 \times 10^3}{20 \times 10^3}\right)\cfrac{(1/1.0 \times 10^3) + (1/20 \times 10^3)}{(1/1.0 \times 10^3) - (1/100 \times 20 \times 10^3)} + \left(\cfrac{100 \times 2.0 \times 10^3}{20 \times 10^3}\right)}$$

$$= \frac{-200 \times 10^3}{1.16 + 10}$$

$$= -17.9 \text{ k}\Omega$$

Equation 14.19 gives

$$G_Z = \frac{-200 \times 10^3}{1 + (200 \times 10^3/20 \times 10^3)}$$

$$= -18.2 \text{ k}\Omega$$

This is an error of 1.7% with respect to Equation 14.18.
The approximation given by Equation 14.23 gives

$$G_Z \simeq -20 \text{ k}\Omega$$

with an error of 11.7% with respect to Equation 14.18.

If the feedback resistor is reduced to 2.0 kΩ, $G_Z = -1.94$ kΩ (Equation 14.18), compared to -1.98 kΩ (Equation 14.19) and -2.0 kΩ (Equation 14.23). The error is less than 3% for both approximations.

The numerical results of this example show that the approximate solutions of Equations 14.19 and 14.23 do not differ appreciably from the exact solution of Equation 14.18. When R_F was lowered from 20 kΩ to 2.0 kΩ, the loop gain $T = h_{fe}R_C/R_F$ increased and brought the results of Equations 14.19 and 14.20 closer to the exact solution. This occurred in spite of the fact that impedance loading and feedthrough effects were more severe due to the decreased value of R_F. We now analyze the effect that loop gain has on both the input and output impedance for this type of amplifier.

INPUT IMPEDANCE FOR THE TRANS-IMPEDANCE AMPLIFIER. The input impedance to this circuit can be found by eliminating V_{out} from Equations 14.16 and 14.17. The result is

$$Z_{in} = \frac{V_x}{I_{in}} = \frac{1}{\cfrac{1}{h_{ie}} + \cfrac{1}{R_F} + \cfrac{h_{fe}[1/h_{ie} - 1/(h_{fe}R_F)]}{1 + (R_F/R_C)}} \qquad (14.24)$$

If $R_F \gg R_C$ and h_{ie}, this can be approximated by

$$Z_{\text{in}} \simeq \frac{h_{ie}}{1 + (h_{fe}R_C/R_F)} \qquad (14.25)$$

This takes the form of

$$Z_{\text{in}} = \frac{Z_{\text{in}}(A)}{1 + AB} \qquad (14.26)$$

$$= \frac{Z_{\text{in}}(A)}{1 + T} \qquad (14.27)$$

where $Z_{\text{in}}(A)$ is the input impedance of the amplifier A without feedback. The qualitative relationship between loop gain T and the input impedance can be observed in Equation 14.27.

For the example above, with $R_F = 20$ kΩ, $R_C = 2.0$ kΩ, $h_{ie} = 1.0$ kΩ, and $h_{fe} = 100$,

$$Z_{\text{in}} = \frac{1.0 \times 10^3}{1 + (100 \times 2.0 \times 10^3/20 \times 10^3)}$$
$$= 90.9 \ \Omega$$

It is a general characteristic of negative feedback amplifiers that the input impedance can be significantly reduced by using a parallel feedback connection.

OUTPUT IMPEDANCE FOR THE TRANS-IMPEDANCE AMPLIFIER.
The output impedance for this circuit configuration can be determined from Figure 14.7, where the independent input signal source has been disabled. A test voltage V_x is applied between terminals A and B, and the resulting current I_x is determined. Then, $Z_{\text{out}} = V_x/I_x$. From the figure, we have

$$I_x = I_b + I_y + h_{fe}I_b = (1 + h_{fe})I_b + I_y$$
$$I_y = \frac{V_x}{R_C}$$

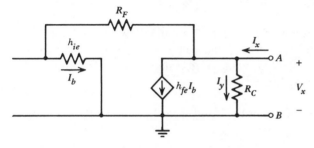

Figure 14.7. The equivalent circuit for the output impedance calculation for the local feedback trans-impedance amplifier.

and

$$I_b = \frac{V_x}{R_F + h_{ie}}$$

Then,

$$I_x = \frac{(1 + h_{fe})V_x}{R_F + h_{ie}} + \frac{V_x}{R_C}$$

or

$$Z_{out} = \frac{V_x}{I_x} = \frac{1}{[(1 + h_{fe})/(R_F + h_{ie})] + (1/R_C)} \tag{14.28}$$

which is reduced from the R_C expected from the amplifier without feedback.

Equation 14.28 can be reduced to a form that expresses Z_{out} as a function of loop gain T, if the feedback network is assumed not to load the amplifier. For $R_F \gg h_{ie}$ and R_C, Equation 14.28 becomes

$$Z_{out} = \frac{R_C}{1 + (h_{fe}R_C/R_F)} \tag{14.29}$$

or

$$Z_{out} = \frac{R_C}{1 + T} \tag{14.30}$$

For $T \gg 1$, Equation 14.30 reduces to

$$Z_{out} \simeq \frac{R_F}{h_{fe}} \tag{14.31}$$

The parallel feedback configuration as shown here and in the case of the series-parallel feedback configuration analyzed later reduces the output impedance of the amplifier by the factor $1 + T$.

If $R_C = 2.0 \text{ k}\Omega$, $h_{fe} = 100$, $R_F = 20 \text{ k}\Omega$, and $h_{ie} = 1.0 \text{ k}\Omega$, then Equation 14.28 provides an exact value for $Z_{out} = 188 \, \Omega$. The loop gain is $T = h_{fe}R_C/R_F = 10$, providing a Z_{out} from Equation 14.30 of 182 Ω. The approximation of Equation 14.31 yields $Z_{out} = 200 \, \Omega$.

Example. In the circuit of Figure 14.7, assume that $R_C = 2.0 \text{ k}\Omega$, $h_{fe} = 100$, and $h_{ie} = 1.0 \text{ k}\Omega$. (a) Find the value of T that provides an output resistance of 50 Ω. (b) What value of R_F satisfies this condition for T? (c) What is the input resistance?

Solution. The value of T can be found from Equation 14.30, where

$$T = \frac{R_C}{Z_{out}} - 1$$

$$= 39$$

The loop gain is seen in Equation 14.29 as $T = h_{fe}R_C/R_F$, allowing a solution for R_F as

$$R_F = \frac{h_{fe}R_C}{T}$$

$$= 5.1 \text{ k}\Omega$$

Then, Z_{in} is estimated from Equation 14.25 as

$$Z_{in} = \frac{h_{ie}}{1 + T}$$

$$= 25 \ \Omega$$

MEASUREMENT OF LOOP GAIN. The open-loop gain T of the feedback amplifier of Figure 14.6 can be measured using the configuration shown in Figure 14.8. Here, the input signal current is reduced to zero and the feedback loop is broken. A test signal is applied to one end of the broken loop. For the condition that R_F does not load the amplifier ($R_F >> h_{ie}$ and R_C), the open-loop gain T becomes

$$T = - \left. \frac{V_{out}}{V_x} \right|_{R_F >> h_{ie}, R_C} \tag{14.32}$$

where

$$V_{out} = - h_{fe}I_bR_C$$

and

$$I_b = \frac{V_x}{R_F + h_{ie}}$$

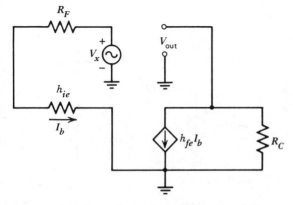

Figure 14.8. A small-signal equivalent circuit for measurement of open-loop gain T for the trans-impedance amplifier.

giving

$$-\frac{V_{out}}{V_x} = \frac{h_{fe}R_C}{R_F + h_{ie}} \qquad (14.33)$$

For $R_F \gg h_{ie}$ and R_C, Equation 14.32 becomes

$$T = \frac{h_{fe}R_C}{R_F} \qquad (14.34)$$

It has been shown that the open-loop gain for all parallel feedback circuits can be measured by breaking the loop and measuring the transfer function $T = -V_{out}/V_x$. Using the values of a previous example where $R_C = 2.0$ kΩ, $h_{ie} = 1.0$ kΩ, $h_{fe} = 100$, and $R_F = 20$ kΩ, Equation 14.33 provides $-V_{out}/V_x = 9.52$ for a measured open-loop transfer function. Equation 14.34 provides the value of open-loop gain that agrees with Equation 14.5 when derived under the assumption that $R_F \gg h_{ie}$ and R_C. Here, Equation 14.34 gives $T = 10$, which is less than a 5% error with respect to the result of Equation 14.33.

14.2.2 The Transfer-Admittance Amplifier

The transfer-admittance or trans-admittance amplifier gain parameter G_Y relates the output current to the input voltage. This feedback connection is also called the series–series connection whose basic configuration is shown in Figure 14.9a. The one-transistor circuit exhibiting this kind of feedback is shown in Figure 14.9b, without the bias components. The small-signal equivalent circuit is shown in Figure 14.10.

The gain parameter $G_Y = I_{out}/V_{in}$ can be determined by analyzing the equivalent circuit. We have

$$I_e = I_b + h_{fe}I_b = (1 + h_{fe})I_b \qquad (14.35)$$

$$I_{out} = -h_{fe}I_b \qquad (14.36)$$

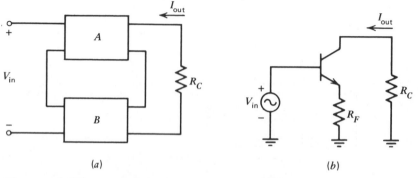

(a)   (b)

Figure 14.9. The trans-admittance or series–series feedback amplifier. (a) Basic configuration. (b) One–transistor implementation.

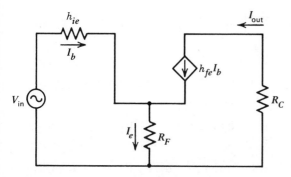

Figure 14.10. The h-parameter small-signal equivalent circuit for the trans-admittance feedback amplifier.

and

$$V_{in} = I_b h_{ie} + I_e R_F \qquad (14.37)$$

Then,

$$V_{in} = -\frac{I_{out}}{h_{fe}} [h_{ie} + (1 + h_{fe})R_F] \qquad (14.38)$$

$$G_Y = \frac{I_{out}}{V_{in}} = \frac{-h_{fe}}{h_{ie} + (1 + h_{fe})R_F} \qquad (14.39)$$

or

$$G_Y = \frac{-h_{fe}/h_{ie}}{1 + (R_F/h_{ie}) + (h_{fe}R_F/h_{ie})} \qquad (14.40)$$

In this series feedback circuit, the feedback element R_F does not load the amplifier A if $R_F \ll h_{ie}$. This constraint reduces Equation 14.40 to the form of Equation 14.4, giving

$$G_Y = \frac{-h_{fe}/h_{ie}}{1 + (h_{fe}R_F/h_{ie})} \qquad (14.41)$$

with

$$A = \frac{-h_{fe}}{h_{ie}}$$

$$B = -R_F$$

and

$$T = \frac{h_{fe}R_F}{h_{ie}}$$

If T is much greater than unity, then

$$G_Y \approx -\frac{1}{R_F} \qquad (14.42)$$

As an example, if $h_{fe} = 100$, $h_{ie} = 1.0 \text{ k}\Omega$, $R_C = 2.0 \text{ k}\Omega$, and $R_F = 50 \text{ }\Omega$, then

$$G_Y = \frac{-100/1.0 \times 10^3}{1 + [(1 + 100)/1 \times 10^3] \times 50}$$

$$= -16.5 \text{ mS}$$

which is comparable to -16.7 mS as predicted by Equation 14.41 and -20 mS from Equation 14.42.

It is important to recognize that R_C is not included in the expressions for the gain parameter of the trans-admittance amplifier. This is because the output is in the form of a current rather than a voltage. The value for R_C is selected in conjunction with the biasing components to establish the operating point for V_{CE} in the forward active region.

INPUT IMPEDANCE FOR THE TRANS-ADMITTANCE AMPLIFIER. The input impedance for the trans-admittance amplifier can be found by substituting Equation 14.35 into Equation 14.37. The result is

$$Z_{\text{in}} = \frac{V_{\text{in}}}{I_b} = h_{ie} + R_F + h_{fe}R_F \qquad (14.43)$$

The assumption of nonloading by R_F on the amplifier is that $R_F << h_{ie}$. This gives

$$Z_{\text{in}} = h_{ie}\left[1 + \left(\frac{h_{fe}}{h_{ie}}\right)R_F\right] \qquad (14.44)$$

This is of the form

$$Z_{\text{in}} = Z_{\text{in}}(A)(1 + AB) \qquad (14.45)$$

$$= Z_{\text{in}}(A)(1 + T) \qquad (14.46)$$

where $Z_{\text{in}}(A)$ is the input impedance of the amplifier A without feedback. For our example, with $h_{ie} = 1.0 \text{ k}\Omega$, $h_{fe} = 100$, and $R_F = 50 \text{ }\Omega$, we find that $Z_{\text{in}} = 6.05 \text{ k}\Omega$ using Equation 14.43 and 6.0 kΩ using Equation 14.46.

A configuration with a feedback voltage connected in series with the input voltage has an input impedance that is significantly increased over the non-feedback circuit. The increase in the input impedance is readily estimated from Equation 14.46.

OUTPUT IMPEDANCE FOR THE TRANS-ADMITTANCE AMPLIFIER. The output impedance for this amplifier can be determined from the circuit in Figure 14.11. The $1/h_{oe}$ parameter is included in this circuit to provide a more meaningful comparison to the nonfeedback case. A test voltage V_x is applied

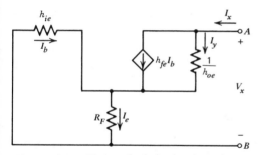

Figure 14.11. The equivalent circuit for determining the output impedance for the one-transistor trans-admittance amplifier.

between the terminals A and B, and the resulting current I_x is found from

$$I_x = h_{fe}I_b + I_y \qquad (14.47)$$

$$I_x = -I_b + I_e \qquad (14.48)$$

$$V_x = I_y \frac{1}{h_{oe}} - I_b h_{ie} \qquad (14.49)$$

From the current divider expression,

$$-I_b = \frac{R_F}{h_{ie} + R_F} I_x \qquad (14.50)$$

Then, combining equations, we get

$$Z_{\text{out}} = \frac{V_x}{I_x} = \frac{1}{h_{oe}} \left(1 + \frac{h_{ie}h_{oe} + h_{fe}}{h_{ie} + R_F} R_F \right) \qquad (14.51)$$

The relation between Equation 14.51 and the open-loop gain T for the circuit of Figure 14.11 can be observed by imposing the nonloading condition $R_F \ll h_{ie}$ and observing that, for the feed forward effect, $h_{ie}h_{oe} \ll h_{fe}$. These conditions give

$$Z_{\text{out}} = \frac{1}{h_{oe}} \left(1 + \frac{h_{fe}R_F}{h_{ie}} \right) \qquad (14.52)$$

The value of $T = h_{fe}R_F/h_{ie}$ can be derived from an analysis of G_Y for the circuit of Figure 14.11 by employing the assumptions used to derive Equation 14.4. Equation 14.52 can then be written as

$$Z_{\text{out}} = Z_{\text{out}}(A)(1 + T) \qquad (14.53)$$

where $Z_{\text{out}}(A)$ is the output impedance of the amplifier A without feedback.
   For example, if $1/h_{oe} = 20$ kΩ, $h_{fe} = 100$, $h_{ie} = 1.0$ kΩ, and $R_F = 50$ Ω,

then Equation 14.53 gives $Z_{out} = 120 \text{ k}\Omega$, compared to 20 kΩ for the amplifier without feedback.

The output impedance is increased when the feedback is in the form of a voltage in series with the output voltage, and the quantitative relation to loop gain is given by Equation 14.53.

14.2.3 **The Voltage Amplifier**

The feedback configuration that is best described as a voltage amplifier is the series–parallel configuration shown in Figure 14.12a. The one-transistor implementation of this configuration is the basic emitter–follower circuit shown without biasing elements in Figure 14.12b. In this circuit, the entire output voltage is fed back in series with the input.

The small-signal equivalent for this circuit is shown in Figure 14.13a. We also include the small-signal equivalent circuit of the common-emitter amplifier in Figure 14.13b for comparison purposes.

The voltage gain for the voltage amplifier can be determined from the equivalent circuit. We have

$$V_{in} = I_b h_{ie} + V_{out} \tag{14.54}$$

$$V_{out} = I_e R_E \tag{14.55}$$

and

$$I_e = I_b + h_{fe}I_b = (1 + h_{fe})I_b \tag{14.56}$$

Then,

$$G_V = \frac{V_{out}}{V_{in}} = \frac{(R_E/h_{ie})(1 + h_{fe})}{1 + (R_E/h_{ie})(1 + h_{fe})} \tag{14.57}$$

This has the same form as Equation 14.4 with

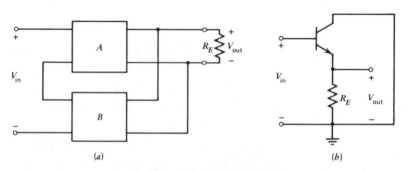

Figure 14.12. The series–parallel or voltage feedback amplifier configuration. (a) Basic configuration. (b) One-transistor implementation.

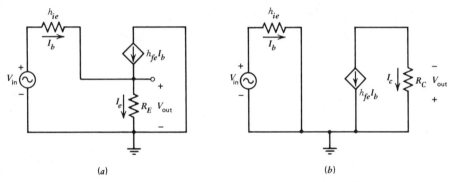

Figure 14.13. Small-signal equivalent circuits for the voltage amplifier and the common-emitter amplifier with the same load. (a) The emitter follower amplifier. (b) The common-emitter amplifier.

$$A = \frac{R_E}{h_{ie}}(1 + h_{fe})$$

(14.58)

$$B = 1$$

and

$$T = \frac{R_E}{h_{ie}}(1 + h_{fe})$$

(14.59)

Analysis of Figure 14.13b yields

$$A_v = \frac{V_{out}}{V_{in}} = \frac{h_{fe}R_E}{h_{ie}}$$

(14.60)

If h_{fe} is large compared to 1, this is the same as A in Equation 14.58.

As an example, if $h_{fe} = 100$, $h_{ie} = 1.0$ kΩ, and $R_E = 500$ Ω, we find $G_V = 0.98$.

INPUT IMPEDANCE FOR THE VOLTAGE AMPLIFIER. The input impedance for this circuit can be found from Equations 14.54 through 14.56. The result is

$$Z_{in} = \frac{V_{in}}{I_b} = h_{ie}\left[1 + \frac{(1 + h_{fe})R_E}{h_{ie}}\right]$$

(14.61)

$$= Z_{in}(A)(1 + T)$$

(14.62)

where $Z_{in}(A)$ is the input impedance of the amplifier without feedback. Equation 14.61 has the same form as Equation 14.44 and has a value $Z_{in} = 51.5$ kΩ for the example circuit. We anticipated this result, since the feedback is in the form of a voltage in series with the input voltage.

OUTPUT IMPEDANCE FOR THE VOLTAGE AMPLIFIER. To determine the output impedance for this configuration, we analyze the circuit in

Figure 14.14. Applying Kirchhoff's current law at the test node results in

$$I_x = I_y - h_{fe}I_b - I_b = I_y - (1 + h_{fe})I_b \tag{14.63}$$

Also,

$$I_y = \frac{V_x}{R_E} \tag{14.64}$$

and

$$I_b = -\frac{V_x}{h_{ie}} \tag{14.65}$$

Then,

$$I_x = V_x \left(\frac{1}{R_E} + \frac{1 + h_{fe}}{h_{ie}} \right) \tag{14.66}$$

or

$$Z_{\text{out}} = \frac{V_x}{I_x} = \frac{R_E}{1 + [(1 + h_{fe})R_E/h_{ie}]} \tag{14.67}$$

which, using the results of the previous derivation, can be written

$$Z_{\text{out}} = \frac{Z_{\text{out}}(A)}{1 + T} \tag{14.68}$$

which is substantially reduced from the output impedance of the amplifier without feedback. This impedance reduction is due to the parallel connection of the feedback on the output. If $h_{fe} = 100$, $h_{ie} = 1.0 \text{ k}\Omega$, and $R_E = 500 \ \Omega$, we find that $Z_{\text{out}} = 9.7 \ \Omega$.

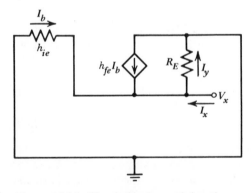

Figure 14.14. Circuit for determining the output impedance for the voltage amplifier.

14.2.4 **The Current Amplifier**

The final possible feedback configuration is the parallel-series or current amplifier connection. This is shown in Figure 14.15a along with the local feedback circuit (without bias components) for this arrangement in Figure 14.15b. A close examination of the circuit in Figure 14.15b indicates that this is the common-base amplifier circuit, as redrawn in Figure 14.15c. It is not customary to think of this circuit as a feedback amplifier, but the total output current, considering the transistor as a common-emitter circuit, is fed back to the input.

The equivalent circuit for this amplifier is shown in Figure 14.16. To determine the closed-loop gain $G_I = I_{out}/I_{in}$ for this circuit, we begin with Kirchhoff's current law at node E:

$$I_{in} = -I_b - h_{fe}I_b = -I_b(1 + h_{fe}) \tag{14.69}$$

but

$$I_{out} = h_{fe}I_b \tag{14.70}$$

Then,

$$I_{in} = -I_{out}\left(\frac{1}{h_{fe}} + 1\right) \tag{14.71}$$

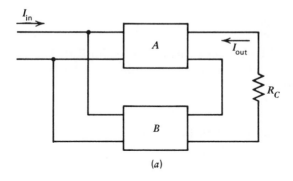

(a)

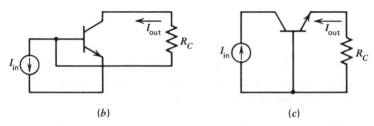

(b)   (c)

Figure 14.15. The parallel–series feedback or current amplifier configuration. (a) Basic configuration. (b) The current amplifier. (c) The common-base amplifier.

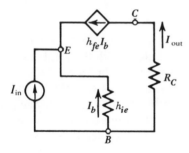

Figure 14.16. The equivalent circuit for the common-base amplifier.

or

$$G_I = \frac{I_{\text{out}}}{I_{\text{in}}} = \frac{-h_{fe}}{1 + h_{fe}} \tag{14.72}$$

This has the same basic form as Equation 14.4 with

$$A = -h_{fe}$$
$$B = -1$$

and

$$T = h_{fe}$$

For $h_{fe} = 100$, we find that $G_I = -0.99$.

INPUT IMPEDANCE OF THE CURRENT AMPLIFIER. The input impedance of the common-base amplifier can be determined by analyzing the circuit in Figure 14.16, where

$$V_{\text{in}} = -V_{be} = -I_b h_{ie} \tag{14.73}$$

Since I_b is related to I_{in} by Equation 14.69, then

$$Z_{\text{in}} = \frac{V_{\text{in}}}{I_{\text{in}}} = \frac{h_{ie}}{1 + h_{fe}} \tag{14.74}$$

and

$$Z_{\text{in}} = \frac{h_{ie}}{1 + T} \tag{14.75}$$

These results are of the same form as Equation 14.26. This form is expected because the feedback connection on the input side is a parallel connection. For $h_{ie} = 1.0 \text{ k}\Omega$ and $h_{fe} = 100$, we have $Z_{\text{in}} = 9.9 \ \Omega$.

OUTPUT IMPEDANCE OF THE CURRENT AMPLIFIER. The output impedance of the common-base amplifier can be determined from the circuit in Figure 14.17, where the parameter $1/h_{oe}$ is included for comparison with the

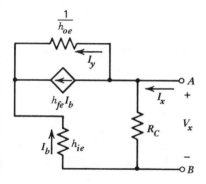

Figure 14.17. The equivalent circuit for the output impedance of the current amplifier.

nonfeedback amplifier. If we had not included h_{oe} in the model, then the output impedance of the common-base amplifier would be infinity and the effect of feedback on the output impedance could not be demonstrated.

Kirchhoff's current law at node A yields

$$I_x = I_y + h_{fe}I_b \tag{14.76}$$

but

$$I_x = -I_b \tag{14.77}$$

Kirchhoff's voltage law results in

$$V_x = I_y \frac{1}{h_{oe}} + I_x h_{ie} \tag{14.78}$$

Then, by substituting Equations 14.76 and 14.77, we get

$$Z_{\text{out}} = \frac{V_x}{I_x} = (1 + h_{fe})\frac{1}{h_{oe}} + h_{ie} \tag{14.79}$$

which can be written in the form

$$Z_{\text{out}} = \left(\frac{1}{h_{oe}} + \frac{h_{ie}}{1 + h_{fe}}\right)(1 + h_{fe})$$

or

$$Z_{\text{out}} = Z_{\text{out}}(A)(1 + T) \tag{14.80}$$

If $h_{fe} = 100$, $1/h_{oe} = 20$ kΩ, and $h_{ie} = 1.0$ kΩ, we get $Z_{\text{out}} = 2.0$ MΩ, which is a greatly enhanced value over that of the nonfeedback configuration.

14.2.5 Summary of Single-Stage Negative Feedback Amplifier Configurations

We have analyzed the four basic feedback configurations and the four local feedback or one-transistor implementations of these connections. The open-loop gain T is the circuit function that describes how the amount of feedback controls the transfer function, the input and output impedances, the bandwidth, and the reduction of distortion. A designer has a primary interest in the analytical expression for the open-loop gain T, since the specific parameters that affect T can be manipulated to achieve the circuit specifications.

The closed-loop gain expression $G = A/(1 + T)$ loses some accuracy when applied to many feedback amplifiers. The loss in accuracy is due mainly to impedance loading of the amplifier by the feedback network and to a lesser extent by the feed forward signal. Table 14.1 summarizes the impedance and gain relationships between the closed- and open-loop gain functions for the four feedback configurations.

14.3 MULTISTAGE FEEDBACK AMPLIFIERS

The previous discussion of single-stage feedback circuit theory is useful to illustrate the properties of the four types of feedback amplifiers. These configurations are quite common, and the analysis also is useful in a practical sense. However, it is equally common to apply negative feedback across several stages of a multistage amplifier. Figure 14.18 shows a parallel–parallel feedback configuration in a two-stage amplifier. We determine the transfer function, input impedance, and output impedance by first deriving the open-loop gain T. It is assumed that $R_F \gg h_{ie1}$ and R_E. In Figure 14.19a, we show the small-signal model of the amplifier assuming that $h_{oe} = h_{re} = 0$. Although h_{oe} is usually significant in an integrated circuit, it is neglected here to simplify the analysis.

Table 14.1. Summary of Feedback Relationships[a]

	Parallel–Parallel	Series–Series	Series–Parallel	Parallel–Series
Gain parameter $G = \dfrac{A}{1 + T}$	Transfer impedance $G_Z = \dfrac{V_{out}}{I_{in}}$	Transfer admittance $G_Y = \dfrac{I_{out}}{V_{in}}$	Voltage $G_V = \dfrac{V_{out}}{V_{in}}$	Current $G_I = \dfrac{I_{out}}{I_{in}}$
Input impedance	Low	High	High	Low
Z_{in}	$\dfrac{Z_{in}(A)}{1 + T}$	$Z_{in}(A)(1 + T)$	$Z_{in}(A)(1 + T)$	$\dfrac{Z_{in}(A)}{1 + T}$
Output impedance	Low	High	Low	High
Z_{out}	$\dfrac{Z_{out}(A)}{1 + T}$	$Z_{out}(A)(1 + T)$	$\dfrac{Z_{out}(A)}{1 + T}$	$Z_{out}(A)(1 + T)$

[a]$Z_{in}(A)$ and $Z_{out}(A)$ refer to the amplifier input and output impedances with the feedback signal disabled.

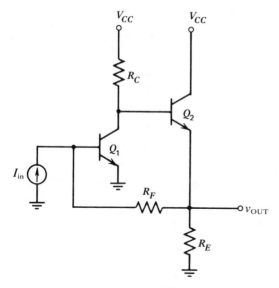

Figure 14.18. Two-stage amplifier using a parallel–parallel feedback configuration.

Figure 14.19*b* shows the equivalent circuit that can be used to derive the
open-loop gain *T*. As previously discussed, $T = -V_{out}/V_x$, where V_x is a test
signal that drives the open-loop configuration. The input signal I_{in} is disabled
for the derivation of *T*.

To derive $T = -V_{out}/V_x$, we first observe that

$$V_{out} = (1 + h_{fe2})R_E I_{b2} \tag{14.81}$$

and

$$I_{b2} = -(h_{fe1}I_{b1})\frac{R_C}{R_C + R_{in2}} \tag{14.82}$$

where R_{in2} is the resistance looking into the base of Q_2. This resistance is given
by

$$R_{in2} = \frac{V_{b2}}{I_{b2}} = h_{ie2} + (1 + h_{fe2})R_E \tag{14.83}$$

I_{b1} can be expressed as

$$I_{b1} = \frac{V_x}{R_F + h_{ie1}} \tag{14.84}$$

Equations 14.84, 14.83, and 14.82 can be substituted into Equation 14.81, with
$R_F \gg h_{ie1}$, to give

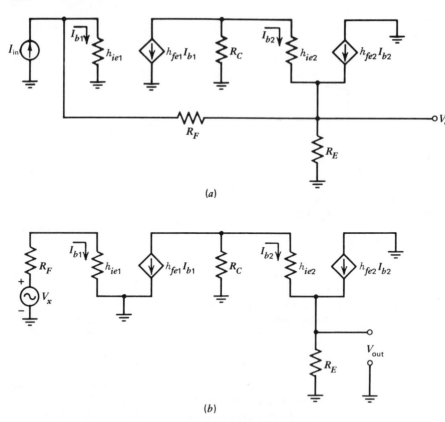

(a)

(b)

Figure 14.19. (a) Small-signal model of parallel–parallel amplifier shown in Figure 14.18. (b) Equivalent circuit to calculate the open-loop gain T.

$$T = -\frac{V_{out}}{V_x} = \frac{(1 + h_{fe2})(h_{fe1})R_E}{R_F} \cdot \frac{R_C}{R_C + h_{ie2} + (1 + h_{fe2})R_E} \quad (14.85)$$

In Figure 14.20, we show the equivalent circuit to calculate the transimpedance transfer function of the amplifier without feedback, $A = V_{out}/I_{in}$. The feedback signal has been disabled, and it is assumed that $R_F \gg h_{ie1}$ and R_E. A can be derived by noting that Equations 14.81, 14.82, and 14.83 apply to Figure 14.20 and that $I_{b1} = I_{in}$. Therefore,

$$A = \frac{V_{out}}{I_{in}} = (1 + h_{fe2})(h_{fe1})R_E \frac{R_C}{R_C + h_{ie2} + (1 + h_{fe2})R_E} \quad (14.86)$$

The closed-loop amplifier transfer impedance function is then

$$G_Z = \frac{A}{1 + T} \quad (14.5)$$

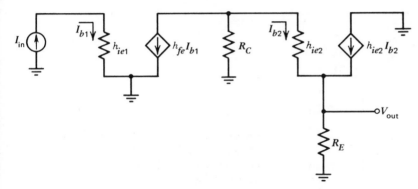

Figure 14.20. Equivalent circuit of Figure 14.18 to calculate the trans-impedance amplifier transfer function $A = V_{out}/I_{in}$ for feedback disabled.

The input resistance of the closed loop amplifier can be calculated from

$$Z_{in} = \frac{Z_{in}(A)}{1 + T} \qquad (14.27)$$

where $Z_{in}(A)$ is the input impedance of the amplifier with the feedback disabled. Assuming that $R_F \gg h_{ie1}$, then

$$Z_{in}(A) = h_{ie1} \qquad (14.87)$$

Similarly, the output impedance is

$$Z_{out} = \frac{Z_{out}(A)}{1 + T} \qquad (14.88)$$

where $Z_{out}(A)$ is the output impedance of the amplifier with the feedback signal disabled. Figure 14.21 shows the small-signal equivalent circuit to derive $Z_{out}(A)$. The output impedance is defined by $Z_{out}(A) = V_x/I_x$ and can be determined as

$$Z_{out}(A) = \frac{V_x}{I_x} = R_E \left\| \left(\frac{h_{ie2} + R_C}{1 + h_{fe2}} \right) \qquad (14.89)$$

Example. For the circuit of Figure 14.18, let $h_{ie1} = h_{ie2} = 1.0$ kΩ, $h_{fe1} = h_{fe2} = 100$, $R_C = 1.5$ kΩ, $R_E = 200$ Ω, and $R_F = 15$ kΩ. Calculate G_Z, Z_{in}, and Z_{out}.

Solution. First, the open-loop gain T is calculated from Equation 14.85 as

$$T = \frac{101 \times 100 \times 200}{15 \times 10^3} \frac{1.5 \times 10^3}{1.5 \times 10^3 + 1.0 \times 10^3 + 101 \times 200}$$

$$= 8.9$$

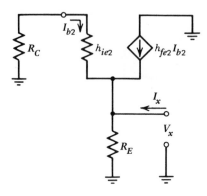

Figure 14.21. Small-signal equivalent circuit to calculate the output impedance $Z_{out}(A)$ of the circuit shown in Figure 14.18.

Then, the amplifier open-loop trans-impedance gain A is calculated from Equation 14.86 as

$$A = \frac{V_{out}}{I_{in}} = -101 \times 100 \times 200 \frac{1.5 \times 10^3}{1.5 \times 10^3 + 1.0 \times 10^3 + 101 \times 200}$$

$$= -133.5 \text{ k}\Omega$$

The closed-loop transfer impedance is then

$$G_Z = \frac{V_{out}}{I_{in}} = \frac{A}{1 + T}$$

$$= -\frac{133.5 \times 10^3}{1 + 8.9}$$

$$= -13.5 \text{ k}\Omega$$

The input impedance is calculated by first obtaining $Z_{in}(A)$ from Equation 14.87 as $Z_{in}(A) = h_{ie1} = 1.0 \text{ k}\Omega$. Then, from Equation 14.27,

$$Z_{in} = \frac{Z_{in}(A)}{1 + T}$$

$$= \frac{1.0 \times 10^3}{1 + 8.9}$$

$$= 101 \text{ }\Omega$$

The output resistance is determined by first obtaining $Z_{out}(A)$ from Equation 14.89:

$$Z_{out}(A) = 200 \| \left(\frac{1.0 \times 10^3 + 1.5 \times 10^3}{101} \right)$$

$$= 22.0 \text{ }\Omega$$

Then,

$$Z_{\text{out}} = \frac{Z_{\text{out}}(A)}{1 + T}$$

$$= \frac{22.0}{1 + 8.9}$$

$$= 2.2 \ \Omega$$

Example. Find a value of R_F so that the input resistance of the amplifier in the previous example is 50 Ω.

Solution. Equation 14.27 can be solved for T to yield

$$T = \frac{Z_{\text{in}}(A)}{Z_{\text{in}}} - 1$$

$$= \frac{1.0 \times 10^3}{50} - 1 \qquad (14.90)$$

$$= 19$$

Then, R_F can be determined from Equation 14.85 as

$$R_F = \frac{(1 + h_{fe2})(h_{fe1})R_E}{T} \frac{R_C}{R_C + h_{ie2} + (1 + h_{fe2})R_E}$$

$$= \frac{101 \times 100 \times 200}{19} \frac{1.5 \times 10^3}{1.5 \times 10^3 + 1.0 \times 10^3 + 101 \times 200} \qquad (14.91)$$

$$= 7.03 \ \text{k}\Omega$$

14.4 SUMMARY

Negative feedback is useful for the control of the gain, the bandwidth, the input impedance, and the output impedance of an amplifier. The distortion is also significantly reduced. In this chapter, we have restricted ourselves to resistive feedback to simplify the analysis, but we consider nonresistive feedback in Chapter 15. The basic concepts of feedback can be examined on single-tran-sistor circuits, including the four fundamental configurations, series–series, parallel–parallel, series–parallel, and parallel–series. Once we see the general effects of the different configurations in terms of the loop gain T and learn to identify the types of connections, we can predict the effects of feedback on multistage amplifiers. As the complexity of the circuits increases, computer simulations are often used to perform the detailed analyses.

REFERENCES

1. P. R. Gray and R. G. Meyer, *The Analysis and Design of Analog Integrated Circuits* (New York: Wiley, 1977).

2. J. Fisher and B. Gatland, *Electronics From Theory Into Practice*, Vol. 2, 2nd ed. (Oxford: Pergamon, 1976).

PROBLEMS

14.1. The feedback amplifier shown in Figure 14.22 has transistor parameter values of $h_{fe} = 50$ and $h_{oe} = h_{re} = 0$. Calculate (a) the input impedance Z_{in}, (b) the output impedance Z_{out}, (c) the closed-loop transfer function $G_Z = V_{out}/V_{in}$, and (d) the loop gain T. Assume that C_C, the coupling capacitor, is a short circuit in the frequency range of interest.

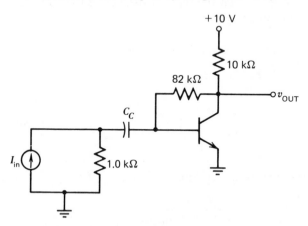

Figure 14.22. Circuit for Problem 14.1.

14.2. (a) Determine the resistor values for the feedback amplifier shown in Figure 14.23 so that $I_{CQ} = 250\ \mu A$, $V_{CEQ} = 6.0\ V$, and the closed loop gain $T = 5.0$. The $h_{fe} = 150$ and $h_{oe} = h_{re} = 0$. (b) Calculate the input resistance. (c) Calculate the input resistance if R_F is removed from the circuit. Assume that C_C and C_B are short circuits for the small-signal calculations.

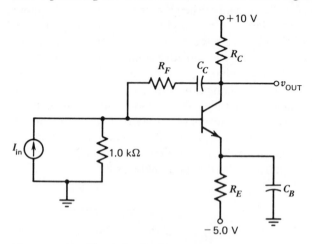

Figure 14.23. Circuit for Problem 14.2.

14.3. Compare the loop gain T for the circuit in Problem 14.2 if h_{oe} is increased from 0 to 20 μS. Use the method of analysis shown in Figure 14.8.

14.4. The transistor in the feedback amplifier in Figure 14.24 has $h_{fe} = 200$ and $h_{oe} = h_{re} = 0$. Calculate (a) the loop gain T, (b) the input resistance, and (c) the output resistance.

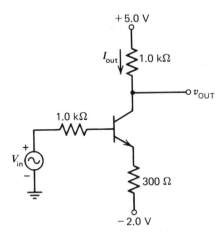

Figure 14.24. Circuit for Problem 14.4.

14.5. (a) Design the resistor values for the circuit in Figure 14.25 so that the loop gain $T = 10$. Keep V_{CEQ} between 3.0 and 7.0 V. The $h_{fe} = 100$ and $h_{oe} = h_{re} = 0$. (b) Calculate the input resistance and the output resistance. Assume that C_C is a short circuit for the small-signal calculations.

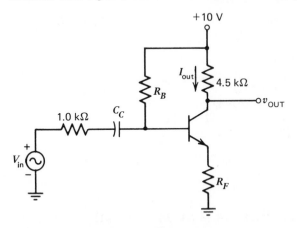

Figure 14.25. Circuit for Problem 14.5.

14.6. For the circuit shown in Figure 14.26, calculate (a) the loop gain, (b) the input impedance, (c) the output impedance, and (d) the transfer function $G_V = V_{out}/V_{in}$. The $h_{fe} = 100$, $h_{re} = h_{oe} = 0$, and the coupling capacitor C_C is assumed to be a short circuit for small-signal analysis purposes.

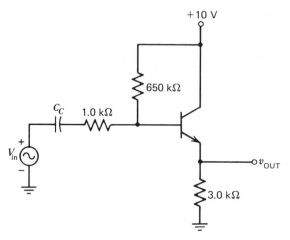

Figure 14.26. Circuit for Problem 14.6.

14.7. Let h_{oe} increase from 0 to 20 μS for the circuit in Problem 14.6. Calculate (a) the loop gain T, (b) the input resistance, (c) the output resistance, and (d) the transfer function. Compare the values with those obtained in Problem 14.6.

14.8. For the circuit shown in Figure 14.27, assume that $h_{fe} = 100$, $h_{oe} = h_{ie} = 0$, and that the capacitors are short circuits in the small-signal analysis. Calculate (a) the loop gain T, (b) the input resistance, (c) the output resistance, and (d) the transfer function $G_I = I_{out}/I_{in}$.

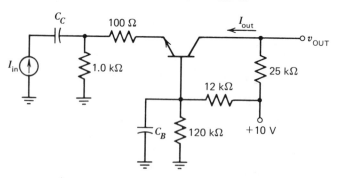

Figure 14.27. Circuit for Problem 14.8.

14.9. The circuit shown in Figure 14.18 has $R_C = 2.0$ kΩ, $R_E = 150$ Ω, $h_{fe1} = h_{fe2} = 200$, $h_{oe} = h_{re} = 0$. (a) Design a value for R_F so that the loop gain $T = 10$. (b) Calculate the transfer function $G_Z = V_{out}/I_{in}$. (c) Calculate the input and output resistances.

14.10. The feedback amplifier shown in Figure 14.28 has transistor parameters whose values are $h_{fe} = 100$, $h_{ie} = 2.0$ k, and $h_{re} = h_{oe} = 0$. The base bias circuitry is omitted in the circuit. Calculate (a) the closed loop transfer function G_Z, (b) the input and output resistances, (c) the open-loop gain T, and (d) the voltage gain $A_v = V_{out}/V_{in}$.

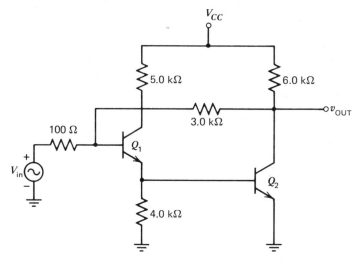

Figure 14.28. Circuit for Problem 14.10.

14.11. The transistor parameters for the feedback circuit shown in Figure 14.29 are $h_{fe} = 100$, $h_{ie1} = 4.0\ k\Omega$, $h_{ie2} = 2.5\ k\Omega$, $h_{ie3} = 600\ \Omega$, and $h_{oe} = h_{re} = 0$. Calculate (a) the closed loop transfer function G_Y, (b) the input impedance, (c) the output impedance, (d) the open-loop gain T, and (e) the voltage gain $A_v = V_{out}/V_{in}$.

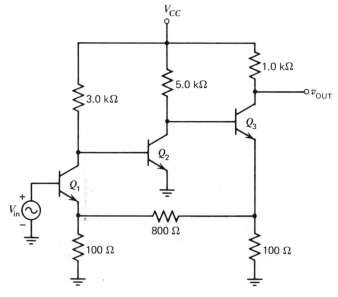

Figure 14.29. Circuit for Problem 14.11.

Note: The following problems are associated with feedback configurations of single-stage MOSFET amplifiers. Employ the feedback principles applied to the bipolar feedback circuits to analyze each of the circuits. Use the MOSFET small-signal model developed in Chapter 11. For each transistor in Problems 14.12 through 14.15, $g_m = 1.0$ mA/V, $r_{ds} = 50$ kΩ, and $r_{gs} = 1.0$ MΩ.

14.12. Calculate the following functions for the circuit in Figure 14.30: (*a*) the closed-loop function G_Z, (*b*) the input impedance, (*c*) the output impedance, and (*d*) the open-loop gain T.

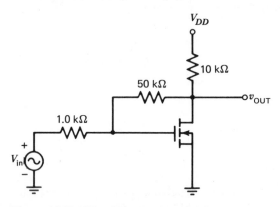

Figure 14.30. Circuit for Problem 14.12.

14.13. Calculate the following functions for the circuit in Figure 14.31: (*a*) the closed loop function G_Y, (*b*) the input impedance, (*c*) the output impedance, and (*d*) the open-loop gain T.

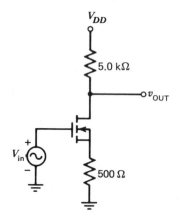

Figure 14.31. Circuit for Problem 14.13.

14.14. For the circuit in Figure 14.32, calculate (*a*) the closed loop function G_V, (*b*) the input and output impedances, and (*c*) the open-loop gain T.

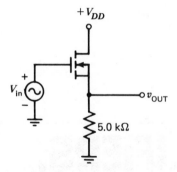

Figure 14.32. Circuit for Problem 14.14.

14.15. For the circuit in Figure 14.33, calculate (*a*) the closed-loop function G_I, (*b*) the input and output impedances, and (*c*) the open-loop gain T.

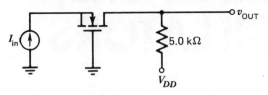

Figure 14.33. Circuit for Problem 14.15.

chapter 15 _____
POWER AMPLIFIERS, TUNED AMPLIFIERS, AND OSCILLATORS

The analog amplifiers discussed in the previous chapters are, in general, small-signal circuits whose purpose is either to increase the amplitude of the signal or to change the impedance level supplying or receiving the signal. In most cases, these amplifiers are expected to accept signals within a broad frequency spectrum and a wide range of amplitudes.

In this chapter, we consider three types of electronic circuits: power amplifiers, tuned amplifiers, and oscillators. These circuits serve different functions, but there are certain features that appear in the analysis of all three, making it logical to consider them together. As we demonstrate below, power amplification is performed most efficiently if the signal is restricted to a single frequency. Tuned circuits are used in communications systems to select a narrow frequency range, centered on a single frequency, from a broad spectrum of similar signals. Oscillators are positive feedback circuits with a gain characteristic that provides a unity open-loop gain at one particular frequency, so that the circuit acts as a source of a single frequency signal. These three circuit types are analyzed and applied in the following sections.

15.1 **POWER AMPLIFIERS**

A circuit that must be capable of supplying a specific amount of power to a load, such as a stereo amplifier to a pair of speakers or a broadcasting amplifier to a transmitting antenna, presents peculiar problems to the designer. Since high power levels may be involved, the efficiency of the circuit becomes quite important. Inefficiency in the circuit gives rise to heat generation in the components that must be minimized. Integrated circuits are especially sensitive to temperature increases on the chip. Heat sinks, forced air cooling, heat pipes, and liquid cooling are popular techniques for transferring heat from electronic components to cooler environments. This heat transfer is essential for the proper operation of the electronic devices. Another significant problem lies in the large-signal nature of power amplifiers. Signals that traverse the extremes in the forward active region of a transistor are subject to distortion. Special techniques are required to minimize this distortion.

The power output stages of integrated circuits are predominantly in the common-collector or common-drain configuration, and this application is examined first. We then examine the power stage designs for radio-frequency (rf) circuits. The rf circuit power stages are usually in the common-emitter or common-source configuration and use resonant circuits for loads.

15.1.1 **Amplifier Classifications**

In our earlier discussions, we have classified amplifiers according to their circuit configurations, such as common-emitter, common-collector, common-base, common-source, common-drain, or common-gate. The need for more efficient circuits has led to another classification scheme that labels circuits according to the portion of the period of the output waveform during which the output transistors conduct current. This designation is usually given in terms of conduction angle, assuming that the input signal is sinusoidal.

Up to this point, the analog circuits that we have presented have had conduction angles of 360° and are called Class A amplifiers. The classifications of amplifiers by conduction angle for sinusoidal input waveforms are listed in Table 15.1.

Table 15.1. Amplifier Classifications

Class	Conduction Angle α (degrees)
A	360
B	180
AB	180–360
C	< 180

In Figure 15.1, the collector current waveshapes for the different classes of bipolar transistor amplifiers are shown.

At this point, it seems peculiar that a circuit could faithfully reproduce the shape of an analog waveform if transistors in the circuit conducted for only a fraction of the input signal period. Class B and AB broadband amplifiers make use of arrangements that allow transistors to share portions of the input signal conduction angle. Class C amplifiers provide single-frequency sinusoidal outputs by injecting energy into resonant circuits for a small part of the cycle. The tuned circuit itself provides the continuity for a sinusoidal output voltage. In the following sections, we examine these circuits in detail.

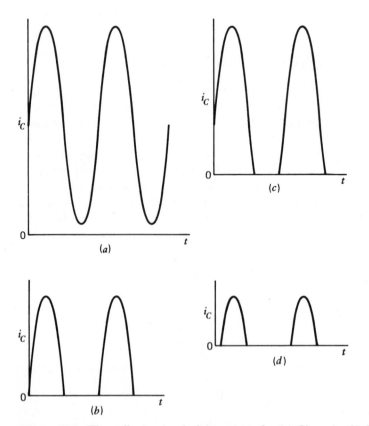

Figure 15.1. The collector (or drain) currents for (*a*) Class A, (*b*) Class B, (*c*) Class AB, and (*d*) Class C bipolar (or MOSFET) amplifiers.

15.1.2 **Common-Collector Output Stage**

In Figure 15.2, we show a common-collector amplifier driven by a sinusoidal voltage source. The power efficiency η of this circuit is defined as

$$\eta = \frac{P_{\text{LOAD}}(\text{ac})}{P_{\text{CIRCUIT}}} \times 100 \tag{15.1}$$

The maximum power efficiency η_{max} is of interest in evaluating the comparative performance of different circuit configurations.

The η_{max} for the circuit of Figure 15.2 may be calculated by obtaining the maximum possible average power delivered to the load. The average load power is related to the peak load current I_M and peak load voltage V_M by

$$P_{\text{LOAD}}(\text{ac}) = \frac{V_M}{\sqrt{2}} \frac{I_M}{\sqrt{2}} \tag{15.2}$$

or

$$P_{\text{LOAD}}(\text{ac}) = \frac{V_M I_M}{2} \tag{15.3}$$

The circuit load line is shown superimposed on the transistor collector characteristic curves in Figure 15.3. If the circuit is biased at Q_1 for maximum possible symmetrical swing, then

$$V_M = \frac{V_{CC} - V_{CE}(\text{sat})}{2} \tag{15.4}$$

or, neglecting the small saturation voltage,

$$V_M \simeq \frac{V_{CC}}{2} \tag{15.5}$$

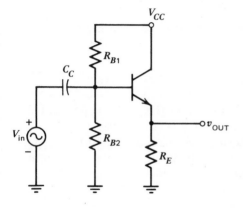

Figure 15.2. A common-collector Class A power amplifier.

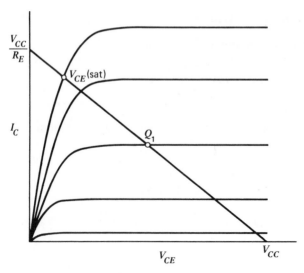

Figure 15.3. The load line of the common-collector power amplifier.

and

$$I_M = \frac{V_{CC}}{2R_E} \tag{15.6}$$

The maximum average power available to the load is obtained by substituting Equations 15.5 and 15.6 into 15.3:

$$P_{\text{LOAD}}(\text{ac}) = \frac{V_{CC}^2}{8R_E} \tag{15.7}$$

If we neglect the small power dissipation in the bias circuit, then

$$P_{\text{CIRCUIT}} = V_{CC}I_{CC} = V_{CC}I_{CQ} \tag{15.8}$$

$$= V_{CC}\frac{V_{CC}}{2R_E}$$

The maximum efficiency of the circuit is then obtained from Equations 15.1, 15.7, and 15.8 as

$$\eta_{\text{max}} = \frac{V_{CC}^2/8R_E}{V_{CC}^2/2R_E} \times 100 = 25\% \tag{15.9}$$

This maximum power efficiency of 25% illustrates the inadvisability of using a Class A configuration as an output stage in a power amplifier.

The power dissipated in the transistor is of particular importance, since transistors are sensitive to heat and it is difficult to remove heat from a semiconductor device. The total average power delivered to the circuit is independent of the signal. When a zero signal is present, half of the power is dissipated in the load as dc power, and the other half is dissipated in the transistor. An increase in the signal power increases the power dissipated in the load and decreases the power dissipated in the transistor. The maximum power dissipated in the transistor occurs for zero input signal and is given by

$$P_Q(NL) = V_{CEQ}I_{CQ} = \frac{V_{CC}}{2}\frac{V_{CC}}{2R_E} = \frac{V_{CC}^2}{4R_E} \tag{5.10}$$

which is two times the maximum power that can be delivered to the load. Thus, a considerable amount of power is dissipated in the transistor when there is no signal present. As shown below, Class B and Class C amplifiers have no power dissipation in the device with zero input signal. The minimum power dissipated in the transistor occurs when the load power is at a maximum. This power is determined by subtracting Equation 15.7 from Equation 15.10, giving

$$P_Q(FL) = \frac{V_{CC}^2}{4R_E} - \frac{V_{CC}^2}{8R_E} = \frac{V_{CC}^2}{8R_E} \tag{15.11}$$

Discrete transistors that dissipate large amounts of heat are constructed with large physical dimensions and often have a large metal heat sink or metal cooling fins attached to the transistor casing. The base width of a power transistor is relatively large, and β values on the order of 10 are typical. Discrete power transistors that are adequately protected with thermal cooling may be capable of power dissipations as large as 80 W.

Example. For the power MOSFET circuit of Figure 15.4, determine the

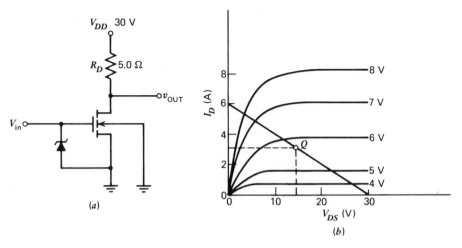

Figure 15.4. (a) A MOSFET Class A common-source power amplifier and (b) its drain characteristics.

power dissipated in the transistor for (a) zero input signal and (b) the maximum possible output signal.

Solution. (a) This situation is very similar to the bipolar transistor circuit discussed above. The quiescent operating point is $V_{DSQ} = 15$ V and $I_{DQ} = 3.0$ A, as indicated in Figure 15.4. The average power dissipated in the transistor under these conditions is

$$P_M = V_{DSQ}I_{DQ}$$
$$= 15 \times 3.0$$
$$= 45 \text{ W}$$

Note that there is also 45 W dissipated in the 5.0-Ω resistor in this case, for a total of 90 W delivered from the supply.

(b) When the input signal is increased to provide an output voltage swing of ± 15 V around the operating point, the signal power delivered to the load is

$$P_{\text{LOAD}}(\text{ac}) = \frac{V_{DD}^2}{8R_D}$$
$$= \frac{30^2}{8 \times 5.0}$$
$$= 22.5 \text{ W}$$

Thus, the power dissipated in the transistor is reduced to 22.5 W when the maximum output signal is present.

15.1.3 Class B Broadband Power Amplifiers

A significant improvement in efficiency and a reduction in transistor dissipation can be achieved by using a pair of transistors in a Class B "push–pull" arrangement. If the transistors are both *npn* or *n*-channel, a center-tapped transformer is used, in the configuration shown in Figure 15.5. There is considerable interest in transistor circuits to eliminate the transformers by using two power supplies and complementary transistors as shown in Figure 15.6*a* for bipolar transistors and in Figure 15.6*b* for MOSFETs. These circuits are emitter–follower and source–follower configurations.

The maximum efficiency of Class B amplifiers can be determined in the following way. The power supply currents in Figure 15.6*a* are half-wave-rectified sine waves, if we neglect distortion. The average value of a half-wave-rectified current is

$$I_{DC} = \frac{I_{CM}}{\pi} \tag{15.12}$$

and the dc power delivered to the circuit is given by

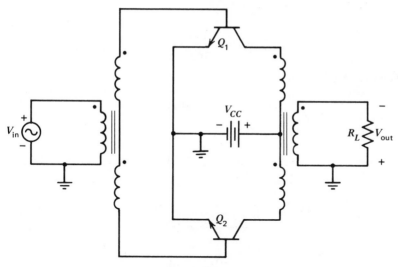

Figure 15.5. A transformer-coupled Class B bipolar transistor push–pull amplifier.

$$P_{\text{CIRCUIT}} = \frac{2V_{CC}I_{CM}}{\pi} \qquad (15.13)$$

but

$$I_{CM} = \frac{V_{CC}}{R_L} \qquad (15.14)$$

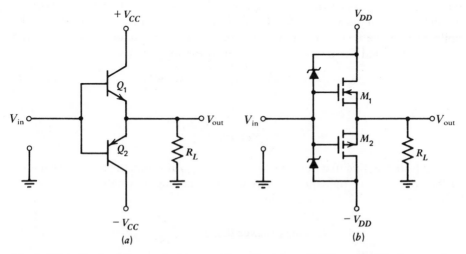

Figure 15.6. Push–pull complementary Class B power amplifiers. (*a*) Bipolar transistor configuration. (*b*) MOSFET transistor-configuration with gate protection diodes.

and

$$P_{\text{CIRCUIT}} = \frac{2V_{CC}^2}{\pi R_L} \qquad (15.15)$$

The ac power delivered to the load is

$$P_{\text{LOAD}}(\text{ac}) = \tfrac{1}{2}I_{CM}^2 R_L$$

or

$$P_{\text{LOAD}}(\text{ac}) = \frac{V_{CC}^2}{2R_L} \qquad (15.16)$$

Then,

$$
\begin{aligned}
\eta_{\text{max}} &= \frac{P_{\text{LOAD}}(\text{ac})}{P_{\text{CIRCUIT}}} \times 100 \\
&= \frac{V_{CC}^2/2R_L}{2V_{CC}^2/\pi R_L} \times 100 \\
&= \frac{\pi}{4} \times 100 \\
&= 78.5\%
\end{aligned}
\qquad (15.17)
$$

This is a considerable improvement over the Class A value of 25%, but it does require two power supplies or a transformer and reasonably well-matched complementary transistors.

Our other concern in the Class A amplifier is the power dissipated in the transistors. In the Class B configurations, there is essentially no power dissipation in the transistors for zero input signal, compared to twice the maximum available output power for the Class A case. The power dissipated in the collectors of the transistors is the difference between the dc power delivered to the circuit and the ac power delivered to the load. Thus,

$$2P_Q = P_{\text{CIRCUIT}} - P_{\text{LOAD}}(\text{ac}) \qquad (15.18)$$

or

$$2P_Q = \frac{2}{\pi}V_{CC}I_{CM} - \frac{R_L I_{CM}^2}{2} \qquad (15.19)$$

It is instructive to determine the value of I_{CM} at which the power dissipated in the collectors of the transistors reaches a maximum. This can be found by setting $dP_Q/dI_{CM} = 0$, resulting in

$$I_{CM}(\text{max power dissip}) = \frac{2V_{CC}}{\pi R_L} \qquad (15.20)$$

Using this value for I_{CM}, the maximum power dissipated in each collector is approximately 20% of the maximum power that can be delivered to the load, compared to 200% for the Class A case.

The major difficulty associated with Class B amplifiers is the distortion due to the transition near zero. Typical transfer characteristics for bipolar and MOS Class B amplifiers are shown in Figures 15.7a and 15.7b. The crossover distortion for the bipolar transistor amplifier is due to the nonlinearity of the base–emitter diode characteristic. The MOSFET amplifier crossover distortion

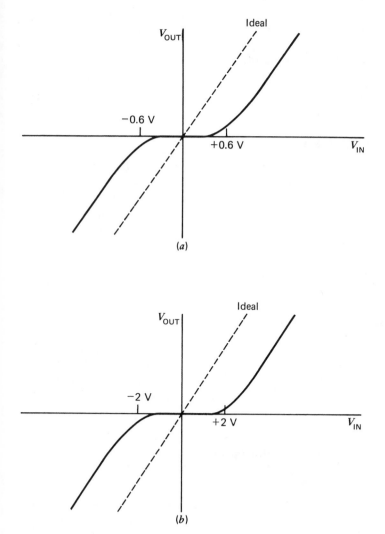

Figure 15.7. Typical transfer characteristics for push–pull amplifiers. (*a*) Bipolar transistors. (*b*) MOSFETs.

is due to the threshold voltages of the transistor pairs. Obtaining matched devices is difficult because the processes for fabricating *npn* and *pnp* or *n*-channel and *p*-channel devices are different and the basic materials properties are different.

Crossover distortion can be eliminated by operating the amplifier in Class AB. If the transistors are biased between the cutoff and Class A modes, a large reduction in distortion can be accomplished and a significant increase in efficiency over that in a Class A configuration can be achieved. An example of a Class AB power stage is given in Chapter 16, where the output stage of an operational amplifier is analyzed. This particular output stage of an integrated circuit does not have high power when compared to the tens of watts delivered to motors or audio speakers from discrete transistor circuits.

In Table 15.2, we summarize the maximum efficiencies for the power amplifier classifications.

15.2 TUNED AMPLIFIERS

The amplifiers described in the previous section are broadband power amplifiers. Reducing the output current conduction angle to below 180° increases the efficiency of a power amplifier but makes sense only if the amplification takes place at a single frequency. This narrow-band amplification is accomplished by using a circuit as a load that is resonant at that frequency. It is also useful to perform narrow-band amplification using Class A or Class B circuits. In this section, we discuss Class C power amplifiers and more general linear tuned amplifiers.

15.2.1 Class C Amplifiers

The broadband amplifiers described above are "linear" amplifiers in that the output signal is approximately linearly related to the input signal. In a tuned amplifier, for a Class A, Class AB, or Class B amplifier, it can be shown [2] that linear amplification is possible, but the fundamental frequency component

Table 15.2. Maximum Efficiencies for Power Amplifiers

Class	Efficiency η (%)
A	25
B	78.5
AB	25–78.5
C	> 78.5

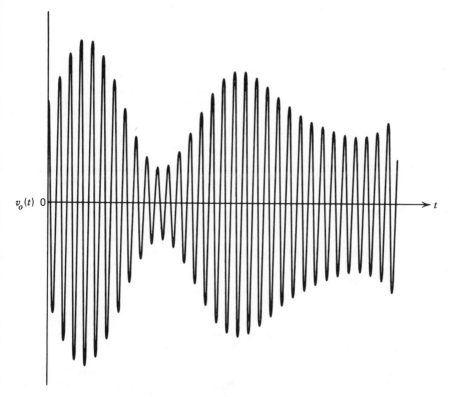

Figure 15.8. An amplitude-modulated signal.

for a Class C amplifier output is not linearly related to the fundamental frequency component for the input signal.

Class C amplifiers are often used in communications systems where the output is an amplitude-modulated signal like that shown in Figure 15.8. It is important to recognize that Class C amplifiers introduce distortion if used to amplify amplitude modulated signals. Therefore, it is necessary to use Class C amplification on a constant-amplitude, single-frequency signal, called the carrier, and provide distortion-free amplitude modulation on the the final stage. The carrier for a commercial broadcast signal must be ultrastable in frequency, which is best accomplished at a low power level, and have output power levels up to 50 kW. This is just one example of the need for the amplification of constant-amplitude, single-frequency sine waves to high power levels that is most efficiently performed by Class C amplification.

A Class C bipolar transistor amplifier is shown in Figure 15.9. The Q of the tuned circuit (see Chapter 12) is

$$Q = \frac{R_L}{\omega_o L} \tag{15.21}$$

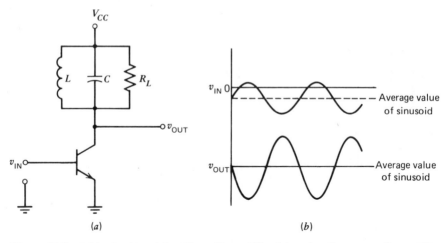

Figure 15.9. A bipolar transistor Class C amplifier (*a*) and voltage waveforms (*b*).

where

$$\omega_o = \frac{1}{(LC)^{1/2}} \tag{15.22}$$

Since the output power P_{LOAD} is, for a fixed power supply voltage, inversely proportional to R_L and the Q is proportional to R_L, a compromise value for R_L must be selected to optimize performance. High efficiency (low conduction angle) and low Q are incompatible since the energy added to the tuned circuit each cycle must be sufficient to provide a satisfactory output waveform.

An interesting application of a Class C amplifier is as a frequency multiplier. If the load is tuned to a frequency either two or three times that of the input frequency and the Q is high enough, the output will "ring" at the higher frequency with an almost constant amplitude. This is a very effective technique for generating the carrier frequencies used for television, broadcast FM, and radar applications.

15.2.2 **Tuned Amplifier Characteristics**

We now consider the small-signal, low-power tuned amplifier that is used as an amplifier in many communications systems. This circuit is usually operated in the Class A common-emitter configuration and can be tuned either on the input, the output, or both.

The circuit for the basic tuned amplifier is shown in Figure 15.10*a* and the admittance (*y*) parameter equivalent circuit is shown in Figure 15.10*b*. We assume that the *y*-parameters have been measured at the frequency ω_o and include the parasitic components associated with the packaging. We represent

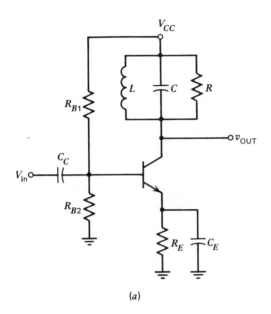

(a)

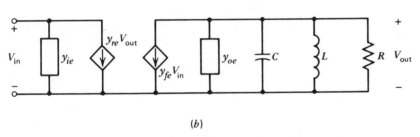

(b)

Figure 15.10. A small-signal tuned amplifier. (a) Circuit. (b) Equivalent circuit.

the y-parameters by a Taylor series expansion about ω_o such that

$$y_{ie} = g_{ie} + j\omega_o c_{ie} + j(\omega - \omega_o)c'_{ie} \tag{15.23}$$

where

$$c'_{ie} = \frac{d[Im(y_{ie})]}{d\omega}\bigg|_{\omega = \omega_o} \tag{15.24}$$

The capacitance c'_{ie} represents the variation in the input admittance as a function of frequency. Some of the parameters have little or no variation with frequency, and it is reasonable to represent them as constant capacitors, but others, particularly y_{ie} and y_{oe}, have a strong frequency dependence. Equation 15.24 is only a first-order approximation to the variation of c_{ie}. In a similar manner,

$$y_{fe} = g_{fe} + j\omega_o c_{fe} + j(\omega - \omega_o)c'_{fe} \tag{15.25}$$

$$y_{re} = g_{re} + j\omega_o c_{re} + j(\omega - \omega_o)c'_{re} \tag{15.26}$$

$$y_{oe} = g_{oe} + j\omega_o c_{oe} + j(\omega - \omega_o)c'_{oe} \tag{15.27}$$

We can calculate the voltage gain from the circuit of Figure 15.10b as

$$A_v = \frac{V_{\text{out}}}{V_{\text{in}}} = \frac{-y_{fe}}{y_{oe} + j\omega C + (1/R) + (1/j\omega L)} \tag{15.28}$$

and expanding y_{oe} in this equation about ω_o results in

$$A_v = \frac{-y_{fe}}{g_{oe} + j\omega(C + c'_{oe}) + j\omega_o(c_{oe} - c'_{oe}) + (1/R) + (1/j\omega L)} \tag{15.29}$$

Note that we have not expanded y_{fe} since it only introduces phase shift into the expression that is almost constant over the frequency range of interest. If we select ω_o such that

$$\omega_o(C + c_{oe}) = \frac{1}{\omega_o L} \tag{15.30}$$

and $c_{oe} \cong c'_{oe}$, then the imaginary part of the denominator goes to zero at $\omega = \omega_o$. Note that ω_o is not $1/(LC)^{1/2}$, the resonant frequency of the tuned circuit external to the transistor but includes the transistor capacitances. Then,

$$\omega_o = \frac{1}{[(C + c_{oe})L]^{1/2}} \tag{15.31}$$

and

$$|A_v(\omega_o)| = A_0 = \frac{|y_{fe}|}{g_{oe} + (1/R)} \tag{15.32}$$

To find the bandwidth, let $\omega' = (\omega - \omega_o) \ll \omega_o$. Then,

$$\frac{V_2}{V_1} = \frac{y_{fe}}{j\omega'C + j\omega_o C + j\omega_o c_{oe} + j\omega' c'_{oe} + g_{oe} + (1/R) + [1/j(\omega_o + \omega')L]} \tag{15.33}$$

The last term of the denominator of this equation can be approximated, using long division, by

$$\frac{1}{j(\omega' + \omega_o)L} \simeq \frac{1}{j\omega_o L} - \frac{\omega'}{j\omega_o^2 L} \tag{15.34}$$

Substituting Equation 15.30, we get

$$\frac{1}{j(\omega' + \omega_o)L} \simeq -j\omega_o(C + c_{oe}) + j\omega'(C + c_{oe}) \tag{15.35}$$

Then, in the vicinity of $\omega = \omega_o$, we can write

$$A_v \approx \frac{-y_{fe}}{j\omega'(2C + c_{oe} + c'_{oe}) + g_{oe} + (1/R)} \tag{15.36}$$

Note that the narrow-band approximation $\omega' \ll \omega_o$ transforms the denominator from a two-pole function to a one-pole function, providing a considerable reduction in the mathematical complexity of the solution. The -3 dB points occur where the real part of the denominator is equal to plus or minus the imaginary part of the denominator, or

$$\omega_1, \omega_2 = \omega_o \pm \frac{g_{oe} + (1/R)}{2C + c'_{oe} + c_{oe}} \tag{15.37}$$

and

$$\omega_2 - \omega_1 = \frac{g_{oe} + (1/R)}{C + [(c_{oe} + c'_{oe})/2]} \tag{15.38}$$

Example. Determine R, C, and $A_v(\omega_o)$ of a tuned amplifier with a center frequency of 100 MHz and a bandwidth of 15 MHz. $= 0.5 \ \mu$H, and the transistor parameters at 100 MHz are:

$$f_{fe} = 30 \text{ mS}, \qquad c_{fe} = -71.5 \text{ pF}, \qquad c'_{fe} = 0$$
$$g_{oe} = 0.4 \text{ mS}, \qquad c_{oe} = 2.0 \text{ pF}, \qquad c'_{oe} = 2.0 \text{ pF}$$
$$g_{ie} = 4.0 \text{ mS}, \qquad c_{ie} = 8.0 \text{ pF}, \qquad c'_{ie} = 2.0 \text{ pF}$$
$$g_{re} = -0.02 \text{ mS}, \qquad c_{re} = -1.4 \text{ pF}, \qquad c'_{re} = 0$$

Solution. To determine C, we use Equation 15.31:

$$C + c_{oe} = \frac{1}{L\omega_o^2}$$

$$= \frac{1}{(0.5 \times 10^{-6})[(2\pi \times 100 \times 10^6)^2]}$$

$$= 5.1 \text{ pF}$$

Since $c_{oe} = 2.0$ pF,

$$C = 3.1 \text{ pF}$$

The value of R necessary to provide the required bandwidth is found from Equation 15.38:

$$R = \frac{1}{(2\pi \times 15 \times 10^6)\ [C + (c_{oe} + c'_{oe})/2] - g_{oe}}$$

$$= \frac{1}{(2\pi \times 15 \times 10^6)(5.1 \times 10^{-12}) - 0.4 \times 10^{-3}}$$

$$= 12.4 \text{ k}\Omega$$

Using Equation 15.36, the gain at the center frequency is found to be

$$A_v = \frac{-y_{fe}}{g_{oe} + (1/R)}$$

$$= \frac{-30 \times 10^{-3} + j\omega_o \times 71.5 \times 10^{-12}}{0.4 \times 10^{-3} + (1/12.4 \times 10^3)}$$

$$= -62 + j93$$

15.2.3 Tuned Amplifier Stability

The input admittance of a tuned amplifier has some interesting frequency characteristics that can influence the stability of the circuit. The feedback parameter y_{re}, which is usually neglected in broadband amplifiers, is a complex, frequency-dependent quantity. It is reasonable to expect that some combination of frequency and driving source admittance Y_s could result in a feedback signal of the proper amplitude and phase to be regenerative and provide a single-frequency output with no input signal. This effect is highly desirable in an oscillator circuit, such as the ones described in the final section of this chapter, but has disastrous consequences when it occurs in an amplifier. In this section, we analyze the input admittance to demonstrate the possibility of instability and investigate methods for preventing oscillation.

INPUT ADMITTANCE OF A TUNED AMPLIFIER. The small-signal equivalent circuit shown in Figure 15.11a can be simplified to that of Figure 15.11b for the analysis of the input circuit. The current I_a in Figure 5.11a is

$$I_a = y_{re}V_{out} \tag{15.39}$$

but

$$V_{out} = -\frac{y_{fe}V_{in}}{y_{oe} + Y_L} \tag{15.40}$$

Then,

$$I_a = -\frac{y_{fe}y_{re}V_{in}}{y_{oe} + Y_L} \tag{15.41}$$

The same current, in the circuit of Figure 15.11b, can be written

$$I_a = y_{eq}V_{in} \tag{15.42}$$

If we compare this result with Equation 15.41, we see that

$$y_{eq} = -\frac{y_{fe}y_{re}}{y_{oe} + Y_L} \tag{15.43}$$

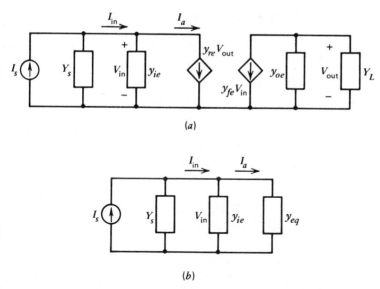

Figure 15.11. Circuits for determining the input admittance of a tuned amplifier. (*a*) The small-signal circuit. (*b*) The simplified circuit.

For a typical transistor, y_{fe} can be assumed to be purely conductive over the frequency range of interest, while y_{re} is purely capacitive. Then,

$$y_{fe} \cong g_{fe} \tag{15.44}$$

and

$$y_{re} = - j\omega c_{re} \tag{15.45}$$

Then, if we substitute these quantities into Equation 15.43, the result is

$$y_{eq} = \frac{j\omega c_{re} g_{fe}}{y_{oe} + Y_L} \tag{15.46}$$

If we substitute Equation 15.27 for y_{oe} and Y_L is the parallel combination of R, L, and C, we have

$$y_{eq} = \frac{j\omega c_{re} g_{fe}}{[g_{oe} + (1/R)] - j[(1/\omega L) - \omega(c'_{oe} + C) - \omega_o(c_{oe} - c'_{oe})]} \tag{15.47}$$

For values of ω less than ω_o, the inductive term is dominant in the denominator, and the imaginary term is a positive number that we designate as B. Then

$$y_{eq} = \frac{j\omega c_{re} g_{fe}}{G_A - jB} \tag{15.48}$$

where $G_A = (g_{oe} + 1/R)$. If we rationalize Equation 15.48, the result is

$$y_{eq} = -\frac{\omega C_{re} g_{fe} B}{G_A^2 + B^2} + j\frac{\omega C_{re} g_{fe} G_A}{G_A^2 + B^2} \qquad (15.49)$$

The importance of this result is that the real part of y_{eq} is negative for the typical conditions of $\omega < \omega_o$, real y_{fe}, and purely reactive y_{re}.

A negative real part for y_{eq} means that it is likely that the input admittance, which is the parallel combination of y_{eq} and y_{ie}, will have a negative real part over some frequency range. This condition can result in self-sustaining oscillations, depending on the value of Y_s in that frequency range. We can write

$$V_{in} = \frac{I_s}{Y_s + y_{ie} + y_{eq}} \qquad (15.50)$$

If we define $Y_D \equiv Y_s + y_{ie} + y_{eq} = G_D + jB_D$, then

$$V_{in} = \frac{I_s}{G_D + jB_D} \qquad (15.51)$$

If G_D is negative at a particular frequency and Y_S is inductive such that $B_D = 0$, then an appropriate value of resistance across the input terminals can reduce G_D to zero. The effect of these conditions is a zero value of Y_D at that frequency, making oscillations occur.

The oscillations can be prevented by assuring that not all of the conditions can be satisfied at any frequency. Note that the negative real part of Y_{in} is not sufficient, in itself, to establish oscillation. An inductive susceptance is also essential, to cancel the capacitive susceptance, or the feedback will not have the proper phase for oscillation. Unfortunately, it is common practice to operate these amplifiers with a tuned input circuit as well as a tuned output circuit. Another way to eliminate oscillations is to decrease the value of the input shunt resistor to make sure that the real part of Y_D is never zero. A reduction in the R of a parallel tuned circuit has the undesirable effect of also reducing the selectivity of the circuit. A final method that assures stability is to make the transistor appear to be a unilateral device, as discussed below.

NEUTRALIZATION. Most bipolar and field effect transistors are potentially unstable over some frequency range due to the feedback parameter y_{12}. A similar effect is observed in vacuum tubes. If the feedback can be canceled by an additional feedback signal that is equal in amplitude and opposite in sign, the transistor becomes unilateral from input to output and oscillation is impossible. This technique for the elimination of potential oscillations is called neutralization.

The small-signal equivalent circuit of a bipolar transistor is shown in Figure 15.12, with an external feedback element Y_N. If we consider this entire circuit as a two-port network, we can determine a new set of admittance parameters (designated a'), with particular interest in the feedback parameter y_{re}'.

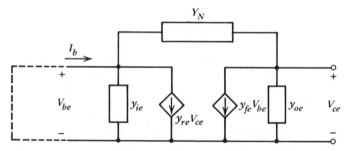

Figure 15.12. The equivalent circuit of a neutralized transistor.

From the definition of the admittance parameters (see Chapter 11),

$$y'_{re} = \frac{I_b}{V_{ce}} \quad \text{with } V_{be} = 0 \qquad (15.52)$$

and with the input terminals short-circuited,

$$I_b = y_{re}V_{ce} - Y_N V_{ce} \qquad (15.53)$$

Then,

$$y'_{re} = y_{re} - Y_N \qquad (15.54)$$

For complete neutralization, $y'_{re} = 0$, which means that

$$Y_N = y_{re} \qquad (15.55)$$

Equation 15.55 indicates that oscillations are impossible if we can find a circuit element that matches y_{re} for all values of frequency and operating conditions. Unfortunately, y_{re} is a nonlinear parameter that is best approximated by

$$y_{re} \simeq - j\omega c_{re} \qquad (15.56)$$

where c_{re} is a slowly varying function of both operating point and frequency. In effect, we are looking for a negative capacitor. An inductor has a negative susceptance, $B = -j(1/\omega L)$, but this component has the inverse of the desired frequency dependence. Inductors are also short circuits at dc and would be useless for this application.

A practical approach to this problem uses a fixed capacitance that is transformer coupled (for 180° phase shift) to provide neutralization over a limited frequency range. A typical circuit is shown in Figure 15.13. It should be recognized that perfect neutralization is impossible and that problems created by limited neutralization may exceed the benefits.

15.2.4 **Multistage Tuned Amplifiers**

In many situations, it is desirable to cascade tuned amplifiers because of insufficient gain in single stage. It is important to recognize that the charac-

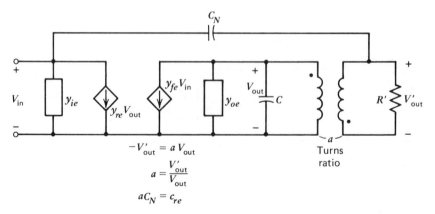

Figure 15.13. A tuned amplifier with neutralization to prevent spontaneous oscillation.

teristics of multistage tuned amplifiers are different from those of single tuned amplifiers.

To begin the analysis, we define

$$G_A \equiv g_{oe} + \frac{1}{R} \qquad (15.57)$$

and

$$C_A = C + \frac{c_{oe} + c'_{oe}}{2} \qquad (15.58)$$

Then, we can rewrite Equation 15.38 as

$$\omega_2 - \omega_1 = \frac{G_A}{C_A} \qquad (15.59)$$

and Equations 15.36 as

$$A_v = \frac{-y_{fe}}{G_A + 2j\omega' C_A} \qquad (15.60)$$

or

$$A_v = \frac{-y_{fe}/G_A}{1 + j\omega'(2C_A/G_A)}$$

and, if $K \equiv y_{fe}/G_A$,

$$A_v = -\frac{K}{1 + j\omega'(2C_A/G_A)} \qquad (15.61)$$

If we have n identical cascaded stages tuned to the same frequency (synchronous tuning), the voltage gain is the product of the gains of the individual stages, resulting in

$$A_v(n\text{-stages}) = (A_v)^n \tag{15.62}$$

$$= (-1)^n \frac{K^n}{[1 + j\omega'(2C_A/G_A)]^n}$$

The magnitude of this gain is

$$|A_v(n\text{-stages})| = \frac{|K^n|}{[1 + (2\omega'C_A/G_A)^2]^{n/2}}$$

but, at the -3 dB points,

$$|A_v(n\text{-stages})| = \frac{|K^n|}{2^{1/2}}$$

Then,

$$2^{1/n} = 1 + \left(2\omega'\frac{C_A}{G_A}\right)^2$$

or

$$\omega_2, \omega_1 = \omega_o + \frac{G_A}{2C_A}(2^{1/n} - 1)^{1/2} \tag{15.63}$$

resulting in a bandwidth of

$$\omega_2 - \omega_1 = \frac{G_A}{C_A}(2^{1/n} - 1)^{1/2} \tag{15.64}$$

We see from this result that the bandwidth is reduced when synchronously tuned circuits are cascaded. The reductions are summarized in Table 15.3. In this configuration, the gain increase is accompanied by a decrease in bandwidth. This change in the frequency characteristics of the amplifier can be critical, since in many communications systems the information is contained in the

Table 15.3. Bandwidths for Synchronously Tuned Cascaded Amplifiers

Number of Stages	Bandwidth
1	$1.0\ G_A/C_A$
2	$0.64\ G_A/C_A$
3	$0.51\ G_A/C_A$
4	$0.43\ G_A/C_A$
5	$0.39\ G_A/C_A$

sidebands and is lost if the bandwidth is too narrow. If the stages are tuned to different frequencies (stagger tuned), it is possible to obtain the gain improvement of a cascaded system with the desired bandwidth.

15.2.5 Graphical Design of Tuned Amplifiers

A useful aid to the understanding and design of tuned circuits for amplifier applications can be developed from phase–plane plots of the characteristics of the parallel RLC circuit. The impedance of the circuit as a function of the Laplace complex frequency s is given by

$$Z(s) = \frac{s/C}{s^2 + (s/RC) + (1/LC)} \tag{15.65}$$

$$= \frac{s/C}{(s - p_1)(s - p_2)} \tag{15.66}$$

where

$$p_1, p_2 = -\frac{1}{2RC} \pm \left[\left(\frac{1}{2RC} \right)^2 - \frac{1}{LC} \right]^{1/2} \tag{15.67}$$

We can also write

$$p_1, p_2 = -\alpha \pm j\beta \tag{15.68}$$

where

$$\alpha \equiv \frac{1}{2RC} \tag{15.69}$$

$$\omega_o \equiv \frac{1}{(LC)^{1/2}} \tag{15.70}$$

and

$$\beta \equiv (\omega_o^2 - \alpha^2)^{1/2} \tag{15.71}$$

Then, Equation 15.65 can be written in the general form

$$Z(s) = \frac{s/C}{s^2 + 2\alpha s + \omega_o^2} \tag{15.72}$$

The pole-zero pattern for a tuned circuit is shown in Figure 15.14.

Operation of this circuit in a tuned amplifier takes place along the $j\omega$ axis. Equation 15.66 for $s = j\omega$ is written

$$Z(j\omega) = \frac{j(\omega/C)}{(j\omega - p_1)(j\omega - p_2)} \tag{15.73}$$

The denominator can be written in polar form if we define

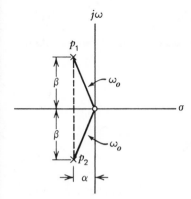

Figure 15.14. The general pole-zero phase-plane plot for a tuned circuit.

$$jw - p_1 = k_1 \angle \tan^{-1}(-\omega/p_1) \tag{15.74}$$

and

$$jw - p_2 = k_2 \angle \tan^{-1}(-\omega/p_2) \tag{15.75}$$

where $k_1 = (p_1^2 + \omega)$ and $k_2 = (p_2^2 + \omega^2)^{1/2}$. These quantities are illustrated in Figure 15.15.

The magnitude of the impedance is

$$|Z(j\omega)| = \frac{\omega/C}{k_1 k_2} \tag{15.76}$$

The area of the triangle in Figure 15.15 is

$$\text{Area} = \tfrac{1}{2}(\alpha \times 2\beta) \tag{15.77}$$
$$= \alpha\beta$$

but the area can also be written as

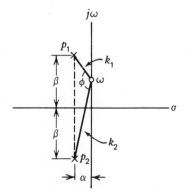

Figure 15.15. The operation of a tuned circuit along the $j\omega$ axis.

$$\text{Area} = \tfrac{1}{2} k_1 k_2 \sin \phi \tag{15.78}$$

where ϕ is the angle between the sides designated k_1 and k_2 in the figure. Equating these two relationships for the area yields

$$\frac{1}{k_1 k_2} = \frac{\sin \phi}{2\alpha\beta} \tag{15.79}$$

If we substitute this result into Equation 15.76, we get

$$|Z(j\omega)| = \frac{(\omega/C) \sin \phi}{2\alpha\beta} \tag{15.80}$$

If we assume that the value of $\omega \simeq \omega_0$, then Equation 15.80 reaches a maximum when $\sin \phi = 1$ ($\phi = 90°$). From basic geometrical concepts, we know that an angle inscribed in a semicircle is a right angle. We conclude that a circle of radius β with a center at $-\alpha$ intersects the $j\omega$ axis at the peak values of $Z_{j\omega}$. This circle is called the "peaking" circle and is illustrated in Figure 15.16. Note that if we decrease the Q of the circuit by increasing α, the circle becomes tangent to the $j\omega$ axis at $\omega = 0$ when $\alpha = \beta$ and there is no intersection when $\alpha > \beta$.

15.2.6 Stagger-Tuned Amplifiers

In this section we apply the graphical technique developed above to a pair of cascaded amplifiers tuned to slightly different frequencies. This stagger tuning provides the opportunity for establishing a more desirable bandpass characteristic.

We assume that the bandwidths of the tuned circuits are narrow. This enables us to simplify the gain expression significantly. If the amplifier voltage gain is given by

$$A_v(j\omega) = \frac{-j(g_m\omega/C)}{(j\omega - p_1)(j\omega - p_2)} \tag{15.81}$$

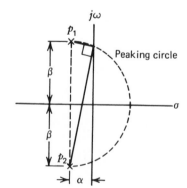

Figure 15.16. The peaking circle for a narrow band-tuned circuit.

and we assume that the Q of the circuit is high enough that we can approximate p_2 as $-j\omega_o$ when ω is in the vicinity of $+\omega_o$, then

$$A_v(j\omega) \simeq \frac{-j(g_m\omega_o/C)}{(j\omega - p_1)(2j\omega_o)} \qquad \text{for } \omega \simeq \omega_o$$

or

$$A_v(j\omega) \simeq - \frac{g_m}{2C(j\omega - p_1)} \qquad \text{for } \omega \simeq \omega_o \qquad (15.82)$$

The approximations derived above apply to both amplifiers.

We assume that the two amplifiers have the same value of $\alpha = 1/2RC$ but different values of $\beta = (\omega_o^2 - \alpha^2)^{1/2}$. The poles can then be designated

$$p_1 = -\alpha + j\beta_1 \qquad (15.83a)$$
$$p_2 = -\alpha - j\beta_1 \qquad (15.83b)$$
$$p_3 = -\alpha + j\beta_3 \qquad (15.83c)$$
$$p_4 = -\alpha - j\beta_3 \qquad (15.83d)$$

as shown in Figure 15.17.

The gain expression for the pair of amplifiers, assuming that the narrow-band approximation is valid, is then

$$A_v(j\omega) \simeq \frac{g_m^2}{4C_1C_3(j\omega - p_1)(j\omega - p_3)} \qquad (15.84)$$

where C_1 and C_3 are the capacitors associated with p_1 and p_3, respectively. We have shown the pole-zero plot for this expression in Figure 15.18. Note the similarity between Equations 15.84 and 15.73. We use the peaking circle concept to estimate the performance of this amplifier. The radius of the circle is $\Delta/2$, where $\Delta = \beta_1 - \beta_3$, and its center is at $-\alpha + j(\beta_3 + \Delta/2)$, assuming that β_1 is greater than β_3. In this case, the peaks are both on the positive $j\omega$ axis, in contrast with the previous case where one was on the positive $j\omega$ axis and

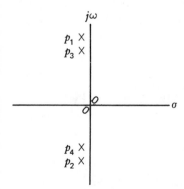

Figure 15.17. The phase-plane plot for the cascaded stagger-tuned amplifiers.

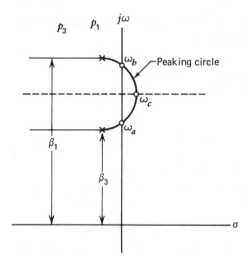

Figure 15.18. The peaking circle applied to a stagger-tuned amplifier pair.

the other was on the negative $j\omega$ axis. The analogous result to Equation 15.80 is

$$|A_v(j\omega)| = \frac{g_m^2 \sin \phi}{4C_1C_3\alpha\Delta} \tag{15.85}$$

but

$$\alpha = \frac{1}{2R_1C_1} = \frac{1}{2R_3C_3}$$

Then,

$$|A_v(j\omega)| = g_m^2 R_1 R_3 \left(\frac{\alpha}{\Delta}\right) \sin \phi \tag{15.86}$$

If $\alpha < \Delta/2$, the peaking circle intersects the imaginary axis at two points and the frequency characteristic is like that shown in Figure 15.19. The peak frequencies can be determined using the geometrical technique shown in Figure 15.20. The peak frequencies ω_a and ω_b are defined by

$$\omega_b, \omega_a = \omega_c \pm \lambda \tag{15.87}$$

where $\omega_c = \beta_3 + \Delta/2$. The quantity λ can be found from the Pythagorean theorem and is given by

$$\lambda = \left[\left(\frac{\Delta}{2}\right)^2 - \alpha^2\right]^{1/2} \tag{15.88}$$

If the peaking circle is tangent to the $j\omega$ axis, the amplifier pair is said to be

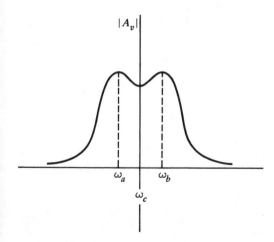

Figure 15.19. The gain-vs.-frequency curve for a stagger-tuned amplifier pair where the peaking circle intersects the $j\omega$ axis at two points.

flat-staggered. Over-staggering occurs if the peaking circuit does not intersect the $j\omega$ axis.

The bandwidth calculation for a stagger-tuned pair is complicated by the three possible cases. For the flat- and over-staggered cases, the conventional definition of bandwidth applies, where $|A_v| = A_p/2^{1/2}$. For the under-staggered or dual peak case, the bandwidth is often specified at the "triple" point, as illustrated in Figure 15.21. It can be shown that this bandwidth is approximated by

$$\text{Bandwidth} \simeq 2^{3/2}\lambda \tag{15.89}$$

As additional stages are added to stagger-tuned cascaded amplifiers, the pass band characteristic can be tailored to suit particular applications. Triple-tuned amplifiers with the third-stage tuned to the center frequency can have nearly flat pass bands.

Example. Determine the peak frequencies and bandwidth for a stagger-tuned amplifier pair with $R_1 = R_3 = 10 \text{ k}\Omega$, $C_1 = C_3 = 100 \text{ pF}$, $L_1 = 82.6 \text{ }\mu\text{H}$, and $L_3 = 123 \text{ }\mu\text{H}$.

Solution. From Equations 15.69, 15.70, and 15.71,

$$\alpha = \frac{1}{2R_1C_2} = \frac{1}{2R_3C_3}$$

$$= \frac{1}{(2 \times 10 \times 10^3)(100 \times 10^{-12})}$$

$$= 500 \times 10^{-3} \text{ s}^{-1}$$

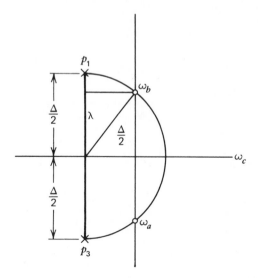

Figure 15.20. A detailed view of the geometry for determining the peak frequencies for a stagger-tuned amplifier pair.

To determine the values for β_1 and β_3, we must first find ω_o for each pole. The results are

$$\omega_{o1} = \frac{1}{(L_1 C_1)^{1/2}}$$

$$= \frac{1}{[(82.6 \times 10^{-6})(100 \times 10^{-12})]^{1/2}}$$

$$= 11 \times 10^6 \text{ rad/s}$$

Similarly,

$$\omega_{o3} = \frac{1}{[(123 \times 10^{-6})(100 \times 10^{-12})]^{1/2}}$$

$$= 9.0 \times 10^6 \text{ rad/s}$$

The respective β values are then

$$\beta_1 = (\omega_{o1}^2 - \alpha^2)^{1/2}$$

$$= [(11 \times 10^6)^2 - (500 \times 10^3)^2]^{1/2}$$

$$= 10.99 \times 10^6 \text{ rad/s}$$

and

$$\beta_3 = [(9.0 \times 10^6)^2 - (500 \times 10^3)^2]^{1/2}$$

$$= 8.99 \times 10^6 \text{ rad/s}$$

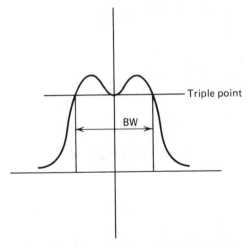

Figure 15.21. The triple-point bandwidth for a dual-peak stagger-tuned amplifier pair.

Then, $\Delta = \beta_1 - \beta_3 = 2.0 \times 10^6$ rad/s and $\Delta/2 = 1.0 \times 10^6$ rad/s. Using Equation 15.88,

$$\lambda = \left[\left(\frac{\Delta}{2}\right)^2 - \alpha^2 \right]^{1/2}$$
$$= [(1.0 \times 10^6)^2 - (500 \times 10^3)^2]^{1/2}$$
$$= 866 \times 10^3 \text{ rad/s}$$

The center frequency is

$$\omega_c = \beta_3 + \frac{\Delta}{2}$$
$$= 8.99 \times 10^6 + 1.0 \times 10^6$$
$$= 9.99 \times 10^6 \text{ rad/s}$$

Then,

$$\omega_a = \omega_c - \lambda$$
$$= 9.99 \times 10^6 - 866 \times 10^3$$
$$= 9.12 \times 10^6 \text{ rad/s}$$

and

$$\omega_b = \omega_c + \lambda$$
$$= 9.99 \times 10^6 + 866 \times 10^3$$
$$= 10.86 \times 10^6 \text{ rad/s}$$

Note that these frequencies do not coincide with ω_{o3} and ω_{o1}. The triple-point bandwidth, from Equation 15.89, is

$$\text{Bandwidth} = (2)^{3/2}\lambda$$
$$= (2)^{3/2}(866 \times 10^3)$$
$$= 2.45 \times 10^6$$

When using stagger-tuned pairs with narrow-band amplifiers, caution must be exercised to ensure that the resonant frequencies of the tuned circuits are close enough to one another so that the center dip is not excessive.

15.3 OSCILLATORS

Oscillators are a class of electronic circuits that include many of the concepts discussed in this and the previous chapters. An oscillator circuit consists of an amplifier, a frequency-selective network, and feedback (positive in this case). In most applications, the result of applying dc power to an oscillator is a steady-state, stable source of constant-frequency sinusoidal output voltage. Oscillator outputs can also be variable frequency or square wave, but we consider only constant frequency sinusoidal circuits here.

In this section, we concentrate on two different types of circuits: the phase-shift oscillator and the Colpitts oscillator. The first of these circuits makes use of resistors and capacitors to establish the frequency of oscillation, while the second uses a more stable arrangement in the form of a resonant circuit. There are numerous other types of oscillator circuits, but the basic principles of these circuits do not differ significantly from the ones described in this section.

15.3.1 The Phase Shift Oscillators

The principle of the phase shift oscillator is to feed the output of a single-stage amplifier into a capacitor–resistor ladder network, whose output becomes in turn the input to the amplifier. A typical MOSFET phase shift oscillator is shown in Figure 15.22. It is assumed that C_S is large enough to bypass the oscillation frequency and that $R_{G1}\|R_{G2}$ is $>> R$.

Analysis of the ladder network yields

$$\frac{V_y}{V_x} = \frac{R^3}{R^3 - j(6R^2/\omega C) - (5R/\omega^2 C^2) + j(1/\omega^3 C^3)} \qquad (15.90)$$

The amplifier stage has a 180° phase shift associated with the gain. For positive feedback, an additional 180° phase shift is required from the ladder network. Since each RC product produces less than 90° phase shift, at least three sections are needed. The ratio of V_y/V_x must be real for a phase shift of 180°. Therefore,

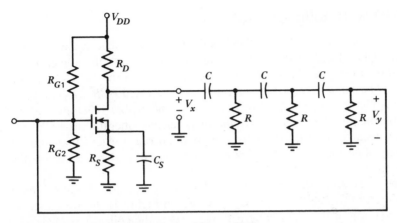

Figure 15.22. A MOSFET phase shift oscillator.

at the frequency of oscillation,

$$\frac{6R^2}{\omega_o C} = \frac{1}{\omega_o^3 C^3} \tag{15.91}$$

or

$$\omega_o = \frac{1}{RC(6)^{1/2}} \tag{15.92}$$

If this value is inserted into Equation 15.90, the result is not only real, which is our condition for the proper phase shift, but also independent of R and C, and is

$$\frac{V_x}{V_y} = -\frac{1}{29} \tag{15.93}$$

The conditions for oscillation are met if the gain of the amplifier is adjusted to -29, resulting in a closed-loop gain of unity, and values of R and C are selected to provide the proper frequency by satisfying Equation 15.92.

Example. Select values for R and C so that a phase shift circuit will oscillate at 1.0 MHz.

Solution. From Equation 15.92, the RC product must be

$$RC = \frac{1}{(6)^{1/2} 2\pi \times 1.0 \times 10^6}$$
$$= 6.5 \times 10^{-8} \text{ s}$$

Since there is no single correct answer for this problem, we must select one quantity and calculate the other. If $R = 1.0$ kΩ, then $C = 65$ pF.

15.3.2 **The Colpitts Oscillator**

A very popular sinusoidal oscillator is the Colpitts oscillator. One version is shown in Figure 15.23. The output voltage is taken from the collector and is a sinusoidal waveform whose frequency is a function of the LC_1C_2 resonant tank circuit. The base current bias circuitry is composed of R_1, R_2, and R_E. Both C_C and C_B are large capacitors presenting negligible reactance at the oscillating frequency. The Colpitts oscillator is usually operated in a Class C mode, which means that the transistor conducts current for less than 180° of the output signal period. The Class C operation imparts a high power efficiency to the circuit.

The circuit sustains the sinusoidal oscillation by a positive feedback path from the collector through the L and C_2 network. In Figure 15.24, we illustrate a simplified version of this path, where V_c represents the collector voltage and V_b represents the voltage that is tapped off to feed the base of the transistor. It is instructive to analyze V_b as a function of V_c. A voltage divider expression for V_b gives

$$V_b = \left(\frac{1}{j\omega C_2}\right)\left[\frac{1}{(1/j\omega C_2) + j\omega L}\right] V_c \qquad (15.94)$$

or

$$V_b = \left(\frac{1}{1 - \omega^2 L C_2}\right) V_c \qquad (15.95)$$

Note that the magnitude of the denominator determines whether V_b is in phase

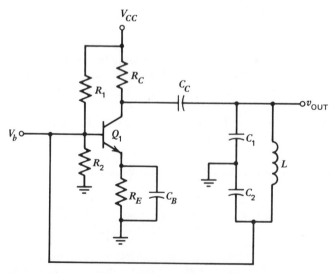

Figure 15.23. Circuit for a Colpitts oscillator.

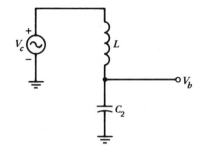

Figure 15.24. Simplified version of positive feedback path in the Colpitts oscillator of Figure 15.23.

or 180° out of phase with respect to V_c. If $\omega^2 L C_2 > 1$, then the bracketed term is negative and V_b is 180° out of phase with respect to V_c. As we shall show, this condition is true for the Colpitts oscillator.

The frequency of oscillation of the Colpitts oscillator is determined by the component values of L, C_1, and C_2. Note that, in going from the collector terminal to the base terminal through the tank circuit, C_1 and C_2 form a series capacitance whose effective value is

$$C' = \frac{C_1 C_2}{C_1 + C_2} \tag{15.96}$$

The resonant frequency of this tank circuit is then

$$\omega_o = \frac{1}{(LC')^{1/2}} \tag{15.97}$$

or

$$\omega_o^2 = \frac{1}{L[C_1 C_2/(C_1 + C_2)]} \tag{15.98}$$

When Equation 15.98 is substituted into Equation 15.95, we get

$$V_b = \left[\frac{1}{1 - \dfrac{(C_1 + C_2)LC_2}{LC_1C_2}} \right] V_c \tag{15.99}$$

$$= \left[\frac{1}{1 - \dfrac{C_1 + C_2}{C_1}} \right] V_c = - \frac{C_1}{C_2} V_c \tag{15.100}$$

Equation 15.100 shows that an oscillating circuit will always feed back a signal to the base that is 180° out of phase with the collector. Equation 15.100 also shows that the relative values of C_1 and C_2 determine the magnitude of the feedback voltage. As the ratio C_1/C_2 increases, the magnitude of feedback voltage increases. The circuit in Figure 15.24 illustrates the phase shift and positive feedback properties inherent in a Colpitts oscillator, but it does not include the resistive and capacitive loading effects of the transistor and bias network.

The Colpitts oscillator in Figure 15.23 will operate in a Class C mode when the oscillating voltage fed back to the base from the collector is large enough to drive the base voltage to a value less than the cut-in voltage of the transistor. Visualize the base voltage as composed of the sum of a dc voltage from V_{CC}, R_1, and R_2 and an oscillating voltage feedback from V_c. If the negative phase of the feedback voltage drives v_{BE} below the cut-in voltage, transistor Q_1 turns off. During the positive portion of the feedback sine wave, the base drive is higher than that due to the bias network alone.

The efficiency of the circuit is improved when the transistor is cut off. In fact, if the tank circuit were to resonate indefinitely ($Q = \infty$), there would be no need to ever turn the transistor on. The main function of the transistor is to supply charge to the tank circuit to compensate for resistive and radiative losses. Figure 15.25 illustrates the waveforms that are present in a Class C Colpitts oscillator. The base and collector waveforms are 180° out of phase. The emitter current is present when v_{BE} is greater than approximately 0.6 V.

The initial turn-on of the oscillator is stimulated by the presence of noise voltages in the circuit. A positive noise voltage on the base induces a drop in collector voltage which in turn induces a further positive voltage on the base

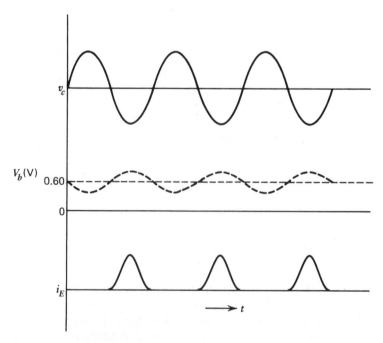

Figure 15.25. Voltage and current waveforms for the Colpitts oscillator in Class C operation.

via the feedback path. The tank circuit filters out all but the resonant frequency as oscillations rapidly build up to a steady-state level.

The bias resistors R_1, R_2, and R_E set a bias current that would occur if the feedback signal were absent. The Class C bias current is a pulse and bears little resemblance to that designed with R_1, R_2, and R_E for dc conditions. However, the bias resistors do influence the amplitude of the emitter current spike. The Colpitts oscillator will be more efficient when the transistor current pulse amplitudes are sufficient for oscillation, and no larger.

A Colpitts oscillator can operate in a Class A mode if the feedback voltage is not large enough to cut off the transistor. This Class A operation is not nearly as efficient as a Class C operation but will allow a linear analysis that illustrates the startup condition.

In Figure 15.26, we depict the Colpitts oscillator redrawn to show it as two two-port networks representing the feedforward path through the transistor and the feedback path through the tank circuit. The bias and load resistors are neglected. The transistor and feedback networks can be described with hybrid (h) parameters, where h refers to the transistor network and h' refers to the feedback network. The h-parameter descriptions for both networks can be written as

$$V_1 = h_{11}I_e + h_{12}V_2 \qquad (15.101)$$

$$I_c = h_{21}I_e + h_{22}V_2 \qquad (15.102)$$

and

$$-V_1 = h'_{11}I_e + h'_{12}(-V_2) \qquad (15.103)$$

$$I_c = h'_{21}I_e + h'_{22}(-V_2) \qquad (15.104)$$

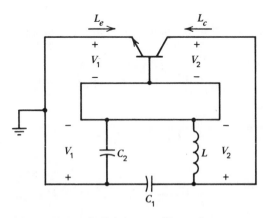

Figure 15.26. A Colpitts oscillator drawn to show the feedback and amplifying device as separate two-port networks. The bias circuitry is not shown.

Substitution yields

$$0 = (h_{11} + h'_{11})I_e + (h_{12} - h'_{12})V_2 \qquad (15.105)$$

$$0 = (h_{21} - h'_{21})I_e + (h_{22} + h'_{22})V_2 \qquad (15.106)$$

Since I_e and V_2 cannot simultaneously be zero, then

$$(h_{11} + h'_{11})(h_{22} + h'_{22}) - (h_{12} - h'_{12})(h_{21} - h'_{21}) = 0 \qquad (15.107)$$

Equation 15.107 contains both real and imaginary terms, each of which must equate to zero. The solution for the imaginary term specifies the frequency of oscillation, while the solution for the real term specifies the starting condition. The starting condition is a measure of the amount of positive feedback necessary to begin oscillation.

The h-parameters of the two networks in Figure 15.26 can be substituted into Equation 15.107, and we can then solve for the real and imaginary parts. Equation 15.107 becomes

$$\left[h_{11} + \frac{1}{j\omega(C_1 + C_2)} \right] \left(h_{22} + j\omega \frac{C_1 C_2}{C_1 + C_2} + \frac{1}{j\omega L} \right)$$
$$- \left(h_{12} - \frac{C_1}{C_1 + C_2} \right) \left(h_{21} + \frac{C_1}{C_1 + C_2} \right) = 0 \quad (15.108)$$

Setting the imaginary terms equal to zero results in

$$\omega_o^2 = \frac{C_1 + C_2}{LC_1 C_2} + \frac{h_{22}}{h_{11} C_1 C_2} \qquad (15.109)$$

or

$$\omega_o^2 \approx \frac{C_1 + C_2}{LC_1 C_2} \qquad (15.110)$$

Note that the h-parameters for the transistor in Figure 15.26 are measured in the common-base configuration.

The real part of Equation 15.108 can be set equal to zero so that we can derive the starting condition. If $\omega^2 = (C_1 + C_2)/LC_1 C_2$, then the real term can be approximated as

$$h_{21} = -h_{fb} = \frac{C_1}{C_1 + C_2} \qquad (15.109)$$

Then,

$$-h_{fb} = \frac{h_{fe}}{1 + h_{fe}} = \frac{C_1}{C_1 + C_2} \qquad (15.110)$$

Equation 15.110 states that $h_{fe}/(1 + h_{fe})$ must be greater than or equal to $C_1/(C_1 + C_2)$ for oscillation to occur.

15.4 **SUMMARY**

Amplifiers designed for the transfer of large quantities of power to loads such as audio speakers also result in the dissipation of power in the amplifying devices. If we use the same techniques in power amplifiers that we had previously introduced for small-signal amplifiers (360° conduction), the maximum efficiency of the power amplifier (Class A) is 25%. Special connections using pairs of transistors biased so that they conduct for only 180° (Class B) can boost the maximum efficiency to 78.5%. Crossover distortion in Class B amplifiers can be eliminated by dissipating more power in a Class AB configuration in which the transistors conduct for more than 180° but less than 360°.

Power amplifier efficiency can be improved significantly if the output signal is restricted to a narrow range of frequencies. An amplifier with a resonant circuit load is one way of accomplishing this narrow-band process. Class B or Class C (conduction less than 180°) amplifiers provide energy to the tuned circuit for part of the cycle, and the resonant circuit makes the current continuous if the frequency of the tuned circuit and the frequency of the input signal are equal.

Tuned amplifiers are also used in small-signal applications in communications systems. Unfortunately, internal feedback in transistors produces unwanted oscillations in tuned amplifiers. Neutralization is one technique that can be used to prevent oscillations. Cascading of tuned amplifiers provides interesting techniques for shaping the pass band of a communications receiver.

Special circuits in which there is positive feedback at a particular frequency are called oscillators. They act as sources of sinusoidal constant-amplitude, constant-frequency signals.

The common property of these different types of circuits is single-frequency operation. Power amplifiers can operate at high efficiency if the load is a resonant circuit. Tuned amplifiers have the possibility of oscillation due to internal feedback, while oscillators require positive feedback at a single frequency for proper operation.

REFERENCES

1. P. R. Gray and R. G. Meyer, *Analysis and Design of Analog Integrated Circuits* (New York: Wiley, 1977).

2. K. K. Clark and D. T. Hess, *Communication Circuits: Analysis and Design* (Reading, Mass.: Addison-Wesley, 1971).

3. E. J. Angello, *Electronic Circuits,* 2nd ed. (New York: McGraw-Hill, 1964).

4. M. V. Joyce and K. K. Clark, *Transistor Circuit Analysis* (Reading, Mass.: Addison-Wesley, 1961).

PROBLEMS

15.1. Calculate the power efficiency η of the circuit in Figure 15.2 if V_{CC} = 10 V, h_{fe} = 100, h_{oe} = h_{re} = 0, R_E = 50 kΩ, R_{B1} = 125 kΩ, and R_{B2} = 83 kΩ. The input signal is v_{IN} = 1.0 sin ωt V.

15.2. Show that the maximum efficiency of a Class A amplifier with a transformer-coupled load is 50%.

15.3. Calculate the maximum power efficiency of the common-emitter circuit shown in Figure 15.27.

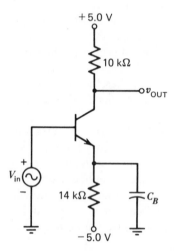

Figure 15.27. Circuit for Problem 15.3.

15.4. The maximum theoretical power efficiency of the circuit shown in Figure 15.2 is 25% (Equation 15.9). Assume that the circuit is biased for maximum symmetrical swing and has V_{CC} = 15 V, h_{fe} = 50, and R_E = 500 Ω. Calculate the maximum theoretical power efficiency for this circuit if the effect of power dissipation in the bias resistors is now included.

15.5. The maximum power rating of a transistor $P_Q(max)$ can be related to the circuit Q-point by a hyperbolic plot of $V_{CE}I_C = P_Q(max)$ on the I_C-versus-V_{CE} plane. Figure 15.28a shows this plot of a locus of transistor operating points that dissipate the maximum rated power of the transistor. Figure 15.28b shows a common-collector circuit with R_E = 8.0 Ω and V_{CC} = 30 V. The maximum collector current rating is $I_C(max)$ = 3.0 A. Find the Q-point of the circuit so that maximum power is delivered to the load.

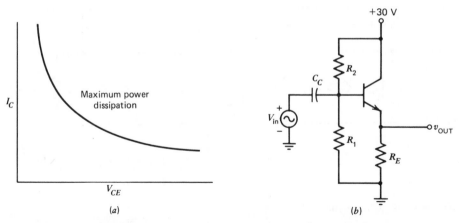

Figure 15.28. (a) Maximum power dissipation locus. (b) Circuit for Problem 15.5.

15.6. A complementary pair Class B amplifier is shown in Figure 15.6a. If V_{CC} = 10 V and V_{BE}(cut-in) = V_{BE}(act) = 0.70 V for Q_1 and V_{EB}(cut-in) = V_{EB}(act) = 0.7 V for Q_2, plot a transfer curve of v_{OUT} versus v_{IN} for $-10 \leq v_{IN} \leq +10$ V. Assume that a 0.1 V difference exists between the collector and emitter terminals at saturation for the two transistors.

15.7. The circuit shown in Figure 15.6a is driven by a sine wave v_{IN} = 2.0 sin ωt V. Assume that V_{BE}(cut-in) = 0.65 V and V_{BE}(act) = 0.70 V. (a) What is the duty cycle of each transistor, and (b) is this "Class B" configuration really a Class B circuit?

15.8. Design a common-emitter tuned load amplifier such as that shown in Figure 15.9. (a) Calculate R, L, and C so that f_o = 10 MHz and Q = 50. (b) If h_{fe} = 100, h_{ie} = 300, and h_{oe} = h_{re} = 0, what is the voltage gain A_v = V_{out}/V_{in} at resonance?

15.9. The transistor parameters for the tuned load common-emitter circuit of Figure 15.10a are given in the numerical example in Section 15.2.2. If A_v = 100 $\angle 115°$, f_o = 150 MHz, and the bandwidth is 15 MHz, find R, L, and C of the circuit.

15.10. The small-signal high-frequency model of a transistor amplifier with an inductively coupled load is shown in Figure 15.29. Find the composite admittance parameters treating the circuit inside the dashed rectangle as a two-port network.

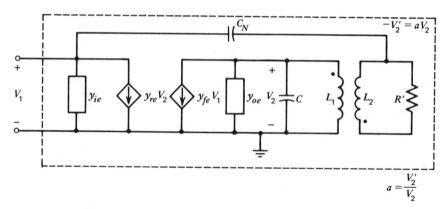

Figure 15.29. Circuit for Problem 15.10.

15.11. Derive an expression for the triple-point bandwidth of a dual-peak stagger-tuned amplifier.

15.12. Design the phase shift network shown in Figure 15.30 to cause oscillation at 100 kHz. Design each section of the RC ladder network to shift the signal by 60°. Scale the magnitudes of the resistors and capacitors in the feedback network such that insignificant loading occurs between RC sections.

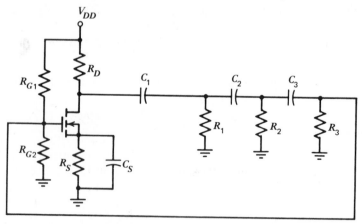

Figure 15.30. Circuit for Problem 15.12.

15.13. In the Colpitts oscillator of Figure 15.23, $C_1 = 1.0$ nF, $C_2 = 10$ nF, $V_{CC} = 10$ V, $L = 100$ μH, $R_C = 3.0$ kΩ, $R_E = 100$ Ω, $R_1 = 180$ kΩ, and $R_2 = 18$ kΩ. The coupling and the bypass capacitor have negligible reactance at the frequency of oscillation. Calculate the oscillation frequency.

15.14. Design a Colpitts oscillator to resonate at 10 MHz. Let the inductor be approximately 100 μH and $V_{CC} = 10$ V. Use the circuit in Figure 15.23.

chapter 16 _____

ANALOG BIPOLAR INTEGRATED CIRCUIT EXAMPLES

The basic theory of analog circuitry has been presented in Chapters 10 through 15. The results of this theory are best illustrated in an integrated circuit (IC) that uses several subcircuits to achieve an overall function. In this chapter, we analyze two important examples of analog integrated circuits and show how large complex circuits can be viewed as the combination of smaller units. In the first example, we examine the bipolar operational amplifier (OP AMP), which is the most popular analog integrated circuit from the standpoint of sales volume. We also consider an integrated circuit version of a voltage regulator.

16.1 A BIPOLAR OPERATIONAL AMPLIFIER

This circuit had its origin in the analog computer. The OP AMP was originally developed to perform a variety of algebraic and calculus operations on electronic waveforms. Properties desired for an ideal OP AMP are infinite input impedance, infinite voltage gain, infinite bandwidth, zero-output impedance and zero dc voltage offset. The commercial OP AMP examined below does not achieve the ideal, but does have a voltage gain in excess of 200,000, input impedance greater than 1 megohm, bandwidth greater than 1 megahertz, output impedance of tens of ohms, and voltage offsets in the millivolt range.

In general, OP AMPs can be considered as consisting of three parts: an input stage, one or more gain stages, and an output stage. The input stage is a differential amplifier with a single-ended output. The gain stages provide power and voltage gain. The output stage adjusts the dc level of the signal and provides a low output impedance. The symbol for the OP AMP is shown in Figure 16.1a. The inverting and noninverting input nodes correspond to those expected for a differential amplifier. In contrast to the differential amplifier, the OP AMP has an input impedance in the megohm range and a much higher voltage gain.

Figure 16.1b shows an OP AMP with a signal source V_{in} and two impedances, Z_1 and Z_2, attached to the external terminals. As we shall show, the OP AMP with negative feedback becomes a circuit dominated by the impedance functions of the feedback and input elements Z_2 and Z_1. The output voltage V_{out} is related to the difference in input voltage ΔV by

$$V_{out} = -A_0 \, \Delta V \qquad (16.1)$$

where A_0 is the relatively large dc or low-frequency voltage gain of the OP AMP. Since values of A_0 are in excess of 200,000, the signals in the microvolt range drive V_{out} to the limits of the circuit power supply.

Negative feedback in the circuit of Figure 16.1b reduces the overall voltage gain V_{out}/V_{in} but does not alter Equation 16.1. The result is that V_{in} and V_{out} may assume nominal operating voltages but ΔV assumes microvolt values. The ΔV is so low that the input terminals are said to be virtually shorted. In Figure 16.1b, the positive or noninverting terminal is grounded, allowing the assumption that the inverting terminal is also at ground potential. This is called a *virtual ground*. The virtual ground concept allows a simple analysis of the voltage gain of the OP AMP circuit of Figure 16.1b.

Since the input impedance at the OP AMP terminal is very large, the current through Z_1 is approximately the same as that through Z_2. We may use this and the virtual ground concept to write

$$I_1 = \frac{V_{in}}{Z_1} = I_2 = -\frac{V_{out}}{Z_2} \qquad (16.2)$$

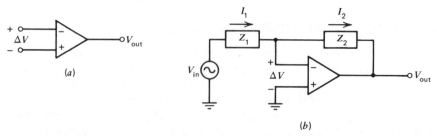

(a)

(b)

Figure 16.1. (a) Operational amplifier symbol. (b) Noninverting operational amplifier configuration with feedback network Z_2 and input network Z_1.

and the voltage gain A_v becomes

$$A_v = \frac{V_{out}}{V_{in}} = -\frac{Z_2}{Z_1} \qquad (16.3)$$

The simplicity of designing with OP AMPS is seen in Equation 16.3. The circuit transfer function is achieved by manipulation of the impedances Z_1 and Z_2. Resistors allow exact voltage scaling between V_{out} and V_{in}. The use of capacitors, inductors, and resistors allows the design of active filters having voltage gains greater than 1. Integral and differential operations are accomplished with simple resistor and capacitor elements, allowing a valuable versatility in signal wave shaping. In this section, our purpose is not to describe the myriad applications of OP AMP circuits but to provide a background for examining the details of the OP AMP itself.

The integrated circuit implementation of the OP AMP represents one of the major advances in the history of analog electronics. Not only does the circuit perform large varieties of commonplace functions, but the designer is freed from bias design and circuit impedance buffering. The integrated circuit of the OP AMP is examined in the following paragraphs.

Integrated circuit versions of the OP AMP have taken slightly different forms using bipolar, JFET, or MOSFET devices. However, all OP AMPs incorporate the basic subcircuits of (1) an input differential amplifier stage, (2) an intermediate gain stage, (3) an output stage, and (4) a biasing network.

The general-purpose OP AMP circuit shown in Figure 16.2 is the most popular analog integrated circuit in history. This particular circuit is designated the 741 OP AMP. The circuit has an apparent complexity that might frustrate a beginning student, but the circuit as shown evolved over a four-year time period and was initially much less complex in appearance. The circuit of Figure 16.2 may be divided into its four functional subunits. In addition to the four basic sections, certain transistors are included that provide overload protection and do not directly support voltage amplification.

The differential input stage consists of transistors Q_1 through Q_8. Transistors Q_1 and Q_2 form composite structures with the lateral *pnp* transistors Q_3 and Q_4. Transistors Q_5 through Q_7 form a current mirror active load, with the output of the differential stage taken from the collectors of Q_4 and Q_6. Transistor Q_8 provides current source biasing to Q_1 and Q_2 in a current mirror configuration with Q_9.

The lateral *pnp* transistors provide some voltage gain as common-base configurations, but, more importantly, provide a positive to negative level shifting of the signal. The dc level of the signal is taken from zero volts on the base of Q_1 or Q_2 to almost $-V_{EE}$ on the collectors of Q_4 and Q_6. Actually, the dc voltage at $V_C(6)$ is approximately two diode voltage drops above $-V_{EE}$, as provided by the base–emitter junctions of Q_{16} and Q_{17}. The negative dc level shifting is required to compensate for the rise in dc voltage when the signal is passed through the intermediate *npn* transistor amplification stage.

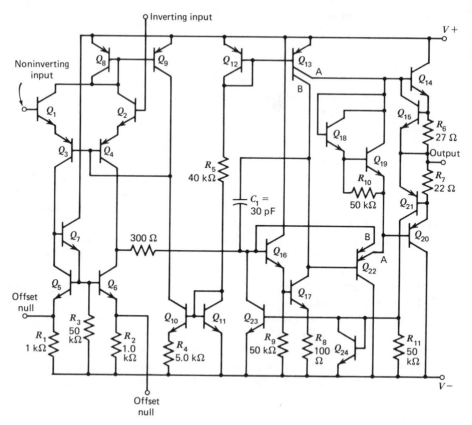

Figure 16.2. Schematic of 741-type operational amplifier.

The intermediate amplification stage contains transistors Q_{16}, Q_{17}, and Q_{13}. Transistor Q_{16} is an emitter–follower that reduces the loading on the collector of Q_6. Transistor Q_{17} is a common-emitter stage with a 100-Ω emitter degeneration resistor. Transistor Q_{13} is a multicollector lateral *pnp* device that acts as an active load for Q_{17} and passes the signal to the output stage.

The output stage consists mainly of the complementary pair Q_{14} and Q_{20}. Transistor Q_{22} is a multiemitter device whose *A* emitter lead performs as an emitter–follower in transferring the signal from the collector of Q_{17} to the base of Q_{20}. Transistors Q_{14} and Q_{20} operate in a Class AB complementary fashion. Both devices are biased just above cutoff by the network formed by Q_{18} and Q_{19}. In this operating region, transistor Q_{14} is cut off by a negative going signal but passes a positive going signal in its forward active region. Transistor Q_{20} operates in a manner opposite to that of Q_{14}, in that it passes negative signals and is cut off for positive signals. This configuration for Q_{14} and Q_{20} is a "push–pull" connection (see Chapter 15) and results in a higher circuit efficiency.

Transistors Q_{18} and Q_{19} represent a series connection between the bases of Q_{14} and Q_{20}. The voltage drop between the two bases is then

$$V_{B14} - V_{B20} = V_{BE18} + V_{BE19} \tag{16.4}$$
$$\simeq 1.4 \text{ V}$$

This voltage is also impressed across the parallel path through Q_{14} and Q_{20}, or

$$1.4 = V_{BE14} + I_{E14}R_6 + I_{E20}R_7 + V_{EB20} \tag{16.5}$$

The voltage drops across R_6 and R_7 are only a few millivolts for nominal conditions, so that

$$V_{BE14} + V_{EB20} \simeq 1.4 \text{ V} \tag{16.6}$$

The voltage excursions at the bases of Q_{14} and Q_{20} are large enough to easily turn these transistors on or off in the push–pull mode. This type of circuit is a Class AB amplifier to denote a circuit that is biased just above cutoff. This is in contrast to a Class A amplifier that always operates in its forward active region and a Class B amplifier that is biased at its cutoff point. Amplifier classification is treated in more detail in Chapter 15.

The transistors in Figure 16.2 that have not been discussed thus far are either for the biasing network or for overload protection. Transistors Q_{15} and Q_{21} provide short-circuit protection for the output stage. These transistors divert current from Q_{14} and Q_{20} when the voltage drop across R_6 or R_7 is large enough to turn on Q_{14} or Q_{20}. The B emitter connection in Q_{22} that is connected to the base of Q_{16} protects Q_{17} under an overdrive condition. If Q_{17} saturates, the B base–emitter circuit of Q_{22} turns on and diverts current from the base of Q_{16} in a negative feedback connection. Transistors Q_{23} and Q_{24} provide negative feedback to Q_{16}, for the condition of strong negative overdriving of the output. Base current from Q_{16} is diverted through Q_{23} when the output current becomes large and negative.

The capacitance C_1 across the base of Q_{16} and the collector of Q_{17} imparts a phase shift to the signal that stabilizes the circuit against oscillations. At high frequencies, the normal phase shift of a signal may become 360°. If the loop gain at that frequency is unity or greater, self-sustained oscillations will occur. The 30-pF capacitor is a value that allows adequate phase margin for stability while providing a gain–bandwidth product near 1 MHz. We now look at more detail of the subunits of the circuit.

16.1.1 DC Analysis

The central biasing network in Figure 16.2 is found in transistors Q_{12}, Q_{11}, Q_{10}, and Q_9 and in resistors R_5 and R_4. Resistor R_5 and transistors Q_{12} and Q_{11} set up a master bias current through R_5. The other subcircuits of the OP AMP scale individual bias currents from that master current, using current mirror circuits. A dc analysis of the 741 amplifier proceeds directly from the current in R_5:

$$I_{R5} = \frac{V_{CC} - (-V_{EE}) - [V_{EB12} + V_{BE11}]}{R_5}$$

$$= \frac{15 + 15 - (0.6 + 0.6)}{40 \times 10^3} \tag{16.7}$$

$$= 720 \ \mu A$$

The differential amplifier subcircuit derives its bias current by scaling I_{R5} through Q_{11}, Q_{10}, and R_4. The relationship of I_{C10} to I_{R5} was derived in Chapter 10 and is

$$I_{C10} = \left(\frac{1}{R_4}\right)\left(\frac{kT}{q}\right) \ln \left(\frac{I_{R5}}{I_{C10}}\right) \tag{16.8}$$

or

$$I_{C10} = \left(\frac{1}{5.0 \times 10^3}\right)(26 \times 10^{-3}) \ln \left(\frac{720 \times 10^{-6}}{I_{C10}}\right) \tag{16.9}$$

An iterative solution of this implicit equation yields

$$I_{C10} \simeq 18.9 \ \mu A$$

Neglecting the base currents out of Q_3 and Q_4 allows us to write

$$I_{C10} \simeq I_{C9} \tag{16.10}$$

Since Q_8 and Q_9 form a mirror,

$$I_{E8} = I_{E9} \simeq I_{C9} \tag{16.11}$$

$$\simeq 18.9 \ \mu A$$

The emitters of Q_1 and Q_2 split the emitter current of Q_8 equally, giving

$$I_{C1} = I_{C2}$$

$$= \frac{I_{E8}}{2} = \frac{18.9}{2} \tag{16.12}$$

$$\simeq 9.46 \ \mu A$$

Neglecting base currents in the differential circuit allows us to write

$$I_{C1} = I_{C3} = I_{C4} = I_{E5} = I_{E6} \tag{16.13}$$

$$\simeq 9.46 \ \mu A$$

The emitter resistors R_1 and R_2 are equal, therefore the emitter currents I_{E5} and I_{E6} are equal.

The emitter current I_{E7} can be estimated by neglecting I_{B5} and I_{B6} and determining the current in R_3

$$I_{R3} \simeq I_{E7}$$

$$= \frac{I_{E6}R_2 + V_{BE6}}{R_5}$$

$$= \frac{(9.46 \times 10^{-6})(1.0 \times 10^3) + 0.6}{50 \times 10^3} \qquad (16.14)$$

$$= 12.2 \ \mu A$$

Resistor R_3 is used to bias Q_7 at a current level approximately equal to that of Q_5 and Q_6.

The biasing of the intermediate gain stage Q_{16}, Q_{17}, and Q_{13} presents an unusual situation. Transistor Q_{16} derives its bias current from the voltage drop across R_9, that is,

$$V_{E16} = V_{BE17} + I_{E17}R_8 \qquad (16.15)$$

Since I_{E17} is set at approximately 540 μA (discussed below), I_{E16} becomes

$$I_{E16} = \frac{V_{BE17} + I_{E17}R_8}{R_9}$$

$$= \frac{0.7 + (0.54 \times 10^{-3} \times 100)}{50 \times 10^3} \qquad (16.16)$$

$$= 15 \ \mu A$$

Transistor Q_{17} obtains its collector bias current from the *pnp* current source (and active load) Q_{13}. Transistor Q_{13} forms a diode mirror current source with Q_{12}. The sum of the collector currents in the A and B collector terminals is equal to I_{C12} if we neglect base currents in Q_{12} and Q_{13}. We should be aware that these lateral *pnp* transistors have low values of β ($\beta \simeq 10$), which may reduce the accuracy of the answer by as much as 10%. The collector currents I_{CA13} and I_{CB13} assume values proportional to the area of each respective collector region in Q_{13}. The area of collector B is approximately three times larger than the area of collector A. Therefore

$$I_{CB13} \simeq 0.75 I_{C12}$$

$$= 0.75 \times 720 \times 10^{-6} \qquad (16.16)$$

$$= 540 \ \mu A$$

and

$$I_{CA13} \simeq 0.25 I_{C12}$$

$$= 0.25 \times 720 \times 10^{-6} \qquad (16.17)$$

$$= 180 \ \mu A$$

Transistor Q_{17} then has a collector current $I_{C17} = I_{CB13} = 540 \ \mu A$.

This dc analysis of the 741 amplifier assumes that the input and offset voltages are zero. In an amplifier with negative feedback, these assumptions are valid. The output stage biasing assumes a zero voltage at the output node. Transistors Q_{18}, Q_{19}, and Q_{22A} take most of the current I_{C17}. The output transistors Q_{14} and Q_{20} take collector current through V_{CC} and V_{EE}, respectively. This collector current in Class AB operation is approximately 150 μA, requiring a more detailed analysis.

The dc node voltages in the circuit can be calculated from the currents in the circuit and by estimating voltages for forward-biased base–emitter junctions.

16.1.2 **AC Analysis**

A detailed manual analysis of a circuit as large as the 741 amplifier would not be worthwhile. Computer models allow accurate calculations but do not provide the analytical expressions necessary for a designer. The analytical expressions of the circuit are handled in two ways. First, the 741 circuit may be broken down into small subcircuits, with analytical expressions written for each subcircuit. The designer may then optimize the performance of each subcircuit. Second, the 741 circuit may be treated as a whole but with simplifying assumptions made for the subcircuits. In this approach, insight may be gained to the overall functioning of the 741 circuit. In this analysis, we will look at the circuit as a whole with simplifying assumptions.

Figure 16.3 illustrates the central circuitry involved in the ac signal amplification. The differential stage consists of a *pnp* differential pair feeding a simple diode mirror active load. The intermediate gain stage is simplified as is the output stage. Figure 16.4 simplifies the circuit further, eliminating the bias-

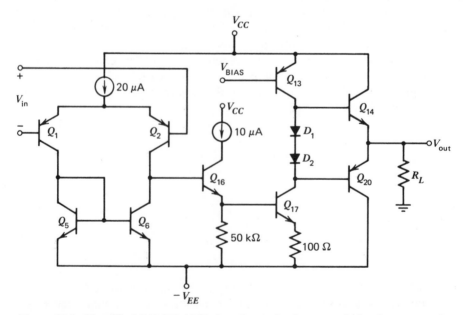

Figure 16.3. Simplified 741 OP AMP showing major features of circuit concerned with ac signal pathways.

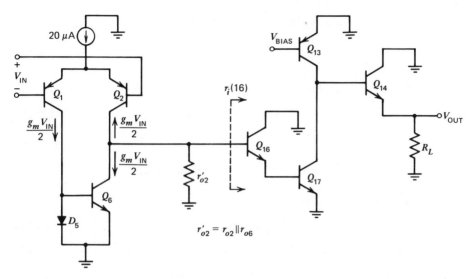

Figure 16.4. Further simplied circuit configuration of Figure 16.3. The biasing diodes D_1 and D_2 and one of the output transistors, Q_{20}, have been removed.

ing diodes and one output transistor, Q_{20}. Only one output transistor, Q_{14}, is shown since the other output transistor, Q_{20}, is the complement of Q_{14}. The effective output resistance of the differential pair is $r_{o2} \| r_{o6}$. We shall use a simplified hybrid parameter model of the transistors as shown in Figure 16.5.

In the circuit of Figure 16.4, the collector currents in Q_1 and Q_2 for a differential input voltage of V_{in} becomes

$$I_{c1} = -I_{c2} = \frac{g_m V_{in}}{2} \tag{16.18}$$

In our assumptions, the output resistances of Q_2 and Q_6 are lumped as r'_{o2}. The output resistance of Q_1 is ignored, although r_{o1} certainly represents a loss factor and must be kept as large as possible. The r'_{o2} is also a loss factor but, as we shall see, lies partly under control of the circuit designer. The I_{c6} is a mirror current of I_{c1} as shown, and the effective input resistance looking into the base

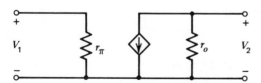

Figure 16.5. Low-frequency hybrid-π used for ac analysis of 741 operational amplifier.

of Q_{16} is designated r_{i16}. The current into the base of Q_{16} is

$$I_{b16} = -g_m V_{in} \frac{r'_{o2}}{r'_{o2} + r_{i16}} \tag{16.19}$$

This current is multiplied as it passes through Q_{16} and Q_{17} to give I_{c17} as

$$I_{c17} \simeq -\beta_{16}\beta_{17}I_{b16} \tag{16.20}$$

or

$$I_{c17} \simeq -g_m V_{in} \frac{r'_{o2}}{r'_{o2} + r_{i16}} \beta_{16}\beta_{17} \tag{16.21}$$

The output voltage then becomes

$$V_{out} = I_{e14}R_L \tag{16.22}$$

$$\simeq \beta_{14}I_{b14}R_L \tag{16.23}$$

If the output resistance of Q_{13} is very high, then we may assume $I_{b14} \cong I_{c17}$ and

$$V_{out} \simeq g_m V_{in} \frac{r'_{o2}}{r'_{o2} + r_{i16}} \beta_{16}\beta_{17}\beta_{14}R_L \tag{16.24}$$

The differential gain of the OP AMP is then

$$A_v = \frac{V_{out}}{V_{in}} \simeq g_m \frac{r'_{o2}}{r'_{o2} + r_{i16}} \beta_{16}\beta_{17}\beta_{14}R_L \tag{16.25}$$

This expression for the OP AMP interprets the differential input stage as a transconductance amplifier delivering an output current of approximately $g_m V_{in}$. The intermediate gain stage and the output transistor multiply this current by approximately the product $\beta_{16}\beta_{17}\beta_{14}$. The transistors Q_{16}, Q_{17}, and Q_{14} simply multiply the signal current by β^3 and then deliver this current to a load resistor to obtain an output voltage.

The calculated voltage gain from Equation 16.25 is quite large. Let us neglect the output resistance of the differential stage and estimate A_v. Let $I_{c1} = I_{c2} = 10 \ \mu A$ and $\beta \simeq 100$. Then

$$g_m \simeq \frac{\beta}{r_\pi} = \frac{\beta}{\beta(kT/q)/I_c} = \frac{I_c}{kT/q}$$

$$= \frac{10 \times 10^{-6}}{26 \times 10^{-3}} \tag{16.26}$$

$$= 385 \ \mu S$$

For a load resistor of 1.0 kΩ, Equation 16.25 becomes

$$A_v = (385 \times 10^{-6})(100^3)(1.0 \times 10^3) \tag{16.27}$$

$$\simeq 385 \ k$$

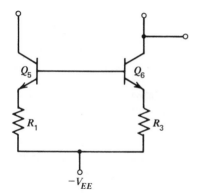

Figure 16.6. Active load portion of differential input stage of 741 operational amplifier.

This value is larger then the nominal 200×10^3 value of voltage gain for a 741 amplifier, but it illustrates the theoretical gain possible. Breadboard rather than IC designs for the 741 have achieved circuit voltage gains close to the theoretical value. The primary degradations to gain in the IC version are thermal gradients.

The output resistance of the differential stage is increased by inclusion of an emitter resistor in Q_6 as shown in Figure 16.6. An equal value of resistance is included in the emitter of Q_5 to balance the bias currents. An analysis of the output resistance for the circuit of Figure 16.6 (see Chapter 10) gives

$$ r_{o6} = \frac{1}{h_{oe}} + R_3 \| h_{ie}' + \frac{R_3}{R_3 + h_{ei}'} \frac{\beta}{h_{oe}} \tag{16.28}$$

where $h_{ie}' = h_{ie6} + r_{i5}$ and r_{i5} is the input resistance looking into the base of Q_5. The presence of R_3 considerably increases the output resistance seen looking into the collector of Q_6. In the 741 amplifier circuit of Figure 16.2, the output resistance of Q_4 (and Q_3) is increased by the presence of the resistance attributed to Q_2 (and Q_1).

The bipolar integrated circuit operational amplifier has made a major contribution to electronics. The 741 OP AMP was the first commercial success in the analog IC industry. Even though more recent circuits have surpassed the performance of the 741, it is instructive to study the circuit design advances that made this circuit possible. The dc bias system with active *pnp* loads and Class AB output stage made possible the high gain stages necessary to allow a single capacitor for frequency compensation. This type of circuit makes use of matched transistor characteristics that are only possible on a mass production basis with integrated circuit processing.

16.2 **VOLTAGE REGULATORS**

Voltage regulators are an integral part of all voltage power supplies. The voltage regulator is specifically designed to hold an output voltage at a specific value. This condition must be held in spite of changes in temperature, input voltage

changes, or changes in the output load. Integrated circuit versions were introduced in the mid-1960s, allowing inexpensive three-terminal voltage regulators
to be part of individual circuit boards of a system. These regulators were
significant accomplishments of integrated circuit technology.

In this section, we first examine the general features of a voltage regulator
and then look at a specific IC version. Figure 16.7 shows the main features of
a voltage regulator. The essence of the regulator is the combination of a voltage
reference, an error amplifier, and a series pass transistor. These elements are
configured in a negative feedback loop through R_A and R_1 to hold the output
voltage constant. In addition, a startup circuit is necessary to assure that the
circuit is placed in its designed operating point and thermal shutdown and output
short-circuit protection are included to protect the circuit against excessive
loading. The pass transistor is designed to handle watts of power and consumes
a large area of the regulator chip. The current source is made of *pnp* transistors.

The voltage reference circuit is carefully designed to provide a voltage
that is quite insensitive to changes in either the temperature or the input voltage
to the regulator. Zener diodes were first used as voltage references, but the
avalanche mode is a strong source of noise and required a breakdown voltage
that was unsuitably high for IC regulators. A very stable voltage reference was
designed by Robert Widlar and is shown in Figure 16.8. The design is called
a band gap reference circuit since its output voltage is that of the silicon band
gap potential. The output reference voltage V_{REF} is

$$V_{REF} = V_{CB3} + V_{BE3}$$
$$= I_{C2}R_2 + V_{BE3}$$

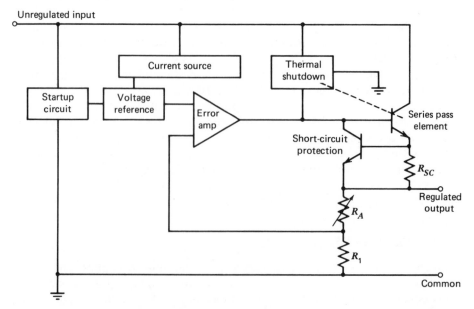

Figure 16.7. Essential features of a voltage regulator circuit.

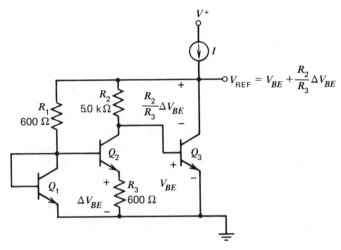

Figure 16.8. Energy band gap voltage reference circuit.

or

$$V_{REF} = \frac{\Delta V_{BE}}{R_3} R_2 + V_{BE3} \qquad (16.29)$$

The reference voltage is then a temperature-dependent function of ΔV_{BE} and V_{BE}. As we shall see, the temperature coefficients of these two voltages are of opposite polarity, and proper adjustment of circuit parameters leads to a circuit operating point that is virtually independent of temperature change. The base–emitter voltage V_{BE} is a complex function of temperature, giving a coefficient of approximately -2 mV/°C. The ΔV_{BE} voltage represents the magnitude of the difference in base–emitter voltages of Q_1 and Q_2. This can be shown by first writing the Kirchhoff voltage law

$$V_{BE1} = V_{BE2} + \Delta V_{BE} \qquad (16.30)$$

or

$$\Delta V_{BE} = V_{BE1} - V_{BE2} \qquad (16.31)$$

Then,

$$\Delta V_{BE} = \frac{kT}{q} \ln\left(\frac{I_{C1}}{I_{s1}}\right) - \frac{kT}{q} \ln\left(\frac{I_{C2}}{I_{s2}}\right) \qquad (16.32)$$

or

$$\Delta V_{BE} = \frac{kT}{q} \ln\left(\frac{I_{C1}}{I_{C2}} \times \frac{I_{s2}}{I_{s1}}\right) \qquad (16.33)$$

Equation 16.31 shows that ΔV_{BE} represents the difference in base–emitter voltages of Q_1 and Q_2 while Equation 16.33 shows the strong positive temperature dependence of ΔV_{BE}. The reverse bias saturation currents are proportional to the area of the base–collector depletion region. Therefore, using current density, $J = I_C/\text{area}$, we may rewrite Equation 16.29 as

$$V_{\text{REF}} = \frac{kT}{q}\frac{R_2}{R_3}\ln\left(\frac{J_1}{J_2}\right) + V_{BE3}(T) \qquad (16.34)$$

Since the temperature coefficients of the two terms on the right-hand side of Equation 16.34 have opposite polarity, it is possible to derive conditions under which V_{REF} is independent of temperature. A detailed analysis, not given here, shows that V_{REF} will be temperature independent at room temperature at a value slightly higher than the band gap voltage of the semiconductor material (silicon). Thus, at room temperature ($T = 25°\text{C}$), output voltage temperature stability will be achieved for $V_{\text{REF}} = 1.262$ V. The variables R_2 and R_3 are adjusted to achieve this voltage. The band gap reference circuit is advantageous in IC form because it is temperature stable, adaptable to relatively low operating voltages, and does not generate excessive noise.

Figure 16.9 illustrates how the output voltage of the reference circuit may be increased to values above the silicon band gap voltage. The reference voltage in Figure 16.9 is

$$V_{\text{REF}} = V_{BE3} + V_{CB3} \qquad (16.35)$$

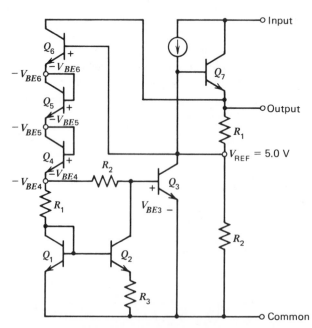

Figure 16.9. Energy band gap reference circuit with the reference voltage scaled higher than the band gap value.

where

$$V_{CB3} = V_{BE4} + V_{BE5} + V_{BE6} + \frac{R_2}{R_3}\frac{kT}{q}\ln\left(\frac{J_1}{J_2}\right) \quad (16.36)$$

Note that

$$I_{C1} \simeq \frac{V_{E4} - V_{B1}}{R_1} \quad (16.37)$$

and

$$I_{C2} \simeq \frac{V_{E4} - V_{B3}}{R_2} \quad (16.38)$$

Since V_{B1} and V_{B3} are both one diode drop above ground, $V_{B1} = V_{B3}$ and

$$V_{REF} = V_{BE3} + V_{BE4} + V_{BE5} + V_{BE6} + \frac{R_2}{R_3}\frac{kT}{q}\ln\left(\frac{J_1}{J_2}\right) \quad (16.39)$$

The room temperature-stable value for V_{REF} in Equation 16.39 can be shown to be a voltage that is four times the value of V_{REF} for the single base–emitter of Equation 16.34. Thus, $V_{REF} = 4 \times 1.262\text{ V} \simeq 5\text{ V}$. The use of three-diode connected transistors (Figure 16.9) has led to a band gap reference circuit voltage of 5 V. Reference voltages above 5 V are easily obtained by scaling the ratio of R_1 and R_2 in Figure 16.9. The output voltage then becomes

$$V_{OUT} = \frac{R_1 + R_2}{R_2}V_{REF} \quad (16.40)$$

Low temperature-coefficient, thin-film resistors are used to minimize the influence of resistors in the circuit on the output voltage.

An IC voltage regulator made by Fairchild Semiconductor Corporation is shown in Figure 16.10. This circuit requires only an input, output and common external pin connections. The three-terminal pin property allowed this circuit to be mounted in transistor packages. This complex circuit may be viewed as a sum of subcircuits, each performing a specific function. The reference voltage V_{REF} appears at the node between R_{19} and R_{20}. The series pass transistor Q_{17} is connected in a Darlington configuration with Q_{16}.

The reference circuit and error amplifier in Figure 16.10 are connected in a negative feedback loop through the pass transistor Q_{17}, R_{20}, and R_{19}. The reference circuit is composed of transistors Q_1 through Q_6 and resistors R_1 through R_3, with Q_3 providing the gain for regulation. The reference voltage is

$$V_{REF} = V_{BE6} + V_{BE5} + I_{C2}R_2 + V_{BE4} + V_{BE3} \quad (16.41)$$

Since

$$I_{C2} \simeq \frac{1}{R_3}\frac{kT}{q}\ln\left(\frac{J_1}{J_2}\right) \quad (16.42)$$

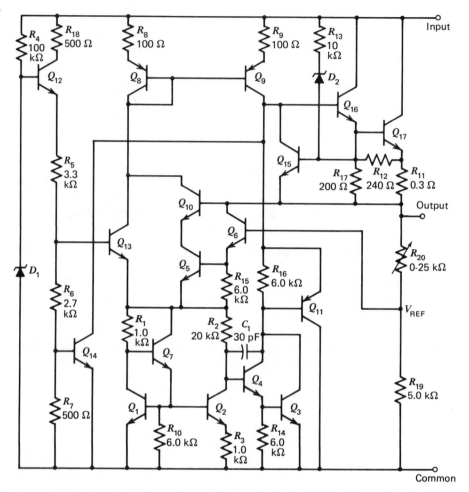

Figure 16.10. Circuit schematic of three terminal voltage regulator from the Fairchild Corporation.

then

$$V_{REF} = 4V_{BE} + \frac{R_2}{R_3}\frac{kT}{q}\ln\left(\frac{J_1}{J_2}\right) \tag{16.43}$$

The reference voltage is then stable at room temperature if $V_{REF} \simeq 5$ V and

$$V_{OUT} = \frac{R_{19} + R_{20}}{R_{19}}V_{REF} \tag{16.44}$$

Transistor Q_7 minimizes the base current drain from I_{R1} with the Q_7 bias current set by R_{10}. Transistor Q_{11} increases the effective gain of Q_3 as the controlled signal is passed to the active load of Q_9 and the base input of Q_{16}.

The negative feedback loop keeps V_{OUT} at a constant value. For example, if V_{INPUT} increased, the current in the bias transistors Q_8 and Q_9 would increase. The base drive to Q_{16} and Q_{17} would subsequently increase and attempt to force the output voltage higher through I_{E17}. As V_{OUT} and V_{REF} attempt to rise, Q_6 and Q_{10} would then increase the drive to Q_3. An increase in I_{C3} would draw current from the base of Q_{16} and hence reduce the attempt for V_{OUT} to increase.

A startup subcircuit includes transistors Q_{12} and Q_{13}, resistors R_4 through R_7, and Zener diode D_1. When power is first applied to the regulator, the Zener diode sets a positive voltage at the base of Q_{12}. Transistors Q_{12} and Q_{13} turn on and supply adequate drive current to the reference voltage transistors. When the regulator reaches its normal operating point, Q_{13} is cut off by the relative voltage values across its base–emitter junction. This cutoff voltage relationship at V_{BE13} may be shown from a dc analysis. The current density ratio J_1/J_2 is set at about 20, so that

$$I_{C2} \simeq \frac{1}{R_3} \frac{kT}{q} \ln \left(\frac{J_1}{J_2} \right)$$

$$= 78 \ \mu A$$

(16.45)

Then,

$$V_{E13} = V_{BE3} + V_{BE4} + I_{C2}R_2$$

$$\simeq 2.96 \ V$$

(16.46)

Transistor Q_{14} is designed to be cut off at room temperature with $V_{BE14} \simeq 0.4$ V. The voltage at V_{B13} is then

$$V_{B13} = \frac{R_6 + R_7}{R_7} V_{BE14}$$

$$\simeq 2.56 \ V$$

(16.47)

Figure 16.10 also contains a number of components that protect the circuit against overload conditions. Transistor Q_{14} acts as a temperature sensor that shuts down the output pass transistor when the chip temperature reaches 175°C. The base–emitter voltage of Q_{14} is set by R_7 to a value $V_{BE14} \simeq 0.4$ V. This is below the cutoff bias for Q_{14} at room temperature. As the chip temperature rises, V_{BE14} remains at about 400 mV, but the collector current rises dramatically. The major source of this current increment is the increase in free carriers within the transistor with elevated temperatures. Base current is then removed from Q_{16}, shutting down the pass transistor Q_{17} and allowing the chip to cool.

Protection against more short time-constant overloads is accomplished by R_{11} and Q_{15}. The regulator in Figure 16.10 has a maximum rated output of about 2.2 A. At this load current, the voltage across R_{11} is $V_{R11} = (0.3)(2.2) = 0.66$ V, which is sufficient to turn on Q_{15}. Base current is then diverted from Q_{16} through the collector of Q_{15}.

Diode D_2 and R_{13} protect the regulator against high input voltages. When the pass transistor Q_{17} exceeds 8 V across the collector–emitter, diode D_2 breaks down and Q_{15} is turned on to divert current from Q_{16} and Q_{17}.

The capacitor C_1 is inserted to stabilize the regulator against high-frequency oscillations. This is necessary since the feedback signal path in the regulator is passed through a high-gain configuration. The compensating capacitor C_1 ensures that any signal frequency that is fed back into the regulator amplifier will be of insufficient phase and amplitude to initiate an oscillatory condition.

The production of the IC voltage regulator removed a significant portion of circuit design for electronic projects. The inclusion of an IC regulator is not only inexpensive but the engineering concern with this problem now is measured in terms of minutes rather than days.

16.3 SUMMARY

Bipolar analog integrated circuits take many forms. In this chapter, we have discussed two representative types of integrated circuits: the operational amplifier and the voltage regulator. These circuits contain specific examples of the design philosophy associated with integrated circuits, and we are able to see major differences between discrete-component electronic design and integrated circuit design. Of particular interest are the *pnp* transistors in the 741 OP AMP, since the geometries of the individual devices differ enough to change their characteristics with respect to each other and to change the operation of the circuit. We also see composite transistor structures with multiple collectors, similar to those in the integrated injection logic circuits described in Chapter 7. The voltage regulator circuits represent a first step toward power integrated circuits, with one transistor representing almost all of the area of the circuit and a package identical to those used for discrete power transistors. Bipolar integrated circuits have dominated the marketplace, but MOS analog circuitry is becoming an important factor. We discuss these circuits in Chapter 17.

REFERENCES

1. B. A. Wooley, S. J. Wong, and D. O. Pederson, *IEEE J Solid-State Circuits* SC-6, 357 (1971).
2. J. E. Solomon, *IEEE J. Solid-State Circuits* SC-9, 314 (1974).
3. R. J. Widlar, *IEEE J. Solid-State Circuits* SC-4, 184 (1969).
4. P. R. Gray and R. G. Meyer, *Analysis and Design of Analog Integrated Circuits* (New York: Wiley, 1977).

chapter 17

ANALOG MOSFET INTEGRATED CIRCUITS

Analog applications of MOSFETs have grown with the need to incorporate analog and digital functions on the same integrated circuit chip. Both NMOS and CMOS analog circuits have been developed in analog-to-digital (A/D) and digital-to-analog (D/A) converters that require OP AMP-type circuits for the conversion process. In the circuits that require analog and digital functions, it is necessary to fabricate the analog and digital circuits from the same technology. While analog and digital functions could be combined on bipolar circuits, MOS technology has developed in this area because of its superior ability to fabricate very large-scale integrated (VLSI) circuitry. In addition, noise margins are better in MOS technology, making the MOSFET especially useful in certain high electrical noise environments, such as automobile engines.

In spite of the MOSFET desirability in VLSI, the MOSFET device has certain disadvantages when compared to the bipolar transistor. As we show below, the transconductance of the MOSFET is significantly lower than that of the bipolar transistor for equal bias currents. The OP AMP circuits constructed with MOSFETs do not compare favorably with the bipolar versions. The MOSFET OP AMPs are characterized by an open-loop gain that is approximately 100 times lower than the bipolar versions with the same number of stages. In addition, offset voltages and noise properties are inferior in the

MOSFET OP AMPs. The MOSFET OP AMP does possess a higher slew rate and bandwidth, but the overall performance of the bipolar analog circuits is better. Nevertheless, one reason for analog MOSFET emergence lies in the fact that MOSFET OP AMPs consume about 20% of the area of bipolar OP AMPs. The inferior performance of the MOSFET circuits is overridden by the ability of the MOSFET to perform a function adequately, even though not the best. MOSFET OP AMPs with open-loop gains of only 1000 are sufficient for many of the voltage conversion and filtering tasks required.

In this chapter, we examine some of the properties of basic analog MOS-FET circuit cells. We then study an OP AMP design that illustrates the similarities and differences between analog MOSFET and bipolar transistor circuit designs. It is advisable to review the MOSFET descriptions in Chapters 5 and 8 before proceeding.

17.1 ANALOG MOSFET DEVICE PROPERTIES

The MOSFET was introduced in Chapter 5, and the MOSFET inverter circuit was examined in Chapter 8. This chapter extends that work. The transistor operating point in analog applications is usually in the saturation region of operation, although occasionally a MOSFET is intentionally biased in the non-saturated state. In Figure 17.1, we show the characteristic curves of a MOSFET having a less than infinite drain-to-source resistance. Several of the device equations are restated here with an emphasis on the nonideal MOSFET.

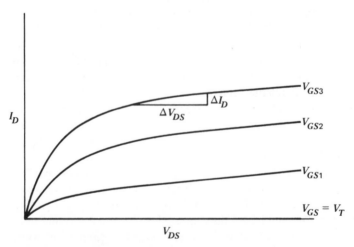

Figure 17.1. MOSFET characteristic curves showing the drain-to-source resistance as $r_{ds} = \Delta V_{DS}/\Delta I_D$ for a constant value of V_{GS}.

The electrical properties of a MOSFET in the saturated state are given by

$$I_D = \left(\frac{k_p}{2}\right)\left(\frac{W}{L}\right)\left(V_{GS} - V_T\right)^2\left(1 + \frac{\lambda}{L}V_{DS}\right) \qquad (17.1a)$$

and in the nonsaturated state by

$$I_D = \left(\frac{k_p}{2}\right)\left(\frac{W}{L}\right)\left[\left(V_{GS} - V_T\right)V_{DS} - V_{DS}^2\right]\left(1 + \frac{\lambda}{L}V_{DS}\right) \qquad (17.1b)$$

Here, the parameter, $k_p = \mu_n C_{ox}$, has been separated from the W/L ratio to preserve the identity of width-to-length ratio in the design of the MOSFET parameters. The finite drain–source resistance r_{ds} is modeled by the inclusion of the $(\lambda/L)V_{DS}$ term, where λ has the units of meters per volt. This resistance can be derived from Equation 17.1a by

$$r_{ds} = \frac{1}{\partial I_D/\partial V_{DS}} = \left(\frac{1}{\lambda}\right)\left(\frac{2}{k_p}\right)\left(\frac{L^2}{W}\right)\left(V_{GS} - V_T\right)^{-2} \qquad (17.2)$$

The r_{ds} term is often referred to as the channel-length modulation resistance and can also be seen as the reciprocal of the slope of the curves in Figure 17.1, or $r_{ds} = \Delta V_{DS}/\Delta I_D$. When the device is in saturation, the pinched-off or high-field portion of the channel increases in length as V_{DS} increases, reducing the length of the channel in the low-field region. The carrier velocity in the pinched-off portion is very high compared to that in the remainder of the channel, resulting in an incremental increase in I_D when the low-field portion is decreased in length. We can think of this situation as a resistor in series with an avalanching diode. The diode supports most of the voltage by widening its space charge region, but this in turn reduces the length and hence the resistance of the resistor. The impact of channel length L upon r_{ds} can be seen from the L^2 term in the numerator of Equation 17.2. The channel modulation coefficient λ can be determined experimentally and has typical values of $\lambda = 20 \times 10^{-6}$ m/V.

The body effect discussed in Chapter 5 is not included in this analysis, but it is significant and represents a degradation in the performance of a MOS-FET. When necessary, this degradation can be included in small-signal analysis by a modification of the transconductance term, g_m[1].

The transconductance is derived from Equation 17.1 as

$$g_m = \frac{\partial I_D}{\partial V_{GS}} = k_p\left(\frac{W}{L}\right)\left(V_{GS} - V_T\right)\left(1 + \frac{\lambda}{L}V_{DS}\right) \qquad (17.3)$$

Using Equation 17.1, this can be rewritten as

$$g_m = \frac{I_D}{(V_{GS} - V_T)/2} \qquad (17.4)$$

Equation 17.4 can be compared to the similar expression for the trans-

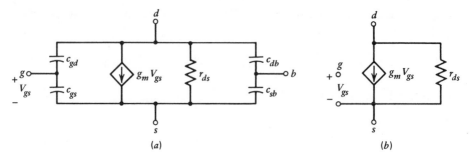

Figure 17.2. (*a*) MOSFET small-signal model. (*b*) Low-frequency small-signal model of MOSFET.

conductance of *bipolar* transistors, which is given by

$$g_m = \frac{I_C}{kT/q} \simeq \frac{I_C}{0.026} \tag{17.5}$$

The magnitude of the denominator in Equation 17.4 ranges from hundreds of millivolts to several volts. For equal drain and collector currents, the MOSFET transconductance is then significantly lower than that of the bipolar transistor. Although it is useful, in some cases, to maximize g_m in a MOSFET, it is never considered that MOSFET transconductance would exceed that of a bipolar transistor under similar bias current conditions.

The small-signal model for a MOSFET is shown in Figure 17.2*a*, where it is assumed that the body effect is incorporated into g_m. The four significant device capacitances are lumped as c_{gd}, c_{gs}, c_{db}, and c_{sb}, where the subscripts *g, d, s,* and *b* refer to gate, drain, source, and body (substrate), respectively. Reduction of these capacitances is a major focus in MOSFET fabrication. To analyze the circuits in this chapter, we use the low-frequency model of Figure 17.2*b*. It is particularly important to relate this figure to the expressions for g_m and r_{ds} (Equations 17.2, 17.3, and 17.4). A designer visualizes these equations at least as often as the small-signal circuit that they represent.

17.2 INVERTER ANALYSIS

The advantage of using a depletion-mode MOSFET load in an inverter circuit was demonstrated in Chapter 8. An ideal enhancement mode MOSFET load with gate connected to drain provides a load resistance of only $1/g_m$. In contrast, an ideal depletion mode MOSFET load with gate connected to source provides an infinite-load resistance. This can be shown using Figures 17.3*a* and 17.3*b*, where the small-signal models for the enhancement and depletion mode MOS-FETs are given. The resistance r_s seen at the source of the enhancement mode MOSFET load device in Figure 17.3*a* is derived from

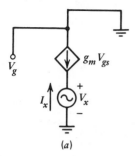 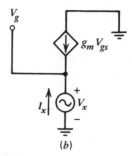

(a)   (b)

Figure 17.3. (a) Enhancement load MOSFET with $V_{gd} = 0$. (b) Depletion load MOSFET with $V_{gs} = 0$.

$$I_x = -g_m V_{gs} \qquad (17.6)$$

where

$$V_{gs} = -V_x$$

Then,

$$r_s = \frac{V_x}{I_x} = \frac{1}{g_m} \qquad (17.7)$$

The resistance seen at the source of the ideal depletion mode MOSFET in Figure 17.3b is infinity, since $V_{gs} = 0$. A nonideal MOSFET would include a parallel r_{ds} component, resulting in a resistance value at the source that is less than the ideal. However, it is clear that a depletion mode MOSFET load provides a much higher resistance than that due to an enhancement mode MOSFET load. The higher load resistance then leads to higher voltage gains in depletion load MOSFET inverters.

The voltage gain of the nonideal depletion mode load inverter shown in Figure 17.4a can be derived from the small-signal equivalent circuit of Figure 17.4b, where

$$V_{\text{out}} = -g_{m1} V_{\text{in}} (r_{ds1} \| r_{ds2}) \qquad (17.8)$$

and

$$A_v = \frac{V_{\text{out}}}{V_{\text{in}}} = -g_{m1}(r_{ds1} \| r_{ds2}) \qquad (17.9)$$

The input resistance of the inverter is in the multimegohm range and is normally neglected in the circuit model. The output resistance of the inverter lies in the tens-of-kilohms range and must always be considered in integrated circuit design.

The bias of a MOSFET requires careful considerations, even though the equations are straightforward. The inverter circuit of Figure 17.4a does not detail the gate–source bias voltage of M_1. In the saturated state, the M_1 transistor bias parameters are defined by

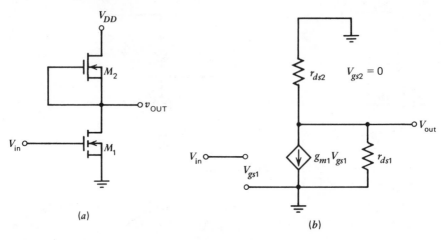

Figure 17.4. (a) MOSFET inverter circuit with depletion load. (b) MOSFET low-frequency small-signal model of part (a).

$$I_{D1} = \left(\frac{k_p}{2}\right)\left(\frac{W}{L}\right)_1 (V_{GS1} - V_{T1})^2 \left(1 + \frac{\lambda_1}{L_1}V_{DS1}\right) \qquad (17.10)$$

Let us neglect the channel length modulation effect ($\lambda_1 = 0$) to observe the interaction between I_{D1}, V_{GS1}, and $(W/L)_1$. Thus,

$$I_{D1} \approx \left(\frac{k_p}{2}\right)\left(\frac{W}{L}\right)_1 (V_{GS1} - V_{T1})^2$$

The MOSFET is in cutoff, saturation, or nonsaturation, depending upon the bias conditions. For $V_{GS1} > V_{T1}$, the transistor is on and is either in the saturation or nonsaturation region of operation. The transition point between saturation and nonsaturation is given by Equation 5.9 and occurs at

$$V_{GS1} = V_{DS1} + V_{T1} \qquad (17.11)$$

The MOSFET is in saturation for $V_{GS1} < V_{DS1} + V_{T1}$ and is in the nonsaturation mode for $V_{GS1} > V_{DS1} + V_{T1}$. These relations are quite important to consider in maintaining the proper bias conditions in an analog MOSFET design.

The inverter of Figure 17.4a is designed so that both transistors are in the saturation mode. This implies that

$$V_{GS1} < V_{DS1} + V_{T1} \qquad (17.12)$$

and

$$0 < V_{DS2} + V_{T2}$$

or

$$V_{DS2} > V_{T2} \qquad (17.13)$$

It is possible that M_1 or M_2 can be driven into their nonsaturated states by either poor bias design or a large input signal. A similar problem occurs in bipolar transistor circuit design, but, in bipolar transistors, it is the base and collector current magnitudes that are critical rather than gate, drain, and source voltages.

The depletion load inverter and its transfer function are shown in Figures 17.5a and 17.5b. It is normally desirable to bias the circuit at the point of maximum slope (Q_1), since $A_v = \partial V_{OUT}/\partial V_{IN}$. An example illustrates the biasing problem.

Example. For Figure 17.5, let $k_p = 20 \times 10^{-6}$ S/V, $\lambda_1 = \lambda_2 = 1.0 \times 10^{-6}$ m/V, $V_{DD} = 10$ V, $W_1 = 50$ μm, $L_1 = 20$ μm, $W_2 = 10$ μm, $L_2 = 100$ μm, $V_{T1} = +1.0$ V, and $V_{T2} = -3.0$ V. (a) Find V_{GS1} such that the circuit is biased at the optimum Q-point. (b) Find the voltage gain at the optimum bias point.

Solution. (a) There are two methods to estimate the value of V_{GS1} that biases the inverter at the midpoint. The first method equates the current in the two transistors, assuming that at the midpoint $V_{DS1} = V_{DS2} = 5.0$ V. Then, with both MOSFETs in saturation, Equation 17.1a provides

$$\left(\frac{k_p}{2}\right)\left(\frac{W}{L}\right)_2 (-V_{T2})^2\left(1 + \frac{\lambda}{L_2} V_{DS2}\right)$$
$$= \left(\frac{k_p}{2}\right)\left(\frac{W}{L}\right)_1 (V_{GS1} - V_{T1})^2\left(1 + \frac{\lambda}{L_1} V_{DS1}\right) \quad (17.14)$$

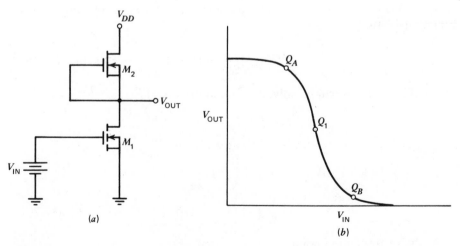

Figure 17.5. (a) Depletion load inverter circuit. (b) Voltage transfer curve of circuit in part (a).

Then,

$$V_{GS1} = \left[\left(\frac{W}{L} \right)_2 \left(\frac{L}{W} \right)_1 (-V_{T2})^2 \frac{1 + (\lambda/L_2)V_{DS2}}{1 + (\lambda/L_1)V_{DS2}} \right]^{1/2} + V_{T1} \quad (17.15)$$

or

$$V_{GS1} = 1.55 \text{ V}$$

The second method is useful when there is uncertainty on the part of the designer concerning the output voltage limits of the inverter. In this method, V_{GS1} is calculated at the two transition points in the transfer curve of Figure 17.5b. As V_{in} increases from zero, the first transition point occurs when M_1 is in saturation and M_2 is leaving the non-saturation (Q_A in Figure 17.5b). The second transition point occurs when M_2 is in saturation and M_1 enters the non-saturation state (Q_B in Figure 17.5b).

At Q_B, Equations 17.1a and 17.1b provide

$$\left(\frac{k_p}{2} \right) \left(\frac{W}{L} \right)_2 (-V_{T2})^2 \left(1 + \frac{\lambda}{L_2} V_{DS2} \right)$$
$$= \left(\frac{k_p}{2} \right) \left(\frac{W}{L} \right)_1 [2(V_{GS1} - V_{T1})V_{DS1} - V_{DS1}^2] \left(1 + \frac{\lambda}{L_1} V_{DS1} \right) \quad (17.16)$$

Also, the transition point relationships for M_1 and M_2 give

$$V_{DS1} = V_{GS1} - V_{T1} \quad (17.17)$$

and the circuit connection of M_2 gives

$$V_{DS2} = V_{DD} - V_{DS1} \quad (17.18)$$

When these relations and the numerical values are substituted into Equation 17.16, we obtain

$$0.9 \left[1 + \frac{(11 - V_{GS1})}{100} \right] = 2.5(V_{GS1} - 1.0)^2[1 + (V_{GS1} - 1.0)/20] \quad (17.19)$$

This cubic equation can be solved by iteration to yield

$$V_{GS1}(Q_B) \cong 1.62 \text{ V}$$

At Q_A, Equations 17.1a and 17.1b also provide

$$\left(\frac{k_p}{2} \right) \left(\frac{W}{L} \right)_2 [2(-V_{T2})V_{DS2} - V_{DS2}^2] \left(1 + \frac{\lambda}{L_2} V_{DS2} \right)$$
$$= \left(\frac{k_p}{2} \right) \left(\frac{W}{L} \right)_1 (V_{GS1} - V_{T1})^2 \left(1 + \frac{\lambda}{L_1} V_{DS1} \right) \quad (17.20)$$

At the Q_A transition point,

$$V_{DS2} = |V_{T2}| \quad (17.21)$$

and

$$V_{DS1} = V_{DD} - V_{DS2} = V_{DD} - |V_{T2}| \tag{17.22}$$

When Equations 17.21 and 17.22 and the numerical values are substituted into Equation 17.20, we obtain

$$0.927 = 3.375(V_{GS1} - 1)^2 \tag{17.23}$$

yielding

$$V_{GS1}(Q_A) = 1.52 \text{ V} \tag{17.24}$$

If we assume that Q_1 lies at the midpoint between Q_A and Q_B, we have

$$V_{GS1} = 1.57 \text{ V}$$

The bias network for this inverter must then provide $V_{GS1} = 1.57$ V in order to achieve an optimal bias point for the inverter.

(b) The voltage gain can be found using Equations 17.2, 17.3, and 17.9, yielding

$$g_{m1} = 3.56 \times 10^{-5} \text{ S}$$
$$r_{ds1} = 2.46 \text{ M}\Omega$$
$$r_{ds2} = 11.11 \text{ M}\Omega$$

and

$$A_v = -g_{m1}(r_{ds1}\|r_{ds2})$$
$$= -71.7$$

The bias statement of the problem asked for a value of V_{GS1}, given all the constants of the transistors. An alternate design problem might give a value for V_{GS1} and demand that one of the W/L ratios be adjusted to provide an optimum bias point. In this case, Equation 17.4 would be the appropriate design aid.

17.3 **VOLTAGE DIVIDERS**

A series of enhancement mode MOSFETs make suitable voltage divider networks for analog design. Figure 17.6 shows three MOSFETs connected as voltage dividers. Certainly, if all transistors were identical, $V_{DS1} = V_{DS2} = V_{DS3} = V_{DD}/3$. However, by adjustment of the individual W/L ratio, each transistor can be made to have a specific voltage drop. Consider the following example.

Example. In Figure 17.6, let $V_T = 1.0$ V, $k_p = 20 \times 10^{-6}$ S/V, $V_{DD} = +10$ V, $\lambda = 1.0$ μm/V, and $M_2 = M_3$, where $W_2 = W_3 = 20$ μm and $L_2 = L_3 = 30$ μm. If $L_1 = 30$ μm, find W_1 such that $V_{D1} = 2.5$ V.

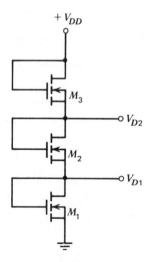

Figure 17.6. MOSFET voltage–divider network.

Solution. Since $M_2 = M_3$, then $V_{DS2} = V_{DS3}$ and $10 = 2V_{DS2} + V_{DS1}$. All three enhancement mode transistors are in the saturated region of operation since their gates are connected to their drains. This allows us to write

$$\left(\frac{k_p}{2}\right)\left(\frac{W}{L}\right)_2 (V_{GS2} - V_{T2})^2 \left(1 + \frac{\lambda}{L_2} V_{DS2}\right)$$

$$= \left(\frac{k_p}{2}\right)\left(\frac{W}{L}\right)_1 (V_{GS1} - V_{T1})^2 \left(1 + \frac{\lambda}{L_1} V_{DS1}\right) \quad (17.25)$$

Since $V_{GS} = V_{DS}$ and $V_{DS2} = (V_{DD} - V_{DS1})/2$, the substitution gives

$$W_1 = 5.0 \times 10^{-6} \left(\frac{8.0 - V_{DS1}}{V_{DS1} - 1.0}\right)^2 \frac{1 + [(10 - V_{DS1})/60]}{1 + (V_{DS1}/30)} \quad (17.26)$$

$$= 69.8 \ \mu m$$

MOSFETs offer an advantage in that there is no loading by stages that use the voltage divider.

17.4   CURRENT SOURCES

A MOSFET current source configuration is shown in Figure 17.7. This circuit uses a voltage divider (M_2, M_3) to control the gate–source voltage across the current source transistor M_1. The two major concerns with the current source are biasing parameters and output resistance. We first examine the bias design problem.

If M_1 and M_2 are identical, the source current I_{SS} exactly mirrors the drain current through M_2 and M_3. In this case, V_{GS1} is calculated from Equation

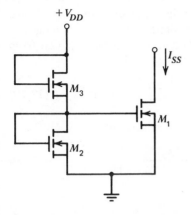

$+V_{DD}$

I_{SS}

M_3

M_1

M_2

Figure 17.7. MOSFET current–mirror network.

17.1a, choosing values of W_1 and L_1 to assist the design conditions. A typical situation keeps V_{GS1} as low as possible so that V_{D1} may drop to a low value without taking M_1 out of saturation. When V_{GS1} is determined and M_2 is made identical to M_1, then the M_2, M_3 voltage divider can be designed to provide the required value of $V_{DS2} = V_{GS1}$. The selection of W_3 and L_3 provides the tools to achieve the desired result.

In addition to exact mirroring of current between I_{SS} and I_{D3}, the current source of Figure 17.7 allows scaling of I_{SS} with respect to I_{D3}. This occurs by altering $(W/L)_1$ and $(W/L)_2$, as shown in the following example.

Example. For Figure 17.7, let $V_T = 1.0$ V, $k_p = 20 \times 10^{-6}$ S/V, $\lambda = 1.0 \times 10^{-6}$ m/V, and $V_{DD} = 10$ V. It is desired to set I_{SS} at 50 μA and to keep V_{GS1} as low as practical. Scale the currents so that $I_{D3} \simeq 20$ μA.

Solution. A value of $V_{GS1} = 1.5$ V is a practical low value and allows V_{D1} to drop to a minimum (before M_1 leaves saturation) of

$$V_{D1} = V_{GS1} - V_T$$
$$= 0.5 \text{ V}$$

We can now calculate a $(W/L)_1$ ratio that will achieve $I_{D1} = 50$ μA at $V_{GS1} = 1.5$ V using Equation 17.1a:

$$I_{D1} = \left(\frac{k_p}{2}\right)\left(\frac{W}{L}\right)_1 (V_{GS1} - V_T)^2 \left(1 + \frac{\lambda}{L_1} V_{DS1}\right) \tag{17.27}$$
$$= 50 \times 10^{-6} \text{ A}$$

A value of V_{DS1} must be selected that represents the quiescent voltage of V_{D1}. If we choose $V_{DS1} = 5.0$ V and substitute the given parameters for this example, then $(W/L)_1$ can be expressed from Equation 17.27 as

$$\left(\frac{W}{L}\right)_1 = \frac{20}{1 + (5.0 \times 10^{-6}/L_1)} \tag{17.28}$$

Equation 17.28 shows that $(W/L)_1$ is a relatively large number. If we choose $L_1 = 10\ \mu m$ as a reasonable lower limit on the channel length of M_1, the result is $W_1 = 133\ \mu m$. Now, M_2 and M_3 can be designed from the information on M_1. We use Equation 17.1a for M_2:

$$I_{D2} = 20 \times 10^{-6} = \left(\frac{k_p}{2}\right)\left(\frac{W}{L}\right)_2 (V_{GS2} - V_T)^2 \left(1 + \frac{\lambda}{L_2} V_{DS2}\right) \quad (17.29)$$

Since $V_{GS2} = V_{DS2} = 1.5$ V, Equation 17.29 can be solved for $(W/L)_2$:

$$\left(\frac{W}{L}\right)_2 = \frac{20 \times 10^{-6}\ (2/k_p)}{(V_{GS2} - V_T)^2[1 + (\lambda/L_2)V_{DS2}]} \quad (17.30)$$

Then,

$$\left(\frac{W}{L}\right)_2 = \frac{8}{1 + (1.5 \times 10^{-6}/L_2)} \quad (17.31)$$

Either W_2 or L_2 must be chosen to complete the design. If L_2 is selected to be equal to $10\ \mu m$, then W_2 is calculated from Equation 17.31 as

$$W_2 = \frac{10^{-5} \times 8}{1 + (1.5 \times 10^{-6}/10 \times 10^{-6})}$$
$$= 69.6\ \mu m$$

The $(W/L)_3$ ratio can now be obtained from

$$I_{D3} = 20 \times 10^{-6} = \left(\frac{k_p}{2}\right)\left(\frac{W}{L}\right)_3 (V_{GS3} - V_T)^2 \left(1 + \frac{\lambda}{L_3} V_{DS3}\right) \quad (17.32)$$

Since $V_{DS2} = 1.5$ V, then $V_{DS3} = V_{DD} - V_{DS2} = 8.5$ V. Equation 17.32 can be reduced by substitution of the known parameters to

$$\left(\frac{W}{L}\right)_3 = \frac{0.036}{1 + (8.5 \times 10^{-6}/L_3)} \quad (17.33)$$

This equation indicates that $(W/L)_3$ is much less than unity. If we choose a minimum width of $W_3 = 10\ \mu m$, then L_3 can be calculated from Equation 17.33 as $L_3 = 289\ \mu m$.

A current source must also have a high output resistance. The output resistance of M_1 is

$$R_{out} = r_{ds1} \quad (17.34)$$

The value is calculated from Equation 17.2 as

$$r_{ds1} = \frac{1}{10^{-6}} \frac{2}{20 \times 10^{-6}} \frac{(10 \times 10^{-6})^2}{133 \times 10^{-6}} \frac{1}{(1.5 - 1.0)^2}$$
$$= 300\ k\Omega$$

This value of output resistance is not particularly high for many applications and can be increased dramatically using a Widlar or Wilson current source design.

17.5 DIFFERENTIAL AMPLIFIER

The small-signal analysis of a differential amplifier was given in Chapter 13 for a load R_D driven by a MOSFET with $r_{ds} = \infty$. Figure 17.8a shows a differential amplifier with depletion loads. Figure 17.8b shows the small-signal equivalent circuit including the r_{ds} channel length modulation resistor elements. The differential mode gain for the single ended output can be derived from the concepts in Chapter 13 as

$$A_d = -\frac{g_{m1}}{2}(r_{ds1} \| r_{ds3}) \tag{17.35}$$

where $g_{m1} = g_{m2}$.

The common mode gain can be derived as

$$A_c = -g_{m1}r_{ds3} \frac{r_{ds1}}{r_{ds1} + r_{ds3} + 2R_S} \frac{1}{1 + g_{m1}(2R_S \| r_{ds1} \| r_{ds3})} \tag{17.36}$$

The common-mode rejection ratio (CMRR) for the single-ended output is then

$$\begin{aligned} \text{CMRR} &= \left| \frac{A_d}{A_c} \right| \\ &= \frac{[1 + g_{m1}(2R_S \| r_{ds1} \| r_{ds3})]}{2} \left(\frac{r_{ds1} + r_{ds2} + 2R_S}{r_{ds1} + r_{ds3}} \right) \end{aligned} \tag{17.37}$$

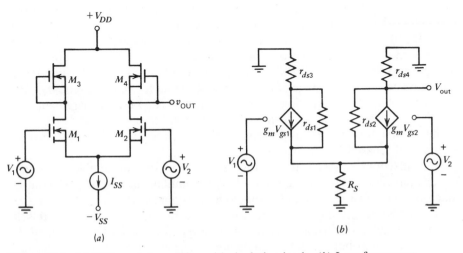

Figure 17.8. (a) Differential amplifier with depletion loads. (b) Low-frequency small-signal model of part (a).

The presence of the channel length modulation term r_{ds1} is seen to degrade the CMRR, while large values of R_S and g_m enhance the CMRR. R_S is increased when present as the output resistance of a current source rather than as a resistor element. The g_m parameter can be manipulated by the designer by observing that

$$g_m = \frac{2I_D}{V_{GS} - V_T} \qquad (17.4)$$

When V_{GS} is minimized, g_m will be maximized. The gate–source voltage can be reduced for a fixed value of I_D by increasing the (W/L) ratio. This is evident when Equation 17.1a is solved for V_{GS}, resulting in

$$V_{GS} = \left[\frac{I_D(2/k_p)(L/W)}{1 + (\lambda/L)V_{DS}} \right]^{1/2} + V_T \qquad (17.38)$$

A MOSFET circuit design will often trade off values of V_{GS} and W/L ratio. This capability is especially useful when circuit dc values affect the transition of a particular transistor into or out of saturation. We examine an example of this next in an NMOS operational amplifier design.

17.6 AN NMOS OPERATIONAL AMPLIFIER

Operational amplifiers provide excellent examples of several aspects of the analog circuit concepts developed thus far. This circuit in MOS has received particular attention because of its role in D/A and A/D converters. We examine an NMOS operational amplifier that illustrates both the differences and similarities in MOSFET and bipolar transistor circuit designs [2].

Figure 17.9 shows a simplified version of this OP AMP. The differential input transistor pair is M_1 and M_2, and a current source is used to bias the two transistors. Depletion mode load devices are used for a double-ended output from M_1 and M_2. A voltage shifter V_{LS} is used to drop the positive dc voltage at the drains of M_1 and M_2 to a value near the negative voltage supply V_{SS}. This negative voltage level shift is necessary to allow adequate biasing levels on the common-source stage of M_{21}. Remember that for saturation of M_{21}, $V_{GS} < (V_{DS} + V_T)$. This condition is extended if V_{DS} can be made large and V_{GS} small. This same voltage level shift was performed on the 741 OP AMP by the emitter–collector voltage drop across the lateral pnp transistors in the input stage, as discussed in Chapter 16.

Transistors M_{17} and M_{18} in Figure 17.9 perform the double- to single-ended conversion of the signal prior to voltage amplification by M_{21}. The output stage is composed of M_{29} and a subcircuit designated B. The capacitor C is necessary to adjust the magnitude and phase of the amplifier so that self-sustained oscillations cannot occur at any frequency. This topic is beyond the scope of this text but is vital to stable amplifier performance [3].

The circuit of Figure 17.9 has several functional similarities to the bipolar

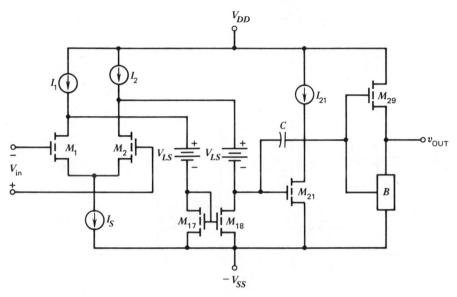

Figure 17.9. Simplified schematic of NMOS operational amplifier. All transistors shown are NMOS enhancement-mode transistors.

741 OP AMP design discussed in Chapter 16. There is, however, one distinct difference in the design that concerns the input resistance of the voltage level shifters. High impedance nodes exist at the drains of M_1, M_2, M_{18}, and M_{21}. A detailed analysis of this situation [5] shows that adequate frequency compensation by capacitor C will occur only if two high impedance nodes are present in the circuit. The voltage level shifter with a low input impedance reduces the drains of M_1 and M_2 to low impedance values, leaving only the drains of M_{18} and M_{21} as the two high impedance nodes. This feature influences the design of the voltage level shifter that we examine below in detail.

17.6.1 **OP AMP Input Stage**

Figure 17.10 shows more detail of the input stage. The bias network for the differential amplifier is formed by M_5, M_6, and M_7. The depletion mode load transistors M_3 and M_4 help feed the signal current to the voltage level shift subcircuit. Current I_6 is designed for approximately 30 μA and V_{GS6} is intentionally kept small. A small value of V_{GS6} allows the common-mode voltage present at the drain of M_6 drop to lower values without causing M_6 to leave saturation. A small value of V_{GS6} is obtained by making $(W/L)_6$ large. Transistors M_5 and M_6 are identical so that $I_{D5} = I_6$ if the effects of transistor drain resistances are neglected (i.e., $\lambda = 0$).

The gate–source voltage of M_6 can be approximated from the transistor parameters $(W/L)_5 = (W/L)_6 = 120\ \mu\text{m}/14\ \mu\text{m}$ and $(W/L)_7 = 30\ \mu\text{m}/100\ \mu\text{m}$.

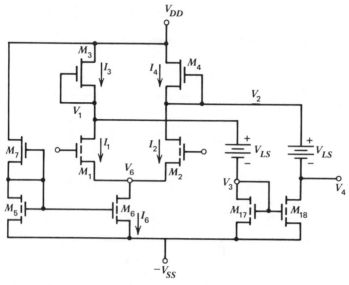

Figure 17.10. Differential input stage of NMOS operational amplifier. Transistors M_3, M_4, and M_7 are depletion-mode MOSFETs while the other transistors are enhancement-mode MOSFETs.

If $V_{TE} = 1.0$ V, $V_{TD} = -3.0$ V, and $\lambda = 0$, then

$$I_{D7} = I_{D5} \tag{17.39}$$

and

$$\left(\frac{k_p}{2}\right)\left(\frac{W}{L}\right)_7 (-V_{TD})^2 = \left(\frac{k_p}{2}\right)\left(\frac{W}{L}\right)_5 (V_{GS5} - V_{TE})^2 \tag{17.40}$$

Substituting the known parameters and solving for V_{GS5} gives

$$V_{GS5} = \left[\frac{30 \times 10^{-6}}{100 \times 10^{-6}}(-3.0)^2 \frac{14 \times 10^{-6}}{120 \times 10^{-6}}\right]^{1/2} + 1 \tag{17.41}$$

Then,

$$V_{GS5} = V_{GS6} = 1.56 \text{ V} \tag{17.42}$$

This value of V_{GS6} allows V_{DS6} to drop to $V_{GS6} - V_{TE} = 0.56$ V before M_6 leaves saturation. This means that the common-mode voltage at the drain of M_6 can drop to $V_6 = 0.56 - V_{SS}$ before M_6 leaves saturation. An important distinction of MOSFET design is apparent in this example in that the designer must be concerned with limits on dc voltage that will cause a transistor to leave the saturated state. This approach contrasts with bipolar transistor design where current levels are set to prevent the bipolar transistor from entering saturation.

The V_{DS7} value is of concern because, for $V_{DS7} > (V_{GS7} - V_{TD})$, transistor M_7 enters the saturated state. Since $V_{GS5} = V_{DS5} = 1.56$ V, $V_{DS7} = (V_{DD} + V_{SS} - 1.56)$. The V_{DD} and V_{SS} values range between 5 and 10 V, keeping M_7 in the saturated state.

The voltage drops across M_3 and M_4 are set at dc values just above $|V_{TD}|$. This allows maximum positive common-mode voltage at V_1 and V_2. Adverse effects of biasing $V_{DS3} = V_{DS4} \simeq |V_{TD}|$ can occur if the ac voltage swings at the drains of M_1 and M_2 are significant. These voltage swings are minimized, however, since the input resistance to the voltage level shifter subcircuits is low.

The common-mode rejection ratio (Equation 17.37) of the input stage and the common-mode voltage swing at the drains of M_1 and M_2 are both increased by assigning low values to V_{GS1} and V_{GS2}. The width and length parameters for M_1 and M_2 are $W/L = 120 \ \mu m/10 \ \mu m$. We can estimate V_{GS1} and V_{GS2} using Equation 17.38 for $\lambda = 0$ and $k_p = 20 \times 10^{-6}$ S/V by

$$
\begin{aligned}
V_{GS1} &= V_{GS2} \\
&= \left[15 \times 10^{-6} \left(\frac{2}{20 \times 10^{-6}} \right) \left(\frac{10 \times 10^{-6}}{120 \times 10^{-6}} \right) \right]^{1/2} + 1 \quad (17.43) \\
&= 1.35 \text{ V}
\end{aligned}
$$

Since V_{GS} must be greater than V_{TE} for conduction, the difference in gate–source voltages on M_1 and M_2 must be less than approximately 0.35 V. This normally does not present a problem for a high-gain OP AMP with negative feedback, since a virtual short condition will exist between the inputs of M_1 and M_2.

The quiescent voltages on M_1, M_2, M_3, and M_4 can be estimated. Since $V_{DS3} = V_{DS4} \simeq |V_{TD}| = 3.0$ V and $V_{S1} = V_{S2} = V_{G1} - V_{GS1} = -1.35$ V, then $V_{DS1} = V_{DS2} = V_{DD} - |V_{TD3}| - V_{S1} = V_{DD} - 4.35$. The power supply V_{DD} is normally greater than 5.0 V, so that M_1 and M_2 are in saturation.

17.6.2 OP AMP Voltage Level Shift Stage

Figure 17.11 shows a detailed schematic of the voltage level shift stage. Transistors M_{13} and M_{14} drop the dc level of V_1 and V_2 (the drain voltages of M_1 and M_2). Voltage V_4 represents the lowered dc level that feeds a common-source voltage amplifier. Transistors M_{11} and M_{12} are shunt feedback elements that lower the input resistance seen by V_1 and V_2.

Transistors M_8, M_9, M_{11}, and M_{12} are designed to ensure that V_1 and V_2 remain at a value of $|V_{TD}|$ below the positive power supply, V_{DD}. This requires that V_9 be set at a value of $|V_{TD}|$ below V_{DD}. Transistors M_{19}, M_{20}, M_{15}, and M_{16} represent a double current mirror, and transistors M_{17} and M_{18} provide the double- to single-ended conversion mentioned previously.

Let us look in more detail at the input resistance of the shunt feedback network of transistors M_{11} and M_{13} (M_{12} and M_{14}). Consider a simplified version of this circuit in Figure 17.12a. The input resistance $R_{in} = \partial V_1 / \partial I_{13}$ may be derived from either a small-signal model or from an analytical expression. Since

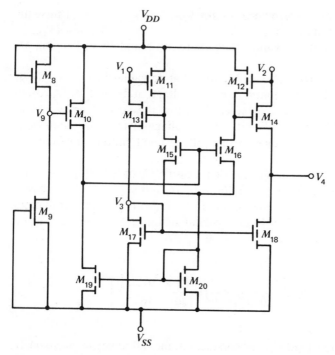

Figure 17.11. Voltage level shift and differential to single-ended converter stage of NMOS operational amplifier. All transistors are enhancement-mode MOSFETs, except depletion-mode MOSFETs M_8 and M_9.

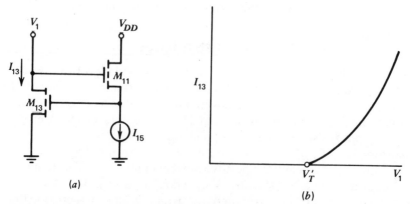

Figure 17.12. (a) NMOS enhancement MOSFETs M_{11} and M_{13} from Figure 17.11, illustrated to show shunt feedback effects at node of V_1. The source of M_{13} is grounded for simplification. (b) The I_{13}-vs.-V_1 relation for circuit in part (a).

the input resistance is sensitive to the applied voltage V_1, we shall derive the input resistance. Figure 17.12b shows the analytical relation between V_1 and I_{13} where the x-axis intercept represents a modified threshold voltage V_T' that must be overcome to allow circuit operation [4]. The current source I_{15} represents that current through M_{15} that is driven by a current source. The source of M_{13} is not grounded in the real circuit, but doing so in Figure 17.12a allows a simpler observation of the feedback properties of M_{11}.

From Figure 17.12a, we may write

$$I_{15} = I_{D11} = \left(\frac{k_p}{2}\right)\left(\frac{W}{L}\right)_{11}(V_{GS11} - V_{T11})^2 \tag{17.44}$$

where $\lambda = 0$. Also,

$$I_{13} = \left(\frac{k_p}{2}\right)\left(\frac{W}{L}\right)_{13}(V_{GS13} - V_{T13})^2 \tag{17.45}$$

and

$$V_1 = V_{GS11} + V_{GS13} \tag{17.46}$$

We can solve for V_{GS13} in Equation 17.46 and substitute this into Equation 17.45 to get

$$I_{13} = \left(\frac{k_p}{2}\right)\left(\frac{W}{L}\right)_{13}(V_1 - V_{GS11} - V_{T13})^2 \tag{17.47}$$

We then solve for V_{GS11} in Equation 17.44 and substitute this into Equation 17.47 to get

$$I_{13} = \left(\frac{k_p}{2}\right)\left(\frac{W}{L}\right)_{13}\left\{V_1 - \left[V_{T11} + V_{T13} + \left[\left(\frac{2}{k_p}\right)\left(\frac{L}{W}\right)_{11}I_{15}\right]^{1/2}\right]\right\}^2 \tag{17.48}$$

Then,

$$I_{13} = \left(\frac{k_p}{2}\right)\left(\frac{W}{L}\right)_{13}(V_1 - V_T')^2 \tag{17.49}$$

where V_T' represents the modified threshold voltage of the circuit.

The resistance can be calculated from Equation 17.49 by

$$\frac{1}{R_{in}} = \frac{\partial I_{13}}{\partial V_1} = k_p\left(\frac{W}{L}\right)_{13}(V_1 - V_T') \tag{17.50}$$

If $(W/L)_{11} = 8.0\ \mu m/100\ \mu m$, $(W/L)_{13} = 120\ \mu m/10\ \mu m$, $k_p = 20 \times 10^{-6}$ S/V, $V_{T11} = V_{T13} = 1.0$ V, and $I_{15} = 20\ \mu A$, then

$$V_T' = 7.0\ V \tag{17.51}$$

and, when $V_1 = 9.0$ V,

$$R_{in} = 2.1\ k\Omega \tag{17.52}$$

This magnitude of resistance is quite low with respect to the resistance at the drains of M_{18} and M_{21}, thereby shunting signal current to the voltage level shift circuit.

Another design feature of the voltage level shifter of Figure 17.11 is the relation between the reference voltage V_9 and V_1 and V_2. As mentioned, V_1 and V_2 are to be approximately $V_{D1} = V_{D2} \leq V_{DD} - |V_{TD}|$. This is accomplished by design techniques that are distinct to MOSFETs. Transistors M_{10}, M_{11}, and M_{12} are made with the same W/L ratio of 8.0 μm/100 μm. The current is M_{11} and M_{12} is made equal to that in M_{10} by using M_{20} to scale up the mirrored current of I_{20} by a factor of 2 over that in M_{10} and M_{19}. The width-to-length ratios are $(W/L)_{19} = 120$ μm/10 μm and $(W/L)_{20} = 240$ μm/10 μm. Since $I_{19} = I_{11} = I_{12}$ and their physical parameters are equal, $V_{GS10} = V_{GS11} = V_{GS12}$. The width-to-length ratios of these transistors are quite low, forcing V_{GS} to be high. If $k_p = 20 \times 10^{-6}$ S/V and $I_{D10} = 20$ μA, then, from Equation 17.38,

$$V_{GS10} = V_{GS11} = V_{GS12}$$
$$= \left[20 \times 10^{-6} \left(\frac{2}{20 \times 10^{-6}} \right) \left(\frac{100 \times 10^{-6}}{8 \times 10^{-6}} \right) \right]^{1/2} + 1 \quad (17.53)$$
$$= 6.0 \text{ V}$$

If $V_{GS10} = V_{GS11}$, then $V_{S10} = V_{S11}$ since V_9 is to be designed equal to V_1. Examination of Figure 17.11 shows that $V_{S10} = V_{S11}$ if $V_{G15} = V_{S15}$. Since

$$V_{G15} = V_{GS15} + V_{GS20} \quad (17.54)$$

and

$$V_{D15} = V_{GS13} + V_{GS17} \quad (17.55)$$

then a condition for $V_{S10} = V_{S11}$ is

$$V_{GS15} + V_{GS20} = V_{GS13} + V_{GS17} \quad (17.56)$$

If $(W/L)_{13} = (W/L)_{15} = (W/L)_{17} = 120$ μm/10 μm, $(W/L)_{20} = 240$ μm/10 μm, $I_{13} = I_{17} = I_{20} = 40$ μA, and $I_{15} = 20$ μA, then Equation 17.38 shows that $V_{G15} = 2.82$ V and $V_{D15} = 3.15$ V. This result gives a voltage difference between V_9 and V_1 of only 0.33 V. It should be emphasized that these calculations are approximate and do not include the effects of channel length modulation or source and drain substrate bias.

17.6.3 OP AMP Voltage Gain Stage

Figure 17.13 shows the voltage gain subcircuit. The capacitor C_1 is placed across the input and output high impedance nodes to provide frequency compensation and stability to the OP AMP. The configuration is also called an integrator since it performs that mathematical function on an input signal. Transistors M_{21}, M_{23}, and M_{24} form a cascode circuit. Transistor M_{24} is the load device, M_{23} is the grounded gate stage, and M_{21} acts as a common-source

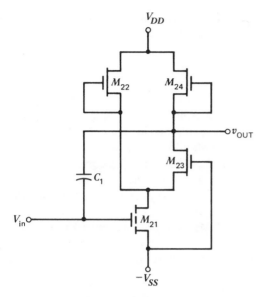

Figure 17.13. Intermediate voltage gain stage with compensating capacitor of NMOS operational amplifier. Transistors M_{22}, M_{23}, and M_{24} are depletion-mode MOSFETs while M_{21} is an enhancement-mode MOSFET.

amplifier. Transistor M_{22} provides more current in M_{21} without affecting the dc current in M_{23} and M_{24}. This arrangement selectively increases the transconductance in M_{21}, as shown in Equation 17.4:

$$g_m = \frac{2I_D}{V_{GS} - V_T}$$

The purpose of the amplifier stage in Figure 17.13 is to achieve maximum voltage gain. The presence of M_{23} provides higher output resistance. The voltage gain analysis of the cascode amplifier is similar to that of the depletion load inverter. This gain is approximately

$$A_v = \frac{V_{out}}{V_{in}} = -g_{m21}(r_{ds21}\|r_{ds23})\delta \tag{17.57}$$

where δ represents the fraction of current delivered to M_{23} from M_{21}. Although M_{22} enhances the transconductance of M_{21}, it also degrades the cascode current gain by bleeding some signal current to ground. The parameter δ represents a current divider action as signal current leaves M_{21}.

The design objective is to maximize A_v as given in Equation 17.57. Equation 17.4 directs us to maximize I_{D21} and minimize V_{GS21}. The actual design specifies $(W/L)_{21} = 657\ \mu m/10\ \mu m$ and $I_{D22} = 159\ \mu A$. The drain current of

M_{24} is $I_{D24} = 96\ \mu A$, making $I_{D21} = 255\ \mu A$. Equation 17.57 also directs that r_{ds21} and r_{ds23} be made as large as possible. Equation 17.2 provides the guidelines for increasing r_{ds}:

$$r_{ds} = \left(\frac{1}{\lambda}\right)\left(\frac{2}{k_p}\right)\left(\frac{L^2}{W}\right)\left(\frac{1}{V_{GS} - V_T}\right)^2$$

The gate–source voltage of M_{21} was made small by the $(W/L)_{21} \approx 65$, thus greatly enhancing r_{ds}. The W/L^2 ratio is also contained in Equation 17.2, causing some degradation in r_{ds21}. Transistor M_{24} should have a large value of r_{ds24}. Decreasing the W/L ratio would increase r_{ds24} but limit the current in M_{24} as dictated by Equation 17.1. In the case of M_{24}, the current is first set at approximately 96 μA, and the $(W/L)_{24}$ is adjusted to provide this current. The final design produced $(W/L)_{24} = 30\ \mu m/27\ \mu m$.

The overall voltage gain of the OP AMP is approximately 1000, with the majority of this gain supplied by M_{21} through M_{24}. The OP AMP is adequate for many applications, but precision converters require special designs to achieve higher overall voltage gain.

17.6.4 OP AMP Output Stage

Figure 17.14 shows the NMOS OP AMP output stage. This subcircuit must provide a low output impedance, zero dc voltage offset, and sufficient current drive for external loads. This design is a Class AB amplifier that combines a cascade of a transconductance and a transresistance amplifier. The composite of these two amplifier sections becomes a voltage amplifier with

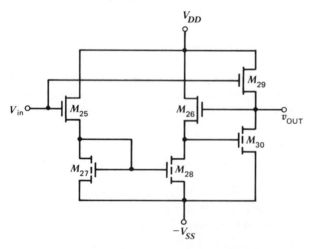

Figure 17.14. Output stage of NMOS operational amplifier. Transistors M_{25}, M_{26}, and M_{29} are depletion-mode while the remainder are enhancement-mode.

nearly unity gain. Transistors M_{25}, M_{27}, and M_{28} form a transconductance amplifier, while M_{29} and M_{30} form a transresistance amplifier. Transistor M_{26} provides the feedback that forces the drain of M_{30} to zero volts in the quiescent condition. The M_{26} shunt feedback connection also lowers the output resistance.

The design of the output stage is such that the gate–source voltages of M_{27} and M_{30} are nearly equal. These two transistors have equal width-to-length ratios ($W/L = 60$ μm/8 μm), forcing $I_{D27} = I_{D28}$. Since transistors M_{25} and M_{26} also have equal width-to-length ratios ($W/L = 8$ μm/80 μm), $V_{DS25} = V_{DS26}$ and $V_{DS27} = V_{DS28}$. This implies that $V_{S25} = V_{S26}$ and $v_{OUT} = v_{IN}$. This condition describes the quiescent condition and partially the dynamic process.

When v_{IN} goes positive, M_{29} conducts more heavily while the gate voltage of M_{30} drops from increased drive to M_{28}. Thus, for positive voltages, M_{29} increased its drain current while M_{30} decreases its drain current. When v_{IN} goes negative, M_{29} shuts down while the gate voltage of M_{30} increases, causing increased I_{D30}. Transistor M_{29} draws a trickle current under zero input condition, therefore classifying this circuit as a Class AB amplifier.

17.7 SUMMARY

Analog MOSFET integrated circuits have found useful applications in areas that require both analog and digital functions on a single chip. The design techniques with MOSFETs require unique insights into MOSFET operation. The MOSFET width-to-length ratio is an integral part of circuit design. In this chapter, some of the basic circuit cells used in analog MOSFET integrated circuit design were analyzed. The depletion load inverter, the voltage divider network, the current source circuits, and the differential amplifier were developed as building blocks for analog MOSFET integrated circuits. An example of an integrated circuit was discussed that illustrated the unique approach taken with analog MOSFET design.

REFERENCES

1. Y. P. Tsividis, *IEEE J. Solid-State Circuits* SC-13, 383 (1978).

2. D. Senderowicz, D. A. Hodges, and P. R. Gray, *IEEE J. Solid-State Circuits* SC-13, 760 (1978).

3. P. R. Gray and R. G. Meyer, *Analysis and Design of Analog Integrated Circuits* (New York: Wiley, 1977), Chapt. 9.

4. P. R. Gray, Basic MOS Operational Amplifier Design—An Overview, in *Analog MOS Integrated Circuits*, P. R. Gray, D. A. Hodges, and R. W. Broderson, Eds. (New York IEEE Press, 1980).

5. J. E. Solomon, *IEEE J. Solid-State Circuits* SC-9, 314 (1974).

PROBLEMS

17.1. Derive Equation 17.9 from Figure 17.4.

17.2. In the example that uses Figure 17.5, find the drain currents at the quiescent points Q_A, Q_1, and Q_B.

17.3. In the example that uses Figure 17.5, how low may the output voltage drop before M_1 enters the nonsaturated state?

17.4. In the example that uses Figure 17.5, suppose that $Y_{GS1} = 2.0$ V. What ratio must $(W/L)_1$ become to keep the inverter biased at the point of maximum voltage gain?

17.5. In Figure 17.6, $V_T = 1.0$ V, $k_p = 10^{-5}$ S/V, $V_{DD} = 15$ V, $\lambda = 10^{-7}$ m/V, and $W/L = 20$ μm/40 μm. Initially, all transistors are equal, and there is a 5.0-V potential drop across each pair of drain-to-source terminals. How would W_2 be changed if it is desired that $V_{D1} - V_{D2} = 3.0$ V, assuming that all other parameters remain the same?

17.6. In Figure 17.15, $V_T = 1.0$ V, $k_p = 10^{-5}$ S/V, and $\lambda = 10^{-7}$ m/V. Keep the minimum length or width of the MOSFETs to 8 μm. If $I_{D4} = 30$ μA, design the dimensions of the transistors such that the current source is $I_{SS} = 100$ μA. Use the design criteria developed in Section 17.4. What is the current source output resistance?

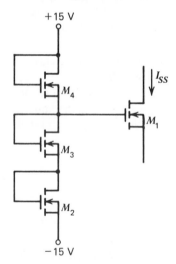

Figure 17.15. Circuit for Problem 17.6.

17.7. Derive (*a*) Equation 17.35 and (*b*) Equation 17.36.

17.8. Derive Equation 17.57.

ANSWERS TO SELECTED PROBLEMS

CHAPTER 2

2.1. (a) n-Type, $n_o = 2 \times 10^{15}$ electrons/cm^3, $p_o = 1.13 \times 10^5$ holes/cm^3. (b) n-Type, $n_o = 2 \times 10^{15}$ electrons/cm^3, $p_o = 2.88 \times 10^{11}$ holes/cm^3.

2.3. (a) $n_o = 7.27 \times 10^{16}$ electrons/cm^3, $p_o = 1.11 \times 10^{-3}$ holes/cm^3. (b) $p_o = 1.3 \times 10^{15}$ holes/cm^3, $n_o = 1.73 \times 10^5$ electrons/cm^3.

2.5. $E = 5.14$ V/cm.

2.7. (a) $R = 2.31$ kΩ. (b) $R = 7.25$ kΩ.

2.9. $J_{\text{DIFF}} = 107$ A/cm^2.

CHAPTER 3

3.1. (a) $V = 0.16$ V. (b) $I = 4.5$ mA.

3.3. 2.25 A, 105 A, 4927 A.

3.5. (a) 59.9 mV. (b) 120 mV.

3.7. $V_{\text{OUT}} = 0.357$ V.

3.9. (a) $V_{\text{IN}} = 1.82$ V. (b)$R_1 = R_2 = 15.7$ Ω, $R_3 = 24.7$ Ω.

3.11. $V_{DC} = 0.652$ V.

3.13. (a) $I_D = 24.8$ mA. (b) $I_D = 5$ μA. (c) Forward bias, $I_D = 24.8$ mA; reverse bias, $I_D = 12.5$ mA.

3.15. (a) $I_D = 25.6$ mA, $V_{D1} = 1.32$ V, $V_{D2} = 1.11$ V. (b) $I_D = 3.87$ mA, $V_{D1} = 0.455$ V, $V_{D2} = 0.677$ V.

3.17. (a) $I_1 = I_2 = I_3 = 0$, $V_{\text{OUT}} = 5$ V. (b) $I_1 = 0$, $I_2 = I_3 = 0.44$ mA, $V_{\text{OUT}} = 0.82$ V. (c) $I_1 = 0$, $I_2 = I_3 = 0.24$ mA, $V_{\text{OUT}} = 2.72$ V. (d) $I_1 = I_2 = 0.225$ mA, $I_3 = 0.45$ mA, $V_{\text{OUT}} = 0.72$ V.

3.19. $V_1 = 1.83$ V.

CHAPTER 4

4.7. $I_{CQ} = 0.74$ mA, $V_{CEQ} = 3.07$ V, $V_E = -1.03$ V, $V_{BC} = -2.37$ V.
4.8. $A_v = -111$, $Z_i = 1.79$ kΩ, $Z_o = 4$ kΩ $A_i = 39.8$.
4.11. $A_v = 91$.
4.15. (a) $R_B = 3.72$ MΩ. (b) $A_v = -17.5$. (c) $A_v = -87.4$. (d) Doubled.
4.17. 388 kΩ.
4.19. $R_C = 21.9$ kΩ.
4.21. $c = 2$ pF, $r = 12.5$ kΩ.

CHAPTER 5

5.1. $I_{DQ} = 0.125$ mA, $V_{DSQ} = 3.75$ V, transistor in saturation region.
5.3. $g_m = 0.42$ mS.
5.5. $I_S = 0.984$ mA, $V_{OUT} = 9.84$ V.
5.7. (a) $V_X = 2.14$ V, $V_{OUT} = 0.64$ V. (c) $V_{OUT} = 0.146$ V. (e) $I_D = 11.2$ mA.
5.9. $V_{OUT} = 8$ V, $V_{DS} = 2$ V.
5.11. (a) $V_{OUT} = 8$ V, $I_D = 0$. (c) $V_{OUT} = 6$ V, $I_D = 80$ μA. (e) $V_{OUT} = 1.78$ V, $I_D = 249$ μA.

CHAPTER 6

6.1. (a) $I_B = I = 0$, $V_{OUT} = 5$ V. (b) $I_B = 0.26$ mA, $I = 2.23$ mA, $V_{OUT} = 0.1$ V.
6.3. For $V_X = V_Y = 0.1$ V; $I_1 = 3.11$ mA, $I_2 = 0.32$ mA, $I_3 = I_4 = 0$, $I_5 = 0.32$ mA, $V_1 = 0.8$ V, $V_B = -0.35$ V, $V_{OUT} = 12$ V. For $V_X = V_Y = 12$ V; $I_1 = I_2 = 1.56$ mA, $I_3 = 4.96$ mA, $I_4 = 1.2$ mA, $I_5 = 0.356$ mA, $V_1 = 6.4$ V, $V_B = 0.8$ V, $V_{OUT} = 0.1$ V.
6.7. (a)(1) $I_1 = 0.7$ mA, $I_2 = 0.093$ mA, $I_3 = I_4 = 0$, $I_5 = 0.093$ mA, $V_1 = 0.8$ V, $V_B = -0.6$ V, $V_{OUT} = 5$ V. (2) $I_1 = I_2 = 0.467$ mA, $I_3 = 2.04$ mA, $I_4 = 0.28$ mA, $I_5 = 0.187$ mA, $V_1 = 2.2$ V, $V_B = 0.8$ V, $V_{OUT} = 0.1$ V. (b) $N = 8$.
6.9. (a)(1) $I_1 = 1.13$ mA, $I_2 = I_3 = I_4 = 0$. (2) $I_1 = 1.49$ mA, $I_2 = 2.73$ mA, $I_3 = 2.23$ mA, $I_4 = 3.1$ mA. (b)(1) 5.54 mW, (2) 32.3 mW. (c) $N = 81$.
6.11. (a) $N = 58$. (b) $N = 46$.
6.13. (a) $I_D = 1.63$ mA. (b) $I_L = 18$ mA.
6.15. (a) 2.25 mW. (b) 7.8 mW.
6.17. Q_6 is active.

CHAPTER 7

7.1. $I_{CQ} = 0.27$ mA, $V_{CEQ} = -2.84$ V.
7.3. $R_E = 41.3$ kΩ, $V_{CE} = -3.2$ V.
7.7. (a) $I_E = 1.0$ mA, $V_{o1} = 0.0$ V, $V_{o2} = -1.5$ V. (b) $I_E = 1.28$ mA, $V_{o1} = -1.4$ V, $V_{o2} = 0.0$ V.
7.9. (a) For $V_{IN} = $ logic 0, $P = 5.2$ mW; for $V_{IN} = $ logic 1, $P = 6.66$ mW.
7.11. (a) $V_R = 2.70$ V. (b) Logic 0 $= 2.4$ V and logic 1 $= 3.0$ V.
7.13. (a) $R_1 = 2.05$ kΩ $R_2 = 3.0$ kΩ. (b) Logic 1 $= +0.24$ V, logic 0 $= -0.24$ V.
7.15. $R_{C1} = 0.12$ kΩ, $R_{C2} = 0.136$ kΩ, $R_E = 0.52$ kΩ, $R_4 = R_5 = 0.66$ kΩ.

CHAPTER 8

8.1. *(a)* Transition point: $V_{OUT} = 2.32$ V, $V_x = 3.32$ V, $V_{OUT} = 1.13$ V when $V_x = 5$ V. *(b)* Transition point: $V_{OUT} = 1.23$ V, $V_x = 2.23$ V, $V_{OUT} = 0.245$ V when $V_x = 5$ V.
8.3. Transition point: $V_{OUT} = 1.71$ V, $V_x = 3.71$ V, $V_{OUT} = 0.70$ V when $V_x = 10$ V.
8.5. *(a)* Load transistor: $V_{OUT} = 3$ V, $V_x = 1.5$ V. Driver transistor: $V_{OUT} = 0.5$ V, $V_x = 1.5$ V. *(b)* $V_{OUT} = 0.031$ V. *(c)* $I_D = 0.4$ mA.
8.7. *(a)* For $V_x = 0.2$ V, $P = 0.0$ mW; for $V_x = 5$ V and $k = 0.1$ mA/V^2, $P = 3.88$ mW; for $V_x = 5$ V and $k = 0.5$ mA/V^2, $P = 4.76$ mW. *(b)* For $V_x = 0.75$ V, $P = 0.0$ mW; for $V_x = 8$ V, $P = 10$ mW. *(c)* For $V_x = 0.5$ V, $P = 0.0$ mW; for $V_x = 5$ V, $P = 2$ mW.
8.11. *(a)* $V_{OUT} = 0.067$ V.
8.13. *(a)* n-Channel transition point: $V_{OUT} = 4$ V, $V_{IN} = 5$ V. p-Channel transition point: $V_{OUT} = 6$ V, $V_{IN} = 5$ V. *(c)* $V_{OUT} = 9$ V for $V_{IN} = 4$ V; $V_{OUT} = 1$ V for $V_{IN} = 6$ V.

CHAPTER 10

10.1. *(a)* For $\beta_{dc} = 50$: $I_{CQ} = 2.14$ mA, $V_{CEQ} = 4.20$ V; for $\beta_{dc} = 200$: $I_{CQ} = 2.88$ mA, $V_{CEQ} = 2.22$ V. *(b)* For $\beta_{dc} = 50$: $I_{CQ} = 0.54$ mA, $V_{CEQ} = 8.54$ V; for $\beta_{dc} = 200$: $I_{CQ} = 1.44$ mA, $V_{CEQ} = 6.11$ V.
10.3. For $\beta_{dc} = 100$: $I_{CQ} = 2.66$ mA, $V_{CEQ} = 4.15$ V; for $\beta_{dc} = 50$: $I_{CQ} = 2.58$ mA, $V_{CEQ} = 4.31$ V.
10.5. $S_\beta = 54.2$ μA.
10.7. $S_I = [(1 + \beta_{dc})(R_E + R_B)]/[R_B + (1 + \beta_{dc})R_E]$.
10.9. $R_1 = 32.1$ kΩ, $R_2 = 17.9$ kΩ.
10.13. $R_1 = R_2 = 0.243$ kΩ, $R_3 = 0.338$ kΩ.
10.15. *(a)* $R_1 = 3.7$ kΩ, $R_2 = 4.39$ kΩ. *(b)* $I_{C4} = 2.04$ mA, $I_1 = 2.12$ mA. For $V_1 = V_2 = -2$ V, $I_{C4} = 1.96$ mA.
10.17. *(a)* $S_\beta = 1.57$ μA. *(b)* $S_\beta = 3.39$ μA for $\beta_{dc} = 10$, $S_\beta = 0.0314$ μA for $\beta_{dc} = 50$. *(c)* $S_V = 40$ μA/V.
10.21. $I_L = 0.544$ mA.

CHAPTER 11

11.1 $z_{11} = h_{11} - (h_{12}h_{21})/h_{22}$, $z_{12} = h_{12}/h_{22}$, $z_{21} = -h_{21}/h_{22}$, $z_{22} = 1/h_{22}$.
11.3. *(a)* $h_{ie} = h_{ic}$, $h_{re} = (1 - h_{rc}$, $h_{fe} = -(1 + h_{fc})$, $h_{oe} = h_{oc}$.
11.5. *(a)* $I_{CQ} = 0.71$ mA, $V_{CEQ} = 5.03$ V. *(c)* $A_V = -97.5$.
11.7. *(a)* $h_{ie} = 2.2$ kΩ, $h_{fe} = 60$, $h_{oe} = 10$ μS. *(b)* $I_{CQ} = 0.71$ mA. *(c)* $V_A = 71$ V.
11.9. $Z_L = 50$ kΩ.
11.11. $Z_{in} = 2.99$ kΩ, $A_v = -148$.

CHAPTER 12

12.1. $f_L = 7.58$ Hz, $f_H = 23.9$ MHz.
12.3. *(a)* $f = 2.65$ Hz. *(c)* $f = 13.1$ Hz.

12.5. (a) $C_M = 0.01$ μF. (b) $A_v = -4.55 \times 10^3$. (c) $f_H = 87.4$ kHz.
12.9. (a) $C_M = 5.86$ pF. (b) $f = 23.9$ MHz.
12.11. (a) $R_E = 2.87$ kΩ, $R_C = 0.47$ kΩ, $C_E = 9.36$ μF.

CHAPTER 13

13.1. (a) $I_E = 0.266$ mA, $I_{C1} = I_{C2} = 0.133$ mA. (b) $V_{CE1} = V_{CE2} = 4.05$ V. (c) $I_{C1} = I_{C2} = 0.147$ mA, $V_{CE1} = 2.35$ V.
13.3. $R_E = 125$ kΩ.
13.5. $R_S = 5$ kΩ.
13.7. (a) $V_{out} = 7.9$ V. (b) $A_v = 29.1$.
13.9. (a) $R_1 = 71.8$ kΩ, $R_C = 28.9$ kΩ. (b) $A_d = -62.6$, $A_c = -0.114$, CMRR = 54.8 dB.
13.11. (a) (1) $v_{o1} - v_{o2} = 0$ V; (2) -0.1 V. (b) (1) 0.768 V; (2) 0.663 V.

CHAPTER 14

14.1. (a) $Z_{in} = 0.257$ kΩ. (b) $Z_{out} = 2.94$ kΩ. (c) $G_z = -54.9$ kΩ. (d) $T = 2.28$.
14.5. (a) $R_F = 0.33$ kΩ, $R_B = 800$ kΩ. (b) $Z_{in} = 36.7$ KΩ $Z_{out} = $ infinity.
14.7. (a) $T = 106$. (b) $Z_{in} = 290$ kΩ. (c) $Z_{out} = 35.7$Ω. (d) $V_{out}/V_{in} = 0.987$.
14.11. (a) $G_Y = 98.3$ mS. (b) $Z_{in} = 1.88$ MΩ. (c) $Z_{out} = 1$ kΩ. (d) $T = 132$. (e) $A_v = -98.3$.
14.13. (a) $G_Y = 0.62$ mS. (b) $Z_{in} = $ infinity. (c) $Z_{out} = 4.69$ kΩ. (d) $T = 0.46$.
14.15. (a) $G_I = -1.0$. (b) $Z_{in} = 1.08$ kΩ, $Z_{out} = 5$Ω. (c) $T = $ infinity.

CHAPTER 15

15.1. $n = 0.87\%$.
15.3. $n_{max} = 4.8\%$.
15.5. $P_{max} = 9$ W.
15.7. Duty cycle = 39.5%.
15.9. $R = 2.96$ kΩ, $C = 5.83$ pF, $L = 14.4$ μH.
15.13 $f_c = 528$ kHz.

CHAPTER 17

17.2. $I_D = 6.12$ μA at Q_A, $I_D = 6.20$ μA at Q_1, $I_D = 6.38$ μA at Q_B.
17.3. $V_{out} = 0.499$ V.
17.5. $W_2 = 126$ μm.

index _____

Acceptor impurities, 19, 123
AC load line, 313, 314, 315
AC resistance of diode, 39, 40
Active load, 434–439, 525, 526, 529–533, 538, 544–547, 553–555, 561–563
  output impedance, 434, 545
Active pull-down/up, 182, 187
Admittance parameters, 340, 341, 357, 361, 362, 494, 495, 500, 501, 521
Alpha (α_F, α_R), 74, 75, 76, 316, 354
Alpha cutoff frequency, 354
Aluminum metallization, 25, 28, 44, 117
Amplification, 3–4, 5
Amplifier, 3, 5, 6, 7, 9
Amplitude modulation (AM), 51–53, 493
AND gate, 145, 183, 188, 231, 232, 234, 273, 284
AND-OR-INVERT gate, 273
Arsenic, 18, 24
Avalanche breakdown, 42, 43, 53, 54

Band diagram, 15–19
Band gap regulator, 534–537
Bandwidth, 367–379, 388–397, 446, 450–451, 455, 470, 475, 496, 497, 503–512, 523, 527, 542
Beta, 75
Beta cutoff frequency, 355, 356
Biasing, diode, 32, 36, 39–49, 54–58
Bias stabilization, 307, 311–323
Bipolar junction transistors, 65–106
  base, 67–75
  base-collector junction, 67–78
  base-emitter junction, 67–79
  base width, 69, 72
  biasing, 65–106, 307–313, 315–318, 319–328, 483–494, 525, 527–530
  current gain, 71, 74–75
  cutoff, 65, 71, 72

equivalent circuits, 80–85, 100–106, 339–357, 494–498
forward-active mode, 68, 71–80
high frequency models, 349–357, 371–379, 521
input impedance, 85, 342, 346
inverse active, 72, 73, 160–166, 169, 173, 174
inverted transistor, 205
large signal models, 100–106
npn, 66
pnp, 67–68, 83, 185–186, 204–208, 210–211, 217, 428, 434–436, 492, 525–534, 540, 554
saturation, 65, 71–74
transconductance, 544
Bode plot, 368, 393, 396, 397
Body effect, 132, 141, 245, 252, 256, 543, 544
Boron, 18, 19, 23–29
Breakdown, 41–43, 49, 53, 57, 60, 183
Bridge rectifier, 51
Broadband amplifier, 367, 484, 488–492, 498
Built-in voltage, 34–36, 42, 352, 353
Bypass capacitor, 310, 312–314, 331, 381, 385, 386, 393, 418, 420, 423, 448, 512, 522

Capacitance, diode, 32, 35–45, 49, 57, 182, 200, 351–353
Capacitor, MOS, 122–124, 293–297
Cascade amplifier, 417–425, 432, 444, 501–502, 519, 562
Cascode amplifier, 560–561
Clamping circuit, 54–56, 146, 150, 153, 165, 167
Classifications of amplifiers, 483–485
  A, 483–486, 490–494, 517–520
  AB, 483–484, 492, 519
  B, 483–484, 487–492, 519, 521
  C, 483–484, 487, 492–494, 514–517
Clipping circuit, 54–56
CMOS, 132, 242, 263–279, 293, 541
Collector dotting, 231–232

Colpitts oscillator, 512, 514–518

Common-base amplifier, 75–77, 114, 204, 210, 211, 349, 371, 376–379, 428, 467–469, 483, 518, 525

Common-base characteristics, 74

Common-base h-parameters, 343, 345, 346, 518

Common-collector (emitter follower) amplifier, 91–93, 343, 348–349, 420–422, 429, 483, 485–488

Common-collector h-parameters, 343, 348, 363

Common-drain amplifier, 358, 422, 483

Common-emitter amplifier, 75, 76, 80, 86–89, 89–91, 94–98, 114, 310–313, 316–319, 349, 354–355, 371–376, 379, 381–386, 390–392, 434–436, 448, 453–460, 464, 467, 483, 494, 520, 526

Common-emitter characteristics, 71, 77, 78, 321, 486

Common-emitter h-parameters, 83, 84, 343, 345, 346, 348, 357, 363, 381, 391

Common-gate amplifier, 358, 361, 483

Common mode, 399–439, 553, 555–557

Common mode rejection ratio (CMRR), 410–411, 416, 553, 554, 557

Common-source amplifier, 313–315, 358, 361–362, 379–381, 392, 436–439, 483, 554–555, 557–562

Compensation, frequency, 533, 540, 555, 560

Complementary MOS (CMOS), 132, 242, 263–279, 293, 541

Conduction band, 15–19

Conductivity, 15, 17, 19, 29, 30

Contact, ohmic, 33, 36, 44, 116, 117

Coupling capacitor, 89–91, 307, 310–313, 346, 368, 370, 381–383, 386, 393, 399, 418, 420

Crossover distortion, 491–492, 519

Current amplifier, 467–470
  input impedance, 468
  output impedance, 468–469

Current mirror, 417, 525, 528–531, 550, 551, 557, 560

Current mode logic, 202–241

Current source, 203–208, 211–213, 217, 230–234, 319–331, 414–417, 428–429, 434–439, 525, 529, 534, 550–553, 554, 559
  output impedance, 325–328, 331, 550, 552

Cut-in voltage, 94, 147, 164, 165, 182, 187, 188, 204, 516, 521

Cutoff frequency, alpha and beta, 354–356

Cutoff frequency (−3 dB):
  upper, 367–379, 393–395, 451
  lower, 367–370, 381–386, 393–396

Darlington connection, 175, 187, 425–427, 429, 433, 439, 444, 537
  input impedance, 426–427

DC load line, 47–48, 78–80, 130–131, 137, 194, 243, 256, 308–310, 312–315, 334, 400–405, 440, 485–486

Demodulation, 52

Depletion mode load, 243, 251–256, 259–262, 274, 275, 283, 291, 298, 301, 302, 544–547, 553–555, 561–563

Depletion region, 34, 352

Detector, 51–53

Differential amplifier, 7, 218–220, 227, 238, 239, 398–417, 429, 433, 439–443, 524–525, 528, 530–533, 553–554, 555, 563
  input impedance, 411–414

Differential mode, 399, 403, 405–416, 433, 439, 532

Diffusion, carrier, 20–22, 29

Diffusion, impurity, 24–29

Diffusion capacitance, 32, 41

Diffusion coefficient, 20–22, 29, 37

Diffusion current, 20, 21, 30, 34–37, 40, 42, 44, 67

Diffusion length, 22, 29, 37, 40, 67, 72

Diodes, 4, 5, 8, 31–64, 145–157, 158, 161, 172–190, 195–200, 222, 223, 231, 232, 286
  equivalent circuit, 45–49

Diode-transistor logic (DTL), 145–157, 158, 161, 164, 167, 183, 188, 189, 194–196

Distortion, 446, 452–453, 470, 475, 483, 488, 491–493, 519

Donor impurity, 18, 19

Doping, 18–29, 66–72

Dotting logic, 231–232

Drift current, 17, 20, 21, 29, 30, 34, 36, 37, 42, 44, 73

Drift velocity, 17

Driver device, 139, 186, 245, 248–261, 275

DTL, 145–157, 158, 161, 164, 167, 183, 188, 189, 194–196

Dynamic memory, 293, 297

Early effect (voltage), 85, 352, 353, 435

Ebers-Moll model, 100

ECL, 218–234

EEPROM, 280, 300–302, 303

Efficiency, emitter, 68, 72, 73, 85

Efficiency, 482–494, 514–520

Einstein relations, 22

Emitter-coupled logic (ECL), 218–234

Emitter efficiency, 68, 72, 73, 85

Emitter follower (common-collector), 91–93, 220–224, 227, 229, 232, 348, 420–422, 429, 464, 488, 526
Energy band, filled, 15–16
Energy gap, 15–16, 534–537
Energy level, 15–16
Epitaxial growth, 28
EPROM, 280, 282, 300–302, 303
Erasable ROM, 280, 282, 300–302, 303
Etching, 23, 25
Excess carrier concentration, 22, 36, 37, 40–43, 70, 104
Excess carrier lifetime, 19, 22, 29, 30, 43, 102, 104
Excess stored charge, 32, 39–45, 58, 104–106, 208
Extrinsic semiconductor, 17–19, 21

Fall time, 156–157
Fan in, 149
Fan out, 151–155, 160–161, 167–175, 193–199, 213, 227–229
Feedback, 342, 372, 374, 446–475, 482, 498, 500, 512–519, 524, 527, 529, 534–540, 557, 559, 563
Field effect transistors, *see* Junction field effect transistor; Metal-oxide-semiconductor field effect transistor
Filled energy band, 15–16
Forward biased junction, 36–41
Frequency response, 340–342, 349–357, 360–363, 367–397, 418, 423, 450, 451, 497, 508, 509
Full-wave rectifier, 49, 51

Gain, 6, 10
Gain-bandwidth product, 355, 356, 375–376, 379, 393, 395, 396, 527
Gallium arsenide, 8, 15–17, 24, 29, 30, 117, 121, 122, 242, 358, 359, 363
Gap, energy, 15, 16, 534–537
Germanium, 6, 8, 15–17, 24, 29, 117
Gunn effect, 358

Half-wave rectifier, 49, 50, 488
High frequency effects, 341, 349–357, 360, 363, 371–381, 527, 540
High frequency FET model, 360–362, 379–381
High frequency h-parameters, 354–355
History of electronics, 4–9
Hybrid-π model, 349–354, 355–357, 371, 372, 373, 374, 392, 394

Hybrid parameter (*h*-parameter) model, 80–93, 326–328, 340–349, 354–355, 381, 391, 397, 405, 418, 419, 421, 453, 517, 518
Hydrofluoric acid, 25

I^2L, 202–214, 216, 217, 224, 227, 234–237
Impedance, effect of feedback on, 446, 455, 456–459, 462–464, 465–466, 468–469, 470, 473–481, 563
Impedance parameters, 340
Impurity profile, 26–27
Input admittance, 361, 495, 498–500
Input impedance, 470
Integrated circuit fabrication, 7, 23–29
Intrinsic semiconductor, 14–17, 19, 22, 30
Inverter (NOT), 145
   bipolar transistor, 100–102
   CMOS, 263–268
   DTL, 145–157
   ECL, 220–229
   I^2L, 202–210
   NMOS, 243–256
   SCTL, 214–216
   STL, 217
   S-TTL, 181–188
   TTL, 157–178
Inverter amplifier, 65, 100, 129–131, 379
Ion implantation, 25, 27, 28, 29

Junction, *pn*, 23–28, 32–43, 65, 66, 71, 72, 80
Junction capacitance, 32, 35, 36, 39–42, 45, 49, 57
Junction field effect transistor (JFET), 8, 114–122
   characteristics, 117–120
   fabrication, 116–117
   frequency characteristics, 121
   input impedance, 114
   models, 357–362
   output impedance, 358
   saturation, 118, 120, 358
   switching, 121
   transconductance, 120–121
Junction transistors, *see* Bipolar junction transistors

Lateral *pnp* transistor, 204, 205, 210, 211, 365, 428, 434–436, 492, 525, 526, 529, 530, 533, 534, 540, 554
Law of the junction, 36
Level shifter:
   bipolar, 417, 427–429, 431, 434, 439, 445, 525
   MOSFET, 554–555, 557–560

Lifetime of excess carriers, 19, 22, 29, 30, 43, 70, 102, 104, 105

Load line, 47, 48, 78–80, 131, 137, 194, 243, 256, 283, 291, 298, 301, 302, 308, 309, 312–315, 334, 400–405, 440, 485, 486

Loop gain, closed, 449, 451, 453, 455, 460, 462, 467, 476, 513

Loop gain, open, 449, 450, 455–460, 463, 464, 470–482, 541, 542

Maximum power rating, 43, 53, 54, 64, 520

Mean-free path, 19

Memory, 280–303

Merged transistor logic (MTL or I^2L), 202–214, 216, 217, 224, 227, 234–237

MESFET, 8, 114–122
 characteristics, 117–120
 enhancement mode, 122
 fabrication, 116–117
 frequency characteristics, 121, 242
 models, 357–362
 switching, 121
 transconductance, 120–121

Metal-oxide-semiconductor field effect transistor (MOSFET), 124–136, 542–544
 biasing, active region, 313–315
 depletion mode, 124, 129, 137, 140
 drain characteristics, 127–129, 132, 243, 245, 256, 358, 542
 drain resistance, 360
 drain-to-source resistance, 542–545, 555, 560, 564
 enhancement mode, 124–129
 input impedance, 114, 545
 inverter amplifier, 129–131
 models, 357–362
 output impedance, 358
 n-channel, 124, 125–129
 p-channel, 124, 131–132
 saturation, 127–129
 switching and frequency effects, 133–136
 theory, 124–129
 transconductance, 129

Metal-semiconductor diode (Schottky), 5, 8, 31, 32, 43–45, 58, 114–117, 122, 178–190, 199, 200, 203, 213–217, 357

Metallization, 28, 29

Miller effect (capacitance), 372–376, 379, 392

Minority carrier lifetime, 19, 22, 29, 30, 43, 70, 102, 104, 105

Mobility (μ), 17, 19, 22, 29, 117, 118, 127, 131, 136

MOSFET, *see* Metal-oxide-semiconductor field effect transistor

MTL, 202–214, 216, 217, 224, 227, 234–237

Multiple-collector transistor, 206, 211–216, 526, 540

Multiple-emitter transitor, 158, 159, 183, 206, 526

NAND gate:
 ALSS-TTL, 185–188
 CMOS, 270–271
 DTL, 145–157
 ECL, 231–233
 LS-TTL, 182–185
 NMOS, 259–260
 STL, 216–217
 S-TTL, 181–182
 TTL, 157–178

Negative feedback, 372, 446–481, 524–527, 529, 534, 537–540, 557, 559, 563

Neutralization, 500–502, 519

NMOS, 242, 243–262, 270, 272, 274

Noise, 229, 230, 534, 536, 541

Noise margin, 155–156, 157, 161, 182–186, 195, 200, 203, 209–210, 221, 226, 227, 230, 235, 265, 267, 541

Nonsaturating logic, 178–188, 218–234

Nonvolatile memory, 297–302

NOR gate:
 CMOS, 269–270
 ECL, 220–230
 I^2L, 210–213
 NMOS, 259–260
 SCTL, 215

Notation, 12–13

n-type impurity, 18, 19, 22–30, 33, 44, 67, 116–117, 123–124

Offset voltage, 523, 529, 541, 562

Ohmic contact, 33, 36, 44, 116, 117

One-sided amplifier, 407, 410–416, 429, 524, 554, 557

Open loop gain, 449, 450, 455–460, 463, 464, 470–482, 541, 542

Operating point, 9–11

Operational amplifier (OP AMP), 7, 398, 429–434, 439, 449, 523–533, 554–563
 bipolar, 429–434, 523–533
 input impedance, 432, 523, 524
 MOSFET, 554–563
 output impedance, 523–524
 741, 523–533

OR gate:
 diode, 57
 domino CMOS, 272
 ECL, 220–231

NMOS, 259
SCTL, 215
Oscillator, 6, 446, 482, 498, 500, 501, 512–518, 519, 522, 527, 540
Output admittance, 83–85, 322, 342, 361, 428

Parallel feedback, 453, 457, 458, 460, 464, 466–468, 470, 475, 557, 563
Parallel-parallel, 453, 470
Parallel resonance, 388–392, 397
Parallel-series, 467, 470
Passive pull-down/up, 150–151, 158, 162, 163, 170, 181, 182, 189, 230
Phase shift oscillator, 512–513, 522
Phase splitter, 185, 186
Phosphorus, 18, 19, 23–29
Photolithography, 23, 25, 28
Piecewise linear model, 48, 59–61, 146, 147, 154–156, 198, 203, 204, 209, 215, 224, 290
Pinch-off, 118–120, 127
Pinch-off voltage, 118–120
Planar process, 158
PMOS, 242, 270, 272, 274, 293
pn junction, 23, 25, 26, 27, 28
Pole, 497, 504, 507, 510
Pole-zero plot, 504–512
Positive feedback, 446, 447, 482, 498, 500, 512–519, 522
Potential barrier, 43
Power amplifier, 65, 68, 482, 483–492, 519
Power-delay product, 208, 237
Power gain, 6, 331
Power supply sensitivity, 315, 316
Propagation delay, 156–157
  CMOS, 265
  ECL, 202, 227, 229
  LS-TTL, 183–185
  NMOS, 261
  S-TTL, 178, 182
  TTL, 161, 170
p-type impurities, 19, 22–23, 44, 67, 116, 117, 122–124
Push-pull amplifier, 488, 489, 526, 527

Quality factor (Q), 388–392, 493, 494, 506, 507, 521
Quiescent (Q) operating point, 9–11

Radio frequency amplifier, 12, 356–357, 483, 492–512
Random access memory (RAM), 280, 282, 287–297, 302, 303
Ratio inverter, 254
RC phase-shift oscillator, 512–513, 522

Read only memory (ROM), 280, 282, 297–302, 303
Recombination, 19, 22, 43, 67, 68, 69, 70, 71
Rectifiers, 49–51
Refresh, 293, 295, 297, 303
Regeneration, 446, 498
Regulated power supply, 53–54, 523, 533–540
Resistivity, 14, 15, 18, 23, 29
Resonance, parallel, 388–392, 397, 492–512
Resonance, series, 386–388, 397
Reverse biased pn junction, 41–43
Reverse breakdown, 42, 43, 53, 54
Reverse saturation current, 37, 45, 49, 58, 60, 155, 158, 160, 161, 174, 218, 230, 295, 535
Rise time, 156, 157
ROM, 280, 282, 297–302, 303

Saturating logic, 145–178, 202–213
Schottky barrier, 5, 8, 31, 32, 43–45, 58, 114–117, 122, 178–190, 199, 200, 203, 213–217, 357
Schottky clamped transistor, 178–181, 182, 184, 198, 203, 204, 214, 216, 218
Schottky TTL gates, 178–188, 216, 218, 224, 227
SCTL, 214–216
Series feedback, 448, 458, 460–462, 464, 465, 467, 470, 475
Series-parallel feedback, 464
Series-series feedback, 460
Shunt feedback, 453, 457, 458, 460, 464, 466–468, 470, 475, 557, 563
Shunt-shunt feedback, 453
Silicon, 6, 8, 15–19, 22–30, 32, 35, 38, 44, 58, 59, 80, 85, 94, 116, 117, 122, 124, 127, 158, 242, 359, 534, 536
Silicon, polycrystalline, 23, 25, 28, 122, 300
Silicon dioxide, 7, 16, 23–28, 114, 115, 122–124, 127, 242
Silicon gate, 122
Silicon nitride, 25, 28, 122
Single-ended amplifier, 407, 410, 411, 415, 416, 429, 524, 554
Source-follower, 422, 488
Space-charge capacitance, 39, 41, 42, 49, 351–353
Space-charge region, 34–43, 49, 68, 72, 115, 117, 118, 122–127
s parameters, 357
SPICE, 12, 89, 362, 447
Stability, bias, 307, 311–319, 321–323, 331–334, 337, 446, 448
Stability, tuned amplifier, 498–501
Stagger tuning, 504, 506–512, 522

Static RAM, 287–293, 302
Storage of excess carriers, 32, 39, 40, 41, 105, 146, 156, 181–182, 213, 227
Storage time, 43, 104–106, 156, 178, 182, 208, 218, 227
STTL, 178–188, 216, 218, 224, 227
Switching effects in diode, 43
Synchronous tuning, 503

Temperature coefficient, 43
Thermal (temperature) stability, 307, 310, 313, 315, 320, 535, 536, 538
Threshold voltage, 123
Totem pole output, 170–176, 185, 189
Transadmittance amplifier, 460–464, 469, 470
  input impedance, 462
  output impedance, 462–464
Transfer characteristics, 149–150, 161, 164–167, 181, 182, 187, 200, 208–209, 215, 219, 224–226, 239, 244, 245, 249, 254, 265–267, 275, 277, 436–439
Transformer-coupled push-pull amplifier, 488–489
Transimpedance amplifier, 453–460, 469, 470, 562
  input impedance, 456–457
  output impedance, 457–459
Transistor-transistor logic (TTL), 145–201, 202, 203, 206, 208, 210, 213, 216, 218, 224, 226, 227, 234, 235, 289–291
Transition region capacitance, 39, 41, 42, 49, 351–353

Transit time, 69, 70, 71, 102, 121, 133, 136
Transmission gate, CMOS, 268, 269, 277
Transmission gate, NMOS, 256–259, 261, 274, 275, 286, 287, 291, 293
Transmission line, 188, 227, 230, 357
Trisate output, 176–178
Tuned circuit, 367, 386–392, 393, 397, 492–512
Two-port network, 83–84
  input impedance, 83–84
  output impedance, 83

Ultraviolet light, 300, 301

Valence band, 15, 16, 17, 18, 19
Varactor diode, 42
Video amplifier, 367
Virtual ground, 524, 557
Voltage amplifier, 464–466, 469, 470
  input impedance, 465
  output impedance, 465–466
Voltage regulator, 53–54, 523, 533–540

Widlar current source, 323–328, 337, 416
  output impedance, 325–328
Widlar voltage reference, 534–537
Wired logic, 188–189

y-parameters, 340, 341, 357, 361, 362, 494, 495, 500, 501, 521

Zener diode, 8, 42, 43, 64, 534